Omics Approaches for Sugarcane Crop Improvement

In this book, the information encompasses various researchable biotechnology aspects of sugarcane, its genomic structure, diversity, comparative and structural genomics, data mining, etc. This book explores both the theoretical and practical aspects of sugarcane crops, focusing on innovative processes. This book argues in favor of developing an integrated research and development system to strengthen the research and development capabilities of all the areas of sugarcane. Further, it covers the recent trends of sugarcane biotechnology, especially in the next-generation sequencing (NGS) era. This book will be very useful for professors and scientists who are working in the area of sugarcane crops by using molecular biology and bioinformatics. It is also useful for students to use as a reference for their classes or thesis projects.

Key features:

- Discusses an integral part of molecular biology and pivotal tools for molecular breeding; enables breeders to design cost-effective and efficient breeding strategies for sugarcane
- Discusses the harnessing genomics technologies for genetic engineering and pathogen characterization and diagnosis of sugarcane
- Provides new examples and problems, added where needed
- Provides insight from contributors drawn from around the globe

Omics Approaches for Sugarcane Crop Improvement

Edited by
Rajarshi Kumar Gaur

CRC Press
Taylor & Francis Group
Boca Raton London New York

CRC Press is an imprint of the
Taylor & Francis Group, an **informa** business

Cover image: ©Shutterstock

First edition published 2023
by CRC Press
6000 Broken Sound Parkway NW, Suite 300, Boca Raton, FL 33487-2742

and by CRC Press
2 Park Square, Milton Park, Abingdon, Oxon, OX14 4RN

CRC Press is an imprint of Taylor & Francis Group, LLC

Library of Congress Cataloging-in-Publication Data
Names: Gaur, Rajarshi Kumar, editor.
Title: Omics approaches for sugarcane crop improvement / Rajarshi Kumar Gaur.
Description: First edition | Boca Raton, FL : CRC Press, 2023. | Includes bibliographical references.
Identifiers: LCCN 2022030713 (print) | LCCN 2022030714 (ebook) | ISBN 9781032273686 (hardback) | ISBN 9781032273723 (paperback) | ISBN 9781003292425 (ebook)
Subjects: LCSH: Sugarcane. | Genomics. | Plant molecular genetics. | Bioinformatics.
Classification: LCC SB231 .O45 2023 (print) | LCC SB231 (ebook) | DDC 633.6/1--dc23/eng/20220822
LC record available at https://lccn.loc.gov/2022030713
LC ebook record available at https://lccn.loc.gov/2022030714

ISBN: 978-1-032-27368-6 (hbk)
ISBN: 978-1-032-27372-3 (pbk)
ISBN: 978-1-003-29242-5 (ebk)

DOI: 10.1201/9781003292425

Typeset in Times
by MPS Limited, Dehradun

Contents

Preface

Biotechnological interventions have produced agronomically improved genotypes, yet scientists are currently working to engineer the sugarcane crop as a platform for large-scale production of chemicals with industrial as well as therapeutic significance. Hence, biotechnological interventions hold great promise to develop a better sugarcane crop with improved agronomic traits, sugar content, and biofuel production.

This book is designed to be self-contained and comprehensive, targeting professors and scientists working on sugarcane biotechnology and its related fields, such as genomics, applied data mining, computational biology, disease management, etc. In addition, we believe that this book will serve as a useful reference for sugarcane and crop improvement in the omics era. The 15 chapters described in this book comprise the following theme:

1. Sugarcane improvement through biotechnology
2. Genomics and transcriptomics in sugarcane
3. Mechanism of sugarcane stress tolerance
4. Pathogen identification
5. Breeding and genomics approaches for sugar productivity
6. Sugarcane as a biofuel

Moreover, we have refined the authors' views in a simpler manner that can be easily understood by the readers.

In this book, the information encompasses various researchable biotechnology aspects of sugarcane, its genomic structure, diversity; comparative and structural genomics; data mining; etc. This book explores both the theoretical and practical aspects of sugarcane crops, focusing on the innovation processes. This book argues in favour of developing an integrated research and development system that will strengthen the research and development capabilities of all areas of sugarcane. Furthermore, it covers the recent trends in sugarcane biotechnology, especially in the next-generation sequencing (NGS) era.

To optimise sugarcane improvement, it is necessary to know the impact a selected trait will have on the general physiology of the plant. However, this is not yet possible because there are too many gaps in our knowledge of the unique development and physiology of sugarcane. Such gaps impair our ability to enhance desired agronomical traits; for example, selection for sugarcane varieties with increased photosynthetic capacity may be useless if sugar accumulation is constrained by temperature, water deficit, or nutrient availability. It may prove difficult to consistently increase sucrose levels in the culm without first knowing the factors that affect sugarcane yield and carbon partitioning. Molecular biology has begun to play an important role in agriculture due to its ability to modify microorganisms, plants, animals, and agricultural processes. In particular, it can aid conventional plant-breeding programmes using molecular markers. Several types of DNA markers and molecular-breeding strategies are now available to plant breeders and geneticists to help overcome many problems faced in conventional breeding. The chapters in this book present an overview of the differentially expressed genes in sugarcane under different biotic and abiotic stressors and the use of new biotechniques to develop tolerance in sugarcane against all stresses. This book will be an important resource for providing an excellent reference book for researchers and students working on sugarcane and its improvement.

Rajarshi Kumar Gaur

Preface

Editor Bio

Prof. (Dr.) Rajarshi Kumar Gaur achieved his PhD in 2005 and now serves as professor for the Department of Biotechnology, Deen Dayal Upadhyaya Gorakhpur University, Gorakhpur, Uttar Pradesh, India. His PhD was on molecular characterization of sugarcane viruses, viz., mosaic, streak mosaic, and yellow luteovirus. He has been the recipient of the MASHAV fellowship of the Israeli government for his post-doctoral studies and joined The Volcani Centre, Israel and BenGurion University, Negev, Israel. In 2007, he received the Visiting Scientist Fellowship from the Swedish Institute Fellowship, Sweden, to work in the Umeå University, Umeå, Sweden. Kumar received his post-doctoral fellowship from ICGEB, Italy, in 2008. He has made significant contributions on plant viruses and published 130 national/international papers, authored 18 edited books, and presented about 50 papers in the national and international conferences.

He has been the recipient of a Fellow of Linnean Society, Fellow of Royal Society of Biology, Fellow of Society of Plant Research, Fellow of Society of Applied Biology (FSAB), and Fellow of International Society of Biotechnology (FISBT). He has been awarded many other honors, including: Prof. B.M. Johri Memorial Award, Society of Plant Research (SPR); Excellent Teaching Award by Astha Foundation, Meerut; UGC-Research Teacher Award; Young Scientist Award, 2012, in Biotechnology by Society of Plant Research (SPR), Meerut; Scientific & Applied Research Centre Gold Medal Award, 2011 for Outstanding Contribution in the Field of Biotechnology. He has visited several laboratories of the United States, Canada, New Zealand, UK, Thailand, Sweden, Italy. He is also associated with several international journals as an academic editor and reviewer: Frontiers in Microbiology, Plose One, Scientific Reports, 3Biotech, J. of Plant Growth Regulators, Molecular Biology Report, Plant Disease, Current Genomic, Scientific African, Indian Phytopathology, Virus Disease, etc. Currently, he is involved in many national & international grants and collaborative projects on plant viruses and disease management.

Contributors

Niaz Ahmad
National Institute for Biotechnology and Genetic Engineering (NIBGE)
Faisalabad, Pakistan
and
Constituent College Pakistan Institute of Engineering and Applied Sciences
Faisalabad, Pakistan

Harinath Babu
Molecular Biology and Genetic Engineering Laboratory
Vasantdada Sugar Institute
Pune, Maharashtra, India

G Boobalakrsihan
Division of Vegetable Science
Indian Agricultural Research Institute, Pusa Campus
New Delhi, India

F. Budeguer
Instituto de Tecnología Agroindustrial del Noroeste Argentino (ITANOA)
Estación Experimental Agroindustrial Obispo Colombres (EEAOC) – Consejo Nacional de Investigaciones Científicas y Técnicas (CONICET)
CCT NOA Sur. Av. William Cross
Las Talitas, Tucumán, Argentina

Rachayya Devarumath
Molecular Biology and Genetic Engineering Laboratory
Vasantdada Sugar Institute
Pune, Maharashtra, India

Suman Devarumath
Vidya Pratishthan's College of Agricultural Biotechnology
Baramati, Maharashtra, India

Abdelaleim Ismail ElSayed
Biochemistry Department
Faculty of Agriculture
Zagazig University
Zagazig, Egypt

R. Enrique
Instituto de Tecnología Agroindustrial del Noroeste Argentino (ITANOA)
Estación Experimental Agroindustrial Obispo Colombres (EEAOC) – Consejo Nacional de Investigaciones Científicas y Técnicas (CONICET)
CCT NOA Sur. Av. William Cross
Las Talitas, Tucumán, Argentina

Zeshan Haider
Functional Genomics and Bioinformatics Program
National Institute for Genomics and Advanced Biotechnology
National Agricultural Research Centre
Islamabad, Pakistan

Somnath Kadappa Holkar
ICAR-National Research Centre for Grapes
Pune, Maharashtra, India

Muhammad Ramzan Khan
Functional Genomics and Bioinformatics Program
National Institute for Genomics and Advanced Biotechnology
National Agricultural Research Centre
Islamabad, Pakistan

Haris Khurshid
Oil Seed Research Program
Crop Sciences Institute
National Agricultural Research Center
Islamabad, Pakistan

S. Radhesh Krishnan
T. Stanes & Company Ltd
Coimbatore, Tamil Nadu, India

Atul Kumar
Amity Institute of Biotechnology
Amity University
Uttar Pradesh Campus, Lucknow, India

Ravindra Kumar
ICAR-Indian Agricultural Research Institute
Regional Station
Karnal, Haryana, India

Sarita Kumari
Department of Agricultural Biotechnology and Molecular Biology
CBS&H, RPCAU-Pusa
Samastipur, Bihar, India

P. Malathi
ICAR-Sugarcane Breeding Institute
Coimbatore, India

Ashutosh Kumar Mall
ICAR-Indian Institute of Sugarcane Research
Lucknow, India

Smriti Mall
Molecular Plant Pathology Lab.
Department of Botany
DDU Gorakhpur University
Gorakhpur, Uttar Pradesh, India

Varucha Misra
ICAR-Indian Institute of Sugarcane Research
Lucknow, India

Zahid Mukhtar
National Institute for Biotechnology and Genetic Engineering (NIBGE)
Faisalabad, Pakistan
and
Constituent College Pakistan Institute of Engineering and Applied Sciences
Faisalabad, Pakistan

Pandiyan Muthuramalingam
Department of Horticultural Science
Gyeongsang National University
Jinju, Korea

Pramod N. Nair
Department of Forestry
Fiji National University
Fiji

Gauri Nerkar
Molecular Biology and Genetic Engineering Laboratory
Vasantdada Sugar Institute
Pune, Maharashtra, India

K. Nithya
ICAR-Sugarcane Breeding Institute
Coimbatore, India

A.S. Noguera
Instituto de Tecnología Agroindustrial del Noroeste Argentino (ITANOA)
Estación Experimental Agroindustrial Obispo Colombres (EEAOC) – Consejo Nacional de Investigaciones Científicas y Técnicas (CONICET)
CCT NOA Sur. Av. William Cross
Las Talitas, Tucumán, Argentina

Muhammad Noman
Functional Genomics and Bioinformatics Program
National Institute for Genomics and Advanced Biotechnology
National Agricultural Research Centre
Islamabad, Pakistan

Ahmad A. Omar
Biochemistry Department
Faculty of Agriculture
Zagazig University
Zagazig, Egypt
and
Citrus Research and Education Center
University of Florida, IFAS
Lake Alfred, FL, USA

S. Ostengo
Instituto de Tecnología Agroindustrial del Noroeste Argentino (ITANOA)
Estación Experimental Agroindustrial Obispo Colombres (EEAOC) – Consejo Nacional de Investigaciones Científicas y Técnicas (CONICET)
CCT NOA Sur. Av. William Cross
Las Talitas, Tucumán, Argentina

A.D. Pathak
ICAR-Indian Institute of Sugarcane Research
Lucknow, India

M.F. Perera
Instituto de Tecnología Agroindustrial del Noroeste Argentino (ITANOA)
Estación Experimental Agroindustrial Obispo Colombres (EEAOC) – Consejo Nacional de Investigaciones Científicas y Técnicas (CONICET)
CCT NOA Sur. Av. William Cross
Las Talitas, Tucumán, Argentina

N. Prabhakaran
T. Stanes & Company Ltd
Coimbatore, Tamil Nadu, India

Madhavi Purankar
Molecular Biology and Genetic Engineering Laboratory
Vasantdada Sugar Institute
Pune, Maharashtra, India

J. Racedo
Instituto de Tecnología Agroindustrial del Noroeste Argentino (ITANOA)
Estación Experimental Agroindustrial Obispo Colombres (EEAOC) – Consejo Nacional de Investigaciones Científicas y Técnicas (CONICET)
CCT NOA Sur. Av. William Cross
Las Talitas, Tucumán, Argentina

Saleem Ur Rahman
National Institute for Biotechnology and Genetic Engineering (NIBGE)
Faisalabad, Pakistan
and
Constituent College Pakistan Institute of Engineering and Applied Sciences
Faisalabad, Pakistan

Hamza Ramzan
National Institute for Biotechnology and Genetic Engineering (NIBGE)
Faisalabad, Pakistan
and
Constituent College Pakistan Institute of Engineering and Applied Sciences
Faisalabad, Pakistan

Ghulam Raza
National Institute for Biotechnology and Genetic Engineering (NIBGE)
Faisalabad, Pakistan
and
Constituent College Pakistan Institute of Engineering and Applied Sciences
Faisalabad, Pakistan

Manuel Díaz de los Ríos
Cuban Research Institute of Sugar Cane By-products
Havana, Cuba

Vimanth S
Department of Agricultural Biotechnology and Molecular Biology
CBS&H, RPCAU-Pusa
Samastipur, Bihar, India

H.K. Sachan
Department of Crop Science
Fiji National University
Fiji

Bilal Saleem
Functional Genomics and Bioinformatics Program
National Institute for Genomics and Advanced Biotechnology
National Agricultural Research Centre
Islamabad, Pakistan

Sumeet Kumar Singh
Department of Seed Science and Technology
PGCA, RPCAU – Pusa
Samastipur, Bihar, India

A. Ramesh Sundar
ICAR-Sugarcane Breeding Institute
Coimbatore, India

M. Swapna
Division of Crop Improvement
ICAR-Indin Institute of Sugarcane Research
Lucknow, Uttar Pradesh, India

Swapan K. Tripathy
Department of Agricultural Biotechnology
Odisha University of Agriculture and Technology
College of Agriculture, Siripur Sq.
Bhubaneswar, Odisha, India

R. Viswanathan
ICAR-Sugarcane Breeding Institute
Coimbatore, India

Kai Wang
College of Life Sciences
Nantong University
Jiangsu, China

Georgina Michelena Álvarez
Cuban Research Institute of Sugar Cane By-products
Havana, Cuba

1 Potential and Advanced Strategies for Sugarcane improvement

M.F. Perera, F. Budeguer, R. Enrique, S. Ostengo, A.S. Noguera, and J. Racedo
Instituto de Tecnología Agroindustrial del Noroeste Argentino (ITANOA), Estación Experimental Agroindustrial Obispo Colombres (EEAOC) – Consejo Nacional de Investigaciones Científicas y Técnicas (CONICET). CCT NOA Sur. Av. William Cross, Las Talitas, Tucumán, Argentina

CONTENTS

1.1 INTRODUCTION

Genetic improvement has played a fundamental role in the development of the sugar industries in almost all the sugarcane-producing countries since it is considered the main method to improve productivity and controls disease (Hogarth et al. 1997). Sugarcane breeding began at the end of the 19th century (Stevenson 1965), and since then, it has been a continuous process mainly oriented toward increasing sugar yield, biomass, resistance to biotic factors, and tolerance to abiotic stresses. Most sugarcane-producing regions of the world have their own breeding programs to develop and improve local varieties adapted to their environments and cultivation practices. Breeding programs make crosses between already improved clones, whose origin usually goes back to a few parents (less than 20 *S. officinarum*; Raboin et al. 2008), whereby the genetic base of modern cultivars is reduced (Arceneaux 1967, Selvi et al. 2006). Conventional breeding applies classical genetic principles based on the phenotype and consists of selecting individuals with desired characteristics to subject them to cycles of crossing and selection. This process is complex and slow, taking each cycle between 10 and 15 years to obtain a new elite cultivar.

Although genetic-improvement programs may differ in strategies and criteria applied, a general scheme is the basis of most current programs (Ming et al. 2006), and it consists of three stages:

1. Selection of parents from a collection or germplasm bank
2. Hybridization using biparental crosses and / or polycrosses
3. Selection of progeny through clonal multiplication steps

The choice of parents depends on the objectives the program pursues and flower availability. For this stage, it is essential to count with a germplasm bank, which constitutes the source of genetic variability for all the traits to be improved. It is important to maximize the genetic distance between the parents for crossings. Genetic differences of parents were estimated considering genetic relationships, geographical origin, and morphological traits. However, pedigree does not always reflect genetic inheritance because of eventual mixed pollen, selfing, and seed admixture (Nair et al. 2002). Since morphology is highly influenced by environment, genetic component of sugarcane accessions could be misestimated from these traits, requiring extensive field determination to gain accuracy. Currently, numerous molecular marker techniques are successfully used to estimate genetic relationships among germplasm bank accessions (see below). Nowadays, the most important criteria for choosing parents are based on either the performance of the parent in the target environment exclusively or additional information on the behavior of their offspring. Moreover, crossing designs have certain limitations to obtain desired combinations and seed amounts, such as flowering timing, categorization of genotypes as male or female, and availability of glasshouse and photo-period facilities (Zhou 2013).

In the second stage, the production of sexual seed is carried out through crosses for the generation of highly variable populations, which will constitute the basis of selection. Regarding the hybridization methods used, biparental crosses are the main strategy employed in most programs, although polycrosses are also carried out to a lesser extent (Ming et al. 2006). An essential requirement to carry out the crosses is that the cane flourishes, a phenomenon that does not occur in all regions; therefore, many breeding programs have adopted the use of photo-periodic and temperature treatments to induce this process. During crossing, temperature and humidity are subject to strict controls to ensure both good pollen viability and survival, as well as seed set. After careful handling, the obtained seeds are stored at -20°C until required, remaining viable for nearly 10 years.

Finally, the third stage consists of identifying and selecting the superior genotypes within the populations generated from crosses. Sugarcane clones are highly heterozigous, allopolyploids and frequently aneuploidy, and, for this reason, the highly variable first generation is subject to selection. It is essential to know the nature and magnitude of the variability present in breeding populations. Selected clones are reproduced and evaluated in different environments and cutting ages, in successive stages of clonal multiplication quantifying environmental effects on the expression of the traits of interest. Parameters such as phenotypic and genotypic variance, and their additive and dominant components, are relevant for establishing a better performance ranking. The large amount of phenotypic data collected for each trait must be subjected to statistical analyses, which provide accurate estimates of the genetic and environmental effects that impact the development and evaluation of cultivars. Statistical analysis based on modern and appropriate models for each stage of selection for sugarcane clones improves characterization of materials and increases the efficiency of the process (Balzarini 2000; Ostengo 2020). After data analysis, the genotypes that met the best characteristics could be released as new commercial varieties.

Although the genetic improvement of sugarcane has been very efficient in obtaining varieties with outstanding characteristics in biomass production, progress in improving industrial performance has been limited in recent years (Jackson 2005; Dal-Bianco et al. 2012). An apparent plateau has been reached in terms of sugar concentration, so the selection of new high-yield varieties has become very difficult (Ming et al. 2002). It must be highlighted that the efficient production of biomass from sugarcane arouses great interest due to its potential to generate ethanol as a biofuel and electricity, reducing the dependence on fossil fuels and the emission of greenhouse gases (Goldemberg 2007). Considering the aforementioned reasons, optimizing the efficiency of breeding programs for sugar and biomass is important for meeting the growing demand for sugarcane.

Molecular knowledge does not compete with classical improvement, but it is intended to improve the quality of the estimates required to carry out each of the breeding stages, adding certainties and accuracy to the determinations. Also, other biotechnological tools, such as *in vitro* propagation of new varieties or quarantine sanitation to introduce new germplasm, can be assisted with molecular markers to increase the sensitivity of disease detection and genetic purity tests. Besides, when an outstanding variety is obtained by conventional breeding, one or few characteristics can be added through transformation and / or gene editing. For this process, numerous omic tools are used in the search for genes and regulatory sequences of interest in databases, the *in silico* design of the constructions for genetic transformation, and the determination of the insertion sites in the genome, among other key steps for the successfully commercial release of the new transgenic variety.

1.2 EXPLORATION, CONSERVATION, AND UTILIZATION OF GENETIC RESOURCES OF SUGARCANE GERMPLASM BANK ASSISTED BY MOLECULAR MARKERS

The *Saccharum* genus traditionally comprised six species: *S. spontaneum*, *S. officinarum*, *S. robustum*, *S. edule*, *S. barberi* and *S. sinense* (Daniels and Roach 1987; D'Hont et al. 1998). However, molecular studies suggested that the genus should be reduced to only two species, *S. officinarum* and *S. spontaneum*, considering the four remaining as hybrids between these two species (Irvine 1999; D'Hont et al. 2002).

Modern sugarcane cultivars are hybrids derived from interspecific crossings performed a century ago between *S. officinarum* L. (2n = 80), the noble sugar-producing species, and the wild species, *S. spontaneum* (2n = 40 – 128) with high fiber content and stress tolerance (Piperidis et al. 2010b; Vieira et al. 2018). The introgression in sugarcane breeding that involved the crossings between these two species and a series of backcrossing is known as nobilization process.

The sugarcane hybrids possess highly polyploid and aneuploidy genomes of 53 to 143 chromosomes (Ingelbrecht et al. 1999), with meiotic abnormalities and an estimated size of >10 gigabases (Gb) (Zhang et al. 2012), i.e., it is one of the most complex genomes of any crop plant (Aitken, 2021). Genetic studies have shown that the alloautopolyploid genome of these hybrids contains about 80% of chromosomes from *S. officinarum*, 10% of chromosomes from *S. spontaneum*, and 10% recombinant chromosomes between the two progenitor species (D'Hont et al. 1996).

The complexity and size of the sugarcane genome is a major limitation in its genetic improvement (Singh et al. 2010). For that reason, in recent years, numerous initiatives to understand the complexity of the sugarcane genome had led to different sequencing projects around the world for both parental species and hybrid genotypes. A minimum tiling path of 4,660 sugarcane BACs of the genome in cultivar R570 that best covers the gene-rich part of the sorghum genome was sequenced and then assembled in a 382 megabases (Mb) single-tiling path (Garsmeur et al. 2018), providing an essential genome template for aligning sequencing data. Also, the assembly of one haploid *S. spontaneum* genome derived from SES208 (Zhang et al. 2018) and gene-space assembly of the Brazilian variety SP80–3280 (Souza et al. 2019), were completed. Besides, genomes of Khon Kaen KK3, a Thailand variety (Shearman et al. 2018) and of the Colombian variety CC01–1940 (www.cenicaña.org 2020) were sequenced.

Increasing sugar productivity is the main concern of sugarcane breeding programs (Singh et al. 2010). To develop new varieties, the breeding programs are assembling germplasm collections to increase the possibilities of novel gene combinations. In that sense, there are two sugarcane global collections held in India and the U.S.A. The Indian collection consists of 1,819 clones, whereas the American collection contains 887 sugarcane clones. Both collections can be used for base-broadening breeding aimed at efficient introgression of desirable alleles (Fickett et al. 2020). Besides, national breeding programs in each country also maintain collections of different numbers of genotypes and exchange their materials to provide germplasm with desirable traits. The safe maintenance of the germplasm collections is a valuable tool for breeders considering the risk of genetic erosion, consisting in a reduction of allelic richness (Van de Wouw et al. 2010).

In terms of diversity, every genus or species is a useful genetic resource for conservation and use, and more efficient introgression programs could be developed with a better knowledge of the donor wild germplasm. Wild sugarcane relatives were employed from the beginning of introgression during the nobilization process, and they are still used in some breeding programs (Cursi et al. 2022). Thus, different breeding programs incorporated genetic material from *Erianthus* species originating from the Old World (*Erianthus Michx*. sect. Ripidium Henrard) and the New World, particularly North America, into *Saccharum* L. (Besse et al. 1998). Among them, *Erianthus arundinaceus* (Retz.) Jeswiet has received special attention due to its high biomass productivity in ratoon crops and its drought tolerance (Jackson and Henry, 2011). However, many sugarcane breeding programs have been crossing it with sugarcane (D'Hont et al. 1995; Besse et al. 1997; Piperidis et al. 2000; Cai et al. 2005a; Nair et al. 2006; Krishnamurthi et al. 2007; Piperidis et al. 2010a; Fukuhara et al. 2013), low production efficiency has been reported among intergeneric hybrids (Piperidis et al. 2000), due to an incompatibility between the *Saccharum* and the *Erianthus* genomes. Several markers were proved useful to follow introgression in *Saccharum* x *Erianthus* hybrids; such as Restriction fragment length polymorphism (RFLP) of the ribosomal DNA (rDNA), (Besse et al. 1996), inter-Alu-like sequences MsCIR2 and EaCIR6 (Alix et al. 1999) and SSR (Chang et al. 2020).

Knowledge of the genetic diversity and population structure among sugarcane parental lines provides valuable information to select the best parental accessions for crossing and broaden the genetic base, i.e., expanding the gene pool, and also contributes to the improved use of genetic resources (Perera et al. 2012a). In this context, molecular markers are powerful tools, both for estimating genetic diversity and understanding the complex genetics of sugarcane (D'Hont et al. 1997). Different types of molecular markers have been used for diversity studies.

In one of the first attempts to estimate diversity, variation of nuclear ribosomal (r) DNA was investigated in cultivated and wild *Saccharum* accessions by digestion with restriction enzymes. Results revealed that clones of the *Erianthus* and *Miscanthus* genera had unique patterns and variation was most extensive in *S. spontaneum*, whereas *S. robustum* was only slightly variable (Glaszmann et al. 1990). After that, 5S rRNA intergenic spacers were amplified and sequenced. Direct sequencing of cloned PCR products revealed different spacer length and sequence polymorphism for *S. officinarum*, *S. spontaneum*, sugarcane cultivars, *S. giganteum*, *Erianthus* spp., *M. sinensis*, Sorghum bicolor, and *Zea mays* (Pan et al. 2000).

RFLP markers with both chloroplast and mitochondrial heterologous probes were useful to study diversity in cultivated (*S. officinarum*, *S. barberi* and *S. sinense*) and wild (*S. spontaneum* and *S. robustum*) genomes in the *Saccharum* complex (D'Hont et al. 1993).

Lu et al. (1994) also performed an RFLP analysis on nuclear DNA of wild and old cultivated sugarcane accessions. Multivariate analyses of the data allowed the separation of the three basic species, *S. spontaneum*, *S. robustum,* and *S. officinarum* and showed that *S. spontaneum* had structure that could be related to the geographic origin of the clones supporting the hypotheses on the origin of secondary species *S. barberi* and *S. sinense*. The authors suggested that RFLPs will be essential in sugarcane for depicting the genomic constitution of modern cultivars.

A collection of *Erianthus* sect. *Ripidium* accessions and *Saccharum* representatives (*S. officinarum* and *S. spontaneum*) were also studied by RFLP analysis using dispersed nuclear single-copy probes from maize (Besse et al. 1997). Results were confirmed by Besse et al. (1998) in an assessment of genetic diversity within and between *Saccharum*, *Erianthus* sect. *Ripidium* and

North American *E. giganteus* (*S. giganteum*) conducted using amplified fragment length polymorphism (AFLP). This study revealed three major groups corresponding to the three genera under study. Within the *Saccharum* group, two subgroups belonged to *S. spontaneum* and *S. officinarum* accessions, respectively, were found.

Moreover, AFLP markers were used to study the genetic relationship among 69 genotypes, representing 16 species from the four genera in the *Saccharum*. Principal component analysis and UPGMA clustering showed that *Saccharum* is closely related to *Miscanthus,* whereas *Narenga* and *Erianthus* were closely related (Cai et al. 2005b).

Selvi et al. (2005) also employed AFLP to understand the genetic constitution of cultivars developed in India suitable for subtropical conditions. Results revealed that the subtropical cultivars retained significantly higher numbers of *S. spontaneum* specific DNA fragments than did the tropical cultivars, reflecting the breeding strategy followed in the development of these cultivars. Additionally, the AFLP-based clustering of the cultivars corresponded well with their pedigree relationships.

Furthermore, AFLP were used to estimate molecular diversity among 270 *S. officinarum* genotypes from all the major regions where *S. officinarum* is grown and 151 cultivars produced by the Australian breeding program or important parents used in it. *S. officinarum* clones from New Guinea displayed greater diversity than *S. officinarum* clones from other regions, confirming the hypothesis that New Guinea is the centre of origin for this species. The *S. officinarum* clones from Hawaii and Fiji formed a separate group that may correspond to introgressed clones with other members of the *Saccharum* complex. Besides, greater diversity was found in the cultivars than in the *S. officinarum* clones due to the introgression of *S. spontaneum* chromatin. It must be highlighted that about 90% of markers present in the *S. officinarum* clone collection were also present in the cultivar collection, suggesting that most of the observed genetic diversity in *S. officinarum* has been captured in Australian sugarcane germplasm (Aitken et al. 2006). Also, in a smaller subset of *S. officinarum*, *S. robustum*, *S. spontaneum*, *S. barberi*, *S. sinense,* and the related genus *Erianthus*, AFLP were employed. The phenetic tree obtained was consistent with the known taxonomical relationships. AFLP gave higher resolution of closely related species into discrete groups than that by RFLP markers, reported earlier. For that, AFLP markers would be an appropriate tool in providing better information about the relationships among the species, estimating diversity, and revealing species and genus specific markers that could be directly applied in sugarcane breeding programs (Selvi et al. 2006).

Random amplified polymorphic DNA (RAPD)-PCR markers were used to characterize a local collection of *S. spontaneum* L. clones and cultivars from the United States. As a substantial degree of genetic diversity was found within the local collection, the information provided would help conserve its genetic diversity (Pan et al. 2005). On the other hand, Indian sugarcane cultivars were characterized using RAPD markers, revealing both the limited genetic base and the need to diversify the genetic base by using new sources from the germplasm (Kawar et al. 2009).

The ability of microsatellite or simple sequence repeat (SSR) markers to determine the genetic diversity between members of the genera *Saccharum* (*S. officinarum*, *S. spontaneum*, *S. sinense*), Old World *Erianthus* Michx. sect. *Ripidium*, North American *E. giganteus* (*S. giganteum*), *Sorghum* and *Miscanthus* was assessed. Clustering revealed a genetic structure for *Saccharum* and *Erianthus* sect. *Ripidium* that reflected closely the relationship previously identified using other marker systems (Cordeiro et al. 2003). It must be highlighted that although SSR are codominant markers in the case of the polyploidy sugarcane, they are scored as present or absent (dominant manner).

Besides, to ensure the genetic identity during shipping and handling, 116 Louisiana commercial sugarcane clones were also genotyped with 21 SSR markers by Pan (2007). After that, recognizing the SSR technique as a valuable tool in characterizing and identifying sugarcane genotypes, Pan (2010) developed the world's first SSR marker-based sugarcane database. For that, 1,025 sugarcane clones were genotyped on a fluorescence-capillary electrophoresis detection platform with 21 highly polymorphic SSR markers. This database, annually updated, provides molecular descriptions for new cultivar registration articles, enables sugarcane breeders to identify mislabeled sugarcane clones in crossing programs, determines the paternity of cross progeny, and ensures that desired cultivars are grown in farmers' fields.

In another study, SSRs were used to access the genetic diversity and interrelationships of *S. barberi, S. spontaneum, S. officinarum,* and commercial cultivars. Clusters revealed groups of individuals having close genetic relationships based on geographical origin, available pedigree information, adaptation zone, and morphological characters (Singh et al. 2010).

Qi et al. (2012) genotyped with SSR 92 elite parents involved in China Mainland sugarcane breeding programs and four wild relatives. Model-based genetic structure, clustering, and principal components analyses consistently revealed five groups. One of the groups was characterized as the introduction group with 46 cultivars predominantly introduced from Australia, Taiwan, India, and the U.S.A.; another group only contained wild germplasm, whereas the three other groups mostly originated from China Mainland.

Furthermore, You et al. (2013) characterized 115 sugarcane genotypes obtained by different Chinese breeding programs and used for parental crossing by SSR. Results revealed that genotypes released by each breeding program showed closer genetic relationships suggesting that breeding programs should increase their genetic diversity of sugarcane varieties to meet the demand of sugarcane cultivation for sugar and bioenergy use. After that, You et al. (2016) obtained similar results when characterizing 181 sugarcane clones by using fluorescence-labeled genomic with SSR and eight expressed sequence tag (EST)-SSR primer pairs.

In China, 150 of the most popular parental lines used in sugarcane breeding programs were analyzed with 21 fluorescence-labeled SSR markers and high-performance capillary electrophoresis (Wu et al. 2019). Based on the population structure analysis, the accessions were clustered into two distinct sub-populations in PCoA: one that contained the majority of clones introduced to China, and the other that grouped accessions native to China. Alternatively AMOVA fixation index value and gene flow value all indicated the very low genetic differentiation between the two groups. This study illustrated that SSR markers are still a very useful tool for genotyping sugarcane.

The genetic structure of Brazilian cultivars contrasted with a group of basic germplasm mainly composed of *S. officinarum* and *S. spontaneum* accessions was assessed through SSR, revealing that this technique is a fast and cost-effective way for routine cultivar identification and management of sugarcane germplasm banks. Authors detected that 97.6% of the cultivar alleles were found in the basic germplasm, while 42% of the basic germplasm alleles were absent in cultivars. Genetic divergence found between cultivars and *S. spontaneum* accessions has practical applications for energy cane-breeding programs because the choice of more divergent parents will maximize the frequency of transgressive individuals in the progeny (Manechini et al. 2018). In another study, Crystian et al. (2018) analyze the genetic base of the Serra do Ouro germplasm bank in Brazil over the past decades using SSR. Results revealed an increase in genetic similarity in the 1970s due to interbreeding with few parents; however, after the 1990s, with the introduction of new parents, there was a decrease in genetic similarity levels.

Even more, SSRs were recently used to characterize the world collection of sugarcane and related grasses maintained at Miami, Florida, U.S.A. that contains nonredundant clones of different species and genera within the *Saccharum* complex. Results revealed average gene diversity of 0.208 among all species groups, where *Erianthus*-like *Saccharum* species, *Miscanthus* spp., and *S. spontaneum* each formed a distinct cluster, *S. robustum*, *S. officinarum*, hybrid cultivars, and *S. edule* grouped together in a major cluster, and *S. sinense* and *S. barberi* formed a distinct grouping (Fickett et al. 2020).

The aforementioned information reflected that among molecular markers, SSRs are advantageous for a variety of applications in sugarcane genetics research and breeding (genetic fingerprinting, kinship analysis, and population genetic studies). Their benefits are being multi-allele, highly informative, having high relative abundance, providing good coverage across the genome, and being experimentally reproducible. In addition, more recent evidence suggests they may play an important role in genome evolution and provide hotspots of recombination. SSRs also have a few advantages even over markers obtained using modern genotyping platforms, though they are low throughput and might cost less per unit (Ahmad et al. 2018).

Besides, inter simple sequence repeats (ISSR) primers designed from SSR regions were useful in clarifying the molecular classification of related *Saccharum* species and characterizing the genuine hybrid from *Saccharum* and *Erianthus* (Zhang et al. 2004).

Also related to SSR markers, 18 sugarcane specific sequence-tagged microsatellite site (STMS) markers developed from Genbank databases were used to analyze the genetic divergence of Indian genotypes. Results revealed that the genetic base of sugarcane germplasm used in varietal improvement in India is reasonably large (Hemaprabha et al. 2005). However, Sindhu et al. (2011) when characterizing cultivated sugarcane hybrids representing all agro-eco climatic regions of tropical India also by using STMS markers found low level of genetic diversity among the commercial hybrids under cultivation.

To determine the utility of single nucleotide polymorphisms (SNPs) as a way to improve the understanding of the complex sugarcane genome, Cordeiro et al. (2006) developed markers to a suite of SNPs identified in a list of sugarcane ESTs. This was the first instance of a technology that has provided an insight into the copy number of a specific gene locus in hybrid sugarcane. The identification of specific and numerous haplotypes/alleles present in a genotype provided the basis for identifying associations between specific alleles and phenotype and between allele dosage and phenotype in sugarcane.

Some authors have suggested that, for breeding purposes, the estimation of genetic diversity should be based on functionally characterized genes or genes of interest since they reflect functional polymorphisms (Alwala et al. 2006). Variability of expressed or regulatory sequences might reflect a selection pressure for these regions, which would be different for each gene. Therefore, the variation detected by functional markers randomly distributed across the genome may detect variability more useful for breeders, since they reflect genetic variations of direct interest. Even more, they can be used to accelerate the breeding process and select important agronomic traits (Marconi et al. 2011). In that sense, target region amplified polimorphism (TRAP) markers amplify functional regions of the genome, providing valuable information related to the variations within coding/functional regions of the plant genome (Li and Quiros, 2001; Singh et al. 2017). TRAP with primers designed from sucrose and cold-tolerance-related EST were used to determine the genetic diversity and interrelationships in sugarcane germplasm collections from the genera *Saccharum* (*S. officinarum*, *S. barberi*, *S. sinense* R, *S. spontaneum*, *S. robustum,* and interspecific hybrids between *S. officinarum* and *S. spontaneum*), *Miscanthus,* and *Erianthus*. Cluster and principal coordinate (PCoA) analyses placed the *Erianthus* spp. and *Miscanthus* spp. genotypes distinctly from each other and from the *Saccharum* species, supporting their taxonomic classification as separate genera. Genotypes of the low sucrose and cold tolerant species, *S. spontaneum*, formed one distinct group, while the rest of the *Saccharum* species formed one interrelated cluster with no distinct subgroups. Results ratify TRAP as a potentially useful marker technique for genetic diversity studies in sugarcane (Alwala et al. 2006). TRAP markers were also used to estimate the genetic similarity among sugarcane varieties and five species of the *Saccharum* complex. Fixed primers designed from candidate genes involved in sucrose metabolism and drought-response metabolism were used. Results suggested that genetic variability in the evaluated genes was lower in the sucrose metabolism genes than in the drought-response metabolism ones (Creste et al. 2009).

Besides, TRAP revealed a higher polymorphism than AFLP markers, indicating that TRAP markers are more effective than AFLP when analyzing the sugarcane genome because they are easier, faster, and more economic than applying AFLP markers (Racedo 2014). In addition, Singh et al. (2017) used TRAP markers to identify promising genotypes bearing enviable agronomic traits (sucrose content and multiple disease resistance) to recommend as proven parents. They found through poly component analysis that 52.48% of the cumulative variation was explained by all the genotypes with respect to the sucrose metabolism-related genes. Moreover, authors reconfirmed that functional TRAP markers would be efficiently useful in genetic studies for sugarcane genetic improvement.

In 2009, Que et al. selected genes related to sucrose metabolism and cold stress to study genetic diversity of important sugarcane clones, including main cultivars, recently released varieties, and germplasms introduced abroad recently through TRAP markers. Although cluster analysis could not contribute completely to parentage grouping of the clones, it could really reflect the variation degree of these clones. Recently, Medeiros et al. (2020) employed TRAP markers to evaluate several accessions of the *Saccharum* complex; they showed that even when considering genes under the selection process, the introgression of new favorable alleles is still possible. Also, TRAP markers revealed that both parental species contributed significantly to the variability in genes involved in sugar metabolism, and considerable variability in the lignin metabolism genes remains underexplored by local Brazilian breeding programs (Junior et al. 2020).

In addition, the potential of the sequence-related amplified polymorphism (SRAP) technique, which preferentially amplifies gene-rich regions, was evaluated to assess the genetic relationships among members of the *Saccharum* species (*S. officinarum*, *S. spontaneum*, *S. robustum*, *S. sinense*, *S. barberi,* and sugarcane hybrids); *Miscanthus* and *Erianthus* were included as outgroups. Dendrogram grouped the genotypes according to their phylogenetic relationships. *Erianthus* and *Miscanthus* were separated as two outgroups, where one clustered *S. robustum*, *S. sinense*, *S. barberi,* and hybrids as different subgroups with each one, including some *S. officinarum* clones, while the second cluster included the *S. spontaneum* clones, exclusively. The high discriminatory power coupled with the possibility that most of the amplicons amplified gene-rich regions of the genome makes SRAP a potentially robust tool for genetic mapping aimed at marker-assisted introgression in sugarcane (Suman et al. 2008).

In another study, the genetic relationships of sugarcane clones were determined based on three types of molecular markers; AFLP, TRAP, and SSR. Although the average genetic similarity estimates based on AFLP (0,675) and TRAP (0,655) was closer to each other than to SSRs (0,522), the correlation between TRAP and SSRs was higher (r = 0.24). Authors suggested that the choice of molecular markers should be considered carefully, based on the purpose of the application in the breeding program, since it is not possible to select a marker system that fits all the requirements for germplasm characterization (Creste et al. 2010). Perera et al. (2012a) also compared AFLP, SSR, and morphological traits to evaluate genetic diversity among the most useful parents of an Argentinean sugarcane breeding program. When morphological data was combined with molecular markers, the dendrogram obtained revealed genetic relationships, and the genetic diversity was better estimated. Authors also suggested that both methods appear to be useful, complementing each other, and should be used together to assist sugarcane breeders in estimating genetic diversity, electing parents for crossings, identifying superior lines, and protecting intellectual property rights. It must be highlighted that, in most cases, a combination of different marker techniques is recommendable for genetic diversity studies since they could reflect different aspects of the genome studied and provide more accurate information (Perera et al. 2016).

EST-SSR markers are preferred to genomic SSR (gSSR) since they represent the functional part of the genome, which can be easily associated with the desired trait. The database of sugarcane EST (SUCEST) offers a great opportunity for developing molecular markers that are directly associated with important agronomic traits. Marconi et al. (2011) employed this database and developed 287 EST-SSRs polymorphic markers that were tested in 18 sugarcane genotypes. Among the markers characterized in this study, some have particular interest, such us those that are related to bacterial-defense responses, generation of precursor metabolites and energy, and involved in carbohydrate metabolic process. Besides, Singh et al. (2013) employed EST-SSR for studying the population structure of sugarcane accessions. The authors found a high level of polymorphism; this not only suggested their utility in sugarcane genetics and genomics, but also enriched the microsatellite marker resources in sugarcane.

Collard and Mackill (2009) developed another DNA marker in rice, the start codon targeted (SCoT) marker. It is based on the short conserved nucleotide sequence that flanks the start codon ATG. SCoT marker involves a single oligonucleotide primer, is PCR based, and, due to the simultaneous binding of the primer on both DNA strands, the sequence between the two binding sites is amplified. SCoT marker presents the following advantages: simple, low-cost, highly polymorphic, gene-targeted, and abundant in the genome (Que et al. 2014). This marker technology was applied in sugarcane to study genetic diversity among 107 sugarcane accessions within a chinese sugarcane germplasm collection. PCA partitioned the 107 sugarcane accessions into two major groups, the domestic group and the foreign introduction group. Each group was further divided based on institutions, where the sugarcane accessions were originally developed. Results provided foundation data for managing sugarcane germplasm, including construction of a core collection and regional variety distribution (Que et al. 2014).

Besides, new technologies are now available. They include high-throughput next-generation sequencing (NGS), currently used to detect SNPs, that have permitted the accumulation of large quantities of molecular data through the analysis of marker regions to make estimates of genetic diversity extremely robust (Goodwin et al. 2016). Although there is a long history of research and development in sugarcane, it is only comparatively recently that marker technology has caught up with the need for discovery of large numbers of single-dose SNP markers; fast, efficient, and cost-effective methods for genotyping can be used for selection in a breeding program (Aitken 2021). In that sense, diversity arrays technology (DArT) provides whole-genome profiling for hundreds to thousands of polymorphic markers in a single assay using a high-throughput microarray platform. However, due to the genome complexity of this sugar-producing crop of high economic importance, an application of DArT genotyping to this species required extensive testing and optimization. The method employs genome complexity reduction using methylation-sensitive restriction enzymes and high-throughput next-generation sequencing (NGS). The sugarcane DArT markers generated with this method identified high genetic differentiation of sugarcane ancestral species from modern cultivars, in agreement with the data available for other types of molecular markers for this crop. Even more, the majority of sugarcane DArT markers segregated in a Mendelian. When 384 clones were sequenced, one-third of sequenced markers came from the transcribed portion of the sugarcane genome. Results obtained validated the potential of DArT technology in providing cost-effective genetic profiles for plants, irrespective of their genome complexity, for effective applications in molecular-assisted breeding, diversity analysis, or genetic identity testing (Heller-Uszynska et al. 2011). In a recent study, the genetic diversity of local and foreign sugarcane varieties selected from the local germplasm in the Philippines was evaluated through SNPs and silico-DArT dominant markers. Cluster analysis and principal coordinate analysis showed low-sucrose varieties grouping separately from commercial hybrids developed in the Philippines. However, genetic distances among varieties genotyped indicate moderate to high genetic relatedness within the local germplasm, especially among commercially-available varieties in the country. Authors suggested that DArT-Seq genotyping was successfully used in the analysis of genetic diversity among current commercial varieties and can be a useful tool in evaluating new breeding materials for the development of more improved varieties (Bello et al. 2019).

To summarize the importance of genetic diversity studies, it must be mentioned that conserving crop biodiversity is an urgent undertaking. Modern cultivars have limited genetic variation (Fickett et al. 2020) since almost 20 *S. officinarum* clones are involved in the genealogy of sugarcane cultivars with only a few being used extensively (Raboin et al. 2008). According to the Intergovernmental Panel on Climate Change, 25–30% of plant species, including crops, will be extinct or endangered in the next century. Hence, safeguarding the existing diversity and producing new diversity in cash crops are of utmost importance to keeping the environment intact (Priyadarshan and Mohan Jain 2022).

In conclusion, different molecular marker systems were useful in studying the genetic diversity of populations of diverse origin, whether wild, domesticated, or modern varieties. Unlike the estimates of diversity made from geographic, pedigree, or morphological data, the genetic parameters of diversity calculated from molecular data that reflect variations at DNA level should better reveal the real relationships. However, it is important to note that a sufficient number of molecular tags distributed throughout the genome should be used, and ideally, the information from different molecular marker systems should be combined to more accurately estimate the diversity present in each population under study.

1.3 MOLECULAR TECHNIQUES TO DETECT SYSTEMIC DISEASES DURING QUARANTINE AND TISSUE CULTURE TO SANITIZE DISEASED PLANTS

Probably due to the genetic bottleneck effect that occurred during the first interspecific hybridizations that gave rise to modern cultivars, the genetic diversity of sugarcane is low. This situation forces the breeding programs to have a germplasm bank that allows them to fully exploit the existing allelic diversity and create new combinations from it. For this process, the exchange of materials between different regions and programs is essential. To effectively use the new genetic resources incorporated, either to introduce them to the germplasm bank or to test them as varieties to be released, it is imperative to subject them to a period of quarantine. During this stage, the plants are inspected and screened for diseases to determine their sanitary status and avoid the introduction of pathogens. The available sugarcane phytopathogen-detection techniques, described below, are varied and also differ in their sensitivity. If the presence of diseases is detected in these imported materials that might have valuable traits for the program, the meristem culture technique offers the possibility of sanitizing them effectively.

Actually, more than 100 pathogens (including bacteria, fungi, viruses, phytoplasmas, and nematodes) have been recognized as causal agents of diseases in sugarcane (Govindaraju et al. 2019). Propagation of sugarcane through vegetative cuttings favors the spread of diseases through planting materials. Most of the breeding programs focus on obtaining varieties resistant to diseases since severe epidemics have threatened sugarcane production throughout its history, and many sugarcane varieties were replaced due to their breakdown to a new disease or to a new pathogenic strain (Viswanathan and Rao 2011).

In general, in modern sugarcane production systems, diseases are predominantly controlled by an integrated approach involving the combination of disease-free planting material obtained by tissue culture, disease-resistant cultivars, applicable farm-management practices, and strict quarantine measures for exchanging foreign materials (Babu et al. 2021). However, to perform any management strategy, a correct diagnosis of the pathogen involved is required. In that sense, regarding available diagnosis techniques, serological and molecular methods are commonly employed.

Among serological methods, ELISA and tissue-blot (or tissue-printing) techniques have become very popular and largely used for routine detection of bacterial, phytoplasma, and viral pathogens in sugarcane. These tools are only specific and sensitive enough to detect moderate to high pathogen titters within infected plants (Amata et al. 2016). However, molecular techniques such as polymerase chain reaction (PCR) and reverse-transcriptase (RT)-PCR assays are more sensitive than the serological techniques, specially when detecting sugarcane viruses and phytoplasmas occurring at low concentrations (Viswanathan and Rao 2011). Besides, PCR offers several advantages over other methods, including versatility, relative simplicity, and specificity, which can be increased by a two-step PCR (nested PCR). Due to the aforementioned reasons, PCR and RT-PCT-based diagnostics are recommended for evaluating quarantine materials and plants obtained through tissue culture. It must be mentioned that most of the available diagnosis protocols for bacteria detection rely on internal transcribed sequences (ITS) regions, while for viruses, the capsid protein is the most commonly used.

In addition, multiplex PCR or RT-PCR allowed, in a single reaction with different pairs of specific primers, the simultaneous detection of various pathogens. This kind of reaction was found to be equally efficient to uniplex-PCR or RT-PCR to amplify the target pathogens from sugarcane (Iglesia et al. 2003) and is very useful in quarantine programs since it saves time and resources.

As PCR and RT-PCR are gel-based systems, they could be laborious and time-demanding for routine diagnosis of a large number of samples. In that sense, quantitative real-time PCR and RT-PCR are gel-free approaches in molecular diagnostics, enabling a sensitive, specific, and fast detection of sugarcane pathogens at very low titters. These techniques could be 100-fold more sensitive than conventional PCR or RT-PCR (Korimbocus et al. 2002).

Quantitative real-time PCR requires using costly instruments that may be unaffordable for some laboratories. To solve this problem, Notomi et al. (2000) developed novel nucleic acid amplification assays termed loop-mediated isothermal amplification (LAMP). This method utilizes auto-cyclic strand displacement at a single, constant temperature in a rapid, simple, and low-cost process. Generally, four to six primers are designed to recognize several regions of the target sequence, which gives LAMP its high specificity; even more, several authors found that LAMP is as sensitive as, or more sensitive than, quantitative real-time PCR assays (Keizerweerd et al. 2015). This diagnosis method was used for detecting bacteria (Liu et al. 2013) and viruses (Keizerweerd et al. 2015) in sugarcane.

A highly sensitive and suitable method for the robust detection of viruses in complex plant tissues, such as sugarcane, is the nucleic acid sequence-based amplification (NASBA) of the target RNA with specific primers and simultaneous real-time detection of the amplification products with molecular beacons (MBs) in a closed tube, called AmpliDet RNA (Gonçalves et al. 2001). MBs are short singled-stranded nucleic acid sequences (30–50 bases) that are designed to have a unique sequence flanked by indirect repeats so that a stem-loop structure is formed.

On the other hand, the detection of pathogens by nucleic acid hybridization tests is not as useful since, many times, it revealed background in healthy plants due to pigment interference. Also, the use of radioactive probes is not always feasible because of the lack of facilities and other inherent limitations in several laboratories (Hema et al. 2003).

Other molecular-diagnostic techniques to detect viruses from leaf extracts and sugarcane juice are immunocapture-reverse transcription-PCR (IC-RT-PCR) and direct-binding RT-PCR (DB-RT-PCR) (Hema et al. 2003). In the first one, viruses are captured with its specific antiserum in antibody-coated tubes and then the RT-PCR is performed, whereas DB-RT-PCR does not require the use of antiserum since it is a nonspecific immobilization system where viruses bound to a solid support can be directly detected by RT-PCR. The second diagnosis method is at least 10-fold less sensitive than IC-RT-PCR.

To provide examples of available diagnostic molecular protocols for detecting the most common bacteria, fungi, and viruses in sugarcane, Table 1.1 is included.

In brief, quarantine is a valuable resource for obtaining foreign material to broaden the genetic diversity of the breeding program. A correct detection and identification of the pathogens present in disease plants avoid introducing and spreading the pathogen in the region. Additionally, due to the difficulties in managing diseases, especially in the case of vegetatively propagated crops like sugarcane, where the planting material is the primary source of the disease in the field, the use of disease-free plants obtained trough tissue culture is recommended to the farmers. In most if not all countries where disease-free sugarcane plants are obtained, molecular diagnosis are routinely applied to ensure the health of the seedlings obtained. The choice of the most appropriate diagnostic method will depend on several factors including sensitivity, number of samples, cost, speed, available equipment, among others. It must be mentioned, that further works are required to develop sensitive diagnostics for field level detection of pathogens to address problems in the field (Viswanathan et al. 2018).

TABLE 1.1
Examples of Available Molecular Diagnosis Protocols for Causal Agents of Sugarcane Diseases

Method	Causal Agent	Disease	Reference
PCR	*Leifsonia xyli susbp. Xyly*	Ratoon stunting disease (RSD)	Pan et al. (1998); Sawazaki et al. (2013)
	Xanthomonas albilineans	Leaf scald (LS)	Fegan et al. (1998); Pan et al. (1997; 1999) Wang et al. (1999)
	Acidovorax avenae subsp. avenae	Red stripe	Song et al. (2003)
	Puccinia melanochepala	Brown rust	Virtudazo et al. (2001) Glynn et al. (2010)
	Puccinia kuehni	Orange rust	Virtudazo et al. (2001) Glynn et al. (2010) Sawazaki et al. (2010a,b, 2013);
	Macruropyxis fulva sp. nov.	Tawny rust	Martin et al. (2015)
	Sporisorium scitaminea	Smut	Sawazaki et al. (2013)
Multiplex PCR	*L. xyli* susbp. *Xyly* and *X. albilineans*	RSD and LS	Iglesia et al. (2003)
Nested PCR	*L. xyli* susbp. *Xyly*	RSD	Lozano et al. (2003)
Multiplex nested PCR	*L. xyli* susbp. *Xyly* and *X. albilineans*	RSD and LS	Davis et al. (1998)
RT-PCR	SCMV	Sugarcane mosaic virus	Yang and Mirkov (1997); Alegria et al. (2003); Viswanathan et al. (2007); Singh et al. (2009)
	SrMV	Sorghum mosaic virus	Yang and Mirkov (1997)
	Fijivirus	Fiji disease virus	Smith and van de Velde (1994)
	SCSMV	Sugarcane streak mosaic virus	Chatenet et al. (2005); Viswanathan et al. (2007; 2008a); Singh et al. (2009)
	SCYLV	Sugarcane yellow leaf virus	Pan (1997); Angel et al. (2001); Chatenet et al. (2001); Gonçalves et al. (2001); Viswanathan et al. (2008c); Singh et al. (2009); Girard et al. (2012)
	SCBV	Sugarcane bacilliform virus	Braithwaite et al. (1995); Singh et al. (2009); Balan et al. (2020)
Multiplex RT-PCR	Fijivirus, SCMV and SCYLV	Fiji disease virus, Sugarcane mosaic virus and Sugarcane yellow leaf virus	Cadavid et al. (2003)
	SCMV, SrMV, SCSMV and SCYLV	Sugarcane mosaic virus, Sorghum mosaic virus, Sugarcane streak mosaic virus and Sugarcane yellow leaf virus	Xie et al. (2009)
	SCMV, SCSMV and SCYLS	Sugarcane mosaic virus, Sugarcane streak mosaic virus and Sugarcane yellow leaf virus	Viswanathan et al. (2010)

TABLE 1.1 (Continued)
Examples of Available Molecular Diagnosis Protocols for Causal Agents of Sugarcane Diseases

Method	Causal Agent	Disease	Reference
DB-RT-PCR	SCSMV	Sugarcane streak mosaic virus	Hema et al. (2003)
IC-RT-PCR	SCSMV	Sugarcane streak mosaic virus	Hema et al. (2003)
	SCBV	Sugarcane bacilliform virus	Matsuoka et al. (2014)
	SrMV	Sorghum mosaic virus	Chen et al. (2020)
Multiplex IC-RT-PCR	SCMV and SCSMV	Sugarcane mosaic virus and Sugarcane streak mosaic virus	Viswanathan et al. (2008b; 2013)
qPCR	*L. xyli* susbp. *Xyly*	RSD	Grisham et al. (2007); Pelosi et al. (2013); Carvalho et al. (2016); Fu et al. (2016); Johnson (2019)
	X. albilineans	LS	Garces et al. (2014); Gutierrez et al. (2016); Wang et al. (2020); Shi et al. (2021)
	SCWL	Sugarcane white leaf phytoplasma	Roddee et al. (2018)
qRT-PCR	SCYLV	Sugarcane yellow leaf virus	Korimbocus et al. (2002); Amata et al. (2016)
	SCSMV	Sugarcane streak mosaic virus	Fu et al. (2015); Wei-Lin et al. (2015)
NASBA	SCYLV	Sugarcane yellow leaf virus	Gonçalves et al. (2001)
LAMP	*L. xyli* susbp. *Xyly*	RSD	Liu et al. (2013)
RT-LAMP	SCMV and SrMV	Sugarcane mosaic virus and Sorghum mosaic virus	Keizerweerd et al. (2015); Viswanathan et al. (2018)

1.4 GENETIC VARIABILITY GENERATION THROUGH HYBRIDIZATION

1.4.1 Selection of Parents

A key factor that determines genetic gain and the success of any breeding program is the selection of parents to carry out the crosses that will generate the clones subjected to rigorous field evaluations during the following 10–12 years (Sanghera and Jamwal 2019, Atkin et al. 2009). However, there is evidence that the selection of parents has received little attention in terms of assistance with molecular markers compared with other stages of the sugarcane improvement scheme (Jackson 2016).

The potential of an individual as a parent can be estimated through its breeding value (BV), which is a measure of the individual's ability to transmit its characteristics to its progeny. BV is affected directly by the estimation of variance components, especially the estimate of additive variance (Falconer and Mackay 1996). At present, the restricted maximum likelihood (REML) and best linear unbiased prediction (BLUP) (Piepho et al. 2008) statistical methods are the most used to obtain BV estimates (Barbosa et al. 2005, Neto et al. 2013, Mbuma et al. 2019). In addition to the data from field trials, there is enough evidence that the incorporation of pedigree information from the coefficient-of-coancestry (pedigree, kinship, consanguinity) matrix has improved the precision of BV estimates in sugarcane (Alwala 2007, Piepho et al. 2008, Atkin et al. 2009, Mendes de Paula et al. 2020). To incorporate genetic correlations between individuals improves the statistical model for estimation of the variance components and BVs. Although pedigree provides extremely valuable information, sometimes it does not necessarily represent the real correlation between the individuals under study since the pedigree records are not always complete and / or may have errors. In addition, the biology of crop reproduction, which presents asymmetric distribution and preferential chromosome pairing in meiosis, added to structural chromosomal differences and frequent aneuploidy (Jannoo et al., 2004), means that not all the assumptions are fulfilled to estimate the kinship coefficients (Lima et al. 2002), which can result in biased estimates of additive variance and BVs (Atkin et al. 2009, Viana et al., 2012). The use of sufficient genotypic data obtained from molecular markers can provide accurate information on shared alleles, better reflecting genetic correlations, and improving estimates of additive variance and genetic merit of individuals (VanRaden 2008, Costa 2015). Few studies in sugarcane incorporated molecular data to improve the estimates of BVs (Jackson 2016, Wei and Jackson, 2017), until the recent development of genomic selection (detailed below) in which high throughput genomic information is incorporated to phenotypic data and appropriate statistical models to accurately estimate BV.

The choice of parents is not an easy task since numerous traits of economic importance are desired to be improved at the same time, and they do not have the same genetic architecture. For example, tons of cane per hectare (TCH) and tons of sugar per hectare (TSH) have been reported as low heritability traits, while Brix presents intermediate heritability (Hoarau et al. 2021, Jackson 2016, Neto et al. 2013; Pisaroglo de Carvalho et al. 2014; Jackson 2018; Wei et al. 2017). This implies that for TCH and TSH, the nonadditive variance components are of great importance (Zhou 2021). In this case, the BVs of the parents will not be very informative about the success of the crosses and the superiority of the clones in the progeny since little of the genetic variance is due to the additive components. This outcome would explain to some extent the low rate of genetic gain obtained in sugarcane breeding compared to other cross-pollinated crops such as maize (Burnquist 2013); traditionally, breeding programs have mostly exploited the additive variance considering BV or general combining aptitude (GCA) of the parents (Zhou 2021). Some recent studies have proposed the exploitation of specific combining aptitude (SCA) and heterosis to exploit the nonadditive genetic effects of important traits (Mendes de Paula 2020, Zhou 2021). This breeding strategy could be assisted by the application of genomics to detect polymorphic molecular markers, which allow the identification of progenitors with different allelic complements to maximize heterosis or hybrid vigour. Some studies have focused on determining genetic distances from different molecular markers systems, suggesting that performing crosses between distant parents would lead to obtaining superior progeny, increasing the genetic gain (Alwala 2007, Lavanya and Hemaprabha 2010, Singh et al. 2017, Wang et al. 2020, Zhou 2021).

In summary, the parent's collection to generate the variability that will be selected for more than a decade is dynamic. New parents must be incorporated while others are discarded so that the available genetic variability does not decrease due to selection. The crosses must be designed to increase the rate of genetic gain; for this, it is essential to know the genetic architecture of the characteristics under selection and the genetic constitution of parents. The precision of estimating BVs can be improved by incorporating genetic information from high-throughput molecular markers. In addition, parents must be studied in their SCA and to exploit the heterosis. For the latter, the aim is to maximize the genetic distance of the parents in a cross, for which the molecular markers distributed throughout the genome provide precise information on the identity between the alleles present in each parent.

1.4.2 Hybridity Determination and Progeny Test Assisted by Molecular Markers

Once progenitors have been chosen, biparental or polycrosses are carried out with genotypes that have enough flowers. At hybridization season, molecular markers are useful tools for the following:

- To verify the hybrid condition of the seeds obtained in a cross since it is very useful to estimate the percentage of intentional paternal hybridizations to better understand and possibly improve the crossing process. As sugarcane inflorescences are hermaphrodite panicles, male or female parental genotypes are classified based on the relative amount of viable pollen. Male parents could be employed as female parents, diminishing the risk of selfing, when an efficient emasculation treatment is used, based on hot water treatment (immersion of the panicle in water at 50°C for 5 minutes) (Perera et al. 2012b, Perera et al. 2013). However, complete pollen sterility is not assured. Mainly, SSR markers were useful to determine the success of enmasculation treatment, identifying true hybrids (McIntyre and Jackson 2001; Manigbas and Villegas, 2004; Pan et al. 2003; Ostengo et al. 2021). Even more, when SSR were used to study the hybrid character of progenies, markers segregated in a Mendelian fashion (Perera et al. 2012b, Pan et al. 2015); however, unexpected segregation patterns that may be associated with the complexity of sugarcane genome were also observed. When SSR were used to assess the fidelity of biparental crosses, three types of progenies were identified: hybrids, selfs, and off-types (had SSR fingerprints that were not found in either parent). Results indicated that selfing may reduce advancement rates more than the off-types (Todd et al. 2020). On the other hand, SSR were useful to confirm true self-fertilized individuals in sugarcane families aimed at the use of self-fertilized plants in an individual reciprocal recurrent selection breeding strategy (Costa 2015).

- To identify the male parent in polycrossings. Although most of the crosses carried out by sugarcane breeding programs are biparental, a small percentage is carried out as polycrossing. This process occurs mainly when little material is available from the female parent, so several male parents are used, maximizing the possible genetic combinations. Numerous authors have reported the success of identifying the progenitors of a cultivar derived from polycrosses using molecular markers, mainly SSR (Pan 2006, Tew and Pan 2010, Xavier et al. 2014) and TRAP (Perera et al. 2020a).
- To reliably identify the pedigree of new varieties or clones. Sugarcane crosses are extremely laborious and errors can occur during the crossing process and subsequent classification and sowing of seeds. When promising clones are in advanced stages, their pedigree can be reconstructed or verified using molecular markers (Perera et al. 2020a).
- To verificate the introgression of wild germplasm into sugarcane as it was aforementioned (D'Hont et al. 1995; Piperidis et al. 2000; Cai et al. 2005; Aitken et al. 2007; Gao et al 2015).

1.5 SELECTING OUTSTANDING CLONES: MOLECULAR MARKERS LINKED TO TRAITS OF INTEREST AND GENOMIC SELECTION

Once the genetic variability has been created through the crosses, the next stages of the breeding scheme involve arduous labour of exhaustive phenotyping in different environments. At the initial stages, there are numerous individuals (reaching up to 200,000–300,000 seedlings in Brazil or China) (Cursi et al. 2021, Qui et al. 2022), so phenotyping cannot be very deep and is carried out in few environments. As the selection progresses, the number of clones under selection is reduced and phenotyping is deepened, improving the determinations and increasing number of environments where genotypes are tested. To increase the selection efficiency of high-yielding clones, several efforts are being conducted to develop molecular marker tools since latest selection stages are expensive, laborious, and time consuming.

Numerous studies aimed to detect genomic regions involved in traits of interest have been developed in sugarcane. The purpose of such studies has been to locate and elucidate the genetic mechanisms of characteristics of interest for breeding to make available molecular markers for marker-assisted selection (MAS). Regarding quantitative inheritance characteristics in sugarcane, quantitative trait loci (QTLs) associated with agronomic traits (TCH, TSH, Brix, Fiber, etc.) were detected in populations derived from biparental crosses (Ming et al. 2002; Aitken et al. 2006, 2008; Pinto et al. 2010, Balsalobre et al., 2017). Subsequently, the genome-wide association studies (GWAS) methodology was also used in sugarcane (Wei et al. 2010; Banerjee et al., 2015, Gouy et al. 2015; Racedo et al. 2016; Siraree et al. 2017) to provide a high resolution evaluation of numerous alleles simultaneously. Although these studies found molecular markers linked to QTLs, implementation remains a challenge in breeding programs since most of the economically important traits are complex and affected by a large number of genes, each with a small effect. However, with markers associated with smut resistance, a successfully implemented MAS was reported in Australian Breeding Program (Wei et al. 2021). Also numerous reports supported the usefulness of diagnostic molecular markers for the presence of the *Bru*1 gene to predict resistance to brown rust in sugarcane breeding programs around the world (Costet et al. 2012; Racedo et al. 2013; Li et al. 2017). Thus, MAS has proven to be suitable for traits controlled by a small number of major genes.

A strategy that is becoming increasingly important as a support for conventional breeding is genomic selection (GS) (Meuwissen et al. 2001). Unlike MAS, which uses a small number of markers that were previously identified as being associated with genes or QTL of interest, in GS all molecular markers are used simultaneously to predict genetic value or phenotypic performance of the individuals of a population, even considering those QTLs of small effect. This genetic merit, called genomic estimated breeding value (GEBV), is the result of the application of a predictive statistical model that relates and integrates the information generated from the molecular markers obtained from high-throughput genotyping with exhaustive phenotypic values of the traits of interest. To successfully implement the SG in a breeding program, once the model is trained, the cross-validation within and between populations is essential. During validation, the predictive capacity of the model calibrated is evaluated by comparing the predicted values to those observed in a new population. As a result of a good calibration, the GEBVs should have sufficient correlation with the breeding value obtained from phenotypic data. Thus, the breeding program will have a powerful tool to design more convenient crosses, selecting parents with high GEBVs, in addition to predicting the genetic value of new individuals under selection only from molecular information, avoiding multiple years of testing and field evaluations, allowing shorten breeding cycles.

The implementation of SG in other crops, such as corn, wheat, barley, and rice (Crossa et al. 2014; Heffner et al. 2011; Spindel et al. 2015; Zhong et al. 2009), has generated interesting results. In the particular case of sugarcane, an initial study carried out by Gouy et al. (2013) demonstrated the potential of the use of SG in complex characters for this polyploid. More recent studies in sugarcane were encouraging and highlight the usefulness of SG in breeding programs (Deomano et al. 2020, Hayes et al. 2021, Voss-Fels et al. 2021).

1.6 MICROPROPAGATION TECHNIQUES TO PROVIDE WITH HEALTHY AND TRUE-TO-TYPE SEED CANE

Several sugarcane producers face the disadvantage of lack of high-quality propagation material (Ali et al., 2008; Rithesh 2013) due to the accumulation of pathogens favored by sugarcane vegetative propagation (Viswanathan 2016). In addition, the major shortcoming of vegetative propagation is limited rate of seed multiplication leading to slow spread of new improved cultivars to growers (Kaur and Sandhu 2014). In that sense, thermotherapy and meristem culture, or a combination of both techniques, emerge as tools to successfully eradicate viruses from infected sugarcane plants (Sreenivasan 1981; Walkey 1980), where micropropagation also enables the massive multiplication of plants to obtain certified vitroplants and increase the sugarcane and sugar productivity per unit area. Plant tissue culture techniques allow the establishment, manipulation and development, under artificial and controlled conditions, of cells, tissues or organs and is very useful for the regeneration of rejuvenated, genetically homogeneous plants free of pests and diseases (Bello-Bello et al. 2018).

Specific *in vitro* sugarcane micropropagation protocols from apical meristems, shoot tips, and leaf whorls to produce seed canes of extremely high phytosanitary quality and similar to mother or stock plants (genetic purity) are available (Kaur and Sandhu 2014). Among them, one of the most safe micropropagation techniques to minimize the genetic variation is the axillary shoot proliferation from shoot tip culture. In brief, for large-scale production, the following stages are involved: i) selection and maintenance of field-grown mother plants, ii) culture establishment, iii) shoot multiplication, iv) rooting and hardening, and v) separation of clumps into single plantlets and transfer to soil (Lal and Krishna 1994, Kaur and Sandhu 2014).

In many projects around the world, the sanitation of the plant material is achieved through *in vitro* cultures of apical meristems from donor plants previously hydro-heat-treated and grown under natural light conditions in greenhouses with anti-aphid screens. It must be mentioned that to guarantee that meristem donor plants and micropropagated seedlings are disease free, both types of plants must be evaluated by molecular diagnosis developed for each important causal agent of disease (Perera et al. 2016). Molecular techniques to detect systemic diseases were commented on in the section 3.

Once *in vitro* seedlings are obtained in the laboratory, all of them are rusticated in special greenhouses with low light and high humidity and thereafter field planted in nurseries for conventionally propagating, before being distributed among sugarcane growers for commercial production (Perera et al. 2016). So, micropropagation allows the massive multiplication of sugarcane plants of optimum quality for the establishment of seed lots (Daíz et al. 2020).

One aspect that must be considered in the massive multiplication of plants through tissue culture is that it can induce the appearance of undesirable mutational events as a consequence of the new conditions to which the cells are subjected. These changes, first described by Larkin and Scowcroft in 1981, known as somaclonal variation, are transmitted to regenerated plants and can affect biochemical and morphological characteristics, of simple or quantitative inheritance. Even if the rate of mutagenesis are the same in cell and tissue cultures as in field-grown plants, the sheer number of occurrences in a cell population (10^6 after 20 cell divisions) would make accumulation of mutants far greater than in field-grown plants. Therefore, somaclonal variants can be detected more frequently in cell cultures than mutations in field-grown populations (Larkin 1998). Although the causes of somaclonal variation have been extensively studied, they remain largely theoretical or unknown (Bairu et al. 2011). The literature indicated that this variation could range from a specific trait to the whole plant genome, and it is associated with changes in the number and structure of chromosomes (translocations, deletions, duplications), point mutations, activation of transposable elements, or changes in organelle DNA (Larkin et al. 1984; Bordallo et al. 2004). So, as the genetic purity of the genotype must be guaranteed, mutational changes are the main drawback of massive micropropagation of plants (Soniya et al. 2001). For that, the optimization of methods that allow the detection of genetic changes caused by *in vitro* culture is essential.

Most of the strategies used to estimate somaclonal variations of micropropagated plants have limitations. Thus, karyotype analysis cannot reveal alterations in genes or small chromosomal rearrangements. Izoenzymatic markers constitute an appropriate method to detect genetic changes, but a low number of bands that only allow the indirect visualization of DNA fragments encoding soluble proteins are obtained (Bordallo et al. 2004). In this sense, molecular markers are useful to evaluate somaclonal variation with greater precision than phenotypic and karyological analyzes (Cloutier and Landry 1994). These techniques allow handling a large number of samples, requiring minimal amounts of material to monitor the entire genome with a potentially unlimited number of markers, and are independent of interaction with the environment (Isabel et al. 1993).

RAPD and AFLP markers have been used for this propose, although the latter present greater reproducibility and allow the detection of higher number of polymorphisms than RAPD (Isabel et al. 1993). Besides, SSR markers were used to evaluate the genetic fidelity in sugarcane plantlets regenerated through micropropagation from leaf whorls (Pandey et al. 2012). Díaz et al. (2020) reported that AFLP markers were replaced by TRAP to evaluate somaclonal variation since the latter involves fewer steps and reagents, and therefore, it is cheaper than the one previously used.

It must be mentioned that comparison of the DNA profiles of the micropropagated plants with the initial genotype profile makes it possible to detect genetic variations before any phenotypic manifestation can be evidenced. In cases where genetic changes are detected, all genetically distinct plants are destroyed to avoid release of inadequate material (Perera et al. 2016). However, these molecular tools complement the routine and irreplaceable field evaluation of phenotypic characters, which enables the detection and elimination of off-type plants.

In summary, regarding productivity, efficiency, and safety, propagated plants from meristems are advantageous; in effect, in the short term, old and/or infected materials could be replaced by these healthy materials of high-yield potential. When field performance of *in vitro* micropropagated and conventionally propagated sugarcane plants was compared, an increase in cane number and cane yield was clearly observed in micropropagated plants (Gosal et al. 1998). Sawant et al. (2014) detected increases in productivity in the field by up to 25% with sugarcane plants obtained through micropropagation. Even more, Jalaja et al. (2008) observed better tillering, juice content, and quality in the micropropagated sugarcane crop as compared to the conventionally raised crop.

1.7 GENETIC ENGINEERING: GENETIC TRANSFORMATION AND GENOME EDITING

As sugarcane production is strongly influenced by biotic and abiotic stress, conventional breeding supported by omics strategies is constantly looking forward to developing new genotypes with increased sugar and biomass yields and higher stress tolerance. However, it is a challenge to find new natural genetic resources to generate varieties with increased tolerance to the above-mentioned production difficulties. For that, elite varieties are the starter material for crop improvement through genetic engineering to add new traits that confer adaptative advantages to overcome the stresses mentioned.

Advances in genetic-engineering techniques allow cloning and characterizing useful genes and also improve commercially important traits of elite sugarcane clones that subsequently lead to the development of an ideal cultivar. Among the available biotechnological tools, the genetic transformation technology, which is capable of incorporating genes from a wide variety of plant and nonplant sources into crops, and the genome editing (GE), capable of editing, inserting, or replacing specific sequences within the genome (Kumar, et al. 2020), are of great importance.

Regarding transgenic sugarcane, particle bombardment and *Agrobacterium*-mediated genetic transformation are the most commonly applied methods in sugarcane due to higher transformation efficiency (Suprasanna 2010, Budeguer et al. 2021). A successful gene-transfer system for plants requires the integration of several proceedings, such as tissue culture to obtain competent cells for genetic transformation and subsequent regeneration of the whole plant, an efficient gene-transfer system to deliver DNA into the genome of the plant, and an efficient selection system that allows the identification of transformed cells (Birch 1997). The first transgenic sugarcane plant with selectable genes was obtained by Bower and Birch (1992) using a simple and efficient system of microprojectile bombardment of embryogenic callus. This procedure was optimized during the following years (Bower et al. 1996), and transgenic plants with a commercial trait (herbicide resistance) were obtained (Gallo-Meagher and Irvine 1996). Numerous examples of transgenic sugarcane expressing traits of commercial interest obtained from both biolistic and *Agrobacterium* transformation are found in the literature (Budeguer et al. 2021).

The success of the genetic transformation in sugarcane was consolidated in the last decade, first, with the commercial release of a drought-tolerant transgenic sugarcane variety. This variety developed by Persero (PT Perkebunan Nusantara XI) expresses a bacterial gene that encodes a choline dehydrogenase; this enzyme participates in the synthesis of glycine betaine, a known osmoregulator that helps maintain water potential and protects cell organization and biological functions during cellular dehydration (Sugiharto 2018). Years later, the events CTB141175 / 01A, CTC91087-6 and CTC93209-4 were commercially released in Brazil. All three events express a toxin from the Cry gene family, which confers resistance to lepidopteran pests (ISAAA 2021).

Regarding omic studies and their support in the development of genetic modified (GM) sugarcane, gene discovery is a key process to identify genes that contribute to the development of a trait or phenotype. It is based on transcriptome studies conducted to the identification of some useful promoters, which can be used to control the tissue-specific transgene expression, and also functional characterization and annotation of genes responsible for important agronomic traits that help in the improvement of sugarcane varieties (Ali et al. 2019). Moreover, the next generation sequencing (NGS) technologies and bioinformatics analyses are suitable tools for the identification and sequence analysis of the transgene(s) insertion sites in the GM genome. Enrique et al. (2021), by using a targeted sequence enrichment approach, coupled with second-generation sequencing (SGS) and bioinformatics analyses, identified and analyzed all flanking sequences for multiple-transgene insertions in the genome of a glyphosate-resistant transgenic sugarcane line TUC 87-3RG (Noguera et al. 2015).

On the other hand, genome editing (GE) with sequence-specific nucleases is revolutionizing crop breeding and has promising applications for sugarcane and other crops with complex genomes (Jung and Altpeter 2016; Weeks 2017). This technology offers advantages on genetic transformation concerning regulatory affairs, time, and costs. For example, in a "prior consultation instance," regulatory agencies of countries such as Argentina (CONABIA) (Lewi and Vicién 2020), Brazil (CTNBIO) and the U.S.A. (APHIS) declared that obtaining products using GE will not be regulated as GM. Hence, if the GE-generated varieties do not contain foreign DNA sequences, they will not be considered transgenic, greatly diminishing the costs and efforts for a commercial release. GE is a technology to generate genetic modifications, such as insertions, deletions, or replacements, within a specific sequence of the genome of an organism, using engineered nucleases. Nowadays, there are four families of engineered nucleases available: meganucleases (MN), zinc-finger nucleases (ZFNs), transcription activator-like effector nucleases (TALENs), and the clustered regularly interspaced short palindromic repeat (CRISPR)-Associated Nuclease 9 (Cas9) technology (Mohan 2016). The technology has found its application in a range of economically important crops providing higher yields, increased nutritional quality, weed protection, and improved abiotic and biotic stress tolerance/resistance (diseases and pests) (Rahman et al. 2019). Concerning GE technology applied to sugarcane genome, there are only a few successful studies published, which is understandable due to the very complex genome of this species and the lack of whole-sequenced reference genomes for varieties. In a first report, Jung and Altpeter (2016) used TALEN nucleases to modify the cell wall composition by reducing lignin content. Field-grown TALEN-mediated mutants of the caffeic acid O-methyl transferase gene, showed a significant reduction in total lignin as well as a change in lignin composition, which resulted in improved saccharification efficiency (up to 43.8%) (Kannan et al. 2018). More recently, Eid et al. (2021) reported efficient multi-allelic editing in the sugarcane magnesium chelatase I subunit (MgCh) gene, coding for a clue enzyme for chlorophyll biosynthesis. The easy detection of chlorophyll depletion during *in vitro* propagation provided a rapidly scorable phenotype for predicting the extent of multiallelic editing in highly polyploid sugarcane. Thus, CRISPR/Cas9-mediated targeted co-mutagenesis of 49 copies/alleles of MgCh resulted in severely reduced chlorophyll content. Subsequently, Tufan Oz et al. (2021) demonstrated the precise co-editing of multiple alleles via template-mediated and homology-directed repair (HDR) of DNA double-strand breaks, induced by the programmable nuclease CRISPR/Cas9. The evaluation resulted in the co-editing of up to three acetolactate synthase (ALS) copies/alleles involved in herbicide resistance. Using a GE approach in sugarcane breeding programs remains as a challenge due to the high polyploidy, genome size, low transformation efficiency, and transgene silencing frequencies reported for this plant species (Mohan 2016). Nevertheless, when more genome sequence information is available, more GE strategies to design specific guided RNAs for targeting specific genes should be readily applicable in sugarcane.

1.8 REMARKED CONCLUSIONS

Classic sugarcane breeding has proven to be successful since its early development; nowadays, it greatly benefits from the incorporation of biotechnological tools. Through the different stages of the breeding scheme, molecular markers and other biotechnological approaches can add certainty and precision to the decisions made by breeders, accelerating the global process.

Numerous germplasm banks had been characterized at molecular level in order to determine genetic diversity of populations of varied origin, whether wild, domesticated, or modern varieties. Ideally, different marker systems should be combined to inspect genome widely. It should be taken into account that as it is intended to analyze the alleles present in around 100 chromosomes per individual, several loci per chromosome will give more reliable information.

When breeding programs incorporate foreign material to broad genetic diversity, the molecular diagnosis of phytopathogens provides certainty about the health of imported material.

From germplasm bank, a dynamic parent's core collection should be defined periodically to perform better combinations for crossings. Little information on the use of biotechnological tools at this stage is found in the literature. However, the recent development of GS in sugarcane including genotyping germplasm bank will allow improving estimates for BVs by incorporating high-throughput genomic information. Further, recent reports suggested taking advantages of SCA and heterosis to exploit the nonadditive genetic effects of important traits in sugarcane. This breeding strategy could be assisted by genomics tools to detect polymorphic molecular markers, which allow the identification of progenitors with different allelic complements in order to maximize heterosis or hybrid vigour.

At hybridization season, molecular markers are useful tools to verify the hybrid condition of the seeds obtained in a cross, to determine the success of enmasculation treatments, to identify the male parent in polycrossings, to reliably identify the pedigree of new varieties or clones and to verify the introgression of wild germplasm into sugarcane.

During the stages of selection of outstanding clones, there are successful reports of the implementation of MAS in breeding programs for traits governed by few genes with a greater effect, such as the case of diagnostic markers for the presence of the *Bru*1 gene that confers resistance to brown rust, or markers associated with smut resistance.

When the program has been successful and new varieties are released to commercial cultivation, biotechnology offers the possibility of multiplying the genotype through the cultivation of meristems. This technique allows the rapid diffusion of the new materials to the producers, guaranteeing the health and genetic purity through the use of the molecular diagnosis of pathogens and molecular markers distributed throughout the genome, respectively.

Genetic engineering offers the possibility of adding one or a few characteristics of interest to the new genotypes obtained, through genetic transformation or gene editing. This allows the genetic repertoire to be expanded for traits that are not available in the germplasm of related species. Several breeding programs have already reported genetically modified varieties that are suitable or are in the process of being commercialized.

1.9 FUTURE PROSPECTS

Sugarcane is one of the most efficient crops for carbon fixation; this makes it especially important not only for food generation but also for the efficient production of biomass due to its potential to generate ethanol as biofuel and electricity, reducing the dependence on fossil fuels and the emission of greenhouse gases. For that, and considering sugarcane as a strategic crop with a growing demand under sustainable framework, it has a highly significant world economic value.

In this regard, classical sugarcane breeding has proven to be successful and played a fundamental role in the development of the sugar industries since it is considered the main method to improve productivity. However, optimizing the efficiency of breeding programs for sugar and biomass is important to meet the growing demand for sugarcane. At present, several biotechnological tools applied at different stages of a classical breeding program allow achieving the required efficiency and sustainability.

Although there are numerous available biotechnological tools, it is imperative to achieve the complete combination and integration of classical breeding assisted by biotechnology to reach high-yielding varieties that constitute the key technology for farmers and industrials.

REFERENCES

Ahmad, A., Wang, J.D., Pan, Y.B., Sharif, R., and S.J. Gao. 2018. Development and use of simple sequence repeats (SSRs) markers for sugarcane breeding and genetic studies. *Agronomy* 8(11): 260.

Aitken, K.S., Li, J.C., Jackson, P., Piperidis, G., and C.L. McIntyre. 2006. AFLP analysis of genetic diversity within *Saccharum officinarum* and comparison with sugarcane cultivars. *Australian Journal of Agricultural Research* 57: 1167–1184.

Aitken, K., Li, J., Wang, L., Qing, C., Fan, Y.H., and Jackson, P. 2007. Characterization of intergeneric hybrids of Erianthus rockii and Saccharum using molecular markers. *Genetic Resources and Crop Evolution* 54(7): 1395–1405.

Aitken, K.S., Hermann, S., Karno, K., Bonnett, G.D., McIntyre, L.C., and P.A. Jackson. 2008. *Theoretical and Applied Genetics* 117(7): 1191–1203.

Aitken, K.S. 2021. History and development of molecular markers for sugarcane breeding. *Sugar Tech.* 10.1007/s12355-021-01000-7

Alegria, O.M., Royer, M., Bousalem, M., Chatenet, M., Peterschmitt, M., Girard J.C., and P. Rott. 2003. Genetic diversity in the coat protein coding region of eighty-six sugarcane mosaic virus isolates from eight countries, particularly from Cameroon and Congo. *Archives of Virology* 148: 357–372.

Ali, A., Naz, Shagufta, Siddiqui, F.A., and J. Iqbal. 2008. An efficient protocol for large scale production of sugarcane through micropropagation. *Pakistan Journal of Botany* 40(1): 139.

Ali, A., Khan, M., Sharif, R., Mujtaba, M., and S.J. Gao. 2019. Sugarcane Omics: An update on the current status of research and crop improvement. *Plants* 8(9): 344.

Alix, K., Paulet, F., Glaszmann, J.C., and A. D'Hont. 1999. Inter-Alu-like species-specific sequences in the Saccharum complex. *Theoretical and Applied Genetics* 99(6): 962–968.

Alwala, S. 2007. Predicting progeny performance and genome analysis in sugarcane (SACCHARUM SPP). 1–173.

Alwala, S., Suman, A., Arro, J.A., Veremis, J.C., and C.A. Kimbeng. 2006. Target region amplification polymorphism (TRAP) for assessing genetic diversity in sugarcane germplasm collections. *Crop Science* 46(1): 448–455.

Amata, R.L., Fernandez, E., Filloux, D., Martin, D.P., Rott, P., and P. Roumagnac. 2016. Prevalence of Sugarcane yellow leaf virus in sugarcane-producing regions in Kenya revealed by Reverse-Transcription Loop-Mediated Isothermal Amplification Method. *Plant Disease* 100(2): 260–268. doi: 10.1094/PDIS-05-15-0602-RE

Angel, J.C., Angel, F.S., and J.I. Victoria. 2001. Diagnóstico molecular de razas del virus del síndrome de la hoja amarilla en caña de azúcar (ScYLV) en Colombia. *Fitopatologia Colombiana* 25(1): 19–22.

Arceneaux, G. 1967. Cultivated sugarcanes of the world and their botanical derivation. *Proceedings of the International Society for Sugar Cane Technology* 12: 844–854.

Atkin, F.C., Dieters, M.J., and J.K. Stringer. 2009. Impact of depth of pedigree and inclusion of historical data on the estimation of additive variance and breeding values in a sugarcane breeding program. *Theoretical and Applied Genetics* 119(3): 555–565.

Babu, K.H., Devarumath, R.M., Thorat, A.S., Nalavade, V.M., Mayur, S., Appunu, C., and P. Suprasanna. 2021. Sugarcane Transgenics: Developments and Opportunities in Genetically Modified Crops. Kavi Kishor, P.B., Rajam, M.V., Pullaiah, T. (eds). Springer, Singapore. 10.1007/978-981-15-5897-9_12

Bairu, M.W., Aremu, A.O., and J. Van Staden. 2011. Somaclonal variation in plants: causes and detection methods. *Plant Growth Regul* 63: 147–173. 10.1007/s10725-010-9554-x

Balagalla, D.N., Wijesuriya, A., Ranathunge, N.P., and A.M.M.S. Perera. 2017. A protocol for in-vitro direct plant regeneration from leaf tissues for micropropagation of sugarcane. *Tropical Agricultural Research* 29 (1): 55–64.

Balan, S., Viswanathan, R., and K.A. Cherian. 2020. Status of leaf fleck caused by Sugarcane bacilliform virus incidence and severity in different sugarcane growing areas of Kerala and Tamil Nadu. *Journal of Sugarcane Research* 10:74–86. 10.37580/JSR.2020.1.10.74-86

Balsalobre, T.W.A., da Silva Pereira, G., Margarido, G.R.A., Gazaffi, R., Barreto, F.Z., Anoni, C.O., et al. 2017. GBS-based single dosage markers for linkage and QTL mapping allow gene mining for yield-related traits in sugarcane. *BMC genomics* 18(1): 1–19.

Balzarini, M.G. 2000. Biometrical models for predicting future performance in plant breeding. PhD Dissertation, Louisiana State University. (Dissertation Abstracts: 99-79242.)

Banerjee, N., Siraree, A., Yadav, S., Kumar, S., Singh, J., Pandey, D.K., and R.K. Singh. 2015. Marker-trait association study for sucrose and yield contributing traits in sugarcane (*Saccharum* spp. hybrid). *Euphytica* 205(1): 185–201.

Barbosa, M.H.P., Resende, M.D.V., Bressiani, J.A., Silveira, L.C.I., and L.A. Peternelli. 2005. Selection of sugarcane families and parents by Reml/Blup. *Crop. Breed. Appl. Biotechnol.* 5, 443–450. doi: 10.12702/1984-7033.v05n04a10

Bello, E.B., Rasco, J.L.S., Sendon, P.M.D., Cueva, F.M.D., Lalusin, A.G., and A.C. Laurena. 2019. Genetic diversity analysis of selected sugarcane (*Saccharum* spp. hybrids) varieties using DArT-Seq technology. *Philippine Journal of Science* 148:103–114.

Bello-Bello, J.J., Mendoza-Mexicano, M., and J.A. Pérez-Sato. 2018. *In Vitro* Propagation of Sugarcane for Certified Seed Production. In *Sugarcane-Technology and Research.* IntechOpen.

Besse, P., McIntyre, C.L., and N. Berding. 1996. Ribosomal DNA variations in *Erianthus*, a wild sugarcane relative (Andropogoneae-Saccharinae). *Theoretical and Applied Genetics* 92: 733–743. 10.1007/BF00226096

Besse, P., McIntyre, C.L., Burner, D.M., and C.G. de Almeida. 1997. Using genomic slot blot hybridization to assess intergeneric *Saccharum*×*Erianthus* hybrids (Andropogoneae -Saccharinae). *Genome* 40: 428–432.

Besse, P., Taylor, G., Carroll, B., Berding, N., Burner, D., and C.L. McIntyre. 1998. Assessing genetic diversity in a sugarcane germplasm collection using an automated AFLP analysis. *Genetica* 104(2): 143–153.

Birch, R.G. 1997. Plant transformation: problems and strategies for practical application. *Annual Review of Plant Biology* 48: 297–326. doi: 10.1146/annurev.arplant.48.1.297

Bordallo, P.N., Silva, D.H., Maria, J., Cruz, C.D., and E.P. Fontes. 2004. Somaclonal variation on *in vitro* callus culture potato cultivars. *Horticultura Brasileira* 22: 300–304.

Bower, R., and R.G. Birch. 1992. Transgenic sugarcane plants via microprojectile bombardment. *The Plant Journal* 2(3): 409–416.

Bower, R., Elliott, A.R., Potier, B.A.M., and R.G. Birch. 1996. High-efficiency, microprojectile-mediated cotransformation of sugarcane, using visible or selectable markers. *Molecular Breeding* 2: 239–249. doi:10.1007/BF00564201

Braithwaite, K.S., Egeskov, N.M., and G.R. Smith. 1995. Detection of sugarcane bacilliform virus using the polymerase chain reaction. *Plant Disease* 79: 792–796.

Budeguer, F., Enrique, R.A., Perera, M.F., Racedo, J., Castagnaro, A.P., Noguera, A.S., and B. Welin. 2021. Genetic transformation of sugarcane, current status and future prospects. *Frontiers in Plant Science*, 2467. Doi: 10.3389/fpls.2021768609

Burnquist, W.L. 2013. Sugarcane research and development: a view from the private sector. *Proceedings of the International Society of Sugar Cane Technologists* 28: 1–7.

Cadavid, M., Angel, J.C., Angel, F., and J.I. Victoria. 2003. Detección simultánea de la enfermedad de Fiji, el mosaico y el síndrome de la hoja amarilla en caña de azúcar utilizando métodos moleculares. *Carta Trimest. Cenicaña* 25(2): 11–15.

Cai, Q., Aitken, K., Deng, H.H., Chen, X.W., Fu, C., Jackson, P.A., and C.L. McIntyre. 2005a. Verification of the introgression of *Erianthus arundinaceus* germplasm into sugarcane using molecular markers. *Plant Breed* 124: 322–328.

Cai, Q.Y., Fan, H., Aitken, K., Piperidis, G., McIntyre, C.L., and P. Jackson. 2005b. Assessment of the phylogenetic relationships within the "*Saccharum* complex" using AFLP markers. *Acta Agronomica Sinica* 31(5): 551–559.

Carvalho, G., Da Silva, T.G.E.R., Munhoz, A.T., Monteiro-Vitorello, C.B., Azevedo, R.A., Melotto, M., and L.E.A. Camargo. 2016. Development of a qPCR for *Leifsonia xyli* subsp. *xyli* and quantification of the effects of heat treatment of sugarcane cuttings on Lxx. *Crop Protection.* 80: 51–55.

Chakravarti, M. 2017. Sugarcane Biotechnology: Challenges and Prospects, Springer, Brazil, p. 2.

Chang, H., Wang, Q., Qiu, Y., Qin, Y., Li, X., Wu, Q., He, W., Guo, Y., Zhang, W., Chen, J., and N. Fang. 2020. Production, identification and characterization of *Erianthus rockii*× *Narenga porphyrocoma* intergeneric hybrids as a new germplasm for sugarcane breeding and genetic research. *Sugar Tech* 22(3): 389–395.

Chatenet, M., Delage, C., and M. Ripolles. 2001. Detection of Sugarcane yellow leaf virus in Quarantine and production of virus-free sugarcane by apical meristem culture. *Plant Disease* 85: 1177–1180.

Chatenet, M., Mazarin, C., Girard, J.C., Fernandez, E., Gargani, D., Rao, G.P., Royer, M., Lockhart, B., and P. Rott. 2005. Detection of Sugarcane streak mosaic virus in sugarcane from several Asian countries. *Proceeding ISSCT* 25: 656–662.

Chen, H., Ali, N., Lv, W., Shen, Y., Qing, Z., Lin, Y., et al. 2020. Comparison of IC-RT-PCR, Dot-ELISA and Indirect-ELISA for the Detection of Sorghum Mosaic Virus in Field-Grown Sugarcane Plants. *Sugar Tech* 22(1): 122–129. 10.1007/s12355-019-00738-5

Cloutier, S., and B.S. Landry. 1994. Molecular markers applied to plant tissue culture. *In Vitro Cellular & Developmental Biology-Plant* 30(1): 32–39.

Collard, B.C.Y., and D.J. Mackill 2009. Start codon targeted (SCoT) polymorphism: a simple, novel DNA marker technique for generating gene-targeted markers in plants. *Plant Molecular Biology Reporter* 27(1): 86–93.

Cordeiro, G.M., Eliott, F., McIntyre, C.L., Casu, R.E., and R.J. Henry. 2006. Characterization of single nucleotide polymorphisms in sugarcane ESTs. *Theoretical and Applied Genetics* 113(2): 331–343.

Cordeiro, G.M., Pan, Y.B., and R.J. Henry. 2003. Sugarcane microsatellites for the assessment of genetic diversity in sugarcane germplasm. *Plant Science* 165(1): 181–189.

Costa, P.M.A., Almeida, C.F., Silveira, G., Soares, B., Baffa, D.C.F., Peternelli, L.A., Bhering. L.L., and M.H.P. Barbosa. 2014. Selfing confirmation in sugarcane by using simple sequence repeat markers: an individual reciprocal recurrent selection scheme. *Genetics and Molecular Research* 13 (4): 8962–8970.

Costa, P.M.D.A. 2015. Prediction of breeding values in sugarcane using pedigree and genomic information. (Thesis) https://www.locus.ufv.br/bitstream/123456789/7539/1/texto%20completo.pdf

Costet, L., Le Cunff, L., Royaert, S., Raboin, L.M., Hervouet, C., Toubi, L., et al. 2012. Haplotype structure around *Bru*1 reveals a narrow genetic basis for brown rust resistance in modern sugarcane cultivars. *Theoretical and applied genetics* 125(5): 825–836.

Creste, S., Sansoli, D.M., Tardiani, A.C.S., Silva, D.N., Gonçalves, F.K., Fávero, T.M., Medeiros, C.N.F., Festucci, C.S., Carlini-Garcia, L.A., and L.R. Pinto. 2010. Comparison of AFLP, TRAP and SSRs in the estimation of genetic relationships in sugarcane. *Sugar Tech* 12(2): 150–154.

Creste, S., Accoroni, K., Pinto, L., Vencovsky, R., Gimenes, M., Xavier, M., and M. Landell. 2009. Genetic variability among sugarcane genotypes based on polymorphisms in sucrose metabolism and drought tolerance genes. *Euphytica* 172: 435–446. 10.1007/s10681-009-0078-2.

Crossa, J., Perez, P., Hickey, J., Burgueno, J., Ornella, L., Cerón-Rojas, J., et al. 2014. Genomic prediction in CIMMYT maize and wheat breeding programs. *Heredity* 112(1): 48–60.

Crystian D., Messias dos Santos, J., Veríssimo, G., de Souza Barbosa, and C. Almeida. 2018. Genetic diversity trends in sugarcane germplasm: Analysis in the germplasm bank of the RB varieties. *Crop Breeding and Applied Biotechnology* 18: 426–431, 10.1590/1984-70332018v18n4n62

Cursi, D.E., Hoffmann, H.P., Barbosa, G.V.S., Bressiani, J.A., Gazaffi, R., Chapola, R.G., et al. 2022. History and current status of sugarcane breeding, germplasm development and molecular genetics in Brazil. *Sugar Tech* 24: 112–133.

Cursi, D.E., Castillo, R.O., Tarumoto, Y., Umeda, M., Tippayawat, A., Ponragdee, W., Racedo, J., Perera, M.F., Hoffmann, H.P., and M. Sampaio Carneiro. 2022. Origin, genetic diversity, conservation and traditional and molecular breeding approaches in sugarcane. In Genetic diversity, erosion, conservation, and utilization - Cash crops. Priyadarshan, P.M., Jain, S. Mohan (Eds.). Springer. 10.1007/978-3-030-74926-2_4

D'Hont, A.Y., Lu, H., Feldmann, P., and J.C. Glaszmann. 1993. Cytoplasmic diversity in sugarcane revealed by heterologous probes. *Sugar Cane* 1: 12–15.

D'Hont, A., Rao, P.S., Feldmann, P., Grivet, L., Islam-Faridi, N., Taylor, P., and J.C. Glaszmann. 1995. Identification and characterization of sugarcane intergeneric hybrids, *Saccharum officinarum* × *Erianthus arundinaceus*, with molecular markers and DNA in situ hybridization. *Theoretical and Applied Genetics* 91: 320–326.

D'Hont, A., Grivet, L., Feldmann, P., Rao, P., Berding, N., and J.C. Glaszmann. 1996. Characterisation of the double genome structure of modern sugarcane cultivars (*Saccharum* spp.) by molecular cytogenetics. *Molecular Genetics and Genomics* 250: 405–413.

D'Hont, A., Seshagiri Rao, P., Alleyne, S., Glaszmann, J.C., and P. Feldmann. 1997. Application of molecular markers in sugarcane breeding. *Proceedings West Indies sugar technologists*. 77–79.

D'Hont, A., Ison, D., Alix, K., Roux, C., and J.C. Glaszmann. 1998. Determination of basic chromosome numbers in the genus *Saccharum* by physical mapping of Ribosomal RNA Genes. *Genome* 41: 221–225.8.

D'Hont, A., Paulet, F., and J.C. Glaszmann. 2002. Oligoclonal interspecific origin of North Indian and Chinese sugarcanes. *Chromosome Research* 10(3): 253–262.

Dal-Bianco, M., Carneiro, M.S., Hotta, C.T., Chapola, R.G., Hoffmann, H.P., Garcia, A.A.F., and G.M. Souza. 2012. Sugarcane improvement: how far can we go?. *Current opinion in biotechnology* 23(2): 265–270.

Daniels, J., and B.T. Roach 1987. Taxonomy and Evolution. In. Sugarcane Improvement through breeding. Heinz, D. (Ed.). Elsevier. 7–84.

Davis, M.J., Rorr, P., and G. Astua-Monge. 1998. Nested, multiplex PCR for detection of both Clavibacter xyli subsp. xyli and Xanthomonas albilineans in sugarcane. *Proceedings of the Cong. Plant Pathology*, ISPP, Edinburgh, 3:3.3.4.

Deomano, E., Jackson, P., Wei, X., Aitken, K., Kota, R., and P. Pérez-Rodríguez. 2020. Genomic prediction of sugar content and cane yield in sugar cane clones in different stages of selection in a breeding program, with and without pedigree information. *Molecular Breeding* 40(4): 1–12.

Díaz, M.E., Perera, M.F., Paz, N.V., Insaurralde Rocco, P., Ovejero, S.N., Cerviño, A.M., Castagnaro, A.P., and yA.S. Noguera. 2020. Proceso de producción de vitroplantas de caña de azúcar de pureza genética y sanidad garantizadas en etapa de laboratorio en la EEAOC. *Revista Industrial y Agrícola de Tucumán* 97 (2): 51–55.

Eid, A., Mohan, C., Sanchez, S., and D. Wang. 2021. Multiallelic, targeted mutagenesis of magnesium chelatase with CRISPR / Cas9 provides a rapidly scorable phenotype in highly polyploid sugarcane. *Frontiers in Genome Editing* 3: 654996. doi:10.3389/fgeed.2021.654996

Enrique, R., Kurth, D., Ibarra-Laclette, E., Noguera, A., Welin, B., and A.P. Castagnaro. 2021. Identification of transgene flanking sequences in a pre-market safety assessed sugarcane in Argentina. *Crop Breeding and Applied Biotechnology* 21(3): e36862133.

Falconer, D.S., and T.E.C. Mackay. 1996. Introduction to quantitative genetics. 4th ed. Longman, New York.

Fegan, M., Croft, B.J., Teakle, D.S., Hayward, A.C., and G.R. Smith. 1998. Sensitive and specific detection of *Clavibacter xyli* subsp. *xyli*, causal agent of ratoon stunting disease of sugarcane, with a polymerase chain reaction-based assay. *Plant Pathology* 47: 495–504.

Fickett, N.D., Ebrahimi, L., Parco, A.P., Gutierrez, A.V., Hale, A.L., Pontif, M.J., et al. 2020. An enriched sugarcane diversity panel for utilization in genetic improvement of sugarcane. *Scientific reports* 10(1): 1–12.

Fu, H.Y., Sun, S.R., Wang, J.D., Ahmad, K., Wang, H.B., Chen, R.K., and S.J. Gao. 2016. Rapid and quantitative detection of *Leifsonia xyli* subsp. *xyli* in sugarcane stalk juice using a real-time fluorescent (TaqMan) PCR assay. *BioMed research international* 10.1155/2016/2681816

Fu, W.L., Sun, S.R., Fu, H.Y., Chen, R.K., Su, J.W., and S.J. Gao. 2015. A one-step real-time RT-PCR assay for the detection and quantitation of Sugarcane streak mosaic virus. *BioMed Research International.* doi: 10.1155/2015/569131

Fukuhara, S., Y. Terajima, S. Irei, T. Sakaigaichi, K. Ujihara, A. Sugimoto, and M. Matsuoka. 2013. Identification and characterization of intergeneric hybrid of commercial sugarcane (*Saccharum* spp. hybrid) and *Erianthus arundinaceus* (Retz.) Jeswiet. *Euphytica* 189: 321–327.

Gallo-Meagher, M., and J.E. Irvine. 1996. Herbicide resistant transgenic sugarcane plants containing the bar gene. *Crop Science* 36: 1367–1374. doi:10.2135/cropsci1996.0011183X003600050047x

Gao, Y.J., Liu, X.H., Zhang, R.H., Zhou, H., Liao, J.X., Duan, W.X., et al. 2015. Verification of progeny from crosses between sugarcane (*Saccharum* spp.) and an intergeneric hybrid (*Erianthus arundinaceus* × *Saccharum spontaneum*) with molecular makers. *Sugar Tech* 17: 31–35. doi:10.1007/s12355-014-0310-9

Garces, F.F., Gutierrez, A., and J.W. Hoy. 2014. Detection and quantification of *Xanthomonas albilineans* by qPCR and potential characterization of sugarcane resistance to leaf scald. *Plant Disease* 98(1): 121–126.

Garsmeur, O., Droc, G., Antonise, R. Grimwood, J., Potier, B., Aitken, K., et al. 2018. A mosaic monoploid reference sequence for the highly complex genome of sugarcane. *Nature Communications* 9(1): 1–10. 10.1038/s41467-018-05051-5

Girard, J.C., Fernandez, E., Daugrois, J.H., Roques, D., Roumagnac, P., and P. Rott. 2012. Genetic diversity of Sugarcane yellow leaf virus in a sugarcane selection plot in Guadeloupe (FWI). *International Sugar Journal* 114(1358): 96–100.

Glaszmann, J.C., Lu, Y.H., and C. Lanaud. 1990. Variation of nuclear ribosomal DNA in sugarcane. *Journal of Genetics and Breeding* 44(3): 191–198.

Glynn, N.C., Dixon, L.J., Castlebury, L.A., Szabo, L.J., and J.C. Comstock. 2010. PCR assays for the sugarcane rust pathogens *Puccinia kuehnii* and *P. melanocephala* and detection of a SNP associated with geographical distribution in *P. kuehnii. Plant Pathology* 59: 703–711.

Goldemberg, J. 2007. Ethanol for a sustainable energy future. *Science* 315(5813): 808–810.

Gonçalves, M.C., Klerks, M.M., Verbeek, M., Vega, J., and J.F.M.J. van den Heuvel. 2001. The use of molecular beacons combined with NASBA for the sensitive detection of Sugarcane yellow leaf virus. *European Journal of Plant Pathology* 108: 401–407.

Goodwin, S., McPherson, J.D., and W.R. McCombie. 2016. Coming of age: ten years of next-generation sequencing technologies. *Nat Rev Genet* 17: 333.

Gosal, S.S., Thind, K.S., and H.S. Dhaliwal. 1998. Micropropagation of sugarcane, an efficient protocol for commercial plant production. *Crop Improvement* 25: 167–171.

Gouy, M., Rousselle, Y., Bastianelli, D., Lecomte, P., Bonnal, L., Roques, D., et al. 2013. Experimental assessment of the accuracy of genomic selection in sugarcane. *Theoretical and applied genetics* 126(10): 2575–2586.

Gouy, M., Rousselle, Y., Chane, A.T., Anglade, A., Royaert, S., Nibouche, S., and L. Costet. 2015. Genome wide association mapping of agro-morphological and disease resistance traits in sugarcane. *Euphytica* 202(2): 269–284.

Govindaraju, M., Li, Y., and M. Zhang. 2019. Emerging Bacterial Disease (Leaf Scald) of Sugarcane in China: Pathogenesis, Diagnosis, and Management. *In Plant Diseases-Current Threats and Management Trends*. IntechOpen.

Grisham, M.P., Pan, Y.B., and E.P. Richard Jr. 2007. Early detection of *Leifsonia xyli* subsp. *xyli* in sugarcane leaves by real-time polymerase chain reaction. *Plant Disease* 91(4): 430–434.

Gutierrez, A., Garces, F.F., and J.W. Hoy. 2016. Evaluation of resistance to leaf scald by quantitative PCR of *Xanthomonas albilineans* in sugarcane. *Plant disease* 100(7): 1331–1338.

Hayes, B.J., Wei, X., Joyce, P., Atkin, F., Deomano, E., Yue, J., et al. 2021. *Theoretical and Applied Genetics* 134(5): 1455–1462.

Heffner, E.L., Jannink, J.L., Iwata, H., Souza, E., and M.E. Sorrells. 2011. Genomic selection accuracy for grain quality traits in biparental wheat populations. *Crop Science* 51(6): 2597–2606.

Heller-Uszynska, K., Uszynski, G., Huttner, E., Evers, M., Carlig, J., Caig, V., et al. 2011. Diversity arrays technology effectively reveals DNA polymorphism in a large and complex genome of sugarcane. *Molecular breeding* 28(1): 37–55.

Hema, M., Savithri, H.S., and P. Sreenivasulu. 2003. Comparison of direct binding polymerase chain reaction with recombinant coat protein antibody based dot-blot immunobinding assay and immunocapture-reverse transcription-polymerase chain reaction for the detection of Sugarcane streak mosaic virus causing mosaic disease of sugarcane in India. *Curr. Sci.* 85: 1774–1777.

Hemaprabha, G., Govindaraj, P., Balasundaram, N., and N.K. Singh. (2005). Genetic diversity analysis of Indian sugarcane breeding pool based on sugarcane specific STMS markers. *Sugar Tech* 7(2): 9–14.

Hoarau, J.Y., Dumont, T., Wei, X., Jackson, P., and A. D'Hont. 2022. Applications of quantitative genetics and statistical analyses in sugarcane breeding. *Sugar Tech* 24 (1): 320–340.

Hogarth, D.M., Cox, M.C., and J.K. Bull. 1997. Sugarcane improvement: past achievements and future prospects. *Crop improvement for the 21st century*. Louisiana State University, Baton Rouge, 29–56.

Iglesia, A., Díaz, M., Alvarez, E., Peralta, E.L., and V. Pazos. 2003. Optimización del diagnóstico múltiple de la escaldadura foliar y el raquitismo de los retoños de la caña de azúcar. *Rev. Protección Veg* 18(1): 15–18.

Ingelbrecht, I.L., Irvine, J.E., and T.E. Mirkov. 1999. Posttranscriptional gene silencing in transgenic sugarcane. Dissection of homology-dependent virus resistance in a monocot that has a complex polyploid genome. *Plant Physiol* 119: 1187–1197. doi:10.1104/pp.119.4.1187

Irvine, J.E. 1999. *Saccharum* Species as Horticultural Classes. *Theoretical and Applied Genetics* 98: 186–194.

ISAAA (2021). Crop Biotech Update, July. Available online at: https://www.isaaa.org/gmapprovaldatabase/crop/default.asp?CropID=27&Crop=Sugarcane(accessed november 15, 2021)

Isabel, N., Tremblay, L., Michaud, M., Tremblay, F.M., and J. Bousquet. 1993. RAPDs as an aid to evaluate the genetic integrity of somatic embryogenesis-derived populations of Picea mariana (Mill.) BSP. *Theoretical and Applied Genetics* 86(1): 81–87.

Jackson, P.A. (2005). Breeding for improved sugar content in sugarcane. *Field Crops Research* 92, 277–290. doi:10.1016/j.fcr.2005.01.024

Jackson, P., and R.J. Henry 2011. *Erianthus*. In Wild Crop Relatives: Genomic and Breeding Resources, Industrial Crops. Kole, C. (Ed.). Springer-Verlag, Berlin, pp. 97–107.

Jackson, P. 2016. Maximising the rate of parental improvement in the Australian Sugarcane Breeding Program Final report submitted to Sugar Research.

Jackson, P.A. 2018. Advances in conventional sugarcane breeding programmes. In Achieving sustainable cultivation of sugarcane Volume 2: Breeding, pests and diseases, ed.Rott, P., Burleigh Dodds Science Publishing, Cambridge, UK, 2018, (ISBN: 978 1 78676 148 4; www.bdspublishing.com).

Jalaja, N.C., Neelamathi, D., and T.V. Sreenivasan 2008. Micropropagation for quality seed production in sugarcane in Asia and the Pacific (Vol. 1). Food and Agriculture Organization of the United Nations.

Jannoo, N., Grivet, L., David, J., D'Hont, A., and J.C. Glaszmann. 2004. Differential chromosome pairing affinities at meiosis in polyploid sugarcane revealed by molecular markers. *Heredity* 93(5), 460–467.

Johnson, A. 2019. Development and validation of a novel quantitative real-time PCR for the detection of Leifsonia xyli subsp. xyli in sugarcane. *Proceedings of the 2019 Conference of the Australian Society of Sugar Cane Technologists* 178–181.

Jung, J.H., and F. Altpeter. 2016. TALEN mediated targeted mutagenesis of the caffeic acid O-methyltransferase in highly polyploid sugarcane improves cell wall composition for production of bioethanol. *Plant Mol. Biol.* 92:131–142.

Junior, C.A.D.K., Manechini, J.R.V., Corrêa, R.X., Pinto, A.C.R., Costa, J.B., Favero, T.M., et al. 2020. Genetic structure analysis in sugarcane (*Saccharum* spp.) using target region amplification polymorphism (TRAP) markers based on sugar- and lignin-related genes and potential application in core collection development. *Sugar Tech.* doi: 10.1371/journal.pone.0233211

Kannan, B., Jung, J.H., Moxley, G.W., Lee, S.M., and F. Altpeter. 2018. TALEN-mediated targeted mutagenesis of more than 100 COMT copies/alleles in highly polyploid sugarcane improves saccharification effciency without compromising biomass yield. *Plant Biotechnol. J.* 16: 856–866.

Kaur, A.P., and Sandhu, J.S. 2014. High throughput in vitro micropropagation of sugarcane (Saccharum officinarum L.) from spindle leaf roll segments: Cost analysis for agri-business industry. *Plant Cell, Tissue and Organ Cultur.* 120: 339–350.

Kawar, P.G., Devarumath, R.M., and Y. Nerkar. 2009. Use of RAPD markers for assessment of genetic diversity in sugarcane cultivars. *Indian Journal of Biotechnology* 8: 67–71.

Keizerweerd, A.T., Chandra, A., and M.P. Grisham. 2015. Development of a reverse transcription loop-mediated isothermal amplification (RT-LAMP) assay for the detection of Sugarcane mosaic virus and Sorghum mosaic virus in sugarcane. *Journal of virological methods* 212: 23–29.

Korimbocus, J., Coates, D., Barker, I., and N. Boonham. 2002.Improved detection of Sugarcane yellow leaf virus using a real-time fluorescent (TaqMan) RT-PCR assay. *J Virol Methods* 103(2): 109–120. doi: 10.1016/s0166-0934(01)00406-2. PMID: 12008005.

Krishnamurthi, M., Sundaram, K.S., Sekar, S., Rajeswari, S., and P.G. Kawar. 2007. Introgression of *Erianthus* for the development of commercial sugarcane cultivars. *Proceedings of the International Society of Sugar Cane Technologists* 26: 678–680.

Kumar, K., Gambhir, G., Dass, A., Tripathi, A.K., Singh, A., Jha, A.K., Yadava, P., Choudhary, M., and Rakshit, S. 2020. Genetically modified crops: current status and future prospects. *Planta* 251(4): 1–27.

Lal, N., and R. Krishna. 1994. Tissue culture for pure and disease-free seed production in sugarcane. *Indian Sugar* 43: 847–848.

Larkin, P.J., and W.R. Scowcroft. 1981. Somaclonal variation—a novel source of variability from cell cultures for plant improvement. *Theoretical and Applied Genetics* 60(4): 197–214.

Larkin, P.J., Ryan, S.A., Brettell, R.I.S., and W.R. Scowcroft. 1984. Heritable somaclonal variation in wheat. *Theoretical and Applied Genetics* 67(5): 443–455.

Larkin, P.J. 1998. Introduction. In Somaclonal variation and induced mutations in crop improvement. Jain, S.M., Brar, D.S., and Ahloowalia, B.S. (Eds). Kluwer Academic Publishers, Dordrecht, pp. 3–13.

Lavanya, L., and G. Hemaprabha. 2010. Genetic diversity within sucrose rich parental pool of sugarcane and its application in sugarcane breeding through hybridization and selection. *Indian Journal of Genetics and Plant Breeding* 70(2): 172–181.

Lewi, D.M., and C. Vicién. 2020. Argentina's local crop biotechnology developments: why have they not reached the market yet?. *Frontiers in bioengineering and biotechnology* 8: 301.

Li, G., and C.F. Quiros. 2001. Sequence related amplified polymorphism (SRAP), a new marker system based on a simple PCR reaction: Its application to mapping and gene tagging in Brassica. *Theoretical and Applied Genetics* 103: 455–461.

Li, W.F., Wang, X.Y., Huang, Y.K., Zhang, R.Y., Shan, H.L., Yin, J., and Z.M. Luo. 2017. Molecular detection of *Bru*1 gene and identification of brown rust resistance in Chinese sugarcane germplasm. *Sugar Tech* 19(2): 183–190.

Lima, M.L.A., Garcia, A.A.F., Oliveira, K.M., Matsuoka, S., Arizono, H., de Souza Jr, C.L., and A.P. De Souza. 2002. Analysis of genetic similarity detected by AFLP and coefficient of parentage among genotypes of sugar cane (Saccharum spp.). *Theoretical and Applied Genetics* 104(1): 30–38.

Liu, J., Xu, L., Guo, J., Chen, R., Grisham, M.P., and Y. Que. 2013. Development of loop-mediated isothermal amplification for detection of *Leifsonia xyli* subsp. *xyli* insugarcane. *Biomed. Res. Int.* 2013 1–8.

Lourens, A.G., and F.A. Martin. 1987. Evaluation of in Vitro Propagated Sugarcane Hybrids for Somaclonal Variation1. *Crop Science* 27: 793–796 cropsci1987.0011183X002700040038x. 10.2135/cropsci1987.0011183X002700040038x

Lozano, A.I., Díaz, M., Peralta, E.L., Pazos, V., and E. Alvarez. 2003. Diagnóstico del raquitismo de los retoños de la caña de azúcar (*Leifsonia xyli* subsp *xyli*) mediante la técnica de la reacción en cadena de la polimerasa anidada (nPCR). *Rev. Protección Veg.* 18(1): 19–22.

Lu, Y.H., D'Hont, A., Walker, D.I.T., Rao, P.S., Feldmann, P., and J.C. Glaszmann. 1994. Relationships among ancestral species of sugarcane revealed with RFLP using single copy maize nuclear probes. *Euphytica* 78(1): 7–18. 10.1007/BF00021393

Manechini, J.R.V., da Costa, J.B., Pereira, B.T., Carlini-Garcia, L.A., Xavier, M.A., Landell, M.G.D.A., and L.R. Pinto. 2018. Unraveling the genetic structure of Brazilian commercial sugarcane cultivars through microsatellite markers. *PloS one* 13(4): e0195623.

Manigbas, N.L., and L.C. Villegas. 2004. Microsatellite markers in hybridity tests to identify true hybrids of sugarcane. *Philipp J Crop Sci* 29(2): 23–32.

Marconi, T.G., Costa, E.A., Miranda, H.R., Mancini, M.C., Cardoso-Silva, C.B., Oliveira, K.M., et al. 2011. Functional markers for gene mapping and genetic diversity studies in sugarcane. *BMC Research Notes* 4(1): 1–9.

Martin, L., Lloyd Evans, D., Castlebury, L., Sifundza, J., Comstock, J., Rutherford, S., and S. McFarlane. 2015. An update on the new species of rust infecting sugarcane in southern Africa. ISSCT XI Pathology and IX Entomology Workshops, Guayaquil, Ecuador.

Marvellous Z. 2020. General and specific combining ability effects for sugarcane yield among South African breeding populations. *South African Journal of Plant and Soil.* doi: 10.1080/02571862.2020.1749318

Matsuoka, S., Kennedy, A.J., Dos Santos, E.G.D., Tomazela, A.L., and L.C.S. Rubio. 2014. Energy cane: its concept, development, characteristics, and prospects. *Adv Bot.* 597275. 10.1155/2014/597275.

Mbuma, N.W., Zhou, M., and R. van der Merwe. 2019. Evaluating parents for cane yield in sugarcane breeding using best linear unbiased prediction analysis of progeny data derived from family plots. *South African Journal of Plant and Soil* 36(1): 21–28.

McIntyre, C.L., and P.A. Jackson 2001. Low level of selfing found in a sample of crosses in Australian sugarcane breeding programs. *Euphytica* 117(3): 245–249.

Medeiros, C., Balsalobre, T.W.A., and M.S. Carneiro. 2020. Molecular diversity and genetic structure of *Saccharum* complex accessions. *Plos One* 15(5): e0233211. 10.1371/journal.pone.0233211.

Mendes de Paula, T.O., Brasileiro, B.P., Cursi, D.E., de Freitas, E.G., dos Santos, J.M., de Resende, M.D.V., Kimbeng, C., and M.H. Pereira Barbosa. 2020. Establishment of gene pools for systematic heterosis exploitation in sugarcane breeding. *Agronomy Journal* 112(5): 3847–3858.

Meuwissen, T.H., Hayes, B.J., and M.E. Goddard. 2001. Prediction of total genetic value using genome-wide dense marker maps. *Genetics* 157(4): 1819–1829.

Ming, R., Liu, S.C., Bowers, J.E., Moore, P.H., Irvine, J.E., and A.H. Paterson. 2002. Construction of a *Saccharum* consensus genetic map from two interspecific crosses. *Crop Science* 42(2): 570–583.

Ming, R., Moore, P.H., Wu, K., D Hont, A., Glaszmann, J.C., Tew, T.L., Mirkov, T., da Silva, J., Lifon, J., Rai, M., Schnell, R.J., Brumbley, S.M., Lakshmanan, P., Comstock, J.C., and A.H. Paterson. 2006. Sugarcane improvement through breeding and biotechnology. *Plant breeding reviews* 27: 15.

Mohan, C. 2016. Genome editing in sugarcane: challenges ahead. *Front. Plant Sci.* 7: 1542. doi:10.3389/fpls.2016.01542

Nair, N. V., and Mary, S. 2006. Rapid analysis reveals the presence of mainland Indian and Indonesian forms of Erianthus arundinaceus (Retz.) Jeswiet in the Andaman–Nicobar Islands. Current, 1118–1122.

Nair, N.V., Selvi, A., Sreenivasan, T.V., and K.N. Pushpalatha. 2002. Molecular diversity in Indian sugarcane cultivars as revealed by randomly amplified DNA polymorphisms. *Euphytica* 127(2): 219–225.

Neto, H.Z., Daros, E., Bespalhok Filho, J.C., Scapim, C.A., Vidigal, M.C.G., and P.S. Vidigal Filho. 2013. Selection of families and parents of sugarcane (*Saccharum* spp.) through mixed models by joint analysis of two harvests. *Euphytica* 193(3): 391–408.

Noguera, A.S., Paz, N.D.V., Díaz, M.E., Perera, M.F., Sepúlveda Tusek, M., Filippone, M.P., and A.P. Castagnaro. 2010. La producción de caña semilla de alta calidad comienza en el laboratorio. Publicación especial vitro plantas 13–20.

Noguera, A., Enrique, R., Perera, M.F., Ostengo, S., Racedo, J., Costilla, D., et al. 2015. Genetic characterization and field evaluation to recover parental phenotype in transgenic sugarcane: a step toward commercial release. *Moecular Breeding* 35: 115. doi:10.1007/s11032-015-0300-y

Notomi, T., Okayama, H., Masubuchi, H., Yonekawa, T., Watanabe, K., Amino, N., and T. Hase. 2000. Loop-mediated isothermal amplification of DNA. *Nucleic Acids Research* 28(12): e63.

Ostengo, S. 2020. Aproximaciones metodológicas para la caracterización de la calidad industrial de genotipos de caña de azúcar. PhD Thesis. National University of Tucumán, Argentina.

Pan, Y.B., Grisham, M.O., and D.M. Burner. 1997. A polymerase chain reaction protocol for the detection of *Xanthomonas albilineans*, the causal agent of sugarcane leaf scald disease. *Plant Disease* 81(2): 189–194.

Pan, Y.B. 1997. A RT-PCR protocol for Sugarcane yellow leaf syndrome virus. USDA, ARS SRRC SRU. www.usda.gov/wps/portal/usda/usdahome (acessed on february, 2021).

Pan, Y.B., M.P. Grisham, D.M., Burner, K.E., Damann, and Q. Wei. 1998. A polymerase chain reaction protocol for the detection of Clavibacter xyli subsp. xyli, the causal bacterium of sugarcane ratoon stunting disease. *Plant Disease* 82: 285–290.

Pan, Y.-B., Burner, D.M., & Legendre, B.L.2000. Genetica 108: 285–295. 10.1023/a:1004191625603

Pan, Y.B., Cordeiro, G.M., Richards, E.P., and R.J. Henry. 2003. Molecular genotyping of sugarcane clones with microsatellite DNA markers. *Maydica*, 48(4): 319–329.

Pan, Y.B., Burner, D.M., Legendre, B., et al. 2005. An assessment of the genetic diversity within a collection of *Saccharum spontaneum* L. with RAPD-PCR. *Genet Resour Crop Evol* 51: 895–903.

Pan, Y.B. 2006. Highly polymorphic microsatellite DNA markers for sugarcane germplasm evaluation and variety identity testing. *Sugar Tech* 8(4): 246–256.

Pan, Y.B. 2007. High throughput genotyping of commercial sugarcane clones with microsatellite (SSR) DNA markers. *Sugar Tech.* 9: 176–181.

Pan, Y.B. 2010. Databasing molecular identities of sugarcane (*Saccharum* spp.) clones constructed with microsatellite (SSR) DNA markers. *The American Journal of Plant Sciences* 1(2): 87–94.

Pan, Y.B., Liu, P., and Y. Que. 2015. Independently segregating simple sequence repeats (SSR) alleles in polyploid sugarcane. *Sugar Tech* 17(3): 235–242.

Pan, Y.B., Grisham, M.P., Burner, D.M., Legendre, B.L., B.and Q. Wei. 1999. Development of a polymerase chain reaction primer highly specific for *Xanthomonas albilineans*, the causal bacterium of sugarcane leaf scald disease. *Plant Disease* 83(3): 218–222.

Pandey, R.N., Singh, S.P., Rastogi, J., Sharma, M.L., and R.K. Singh. 2012. Early assessment of genetic fidelity in sugarcane (Saccharum officinarum) plantlets regenerated through direct organogenesis with RAPD and SSR markers. *Aus. J. Crop Sci.* 6: 618–624.

Pelosi, C.S., Lourenço, M.V., Silva, M., Santos, A.Z., França, S.C., and M. Marins. 2013. Development of a Taqman real-time PCR assay for detection of *Leifsonia xyli* subsp *xyli. Tropical Plant Pathology* 38: 343–345.

Perera, M.F., Arias, M.E., Costilla, D., Luque, A.C., García, M.B., Cuenya, M.I., Racedo, J., Ostengo, S., Filippone, M.P., and A.P. Castagnaro. 2012a. Genetic diversity assessment and genotype identification in sugarcane based on DNA markers and morphological traits. *Euphytica* 185(3): 491–510. DOI: 10.1 007/s10681-012-0661-9.

Perera, M.F., García, M.B., Díaz Romero, C., Cuenya, M.I., Filippone, M.P., and A.P. Castagnaro, 2012b. Biparental crosses confirmed by SSR with mendelian inheritance in sugarcane breeding. *Revista Industrial y Agrícola de Tucumán* 89(2): 1–7.

Perera, M.F., García, M.B., Díaz, R., Cuenya, M.I., Filippone, M.P., and A.P. Castagnaro. 2013. Validation of the effectiveness of an emasculation treatment in sugarcane by using molecular markers. In *International Society of Sugar Cane Technologists: Proceedings of the XXVIIIth Congress, June 24 to June 27, 2013, São Paulo, Brazil.* Sociedade dos Técnicos Açucareiros e Alcooleiros do Brasil (STAB) & The XXVIIIth ISSCT Organising Committee.

Perera, M.F., Racedo, J., García, M.G., Pardo, E.M., Rocha, C.M.L., Orce, I.G., Chiesa, M.A., Filippone, M.P., Welin, B., and A.P. Castagnaro. 2016. Use of molecular markers to improve the agro-industrial productivity in the North West of Argentina. *Molecular Biology* 5: 153. doi 10.4172/2168-9547. 1000153

Perera, M.F., Costilla, D.D., Ovejero, N.S., Aybar Guchea, M., Figueroa, M.F., Noguera, A.S., Cuenya, M.I., and A.P. Castagnaro. 2020a. Caracterización botánica mediante marcadores fenotípicos y moleculares de variedades de caña de azúcar obtenidas por el Programa de Mejoramiento Genético de la EEAOC. *Revista Industrial y Agrícola de Tucumán* 97(2): 51–55.

Perera, M.F., Ovejero, N.S., Racedo, J., Noguera, A.S., Cuenya, M.I., and A.P. Castagnaro. 2020b. TRAP markers allow identifying transgenic events genetically close to their parental genotype. *Sugar Tech.* DOI 10.1007/s12355-020-00836-9

Piepho, H.P., Möhring, J., Melchinger, A.E., and A. Büchse. 2008. BLUP for phenotypic selection in plant breeding and variety testing. *Euphytica* 161: 209–228. doi:10.1007/s10681-007-9449-8

Pinto, L.R., Garcia, A.A.F., Pastina, M.M., Teixeira, L.H.M., Bressiani, J.A., Ulian, E.C. et al. 2010. *Euphytica* 172(3): 313–327

Piperidis, G., Christopher, M.J., Carroll, B.J., Berding, N., and A. D'Hont. 2000. Molecular contribution to selection of intergenic hybrids between sugarcane and the wild species *Erianthis arundinaceus. Genome* 43: 1033–1037.

Piperidis, N., Chen, J.W., Deng, H.H., Wang, L.P., Jackson, P., and G. Piperidis. 2010a. GISH characterization of *Erianthus arundinaceus* chromosomes in three generations of sugarcane intergeneric hybrids. *Genome* 53: 331–336.

Piperidis, G., Piperidis, N., and A. D'Hont. 2010b. Molecular cytogenetic investigation of chromosome composition and transmission in sugarcane. *Molecular Genetics and Genomics* 284(1): 65–73. 10.1007/ s00438-010-0546-3

Pisaroglo de Carvalho, M., Gezan, S.A., Peternelli, L.A., and M.H. Pereira Barbosa. 2014. Estimation of additive and nonadditive genetic components of sugarcane families using multitrait analysis. *Agronomy Journal* 106(3): 800–808.

Priyadarshan P.M., and S. Mohan Jain. 2022. Cash Crops: An Introduction. In *Genetic diversity, erosion, conservation, and utilization - Cash crops*. Priyadarshan, P.M., and Jain, S. Mohan (Eds.). Springer. 10.1007/978-3-030-74926-2

QI, Y.-W., Pan, Y.-B., Lao, F.-Y., Zhang, C.-M., Fan, L.-N., He, H.-Y., Liu, R., Wang, Q.-N., Liu, S.-M., Liu, F.-Y., Li, Q.-W., & Deng, H.-H.2012. Genetic structure and diversity of parental cultivars involved in China mainland sugarcane breeding programs as Inferred from DNA Microsatellites. *Journal of Integrative Agriculture*, 11:1794–180310.1016/s2095-3119(12)60184-1.

Qi, Y., Gao, X., Zeng, Q., Zheng, Z., Wu, C., Yang, R., et al. 2022. Sugarcane Breeding, Germplasm Development and Related Molecular Research in China. *Sugar Tech* 24(1): 73–85.

Qi, Y.W., Pan, Y.B., Lao, F.Y., Zhang, C.M., Fan, L.N., He, H.Y., et al. 2012. Genetic structure and diversity of parental cultivars involved in China mainland sugarcane breeding programs as inferred from DNA microsatellites. *Journal of Integrative Agriculture*, 11(11), 1794–1803.

Que, Y., Pan, Y., Lu, Y., Yang, C., Yang, Y., Huang, N., and Xu, L. 2014. Genetic analysis of diversity within a Chinese local sugarcane germplasm based on start codon targeted polymorphism. *BioMed research international*. 10.1155/2014/468375

Que, Y.X., Chen, T.S., Xu, L.P., and R.K. Chen. 2009. Genetic diversity among key sugarcane clones revealed by TRAP markers. *Journal of Agricultural Biotechnology* 17(3): 496–503.

Raboin, L.M., Pauquet, J., Butterfield, M., D'Hont, A., and J.C. Glaszmann 2008. Analysis of genome-wide linkage disequilibrium in the highly polyploid sugarcane. *Theoretical and Applied Genetics*. 116(5): 701–714.

Racedo, J., Perera, M.F., Bertani, R., Funes, C., González, V., Cuenya, M.I., et al. 2013. *Bru*1 gene and potential alternative sources of resistance to sugarcane brown rust disease. *Euphytica* 191(3): 429–436.

Racedo, J. 2014. Aproximaciones biotecnológicas aplicadas al mejoramiento de caña de azúcar para la obtención de variedades diferenciadas en su potencial industrial: azucareras y alcoholeras. PhD Thesis. National University of Tucumán, Argentina.

Racedo, J., Gutiérrez, L., Perera, M.F., Ostengo, S., Pardo, E.M., Cuenya, M.I., et al. 2016. Genome-wide association mapping of quantitative traits in a breeding population of sugarcane. *BMC plant biology* 16(1): 1–16.

Rahman, M., Sultana, S., Nath, D., Kalita, S., Chakravarty, D., Mehta, S., et al. 2019. Molecular breeding approaches for disease resistance in sugarcane. in Disease Resistance in Crop Plants. Wani, S. (ed.). Springer: Cham. doi: 10.1007/978-3-030-20728-1_7

Rithesh, S. 2013. *Comparative economics of sugarcane processed for sugar and jiggery* (Doctoral dissertation, UASD).

Roddee, J., Kobori, Y., and Y. Hanboonsong. 2018. multiplication and distribution of sugarcane white leaf phytoplasma transmitted by the leafhopper, matsumuratettix hiroglyphicus (matsumura) (hemiptera: cicadellidae), in infected sugarcane. *Sugar Tech* 20: 445–453. 10.1007/s12355-017-0559-x

Salokhe, S. 2021. Development of an efficient protocol for production of healthy sugarcane seed cane through Meristem culture. *Journal of Agriculture and Food Research* 4: 100126

Sanghera, G.S., and N.S. Jamwal. 2019. Identification of potential crosses based on vigour, cane characteristics and HR Brix for first clonal selection in sugarcane. *Indian Journal of Sugarcane Technology* 34(01): 12–16.

Sawant, R.A., Tawar, P.N., Meti, N.T., and P.K. Ranjekar. 2014. Role of sugarcane micropropagation for production of quality seed. *International Journal of Recent Biotechnology* 2: 34–41.

Sawazaki, H.S., Gonçalves, C.R.N.C.B, Polez, V.L.P., Ribeiro, C., Martins, M.C., Carvalho, V.C., Senger, M.M.M., and R.A. Veiga. 2010a. Infecção da ferrugem laranja na região de Piracicaba, SP. *Summa Phytopathol*. 36: 168.

Sawazaki, H.E., Gonçalves, C.R.N.C.B., Sa, L.A.N., Ribeiro, C., Martins, M.C., Carvalho, V.C., Senger, M.M.M., and R.F.A. Veiga. 2010b. Análises moleculares de bactérias e fungo em cana-de-açúcar. *Summa Phytopathol* 36: 217.

Sawazaki, H.E., de SA, L.A.N., Gonçalves, C.R.N.C.B., Veiga, R., and C.A. Colombo. 2013. Molecular diagnosis optimization of virus, bacteria and fungi in sugarcane. *Embrapa Meio Ambiente-Artigo* em periódico indexado (ALICE).

Selvi, A., Nair, N.V., Noyer, J.L., Singh, N.K., Balasundaram, N., Bansal, K.C., et al. 2005. Genomic constitution and genetic relationship among the tropical and subtropical Indian sugarcane cultivars revealed by AFLP. *Crop Science* 45(5): 1750–1757.

Selvi, A., Nair, N.V., Noyer, J.L., Singh, N.K., Balasundaram, N., Bansal, K.C., et al. 2006. AFLP analysis of the phenetic organization and genetic diversity in the sugarcane complex, Saccharum and Erianthus. *Genetic Resources and Crop Evolution* 53(4): 831–842.

Shearman, J.R., Chutima, S., C. Naktang, et al. 2018. Assembling a Hybrid Sugarcane Genome. PAG Conference.

Shi, Y., Zhao, J.Y., Zhou, J.R., Ntambo, M.S., Xu, P.Y., Rott, P.C., and S.J. Gao. 2021. Molecular detection and quantification of *Xanthomonas albilineans* in juice from symptomless sugarcane stalks using a real-time quantitative PCR assay. *Plant Disease* 10.1094/PDIS-03-21-0468-RE

Sindhu, R., Govindaraj, P., Balamurugan, A., and C. Appunu 2011. Genetic diversity in sugarcane hybrids (*Saccharum* spp complex) grown in tropical India based on STMS markers. *Journal of plant biochemistry and biotechnology* 20(1): 118–124.

Singh, D., Tewari, A.K., Rao, G.P., Karuppaiah, R., Viswanathan, R., Arya, M., and V.K. Baranwal. 2009. RT-PCR/PCR analysis detected mixed infection of DNA and RNA viruses infecting sugarcane crops in different states of India. *Sugar Tech* 11(4): 373–380.

Singh, R., Mishra, S.K., Singh, S.P., Mishra, N., and M. Sharma. 2010. Evaluation of microsatellite markers for genetic diversity analysis among sugarcane species and commercial hybrids. *Australian Journal of Crop Science* 4(2): 116–125. https://search.informit.org/doi/10.3316/informit.982597773668412

Singh, R.K., Jena, S.N., Khan, S., Yadav, S., Banarjee, N., Raghuvanshi, S., et al. 2013. Development, cross-species/genera transferability of novel EST-SSR markers and their utility in revealing population structure and genetic diversity in sugarcane. *Gene* 524(2): 309–329.

Singh, R.B., Singh, B., and R.K. Singh. 2017. Study of genetic diversity of sugarcane (Saccharum) species and commercial varieties through TRAP molecular markers. *Ind J Plant Physiol.* 22: 332–338. 10.1007/s40502-017-0314-z

Siraree, A., Banerjee, N., Kumar, S., Khan, M.S., Singh, P.K., Sharma, S., et al. 2017. Identification of marker-trait associations for morphological descriptors and yield component traits in sugarcane. *Physiology and molecular biology of plants* 23(1): 185–196.

Smith, G.R., and R. van de Velde. 1994. Detection of sugarcane mosaic virus and Fiji disease virus in diseased sugarcane using the polymerase chain reaction. *Plant Disease* 78: 557–561.

Song, W.Y., Sechler, A.J., Hatziloukas, E., Kim, H.M., and N.W. Schaad. 2003. Use of PCR for rapid identification of *Acidovorax avenae* and *A. avenae* subsp. citrulli. In *Pseudomonas syringae and related pathogens* (pp. 531–544). Springer, Dordrecht.

Soniya, E.V., Banerjee, N.S., and M.R. Das. 2001. Genetic analysis of somaclonal variation among callus-derived plants of tomato. *Current science* 80(9): 1213–1215.

Souza, G.M., Van Sluys, M.A., Lembke, C.G., Lee, H., Margarido, G.R.A., Hotta, C.T. et al. 2019. Assembly of the 373k gene space of the polyploid sugarcane genome reveals reservoirs of functional diversity in the world's leading biomass crop. *GigaScience* 8(12): 10.1093/gigascience/giz129.

Spindel, J., Begum, H., Akdemir, D., Virk, P., Collard, B., Redona, E., et al. 2015. Genomic selection and association mapping in rice (Oryza sativa): effect of trait genetic architecture, training population composition, marker number and statistical model on accuracy of rice genomic selection in elite, tropical rice breeding lines. *PLoS genetics* 11(2): e1004982.

Sreenivasan, T.V., and N.C. Jalaja. 1981. Utilization of tissue culture technique in sugarcane improvement. *C. Meristem culture. Annual Report. Sugarcane Breeding Institute, Coimbatore*, 68.

Stevenson, G.C. 1965. Genetics and Breeding of Sugarcane. pp. 284. Tropical Science Series. Longmans, Green y Co Ltd., Londres

Sugiharto, B. 2018. Biotechnology of drought-tolerant sugarcane. In Sugarcane–Technology and Research. de Oliveira, A.B. (ed.). InTech, London.

Suman, A., Kimbeng, C.A., Edmé, S.J., and J. Veremis. 2008. Sequence-related amplified polymorphism (SRAP) markers for assessing genetic relationships and diversity in sugarcane germplasm collections. *Plant Genetic Resources* 6(3): 222.

Suprasanna, P. 2010. Biotechnological interventions in sugarcane improvement: Strategies, methods and progress. *BARC Newsletter* 45(11): 47–53.

Tew, T.L., and Y.B. Pan. 2010. Microsatellite (simple sequence repeat) marker-based paternity analysis of a seven-parent sugarcane polycross. *Crop Science* 50(4): 1401–1408. 10.2135/cropsci2009.10.0579

Todd, J., Pan, Y.B., and D. Boykin. 2020. Fidelity of sugarcane crosses assessed with SSR markers. *Agronomy* 10(3): 386.

Tufan Oz, M., Altpeter, A., Karan, R., and A. Merotto. 2021. CRISPR / Cas9mediated multi-allelic gene targeting in sugarcane confers herbicide tolerance. *Front. Genome Ed.* 3: 673566. doi:10.3389/fgeed.2021.673566

VanRaden, P.M. 2008. Efficient methods to compute genomic predictions. *Journal of Dairy Science* 91: 4414–4423. (Available online at: 10.3168/jds.2007-0980) (verified 29 Feb 2012).

Vázquez, A.M. 2001. Insight into somaclonal variation. *Plant Biosystems-An International Journal Dealing with all Aspects of Plant Biology* 135(1): 57–62.

Van de Wouw, M., Kik, C., van Hintum, T., van Treuren, R., and B. Visser. 2010. Genetic erosion in crops: concept, research results and challenges. *Plant genetic resources* 8(1): 1–15.

Viana, J.M.S., DeLima, R.O., Faria, V.R., Mundim, G.B., de Resende, M.D.V., and F.F. de Silva. 2012. Relevance of pedigree, historical data, dominance, and data unbalance for selection efficiency. *Agronomy Journal* 104(3): 722–728.

Vieira, M.L.C., Almeida, C.B., Oliveira, C.A., Tacuatiá, L.O., Munhoz, C.F., Cauz-Santos, L.A., Pinto, L.R., Monteiro-Vitorello, C.B., Xavier, M.A., and E.R. Forni-Martins. 2018. Revisiting meiosis in sugarcane: chromosomal irregularities and the prevalence of bivalent configurations. *Front. Genet.* 9: 213. doi: 10.3389/fgene.2018.00213.

Virtudazo, E.V., Nakamura, H., and M. Kakishima. 2001. Phylogenetic analysis of sugarcane rusts based on sequences of ITS, 5.8 S rDNA and D1/D2 regions of LSU rDNA. *Journal of General Plant Pathology* 67: 28–36.

Viswanathan, R., Balamuralikrishnan, M., and R. Karuppaiah. 2007. Sugarcane mosaic in India: A cause of combined infection of sugarcane mosaic virus and sugarcane streak mosaic virus. *Sugar Cane Intern* 25: 6–14.

Viswanathan, R., Balamuralikrishnan, M., and R. Karuppaiah. 2008a. Characterization and genetic diversity of sugarcane streak mosaic virus causing mosaic in sugarcane. *Virus Genes* 36: 553–564.

Viswanathan, R., Balamuralikrishnan, M., and R. Karuppaiah. 2008b. Duplex-reverse transcription-polymerase chain reaction (D-RT-PCR) a technique for the simultaneous detection of viruses causing sugarcane mosaic. *Sugar Tech* 10: 81–86.

Viswanathan, R., Balamuralikrishnan, M., and R. Karuppaiah. 2008c. Identification of three genotypes of Sugarcane yellow leaf virus causing yellow leaf disease from India and their molecular characterization. *Virus Genes* 37: 368–379.

Viswanathan, R., Karuppaiah, R., and M. Balamuralikrishnan. 2010. Detection of three major RNA viruses infecting sugarcane by multiplex reverse transcription-polymerase chain reaction (multiplex-RT-PCR). *Australas. Plant Pathol* 39: 79–84.

Viswanathan, R., and G.P. Rao, G. 2011. Disease scenario and management of major sugarcane diseases in India. *Sugar Tech* 13(4): 336–353.

Viswanathan, R., Ganesh Kumar, V., Karuppaiah, R. et al. 2013. Development of Duplex-Immunocapture (Duplex-IC) RT-PCR for the Detection of Sugarcane streak mosaic virus and Sugarcane mosaic virus in Sugarcane. *Sugar Tech* 15:399–405. 10.1007/s12355-013-0216-y

Viswanathan, R. 2016. Varietal degeneration in sugarcane and its management in India. *Sugar Tech* 18(1): 1–7.

Viswanathan, R., Parameswari, B., and K. Nithya. 2018. Molecular characterization of sugarcane viruses and their diagnostics. In Crop improvement through microbial biotechnology Elsevier. pp. 175–193.

Voss-Fels, K.P., Wei, X., Ross, E.M., Frisch, M., Aitken, K.S., Cooper, M., and B.J. Hayes. 2021. Strategies and considerations for implementing genomic selection to improve traits with additive and non-additive genetic architectures in sugarcane breeding. *Theoretical and Applied Genetics*, 134(5): 1493–1511.

VV Tew, T.L., and Y.B. Pan. 2010. Microsatellite (simple sequence repeat) marker-based paternity analysis of a seven-parent sugarcane polycross. *Crop Science* 50: 1401–1408. doi:10.2135/cropsci2009.10.0579.

Walkey, D.G.A., and W. DGA. 1980. Production of virus-free plants by tissue culture.

Wang, H.B., Xiao, N.Y., Wang, Y.J., Guo, J.L., and J.S. Zhang. 2020. Establishment of a qualitative PCR assay for the detection of *Xanthomonas albilineans* (Ashby) Dowson in sugarcane. *Crop Protection* 130: 105–153.

Wang, Z.K., Comstock, J.C., Hatziloukas, E., and N.W. Schaad. 1999. Comparison of PCR, Bio-PCR, DIA, ELISA, and isolation on semiselective medium for detection of *Xanthomonas albilineans*, the causal agent of leaf scald of sugarcane. *Plant Pathology* 48: 245–252.

Wang, Z., Pan, Y., Luo, J., You, Q., Xu, L., Zhang, H., and Y. Que. 2020. SSR-Based Genetic Identity of Sugarcane Clones and its Potential Application in Breeding and Variety Extension. *Sugar Tech* 22(3): 367–378. 10.1007/s12355-019-00788-9

Weeks, D.P. 2017. Gene editing in polyploid crops: wheat, camelina, canola, potato, cotton, peanut, sugar cane, and citrus. *Progress in molecular biology and translational science* 149, 65–80.

Wei, X. 2016. Maximising the rate of parental improvement in the Australian Sugarcane Breeding Program: final report 2008/319.

Wei, X., Jackson, P.A., Hermann, S., Kilian, A., Heller-Uszynska, K., and E. Deomano. 2010. Simultaneously accounting for population structure, genotype by environment interaction, and spatial variation in marker–trait associations in sugarcane. *Genome* 53(11): 973–981.

Wei, X., Eglinton, J., Piperidis, G., Atkin, F., Morgan, T., Parfitt, R., and F. Hu. 2022. Sugarcane Breeding in Australia. *Sugar Tech* 24(1): 151–165.

Wei-Lin, F., Sheng-Ren, S., Hua-Ying, F., Ru-Kai, C., Jin-Wei, S., and G. San-Ji. 2015. A One-Step Real-Time RT-PCR Assay for the Detection and Quantitation of Sugarcane Streak Mosaic Virus. *BioMed Research International* 10.1155/2015/569131

Wei, X., and Jackson, P. 2017. Addressing slow rates of long-term genetic gain in sugarcane. Pakistan Sugar Journal 32(4): 23.

Wu, J., Wang, Q., Xie, J., Pan, Y.B., Zhou, F., Guo, Y., et al. 2019. SSR marker-assisted management of parental germplasm in sugarcane (*Saccharum* spp. hybrids) breeding programs. *Agronomy* 9(8):449.

Xavier, M.A., Pinto, L.R., Fávero, T.M., Perecin, D., Carlini-Garcia, L.A., and M.G.A. Landell. 2014. Paternity identification in sugarcane polycrosses by using microsatellite markers. *Genetics and Molecular Research* 13(1): 2268–2277. 10.4238/2014.March.31.7

Xie, Y., Wang, M., Xu, D., Li, R., and G. Zhou. 2009. Simultaneous detection and identification of four sugarcane viruses by one-step RT-PCR. *J. Virol. Methods* 162(1-2):64–68.

Yang, Z.N., and T.E. Mirkov. 1997. Sequence and relationships of sugarcane mosaic and sorghum mosaic virus strains and development of RT-PCR-based RFLPs for strain discrimination. *Phytopathology* 87(9): 932–939.

You, Q., Xu, L., Zheng, Y., and Y. Que. 2013. Genetic diversity analysis of sugarcane parents in Chinese breeding programmes using gSSR Markers. The Scientific World Journal, 2013:1–1110.1155/2013/613062

You, Q., Pan, Y.B., Xu, L.P., Gao, S.W., Wang, Q.N., Su, Y.C., et al. 2016. Genetic diversity analysis of sugarcane germplasm based on fluorescence-labeled simple sequence repeat markers and a capillary electrophoresis-based genotyping platform. *Sugar Tech* 18(4), 380–390.

Zhang, J., Nagai, C., Yu, Q., Pan, Y.B., Ayala-Silva, T., Schnell, R.J., et al. 2012. Genome size variation in three Saccharum species. *Euphytica* 185(3): 511–519.

Zhang, J., Zhang, X., Tang, H., Zhang, Q., Hua, X., Ma, X., et al. 2018. Allele-defined genome of the autopolyploid sugarcane Saccharum spontaneum L. *Nature genetics* 50(11): 1565–1573.

Zhang, M.Q., Zheng, X.F., Yu, A.L. *et al.* 2004. Molecular marker application in sugarcane. *Sugar Tech* 6: 251–259. 10.1007/BF02942505

Zhong, S., Dekkers, J.C., Fernando, R.L., and J.L. Jannink. 2009. Factors affecting accuracy from genomic selection in populations derived from multiple inbred lines: a barley case study. *Genetics* 182(1): 355–364.

Zhou, M. 2013. Conventional sugarcane breeding in South Africa: progress and future prospects. *American Journal of Plant Sciences* 4: 189–196 10.4236/ajps.2013.42025

Zhou, M. 2021. General and specific combining ability for cane yield and implications for sugarcane breeding. *Crop Science* 61(1): 539–551.

2 Role of Biotechnology for Improvement of Sugarcane

Saleem Ur Rahman and Hamza Ramzan
National Institute for Biotechnology and Genetic Engineering (NIBGE), Faisalabad, Pakistan
Constituent College Pakistan Institute of Engineering and Applied Sciences, Faisalabad, Pakistan

Haris Khurshid
Oil Seed Research Program, Crop Sciences Institute, National Agricultural Research Center Islamabad, Pakistan

Niaz Ahmad
National Institute for Biotechnology and Genetic Engineering (NIBGE), Faisalabad, Pakistan
Constituent College Pakistan Institute of Engineering and Applied Sciences, Faisalabad, Pakistan

Kai Wang
College of Life Sciences, Nantong University, Jiangsu, China

Zahid Mukhtar and Ghulam Raza
National Institute for Biotechnology and Genetic Engineering (NIBGE), Faisalabad, Pakistan
Constituent College Pakistan Institute of Engineering and Applied Sciences, Faisalabad, Pakistan

CONTENTS

DOI: 10.1201/9781003292425-2

2.1 INTRODUCTION

Sugarcane (*Saccharum officinarum* L.) is a tropical C4 grass that is interspecific hybridized and can store up to 50% of its dry weight of sucrose in its stem (McCormick, Watt, and Cramer 2009). Its crushable stems are grown on more than 26 million hectares of land in tropical and subtropical parts of the globe, yielding up to 1.94 billion metric tons of crushable stems annually. Currently, Brazil is the world's leading sugarcane producer, followed by India, China, and Pakistan. This crop contributes nearly two-thirds of global sugar production and has gained increased attention because ethanol derived from cane sugar represents an important renewable biofuel source, with the potential to transform it into a global commodity and significant energy source (D'Hont et al. 2008). Besides ethanol production, sugarcane is widely used as feed for animals (Mordenti et al. 2021) and humans (Ebadi and Azlan 2021). These products have also an important role in the treatment of many diseases; for instance, the polyphenols present in sugarcane help to treat diabetes and are a strong antioxidant that plays a major role in cancer treatment (Prakash et al. 2021). It has been found that sugarcane bagasse helps to clean the environment because they mitigate the heavy metals from the environment (Iwuozor et al. 2021).

Hence, sugarcane plays a huge role in decreasing malnutrition, which is the main problem currently of the increasing population across the globe. However, the sugarcane crop faces various biotic and abiotic challenges from the time of emergence to harvesting; they have a huge impact on the performance, quality, and yield. To deal with these challenges, stress-tolerant sugarcane is direly needed to increase the yield and decrease the cost of production by using cost-effective and environmentally friendly strategies. However, sugarcane's 1) heterogeneous character, 2) vegetative propagation, 3) poly-aneuploid genome, 4) less fuzz viability, and 5) long breeding cycles make the research advancements less effective through normal breeding (Scortecci et al. 2012) and demand for modern breeding programs to overcome these challenges.

Currently, advancements in genetic improvement techniques using various biotechnological methods have opened new and distinctive research avenues for the improvement of sugarcane (Mustafa et al. 2018). The recombinant DNA technology allows the incorporation of one or more genes (gene stacking) via genetic engineering (GE) and regulation of single or many features; this has tremendous promise for the development of improved sugarcane clones. Recently, advances in the GE have enabled the editing of the plant genome using new plant-breeding technologies (NPBTs) such as zinc-finger nucleases (ZFNs), transcription activator-like effector nucleases (TALENs), and clustered regularly interspaced short palindromic repeats (CRISPR/Cas9), which paved the way for enhanced genetic modification of sugarcane. These nucleases edit the genome(s) by replacement, deletion, and insertion of genes/DNA fragments into a newly edited genome (Gao 2021).

For the identification of diverse genotypes in the sugarcane gene pool, several molecular markers linked with important regions of the genome have also been reported (Raza et al. 2019; Mahadevaiah et al. 2021; Aitken 2021). Furthermore, a wide range of genomics and transcriptomics technologies, especially 1) bioinformatics, 2) proteomics, and 3) metabolomic approaches, have revealed complex relationships between metabolites, proteins, and genes that may aid in the understanding of gene expression and molecular pathways, which have a major role in stress responses (Ali et al. 2019). Despite this possibility, the genetic study of sugarcane cultivars is still necessary to find genes linked with high yield and tolerance to multiple biotic and abiotic stresses.

The present chapter aims to enlighten the comprehensive scenario regarding genetic improvement in sugarcane through modern biotechnology. Moreover, it also discussed the different aspects of biotechnology like NPBTs and the role of advanced technologies (bioinformatics and functional genomics) in the development of new cultivars that can cope with the ongoing challenges to the sustainable production of sugarcane.

2.2 ECONOMIC IMPORTANCE OF SUGARCANE

Sugarcane is grown as a cash crop in several tropical and subtropical regions of the world and is widely grown as a cane crop in almost 105 countries. The production of sugarcane always plays a

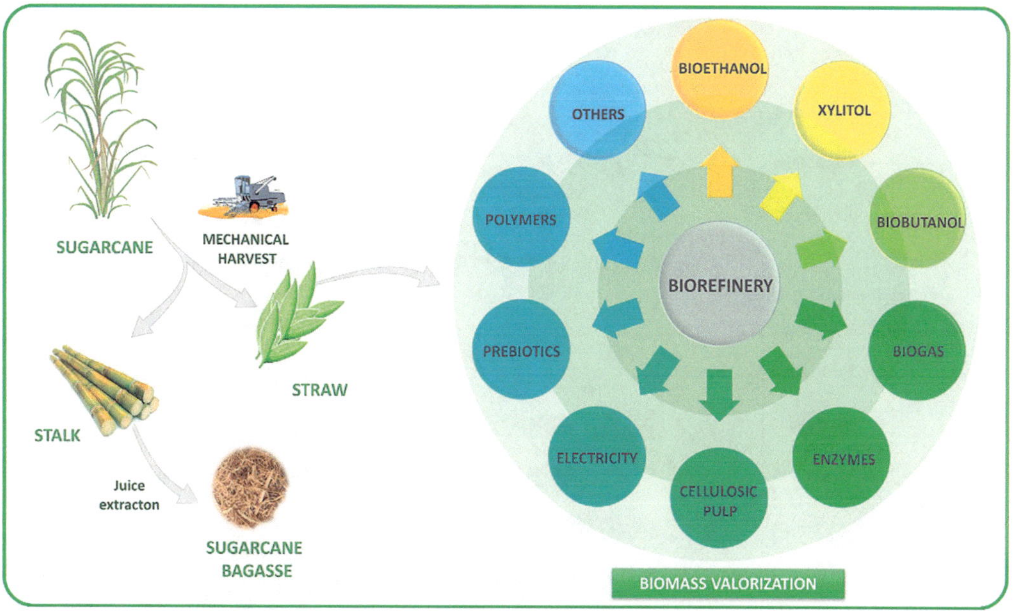

FIGURE 2.1 Systematic view of the process of sugarcane straw and its applications in the market (Aguiar et al. 2021).

crucial role in the uplifting of socioeconomic conditions of sugarcane-growing countries. Its product, sugar (10–13%), contributes significantly to the import and export of any country. Hence, sugarcane production has a direct impact on the economy of any country. For example, in China, it contributes ~90% share to overall sugar production (Li and Yang 2015). Furthermore, this crop produces many products other than sugar, like fiber, paper, biofuel, and organic fertilizers. It has a crucial role in the reduction of pollution since it acts as a replacement for fossil fuels in the vehicle industry (Tew et al. 2008).

Bioethanol as biofuel in Brazil and Argentina has been extensively used (Waclawovsky et al. 2010). In 2019–2020, Brazil produced 642.7 million tons that were cultivated on an area of 8.44 million hectares, which generate an income and made Brazil sufficient in ethanol for the country's use (Cursi et al. 2021). The products of sugarcane, such as sugarcane juice, provide many useful by-products like molasses and bagasse that are further used in the making of fiberboards, plastics, and papers (Figure 2.1). Molasses is used as a vital part in the manufacturing of citric acid, butyl alcohol, and ethyl alcohol. It also provides rum, which is known as a portable spirit (Costa et al. 2021). Animal feed is also one of the products that are made of molasses. The green parts of sugarcane are served as fodder for animals' feed. Its biomass contains a higher level of cellulose, lignin, and hemicellulose than any other C4 crops, like maize and sorghum (Botha 2010). Sugarcane and its by-products are also a source of income and foreign exchanges and one of the important commodities of international trade. The extensive use of sugarcane and its by-products forced the farmers and consumers to cultivate more sugarcane and produce sugarcane by adapting new varieties having different valuable agronomic traits to meet their demand.

2.3 GENETICS OF SUGARCANE

Sugarcane belongs to the family *Poaceae*. It has a very close resemblance with maize and sorghum as they all come under the umbrella of the *Andropogoneae* tribe and *Saccharum* subtribe. Six species have been recognized within the *Saccharum* subtribe (Figure 2.2). Today's commercial sugarcane hybrids are the products of breeding over 100 years, with the starting clones produced

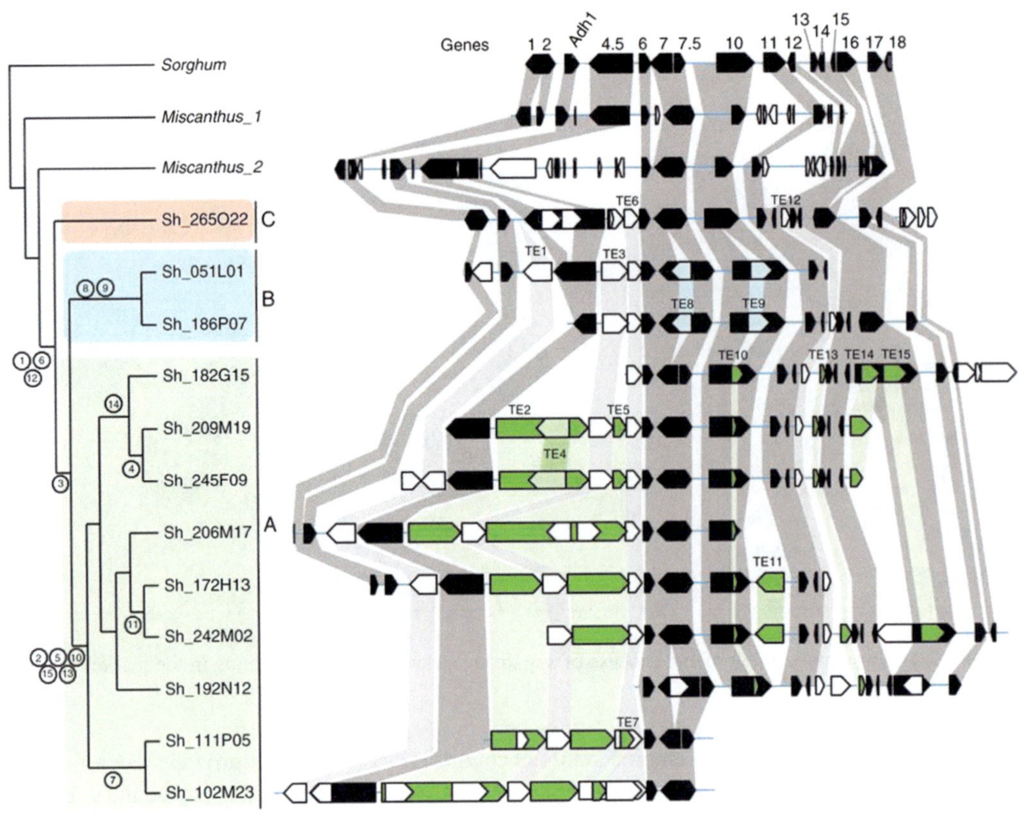

FIGURE 2.2 Comparison of sugarcane with sorghum and miscanthus. Comparison of the 12 sugarcane hom (oe)ologous haplotypes (BACs) of the Adh1 region, together and with sorghum and miscanthus orthologues. Genes are represented by black boxes, and collinear genes are connected in dark grey. TEs are represented by white boxes; collinear TEs are connected in light grey or, when conserved within haplotype group A or B, in green or blue, respectively. Phylogenetic relationships among haplotypes are represented on the left with haplotypes from groups A, B, and C highlighted in green, blue, and red, respectively, and with circles positioning TE insertion times (Pompidor et al. 2021).

by crossing two founding species (*Saccharum officinarum* and *S. spontaneum*) in India and Indonesia. Current sugarcane varieties have a highly complex and large genome with 100–130 chromosomes (Wang, Fang, and Zhang 2021). However, the introgression of the *Saccharum spontaneum* to domesticated *Saccharum officinarum* has proved a big source of modern varieties with many distinct traits. These species are widely regarded as the originators of cultivated sugarcane and are widely dispersed around the world (Cordeiro et al. 2007). However, the history of sugarcane's domestication is not completely established; New Guinea and Southeast Asia are the regions where sugarcane was initially domesticated. The *S. officinarum* was primarily cultivated in New Guinea by crossing it with a wild species: *S. robustum* (D'Hont et al. 2008).

High throughput next-generation (NGS) sequencing made it easy to sequence plant genomes as many plant genomes have been sequenced such as rice, grape, and sorghum (Paterson et al. 2009). The sorghum genome helped sugarcane scientists to study sugarcane crops at the genomic level because the sorghum genome has a tight genetic link with the sugarcane genome. But the availability of mosaic monoploid reference sequence for the highly complex genome of sugarcane made it easy further (Garsmeur et al. 2018; Riaño-Pachón and Mattiello 2017). The genome sequence of sugarcane showed that *S. officinarum* and *S. spontaneum* are distinct by transposable elements and a few large chromosomal rearrangements, suggesting their different size of genome and numbers of the basic chromosome. Moreover, 80%, 10%, and 10% of the chromosomes can be identified based on their derivation from *S. officinarum, S. spontaneum,* and their recombination, respectively (Garsmeur et al. 2018). Due to the high ploidy level, the genome size of the sugarcane is large, and it varies between different genotypes according to their chromosome number. The availability of a complete genome sequence is important for identifying different sources of genes and fitting those cultivars in a suitable environment. Recently, 12 sugarcane cultivars were compared with sorghum and miscanthus and found that the three genomes of sugarcane (A, B, and C) diverged about 0.8 and 1.3 million years ago from modern cultivars cultivated (Figure 2.2) (Pompidor et al. 2021). However, two of the genomes (A and B) still need to be discovered.

2.4 CHALLENGES FOR SUSTAINABLE SUGARCANE PRODUCTION

Plants are subjected to a variety of environmental stresses, and they have evolved a diverse range of approaches to either avoid or deal with the stressful environment. Different environmental stresses have different effects on the expression of genes, physiology, and metabolism of plants, which all differ from one another. It has been found that plants can sense any difference in their surrounding environmental signals; nevertheless, after several research, only a few well-known sensors have been identified (Zhu 2016).

The major challenges for sugarcane productivity per unit area are biotic and abiotic stresses. In biotic stresses, bacteria, fungus, viruses, nematodes, and pests such as borer, billbug, etc., are the main hindrance to sugarcane production (Jones 2021). Reports suggest that about 64 diseases are the causative agents of widespread damage throughout the growing season, resulting in numerous outbreaks. The majority of these diseases are caused by fungal pathogens. The most frequent diseases are rust, wilt, red rot, and smut, which have a serious effect on sugarcane productivity. Chewing and sucking insects are also the major pests of this crop, causing a significant decline in the yield (Goebel and Sallam 2011). This scenario demands the development of disease-resistant cultivars to cope with these problems and to enhance sugarcane productions (Manavalan 2021).

Among abiotic stress, drought stress is a significant contributor, with rising water scarcity being seen throughout the globe. This stress is predicted to continue to intensify and become more severe by the end of the 20th century (Sun et al. 2020) and demand more water. It has been found that the use of freshwater is increasing in irrigation in the agriculture sector, which is an alarming situation. It has been shown that about 65% of worldwide freshwater is used, suggesting the dire need for plant genotypes that could survive in a water-scarcity environment (Chen et al. 2018; Kates and Parris 2003; Gomathi et al. 2015). Furthermore, cold is also significant stress for sugarcane that is produced by either freezing or chilling temperatures. Although cold stress is uncommon in tropical locations, where most of the world's sugarcane is produced, it sometimes has devastating effects on this crop in these areas. This results because most tropical plants have not evolved a mechanism to escape the damaging effects of the cold (Su et al. 2020). Hence, it is obvious that sugarcane varieties vary in their susceptibility to cold (Thiebaut et al. 2012), indicating the existence of alleles that may assist this tropical crop in coping with this stress. The cold-tolerant genes have the potential to cope with cold stress in a future breeding program of sugarcane. Among abiotic stresses, the low fertility soils are also a factor that reduces the yield of sugarcane (Meyer and Wood 2001).

2.5 STRATEGY FOR ENHANCING SUGARCANE PRODUCTIVITY

Sugarcane faces a lot of challenges (biotic and abiotic) that deeply impact its productivity and demand a robust and cost-effective strategy to cope with it as early as possible. So, it is highly needed to produce stress-tolerant sugarcane and analyze its sensitivity to varied stresses to provide long-term agricultural sustainability and food security for the world's rapidly growing population (Zhu 2016). Conventional breeding technology has played a significant role in the improvement of many crop plants. Likewise, in sugarcane, this breeding method has been used to develop several varieties (Zhou 2013). However, due to a long breeding time, conventional breeding has some limitations that have been complemented with advanced biotechnological methods. Transgenic approaches have the potential to develop transgenic plants with less time with required characters. However, due to regulations and consumers' interests, this approach is limited (Trumbo and Powell 2016). But, recently, the advent of the NPBTs platform made it easy for the breeder to produce sugarcane cultivars with their desired character(s) (Eid et al. 2021). So, it will be very helpful to use conventional breeding coupled with NPBTs to uplift sugarcane productivity to meet the demand of the growing population. Different strategies and their use for sugarcane improvement have been discussed such as traditional breeding and biotechnological approaches:

2.5.1 Traditional Breeding

Plant breeding provides dynamic strategies in crop growth and improvement when subjected to a variety of environmental challenges. It provides a way of possibly ensuring food security and safety in extreme weather conditions, as well as assisting plants in escaping from different stressors at a critical period of plant growth, via the development of stress-resistant cultivars (Singh 1981).

In conventional breeding, the genes pre-existing within species, or their close relatives, are brought together by sexual crossing, and plants with desired characteristics are then selected. However, the recombination of genetic elements through mating and selection is not precise, and the newly developed crop contains some undesirable genes that make conventional breeding less effective. Moreover, the success of the breeding process is limited by the lack of germplasm with desired traits, difficulties in establishing selection conditions, and the complexity of resistance mechanisms (Flowers and Yeo 1995).

In sugarcane, most of the disease control was accomplished by natural selection of its resistant clones before the introduction of sugarcane farming and breeding programs. Modern sugarcane varieties were generated via breeding activities that are cultivated under a variety of environmental conditions. Because of the frequent intercrossing program in sugarcane aimed at improving commercial features, there has been genetic drift, which has made the development of clones through conventional breeding more challenging. Nonetheless, extensive interspecific crossing projects carried out across the globe have resulted in relative resistance to gumming disease, as well as an increase in sugar yields in certain varieties. These late interspecific crossings have given a crucial foundation for today's germplasm, which is used in breeding programs that rely on interspecific hybridizations for their breeding success (Gallacher and Berding 1997). However, the complex nature and size of the sugarcane genome provide a significant barrier to the advancement of traditional breeding approaches. Although continued selective breeding for increased sucrose accumulation has been responsible for more than half of the yield growth over the last 50 years, it has likely reached its maximum potential. This is because there is a strong correlation between various features. Apart from this fact, sugarcane breeding is a time-consuming process since it takes 12 to 14 years to produce new elite clones from 100000 or more seedlings, at a cost of millions of dollars for each clone. Furthermore, sugarcane is a poor option for classical genetic alterations since it features one nonsynchronous blooming and low sexual seed viability, and tends to blossom just once per year

2.5.2 Role of Biotechnology in Improving Sugarcane Productivity

Biotechnology is the study of genetic engineering (GE). The GE allows gene transfer between unrelated genera or species. This technique has been widely used for crop improvement. Compared with conventional breeding, GE has its advantages. First, it extends the genetic base because, theoretically, genetic engineering can transfer agronomically useful genes from a different organism. Second, it is more effective because it avoids the problem of linkage drag associated with conventional breeding and is less time-consuming. Through GE, several transgenic plants have been produced to fight biotic and abiotic stresses as well as to increase their nutritional value. When compared to plants produced via conventional breeding, these genetically altered plants exhibit remarkable tolerance to climate fluctuations (Sharoni et al. 2011). Contrary to the classical breeding and marker-assisted selection approaches, the direct introduction of genes by GE is a more attractive and quicker solution for improving stress tolerance (Dunwell 2000; Wang, Vinocur, and Altman 2003). Hence, GE will also have a dramatic impact on the production of sugarcane in the future.

PEG, particle bombardment, electroporation, and *Agrobacterium*-mediated gene transformation are some of the physical and biological transformation strategies that have been used in sugarcane, with varying degrees of effectiveness. Particle bombardment and *Agrobacterium*-mediated gene transformation are the most widely employed methods in sugarcane GE due to their higher transformation efficiency (Suprasanna 2010). Given the heterozygosity of the plant and the lengthy generation time necessary for vegetative propagation, bio-ballistics, also known as DNA-gun technology, is the most stable and durable technology for gene transfer in sugarcane (Arencibia et al. 1995). The biolistic bombardment has altered the genetic makeup of sugarcane cultivars that are suitable to produce embryonic callus cultures. As a result of the Emu promoter controlling the *npt-II* gene, the first successful transgenic sugarcane plant production took place in Australia (Guo et al. 2019; Herrera-Estrella et al. 1983).

The *Agrobacterium*-mediated transformation is also used as a result of the tumor-inducing T-DNA sequences being transmitted to plant cells by the gram-negative soil bacteria *Agrobacterium tumefaciens.* These sequences are now part of the plant genome (Herrera-Estrella et al. 1983), which make this transformation less effective; however, gene integration and other methods make this technology useable for genetic transformation. The transformation in sugarcane has been done (Qamar et al. 2021), and several characters have been introduced, such as insect resistance (Iqbal et al. 2021), herbicide resistance, etc. Several cry genes from *Bacillus thuringiensis*, including cry1A, cry1Ab, cry1Ac, and cry1Aa3, have been introduced in sugarcane, enhancing the plant's resistance to borers and other pests (Arencibia et al. 1999; Braga et al. 2003; Xu et al. 2008; Guo et al. 2019). Herbicide resistance is a well-known example of a trait induced by this genetic modification in sugarcane. Herbicide resistance coupled with insect resistance is the example of such modification as cry1Ab, EPSPS, and PMI, three tolerant genes in transgenic sugarcane, which display substantial levels of herbicide resistance and insect resistance in the laboratory and the field (Gallo-Meagher and Irvine 1996).

To boost the resilience of sugarcane to drought stress, numerous research has been undertaken, for example, by transferring two genes, i.e. Trehalose synthase (TSase) and the vacuolar pyrophosphatase (*AVP1*) genes to the tissues of the apical meristem (Wang et al. 2005). Expression profiling was used to identify genes in sugarcane that were expressed in response to abiotic stressors. The overexpression of *EaDREB2* (the E. arundinaceus *DREB2* gene) and pyramiding with PDH45 (a pea DNA helicase gene) in transgenic sugarcane cultivars increased salinity and drought tolerance. However, it has been shown that sugarcane plants transformed with the PDH45 gene under the control of PortUbi2.3 promoter are successful both under drought and salinity stress (Augustine et al. 2015). As a result of these advances, the transgenic approach utilized in sugarcane is more effective than previous methods since it uses the standard *in vitro* plant-regeneration systems, which are virulent, and drugs that activate the via genes. The traits improved through GE have been presented in Table 2.1. However, this technology needs further advancement to minimize the limitations inherent with using this technology.

Many countries have approved transgenic sugarcane. Indonesia approved drought-tolerant sugarcane expressing choline dehydrogenase (betA), which accumulates membrane-protectant glycine betaine. Other drought-resistant varieties of sugarcane in field trials in Argentina, Brazil, India, Egypt, South Africa, Kenya, and Uganda (Marshall 2014). The transgenic events approved are presented that were retrieved from ISAA (https://www.isaaa.org/gmapprovaldatabase/crop/default.asp?CropID=27&Crop=Sugarcane). Apart from these, about 6000 transgenic lines of sugarcane have been approved by Australia's gene technology that is resistant to the herbicide.

(https://www.isaaa.org/kc/cropbiotechupdate/article/default.asp?ID=5011).

TABLE 2.1
Approved Genetically Modified Sugarcane

Event Name	Introduced Gene/s	Gene Source	GM Trait	Method of Transformation	Company/ Developer	Country where approved
CTB141175/ 01-A (CTC175-A)	*Cry1Ab*	*Bacillus thuringiensis subsp. kurstaki*	Lepidopteran insect resistance	Microparticle Bombardment	Centro de Tecnologia Canavieira	Brazil, Canada, United States
CTC91087-6	*Cry1Ac*	*Bacillus thuringiensis subsp. Kurstaki strain HD73*	Lepidopteran insect resistance	Not available	Centro de Tecnologia Canavieira (CTC)	Brazil, United States
CTC93209-4	*Cry1Ac*	*Bacillus thuringiensis subsp. Kurstaki strain HD73*	Lepidopteran insect resistance	Microparticle Bombardment	Centro de Tecnologia Canavieira	Brazil
NXI-1T	*EcBetA* *nptII** *Aph4 (hpt)**	*Escherichia coli* *Escherichia coli Tn5 transposon* *Escherichia coli*	Drought stress tolerance , Antibiotic resistance	*Agrobacterium tumefaciens-mediated plant transformation*	PT Perkebunan Nusantara XI	Indonesia
NXI-4T	*RmBetA*	*Rhizobium meliloti*	Abiotic Stress Tolerance	*Agrobacterium tumefaciens-mediated plant transformation*	PT Perkebunan Nusantara XI	Indonesia
NXI-6T	*RmBetA*	*Rhizobium meliloti*	Drought stress tolerance	*Agrobacterium tumefaciens-mediated plant transformation*	PT Perkebunan Nusantara XI	Indonesia

2.5.2.1 Marker-Assisted Breeding

Through conventional breeding, the development of crop varieties is a long process that takes about 20–25 years. Biotechnology plays an important role in the development of these cultivars in the less than 10 years it takes to bring it to market. Among these, one such method is marker-assisted selection (MAS), which is an easy and fast technique for selecting any traits in plants. Marker-assisted selection uses DNA markers linked with desirable genes in plants or any organism by which these organisms are selected. These markers are either specific sequences of a gene that recognize it or a DNA sequence that is closely linked with a particular gene and is inherited with this gene. The desired genes are involved in disease resistance, high yield, and adaptation, etc.

Sugarcane's limited genetic base and poly-aneuploid genome make commercial cultivar development difficult. Although sugarcane has been nobilized, little progress has been made in increasing sugar content. When it comes to the characterization of Saccharum species' germplasm, molecular markers have shown to be the most effective method so far. Molecular markers and tiny regions/pieces of DNA sequence that differ across various species also display polymorphism. The genetic study of sugarcane has made extensive use of molecular markers, such as RAPDs (random amplified polymorphic DNA), AFLPs (amplified fragment length polymorphisms), RFLPs (restriction fragment length polymorphisms), SSRs (simple sequence repeats), TRAPs (target region amplification polymorphisms), and ISSRs (inter simple sequence repeats) (Manimekalai et al. 2020). Using microsatellite DNA markers in association with phenotypic characteristics and pedigree data, genetic diversity in sugarcane germplasm has been effectively discovered. In sugarcane, DNA markers have been used to determine the genetic diversity and phylogenetic relationships, which is useful in germplasm categorization and selection of breeding strategies (Cordeiro, Pan, and Henry 2003). Sugarcane genetic markers, including SSR, AFLP, and TRAPs, have been successfully utilized to investigate genetic differences without the influence of environmental circumstances (Singh et al. 2015). About 36 SSR markers were used to examine the genetic diversity of 1002 sugarcane and related grass accessions, and the results revealed that 209 alleles were associated with different agronomic traits (Nayak et al. 2014). An additional genetic diversity study was performed on 25 sugarcane genotypes using TRAP markers, which revealed that all accessions shared a 70% genetic similarity (Singh, Singh, and Singh 2017), and the genetic variation across accessions was low (Nayak et al. 2014). Using 21 fluorescent SSR markers and HPCE (high-performance capillary electrophoresis) to conserve germplasm, recent research revealed that 150 of China's most popular sugarcane accessions successfully underwent genetic diversity analysis, resulting in a considerable increase in genetic variety. By identifying quantitative trait loci (QTLs) and generating genomic maps, molecular markers aid in the understanding of genetic architecture (Pinto et al. 2010). In two species of Saccharum, i.e., *officinarum* and *spontaneum*, partial genetic maps and linkages have been constructed using F1 data (Zhou et al. 2021). *S. officinarum* was genetically mapped using 72 SSR and 40 AFLP primer pairs, which indicated 28 repulsion connections. It has been shown that by using AFLP markers rather than SRAP and TRAP markers for genetic maps, *S. officinarum* and *S. spontaneum* can be distinguished from other sugarcane species in genetic mapping analysis. Three hundred and thirty-three sugarcane accessions were genotyped using 3327 DArT (diversity arrays technology), AFLP, and SSR markers to investigate 13 variables linked with morphology, residual bagasse content, disease resistance, and yield (Ahmad et al. 2018; Gouy et al. 2015). Using 174 SSR primers, researchers were able to determine the phenotypic characteristics of 32 variables in 92 sugarcane cultivars. QTL mapping was used to generate linkage maps and identify genomic areas associated with characteristics of interest in the biparental population (Collard et al. 2005).

The use of molecular markers in plant breeding is very common, especially through QTL mapping, which leads to marker-assisted selection (Takeda and Matsuoka 2008). A treasure of knowledge regarding the genotypic behavior of a plant's quantitative traits is a very important component in most breeding programs fulfilled by QTL mapping. QTL mapping helps breeders to gain a proper understanding of the basic design of quantitative traits at the genetic level, which is indispensable for modern plant breeding. At some point, QTL mapping gives the tools to construct breeding procedures that include both marker information and phenotypic information into their design.

To monitor genetic loci associated with salt resistance, molecular markers may be employed instead of time-consuming phenotypic assessments, saving time and money. PCR-based markers have been used to assess sugarcane varieties' relative salt tolerance and yield potential. RAPD markers, for example, demonstrated genetic variation in both control and salt-tolerant, *S. officinarum* lines, when used in vitro mutagenesis, and *S. officinarum* salt-tolerant lines were molecularly characterized (Suprasanna 2010; Suprasanna et al. 2009). It is also possible to screen sugarcane lines for salt tolerance and susceptibility using 15 ISSR markers chosen based on the similarity index between the cultivars tested. This work showed that the ISSR markers helped determine the genetic diversity of sugarcane cultivars. Researchers also carried out molecular characterization of 18 sugarcane genotypes under salinity; they used five TRAP markers in this investigation. Additional discoveries include the following: When polymorphisms ranging from 25%–100% were used, the genotypes studied under salinity were discovered to have a similarity coefficient of 0.72, indicating little variance among them. Everything is considered; the genotypic and phenotypic behavior of the sugarcane can easily be exploited with molecular markers. They also serve as a basis for the discovery of modern desired characteristics that are resistant to environmental stresses such as drought.

2.5.2.2 Somaclonal Variations

Somaclonal variation is the variation in the regenerated plant due to the tissue-culture technique. It is regarded as a benefit since it allows the development of novel genotypes that have been successfully used in agriculture. Here are some features of the somaclonal variation.

- In high frequencies, somaclonal changes may be seen.
- Some alterations may be innovative and not reproducible using normal breeding methods.
- *In vitro* screening decreases the time it takes to isolate a somaclone with a particular characteristic.
- New desirable characteristics may arise that were not previously present in the germplasm.
- It is far less expensive than other genetic-engineering technologies.

Heinz et al. (1973), for the first time, describe crop improvement by somaclonal variation in the field crops. The presence of somaclonal variants has been reported in many crops, including sugarcane (Burner and Grisham 1995), and a resistant variety of sugarcane against eyespot and red rot disease was developed (Rastogi, Siddhant, and Sharma 2015). Sugarcane cultivar CP65–357 developed, being derived through callus culture; it is a salt (NaCl) tolerant cultivar (Gandonou et al. 2006). Wagih, Ala, and Musa (2004) generated eight drought-resistant varieties of sugarcane (*Saccharum* hybrids) from embryogenic callus, which were then cultivated in the greenhouse for additional testing under water stress. Four salt-resistant somaclonal variants of the sugarcane variety CP48–103 were generated from embryogenic calli (Baer et al. 2007). This shows that somaclonal variation is one of the approaches that help in sugarcane improvement. The technology can be further improved for sugarcane variety production for nutritional as well as for stress responses that will help in sugarcane production and breeding programs.

2.5.3 Genome Editing (CRISPR/Cas9 Technology)

Genome editing is the process of altering an organism's genetic makeup via DNA insertions and deletions. Mega nucleases, zinc-finger nuclease (ZFN), transcription activator-like effector nuclease (TALEN), and clustered regularly interspaced short palindromic repeats (CRISPR)-associated nuclease (CRISPR/Cas9 technologies) are four families of nuclease that are used for the

TABLE 2.2
Genome-Edited Sugarcane for Different Traits

Target Gene (s)	The Function of the Targeted Gene (s)	Delivery Method/ Transformation Method	Technique/ system Used	Reference
Caffeic acid O methyltransferase (*COMT*)	Cell wall composition	*Agrobacterium-mediated* transformation	TALENs	(Jung and Altpeter 2016)
Magnesium chelatase subunit I (*MgCh*)	Chlorophyll biosynthesis	*Biolistic transformation*	CRISPR/Cas9	(Eid et al. 2021)
Acetolactate synthase (ALS)	Herbicide resistance	*Biolistic transformation*	CRISPR/Cas9	(Oz et al. 2021)
Cre/*lox*	Codon optimization	*Biolistic transformation*	CRISPR/Cas9	(Zhao, Karan, and Altpeter 2021)
Caffeic acid O methyltransferase (COMT)	Cell wall composition	*Agrobacterium-mediated* transformation	TALENs	(Kannan et al. 2018)

genetic improvement of many crops (Kathiria and Eudes 2014). The genome editing of sugarcane is very difficult because of its high genome size and polyploidy level, but efforts have been made to modify the sugarcane genome by NPBTs. The success of NPBTs depends upon the availability of whole-genome sequences (WGS). The WGS of sugarcane is available, which made it is easy to target it for desired traits/characters (Garsmeur et al. 2018). However, there is very limited information available regarding NPBTs in sugarcane (Table 2.2). The study of genome editing in sugarcane was performed for the first time by Jung and Altpeter (2016), who targeted multiple sites in the caffeic acid O-methyltransferase (COMT) gene and led to its improvement. They used TALENs by *Agrobacterium*-mediated transformation and improved the cell wall. This study was further proceeded by (Kannan et al. 2018), generating mutant lines having high yields. However, there are no such reports on genome editing of sugarcane for resistance to pathogens (viruses, fungus, bacteria, etc) and nutritional improvements. The genome editing in sugarcane has been discussed by (Mohan 2016) in detail. However, more efforts are needed to target the sugarcane genome for the modification of different traits. It has been found that host genes that interact with different pathogens, which are known as susceptible (S) genes involved in stress response, have been targeted in many crops (Pyott, Sheehan, and Molnar 2016; Chandrasekaran et al. 2016). Identification of these “S” genes in sugarcane for various pathogens and targeting them will be important in the improvement of sugarcane. Apart from this idea, if the S gene(s) are not known in sugarcane, targeting different genes of the pathogen (s) instead of host genes, such as for viruses, fungus, etc., could be an option since this approach has been used in multiple crops for virus resistance (Ali et al. 2015; Iqbal, Sattar, and Shafiq 2016). Furthermore, it is highly important to optimize the conditions for CRISPRcas9 vector systems, such as *Agrobacterium-mediated transformation*, biolistic, etc. Recently, a nanoparticle-based delivery system has been used in crops that are recalcitrant to *Agrobacterium* and tissue culture; *Cannabis sativa* is one (Ahmed et al. 2021). Hence, using the nanoparticle-based delivery system for genome editing of sugarcane will revolutionize the breeding program further.

2.6 ROLE OF BIOINFORMATICS TOOLS AND FUNCTIONAL GENOMICS IN SUGARCANE PRODUCTIVITY

In the present era, the whole world is headed toward using artificial intelligence and big data science for precision agriculture (Zhang et al. 2021). Because of technological advancements, there has been an explosion of biological data, which must be organized and analyzed before it can be used to accelerate future studies and provide advantages. It is the goal of bioinformaticians to classify and logically present this data for accessibility or to design tools that enable further use of the data by uncovering the previously undiscovered relationship between the data.

Currently, the big available resource of genomic data for sugarcane is 250000 expressed sequence tags (ESTs) analyzed by some research groups (Butterfield et al. 2004; Vettore et al. 2003). There is also a lot of information accessible in the form of genetic mapping, but the sugarcane genome is still unknown (Reffay et al. 2005; Raboin et al. 2006). Modern plant science advances, especially developing genomic technologies, hold the great possibility to advance knowledge of sugarcane genetics and breeding. The NGS technology has made important contributions to the advancement of agricultural genetics and breeding. Unique genotypes for breeders, such as the parental genotypes of populations used for mapping, could be sequenced using NGS to obtain genome or transcriptome data in commercial programs. Using this information, it is possible to compare it to those of well-characterized model species or closely related and widely cultivated species; including the use of NGS, sequencing big genomes remains a costly procedure (Cardoso-Silva et al. 2014).

Studies of an organism's full protein and metabolite complements are the current associations in the chain-linking phenotype to genotype, according to proteomics and metabolomics (Barnabas et al. 2015). Many proteomics-related datasets have been collected at this site, which provides a model of prospective data organization when sugarcane proteomics research progresses to a more advanced stage (van Heerden et al. 2010). Even though sugarcane metabolic omics research has only just begun, great consideration must be paid to the arrangement and display of data that will be created in the future. The Golm Metabolome Database (http://csbdb.mpimp-golm.mpg.de/csbdb/gmd/gmd.html) is one such database (accessed at http://csbdb.mpimp-golm.mpg.de). Mass spectrum libraries, metabolite profiles, and other important information may be found in this database (Kopka et al. 2005). Other options include the metabolite analysis tool KNAPSACK (http://kanaya.naist.jp/KNApSACK), which provides information on 17 different species of the *Saccharum* genus and is downloaded for free from the NAIST website. A focus on the biological underpinnings of the chemicals included within this database is emphasized, and the data may be retrieved in several methods from the database. In addition, it offers a tool for analyzing data from the mass spectral domain.

Interaction in the evaluation and identification of genetic markers for future development and the ability to use genetic maps and resources interactively is critical for the effective utilization of genetic maps and resources. For other crop species, outstanding examples are already present, e.g., wheat and rice. GrainGenes (http://wheat.pw.usda.gov/GG2/index.shtml) is a USDA collection of materials aimed at the *Triticeae* and oat research communities. Gramene, on the other hand, includes map data for a wide range of plants, with a focus on rice (Liang et al. 2007). The sugarcane sector has much fewer readily available maps than the average industry. GrainGenes has two *Saccharum spontaneum* maps, one from 1993 and the other from 1995, although neither map is still in use by the grain business. CIRAD has created a freely accessible sugarcane genetic mapping database (Liang et al. 2007). The integration of marker data and genetic maps into existing databases that formulate a stream of trait data for future breeding programs is another method that can be used to increase the usefulness of marker data and genetic maps. One such example is SPIDNet, a database of plant breeding that serves as the foundation for the variety development program of Australia named as BSES-CSIRO Joint Venture Program. Marker data are now being added to this database to help with the implementation of marker-assisted selection, which will increase the database's capacity; however, it is not yet accessible to the general public (Casu 2010).

It will be feasible to examine sugarcane allelic diversity in detail and boost gene discovery projects with the arrival of high-throughput resequencing technologies in the future. In this complex grass, after gene promoters for a large proportion of the genes have been located and a complete genome assembly has been completed, regulatory network research and the development of systems biology tools might be considered. Combining transcriptome research with the improvement in bioinformatics approaches that can exploit the information more efficiently will provide a starting point for describing complicated aspects such as yield. The application of sugarcane ChIP-seq technology to identify transcription factor (TF) sites and gene promoters are being researched. This study's results will have several direct implications in plant breeders' efforts to enhance agronomic performance by selecting cis-regulatory elements and TF changes in their quest for better-suited genotypes to their surroundings. Whole-genome sequencing will also result in the creation of a catalog of entire ORFeomes, which will be useful in the production of transgenics. It needs a strong computational infrastructure to discover novel genes and test their functionality in models.

2.7 FUTURE PROSPECTIVE

Advancements including genomics, genetic mapping, the use of molecular markers, and NPBTs opened ways for sugarcane improvement. This crop has a huge potential in energy generation through biomass of feedstock and its further processing in terms of biorefinery. An increase in understanding and the use of new technologies greatly aid in achieving the objective of sustainable productions. Sugarcane has a lot of genetic potentials that are now untapped and might be leveraged to boost biomass production even more. Selection for the traits related to the biomass would not affect the total quantity of sugar of the crop to make future advances in sugarcane genetic improvement; for this purpose, a sound understanding of feedback inhibition is mandatory, which is almost certainly mediated by sugar signaling. Biotechnological approaches such as genome editing have a huge potential to develop new cultivars of sugarcane with enough resistance for abiotic and biotic stresses as well as to improve it for nutritional purposes. Targeting host genes by these advanced methods will be essential in sugarcane breeding and its improvement. The use of nanotechnology and new bioinformatic tools for sugarcane genetic improvement is highly needed. The susceptible genes and their interaction with foreign pathogens have not been explored in sugarcane, so in the future, finding "S" genes will make the research in sugarcane easier and understating their genomic selection for different environmental stresses easier, too.

REFERENCES

Aguiar, Andre, Thais Suzane Milessi, Daniella Regina Mulinari, Melina Savioli Lopes, Sirlene Maria da Costa, and Rafael Garcia Candido. 2021. Sugarcane straw as a potential second generation feedstock for biorefinery and white biotechnology applications. *Biomass and Bioenergy* 144:105896.

Ahmad, Ali, Jin-Da Wang, Yong-Bao Pan, Rahat Sharif, and San-Ji Gao. 2018. Development and use of simple sequence repeats (SSRs) markers for sugarcane breeding and genetic studies. *Agronomy* 8 (11):260.

Ahmed, Sajjad, Xuefei Gao, Md Asraful Jahan, Maxwell Adams, Nianqiang Wu, and Nik Kovinich. 2021. Nanoparticle-based genetic transformation of Cannabis sativa. *Journal of Biotechnology* 326:48–51.

Aitken, K.S. 2021. History and development of molecular markers for sugarcane breeding. *Sugar Tech* 24:341–353.

Ali, Ahmad, Mehran Khan, Rahat Sharif, Muhammad Mujtaba, and San-Ji Gao. 2019. Sugarcane Omics: An update on the current status of research and crop improvement. *Plants* 8 (9):344.

Ali, Zahir, Aala Abulfaraj, Ali Idris, Shakila Ali, Manal Tashkandi, and Magdy M. Mahfouz. 2015. CRISPR/Cas9-mediated viral interference in plants. *Genome biology* 16 (1):1–11.

Arencibia, Ariel D., Elva R. Carmona, Maria T. Cornide, et al. 1999. Somaclonal variation in insect-resistant transgenic sugarcane (Saccharum hybrid) plants produced by cell electroporation. *Transgenic Research* 8 (5):349–360.

Arencibia, Ariel, Pedro R. Molina, Gustavo de la Riva, and Guillermo Selman-Housein. 1995. Production of transgenic sugarcane (Saccharum officinarum L.) plants by intact cell electroporation. *Plant cell reports* 14 (5):305–309.

Augustine, Sruthy Maria, J. Ashwin Narayan, Divya P. Syamaladevi, et al. 2015. Overexpression of EaDREB2 and pyramiding of EaDREB2 with the pea DNA helicase gene (PDH45) enhance drought and salinity tolerance in sugarcane (Saccharum spp. hybrid). *Plant cell reports* 34 (2):247–263.

Baer, G Ya, A.I. Yemets, N.A. Stadnichuk, D.B. Rakhmetov, and Ya B. Blume. 2007. Somaclonal variability as a source for creation of new varieties of finger millet (Eleusine coracana (L.) Gaertn.). *Cytology and Genetics* 41 (4):204–208.

Barnabas, Leonard, Ashwin Ramadass, Ramesh Sundar Amalraj, Malathi Palaniyandi, and Viswanathan Rasappa. 2015. Sugarcane proteomics: An update on current status, challenges, and future prospects. *Proteomics* 15 (10):1658–1670.

Botha, Frederik C. 2010. Future prospects. *Genetics, Genomics and Breeding of Sugarcane* 31:249–264.

Braga, Daniella P.V., Enrico D.B. Arrigoni, Marcio C. Silva-Filho, and Eugênio C Ulian. 2003. Expression of the Cry1Ab protein in genetically modified sugarcane for the control of Diatraea saccharalis (Lepidoptera: Crambidae). *Journal of New Seeds* 5 (2-3):209–221.

Burner, D.M., and M.P. Grisham. 1995. Induction and stability of phenotypic variation in sugarcane as affected by propagation procedure. *Crop science* 35 (3):875–880.

Butterfield, M.K., R.S. Rutherford, D.L. Carson, B.I. Huckett, and C.H. Bornman. 2004. Application of gene discovery to varietal improvement in sugarcane. *South African Journal of Botany* 70 (1):167–172.

Cardoso-Silva, Claudio Benicio, Estela Araujo Costa, Melina Cristina Mancini, et al. 2014. De novo assembly and transcriptome analysis of contrasting sugarcane varieties. *PloS one* 9 (2):e88462.

Casu, Rosanne. 2010. Role of bioinformatics as a tool for sugarcane research. *Genetics, Genomics and Breeding of Sugarcane*1:229–248.

Chandrasekaran, Jeyabharathy, Marina Brumin, Dalia Wolf, et al. 2016. Development of broad virus resistance in non-transgenic cucumber using CRISPR/Cas9 technology. *Molecular plant pathology* 17 (7):1140–1153.

Chen, Bin, M.Y. Han, Kun Peng, et al. 2018. Global land-water nexus: Agricultural land and freshwater use embodied in worldwide supply chains. *Science of the Total Environment* 613:931–943.

Collard, Bertrand C.Y., M.Z.Z. Jahufer, J.B. Brouwer, and E.C.K. Pang. 2005. An introduction to markers, quantitative trait loci (QTL) mapping and marker-assisted selection for crop improvement: the basic concepts. *Euphytica* 142 (1):169–196.

Cordeiro, Giovanni M., Yong-Bao Pan, and Robert J. Henry. 2003. Sugarcane microsatellites for the assessment of genetic diversity in sugarcane germplasm. *Plant science* 165 (1):181–189.

Cordeiro, G.M., O. Amouyal, F. Eloitt, and R.J. Henry. 2007. Kole genome mapping and molecular breeding in plants, pulses, sugar and tuber crops. *Springer* 3:175–204.

Costa, Ane Catarine Tosi, Jacob Hornick, Tathiana Ferreira Sá Antunes, et al. 2021. Complete genome sequence and analysis of a Saccharomyces cerevisiae strain used for sugarcane spirit production. *Brazilian Journal of Microbiology* 52(3): 1087–1095.

Cursi, Danilo Eduardo, Herman Paulo Hoffmann, G.V.S. Barbosa, et al. 2021. History and current status of sugarcane breeding, germplasm development and molecular genetics in Brazil. *Sugar Tech* 24:112–133.

D'Hont, Angélique, Glaucia Mendes Souza, Marcelo Menossi, et al. 2008. Sugarcane: a major source of sweetness, alcohol, and bio-energy. In *Genomics of tropical crop plants*: Springer.

Dunwell, Jim M. 2000. Transgenic approaches to crop improvement. *Journal of experimental botany* 51 (suppl_1):487–496.

Ebadi, Samarghand, and Azrina Azlan. 2021. Nutritional Composition and Role of Non-centrifugal Sugar (NCS) in Human Health. *Current Nutrition & Food Science* 17 (3):249–257.

Eid, Ayman, Chakravarthi Mohan, Sara Sanchez, Duoduo Wang, and Fredy Altpeter. 2021. Multiallelic, Targeted Mutagenesis of Magnesium Chelatase With CRISPR/Cas9 Provides a Rapidly Scorable Phenotype in Highly Polyploid Sugarcane. *Frontiers in Genome Editing* 3:10.

Flowers, T.J., and A.R. Yeo. 1995. Breeding for salinity resistance in crop plants: where next? *Functional Plant Biology* 22 (6):875–884.

Gallacher, D.J., and N. Berding. 1997. Purpose, selection, and application of descriptors for sugarcane germplasm. *Australian journal of agricultural research* 48 (6):759–768.

Gallo-Meagher, M., and J.E. Irvine. 1996. Herbicide resistant transgenic sugarcane plants containing the bar gene. *Crop Science* 36 (5):1367–1374.

Gandonou, Christophe B., Tomader Errabii, Jamal Abrini, Mohamed Idaomar, and Nadia Skali Senhaji. 2006. Selection of callus cultures of sugarcane (Saccharum sp.) tolerant to NaCl and their response to salt stress. *Plant Cell, Tissue and Organ Culture* 87 (1):9–16.

Gao, Caixia. 2021. Genome engineering for crop improvement and future agriculture. *Cell* 184(6): 1621–1635.

Garsmeur, Olivier, Gaetan Droc, Rudie Antonise, et al. 2018. A mosaic monoploid reference sequence for the highly complex genome of sugarcane. *Nature communications* 9 (1):1–10.

Goebel, François-Régis, and Nader Sallam. 2011. New pest threats for sugarcane in the new bioeconomy and how to manage them. *Current Opinion in Environmental Sustainability* 3 (1-2):81–89.

Gomathi, R., P.N. Gururaja Rao, K. Chandran, and A. Selvi. 2015. Adaptive responses of sugarcane to waterlogging stress: An over view. *Sugar Tech* 17 (4):325–338.

Gouy, Matthieu, Yves Rousselle, A. Thong Chane, et al. 2015. Genome wide association mapping of agro-morphological and disease resistance traits in sugarcane. *Euphytica* 202 (2):269–284.

Guo, Minliang, Jingyang Ye, Dawei Gao, Nan Xu, and Jing Yang. 2019. Agrobacterium-mediated horizontal gene transfer: Mechanism, biotechnological application, potential risk and forestalling strategy. *Biotechnology advances* 37 (1):259–270.

Heinz, D.J. 1973. Sugar-cane improvement through induced mutations using vegetative propagules and cell culture techniques. In *Induced mutations in vegetatively propagated plants*. International Atomic Energy Agency (IAEA), Austria.

Herrera-Estrella, Luis, Ann Depicker, Marc Van Montagu, and Jeff Schell. 1983. Expression of chimaeric genes transferred into plant cells using a Ti-plasmid-derived vector. *Nature* 303 (5914):209–213.

Iqbal, Aneela, Raham Sher Khan, Mubarak Ali Khan, et al. 2021. Genetic Engineering Approaches for Enhanced Insect Pest Resistance in Sugarcane. *Molecular Biotechnology* 63:557–568.

Iqbal, Zafar, Muhammad N. Sattar, and Muhammad Shafiq. 2016. CRISPR/Cas9: a tool to circumscribe cotton leaf curl disease. *Frontiers in plant science* 7:475.

Iwuozor, Kingsley O., Ifeoluwa Peter Oyekunle, Iyiola Olatunji Oladunjoye, Enoch Mayowa Ibitogbe, and Temidayo Samuel Olorunfemi. 2021. A review on the mitigation of heavy metals from aqueous solution using sugarcane bagasse. *Sugar Tech* 24:1167–1185.

Jones, Roger A.C. 2021. Global plant virus disease pandemics and epidemics. *Plants* 10 (2):233.

Jung, Je Hyeong, and Fredy Altpeter. 2016. TALEN mediated targeted mutagenesis of the caffeic acid O-methyltransferase in highly polyploid sugarcane improves cell wall composition for production of bioethanol. *Plant molecular biology* 92 (1):131–142.

Kannan, Baskaran, Je Hyeong Jung, Geoffrey W. Moxley, Sun-Mi Lee, and Fredy Altpeter. 2018. TALEN-mediated targeted mutagenesis of more than 100 COMT copies/alleles in highly polyploid sugarcane improves saccharification efficiency without compromising biomass yield. *Plant biotechnology journal* 16 (4):856–866.

Kates, Robert W., and Thomas M. Parris. 2003. Long-term trends and a sustainability transition. *Proceedings of the National Academy of Sciences* 100 (14):8062–8067.

Kathiria, Palak, and François Eudes. 2014. Nucleases for genome editing in crops. *Biocatalysis and Agricultural Biotechnology* 3 (1):14–19.

Kopka, Joachim, Nicolas Schauer, Stephan Krueger, et al. 2005. GMD@ CSB. DB: the Golm metabolome database. *Bioinformatics* 21 (8):1635–1638.

Li, Yang-Rui, and Li-Tao Yang. 2015. Sugarcane agriculture and sugar industry in China. *Sugar Tech* 17 (1):1–8.

Liang, Chengzhi, Pankaj Jaiswal, Claire Hebbard, et al. 2007. Gramene: a growing plant comparative genomics resource. *Nucleic Acids Research* 36 (suppl_1):D947–D953.

Mahadevaiah, C., Prakash Hapase, V. Sreenivasa, et al. 2021. Delineation of genotype× environment interaction for identification of stable genotypes for tillering phase drought stress tolerance in sugarcane. *Scientific reports* 11 (1):1–13.

Manavalan, R. 2021. Efficient Detection of Sugarcane Diseases through Intelligent Approaches: A Review. *Asian Journal of Research and Review in Agriculture* 3(4):27–37.

Manimekalai, Ramaswamy, Gayathri Suresh, Hemaprabha Govinda Kurup, Selvi Athiappan, and Mallikarjuna Kandalam. 2020. Role of NGS and SNP genotyping methods in sugarcane improvement programs. *Critical Reviews in Biotechnology* 40 (6):865–880.

Marshall, Andrew. 2014. Drought-tolerant varieties begin global march. *Nature* 201:4.

McCormick, A.J., D.A. Watt, and M.D. Cramer. 2009. Supply and demand: sink regulation of sugar accumulation in sugarcane. *Journal of experimental botany* 60 (2):357–364.

Meyer, J.H., and R.A. Wood. 2001. The effects of soil fertility and nutrition on sugarcane quality: a review. Paper read at Proc S Afr Sug Technol Ass.

Mohan, Chakravarthi. 2016. Genome editing in sugarcane: challenges ahead. *Frontiers in plant science* 7:1542.

Mordenti, Attilio Luigi, Elisa Giaretta, Luca Campidonico, Paola Parazza, and Andrea Formigoni. 2021. A review regarding the use of molasses in animal nutrition. *Animals* 11 (1):115.

Mustafa, Ghulam, Faiz Ahmad Joyia, Sultana Anwar, Aqsa Parvaiz, and Muhammad Sarwar Khan. 2018. Biotechnological interventions for the improvement of sugarcane crop and sugar production. *Sugarcane-Technology and Research; IntechOpen: London, UK*:113–138.

Nayak, Spurthi N., Jian Song, Andrea Villa, et al. 2014. Promoting utilization of Saccharum spp. genetic resources through genetic diversity analysis and core collection construction. *PloS one* 9 (10):e110856.

Oz, Mehmet Tufan, Angelika Altpeter, Ratna Karan, Aldo Merotto, and Fredy Altpeter. 2021. CRISPR/Cas9-mediated multi-allelic gene targeting in sugarcane confers herbicide tolerance. *Frontiers in genome editing* 3:1–15.

Paterson, Andrew H., John E. Bowers, Remy Bruggmann, et al. 2009. The Sorghum bicolor genome and the diversification of grasses. *Nature* 457 (7229):551–556.

Pinto, L.R., Antonio Augusto F Garcia, Maria Marta Pastina, et al. 2010. Analysis of genomic and functional RFLP derived markers associated with sucrose content, fiber and yield QTLs in a sugarcane (Saccharum spp.) commercial cross. *Euphytica* 172 (3):313–327.

Pompidor, Nicolas, Carine Charron, Catherine Hervouet, et al. 2021. Three founding ancestral genomes involved in the origin of sugarcane. *Annals of botany* 127 (6):827–840.

Prakash, Monica D., Lily Stojanovska, Jack Feehan, et al. 2021. Anti-cancer effects of polyphenol-rich sugarcane extract. *Plos one* 16 (3):e0247492.

Pyott, Douglas E., Emma Sheehan, and Attila Molnar. 2016. Engineering of CRISPR/Cas9-mediated potyvirus resistance in transgene-free Arabidopsis plants. *Molecular plant pathology* 17 (8):1276–1288.

Qamar, Zahida, Idrees Ahmad Nasir, Mounir G. Abouhaidar, et al. 2021. Novel approaches to circumvent the devastating effects of pests on sugarcane. *Scientific reports* 11 (1):1–25.

Raboin, L-M, Karine Miranda Oliveira, Loïc Lecunff, et al. 2006. Genetic mapping in sugarcane, a high polyploid, using bi-parental progeny: identification of a gene controlling stalk colour and a new rust resistance gene. *Theoretical and Applied Genetics* 112 (7):1382–1391.

Rastogi, Jyoti, P Bubber Siddhant, and Brij Lal Sharma. 2015. Somaclonal variation: A new dimension for sugarcane improvement. *GERF Bull Biosci* 6 (1):5–10.

Raza, Ali, Ali Razzaq, Sundas Saher Mehmood, et al. 2019. Impact of climate change on crops adaptation and strategies to tackle its outcome: A review. *Plants* 8 (2):34.

Reffay, Nathalie, Phil A. Jackson, Karen S. Aitken, et al. 2005. Characterisation of genome regions incorporated from an important wild relative into Australian sugarcane. *Molecular Breeding* 15 (4):367–381.

Riaño-Pachón, Diego Mauricio, and Lucia Mattiello. 2017. Draft genome sequencing of the sugarcane hybrid SP80-3280. *F1000Research* 6.

Scortecci, Katia C., Silvana Creste, Tercilio Calsa Jr, et al. 2012. Challenges, opportunities and recent advances in sugarcane breeding. *Plant breeding* 1:267–296.

Sharoni, Akhter Most, Mohammed Nuruzzaman, Kouji Satoh, et al. 2011. Gene structures, classification and expression models of the AP2/EREBP transcription factor family in rice. *Plant and cell physiology* 52 (2):344–360.

Singh, Amandeep, Uma Ranjan Lal, Hayat Muhammad Mukhtar, Prabh Simran Singh, Gagan Shah, and Ravi Kumar Dhawan. 2015. Phytochemical profile of sugarcane and its potential health aspects. *Pharmacognosy reviews* 9 (17):45.

Singh, Daljit. 1981. The relative importance of characters affecting genetic divergence. *The Indian Journal of Genetic and Plant Breeding* 41 (2):237–245.

Singh, R.B., Balwant Singh, and R.K. Singh. 2017. Study of genetic diversity of sugarcane (Saccharum) species and commercial varieties through TRAP molecular markers. *Indian Journal of Plant Physiology* 22 (3):332–338.

Su, Weihua, Yongjuan Ren, Dongjiao Wang, et al. 2020. The alcohol dehydrogenase gene family in sugarcane and its involvement in cold stress regulation. *BMC genomics* 21 (1):1–17.

Sun, Yuan, Cuiting Wang, Han Y.H. Chen, and Honghua Ruan. 2020. Response of plants to water stress: a meta-analysis. *Frontiers in plant science* 11:978.

Suprasanna, P. 2010. Biotechnological interventions in sugarcane improvement: Strategies, methods and progress. *BARC Newsletter* 47–53.

Suprasanna, P., V.Y. Patade, E.R. Vaidya, and V.D. Patil. 2009. Radiation induced in vitro mutagenesis, selection for salt tolerance and characterization in sugarcane. Paper read at Induced plant mutations in the genomics era. Proceedings of an International Joint FAO/IAEA Symposium. Rome: Food and Agriculture Organization of the United Nations.

Takeda, Shin, and Makoto Matsuoka. 2008. Genetic approaches to crop improvement: responding to environmental and population changes. *Nature Reviews Genetics* 9 (6):444–457.

Tew, Thomas, Anna Hale, Sarah Lingle, and R. Cobil. 2008. Improving sugarcane as a bioenergy crop in the US. Paper read at Agronomy Abstracts.

Thiebaut, Flavia, Cristian A. Rojas, Karla L. Almeida, et al. 2012. Regulation of miR319 during cold stress in sugarcane. *Plant, cell & environment* 35 (3):502–512.

Trumbo, Jennifer, and Douglas Powell. 2016. Why transgenic plants are so controversial. In *Plant biotechnology and genetics: principles, techniques, and applications*. John Wiley & Sons, Hoboken, New Jersey, USA 366–382.

van Heerden, Philippus D.R., Robin A. Donaldson, Derek A. Watt, and Abraham Singels. 2010. Biomass accumulation in sugarcane: unravelling the factors underpinning reduced growth phenomena. *Journal of experimental botany* 61 (11):2877–2887.

Vettore, André L, Felipe R. da Silva, Edson L. Kemper, et al. 2003. Analysis and functional annotation of an expressed sequence tag collection for tropical crop sugarcane. *Genome research* 13 (12):2725–2735.

Waclawovsky, Alessandro J., Paloma M. Sato, Carolina G. Lembke, Paul H. Moore, and Glaucia M. Souza. 2010. Sugarcane for bioenergy production: an assessment of yield and regulation of sucrose content. *Plant Biotechnology Journal* 8 (3):263–276.

Wagih, Mohamed E., A. Ala, and Y. Musa. 2004. Regeneration and evaluation of sugarcane somaclonal variants for drought tolerance. *Sugar Tech* 6 (1):35–40.

Wang, Tianyou, Jingping Fang, and Jisen Zhang. 2021. Advances in Sugarcane Genomics and Genetics. *Sugar Tech* 24:354–368.

Wang, Wangxia, Basia Vinocur, and Arie Altman. 2003. Plant responses to drought, salinity and extreme temperatures: towards genetic engineering for stress tolerance. *Planta* 218 (1):1–14.

Wang, Zi-Zhang, Shu-Zhen Zhang, Ben-Peng Yang, and Yang-Rui Li. 2005. Trehalose synthase gene transfer mediated by Agrobacterium tumefaciens enhances resistance to osmotic stress in sugarcane. *Sugar Tech* 7 (1):49–54.

Xu, Jing-Sheng, Shiwu Gao, Liping Xu, and Rukai Chen. 2008. Construction of expression vector of CryIA (c) gene and its transformation in sugarcane. *Sugar tech* 10 (3):269–273.

Zhang, Peng, Zhiling Guo, Sami Ullah, Georgia Melagraki, Antreas Afantitis, and Iseult Lynch. 2021. Nanotechnology and artificial intelligence to enable sustainable and precision agriculture. *Nature Plants* 7 (7):864–876.

Zhao, Yang, Ratna Karan, and Fredy Altpeter. 2021. Error-free recombination in sugarcane mediated by only 30 nucleotides of homology and CRISPR/Cas9 induced DNA breaks or Cre-recombinase. *Biotechnology Journal* 16 (6):2000650.

Zhou, Jing-Ru, Hui-Dong Sun, Ahmad Ali, et al. 2021. Quantitative proteomic analysis of the sugarcane defense responses incited by Acidovorax avenae subsp. avenae causing red stripe. *Industrial Crops and Products* 162:113275.

Zhou, Marvellous. 2013. Conventional sugarcane breeding in South Africa: progress and future prospects.

Zhu, Jian-Kang. 2016. Abiotic stress signaling and responses in plants. *Cell* 167 (2):313–324.

3 Biotechnological Intervention in Sugarcane: Progress Made So Far

Gauri Nerkar and Madhavi Purankar
Molecular Biology and Genetic Engineering Laboratory, Vasantdada Sugar Institute, Pune, Maharashtra, India

Suman Devarumath
Vidya Pratishthan's College of Agricultural Biotechnology, Baramati, Maharashtra, India

Rachayya Devarumath and Harinath Babu
Molecular Biology and Genetic Engineering Laboratory, Vasantdada Sugar Institute, Pune, Maharashtra, India

CONTENTS

3.1 INTRODUCTION

Sugarcane (*Saccharum officinarum* hybrids) is an important sugar and biofuel-producing crop contributing to more than 80% of the world sugar production and approximately 40% ethanol production. It is C4 grass with high biomass content, growing in tropical and subtropical regions of the world. Sugarcane has a polyploid-aneuploid genome (2n= 100–130), and the cultivated sugarcane varieties are *Saccharum officinarum* hybrids arising from interspecific crosses between *Saccharum officinarum* and *Saccharum spontaneum*; with major contribution from the *Saccharum officinarum* (D'Hont et al. 1996). Sugarcane genome also has the contribution from its wild relatives *viz.S. robustum*, *S. sinense*, *S. barberi* and related grass genera such as *Erianthus*, *Miscanthus*, and *Narenga* (Daniels and Roach 1987). The conventional breeding programs have successfully developed elite cultivars with superior agronomic traits, such as high sugar content, high yield, resistance to pests and diseases, as well as improved ratooning ability (Ming et al. 2006). But, the breeding programs are often difficult and time-consuming due to the complex nature of sugarcane genome, low fertility, and narrow genetic base (Altpeter and Karan 2018; Nerkar et al. 2018). The biotechnological tools, like mutation breeding, marker-assisted selection,

DOI: 10.1201/9781003292425-3

genetic engineering, genome editing, genomics, and transcriptomics can, speed up the process of the development of new sugarcane varieties. During the past two decades, a recalcitrant monocot crop like sugarcane has witnessed tremendous progress in the efforts leading toward its genetic improvement.

After the first report of callus production from sugarcane (Heinz and Mee 1969), the clonal propagants raised from tissue culture were thought to be an efficient tool for the rapid production and multiplication of true-to-type plants. However, the reports on somaclonal variation offered a new way of developing genetically improved sugarcane varieties. Induced mutagenesis and somaclonal variation have been extensively applied for sugarcane improvement, and they offer a simpler, faster, and cost-effective way of developing new sugarcane varieties. With the availability of improved tissue culture and regeneration protocols, successful genetic transformation in sugarcane has been reported using the biolistic as well as the *Agrobacterium*-mediated transformation methods (Lakshmanan et al. 2005; Altpeter and Oraby 2010; Nerkar et al. 2018; Budeguer et al. 2021). Being a largely vegetative propagated crop, there is a lesser chance for the escape of the transgene to the wild relatives of sugarcane. Also, the lack of segregation of genes in the progeny opens up the possibility for the gene stacking of several unlinked target genes, making sugarcane an ideal crop for metabolic engineering (Altpeter and Karan 2018). The nuclear transformation technologies involve random insertion of the transgene, affecting the transgene expression in the genetically transformed plants (Dermawan et al. 2016). The recent emergence of the genome-editing technologies has revolutionized the field of agriculture with their ability to precisely modify the plant genomes. Genome editing encompasses designer nucleases such as mega nuclease, zinc finger nucleases (ZFNs), transcription activator–like effector nucleases (TALEN), and CRISPR (clustered regularly interspaced short palindromic repeats)-Cas9 (CRISPR associated), which facilitate target gene insertionat a predetermined location in the genome. These technologies, in combination with the data obtained from genomic and transcriptomic studies, will surely improve our understanding of the sugarcane genome gene function and accelerate the development of improved varieties.

3.2 MUTATION BREEDING

Although plants derived from tissue culture are the clonal propagants or the true-to-type plants, under the *in vitro* conditions, plants do exhibit a high frequency of genetic variations, which are referred to as somaclonal variations. Hence, researchers often check the micropropagated progeny for genetic variation when the production of true-to-type clones is the main objective (Devarumath et al. 2007).

Breeders have for a long time identified somaclonal variations as a tool to create variability. The use of plant-growth regulators affects the cell division, the callus formation, and selective proliferation of specific cell types (Rutherford et al. 2014). In sugarcane, the somaclonal variation and induced mutation have been employed to induce tolerance against several abiotic stress factors. Induced mutagenesis in sugarcane by gamma radiation (Patade et al. 2008; Nikam et al. 2014, 2015) and EMS (Gadakh et al. 2015) have been used for inducing salt tolerance in sugarcane. Chaum et al (2013) checked field performances of six mutant clones derived through gamma radiation and EMS on saline-sodic soil. Three clones 'K88–92', '(A3)AE1–18' and 'KK3 were found to be salt tolerant in terms of the number of internodes, polarization, BRIX, fibre percent, juice purity, and CCS % (commercial cane sugar). The physiological attributes Fv/Fm, the quantum efficiency of PSII (ΦPSII), and water-use efficiency (WUE) were not found to be affected by salinity in these clones. Nikam et al. (2014) performed gamma radiation-induced mutagenesis and NaCl selections on Co 740 variety. Out of 214 putative salt tolerant clones, 24 plants were checked on saline fields. An increase in girth, number of canes, and yield was observed in the mutant clones. A similar experiment on variety Co 86032 showed improvement in agronomic characteristics of the canes under saline conditions (Nikam et al. 2015).

PEG has been frequently used as a primary selecting agent for obtaining drought-tolerant putative callus/ plantlets (Suhesti et al. 2015). Both physical mutagen: gamma radiation and chemical mutagen *viz*. sodium nitrite (Wagih et al. 2004), EMS (Masoabi et al. 2018) have been employed for mutagenesis. Rao and Jabeen (2013) studied the effect of PEG on catalase, peroxidase, SOD, ascorbate, and proline content. Masoabi et al. (2018) used 16 mM EMS on callus of sugarcane variety NCo 310. Shamansori et al. (2018) developed putative drought-tolerant somaclones from Iranian sugarcane varieties CP48–103, CP69–1062 and CP57–614, which were observed to produce more proline when drought stress implemented *in vitro*.

Somaclonal variation has also been used for developing stress-tolerant somaclonal variants. Prasad and Naik (2000) developed eye spot disease-resistant somaclones from variety Co 419. Kwon-Ndung and Ifenkwe (2000) isolated eight smut disease-resistant mutant clones from B47419, BJ6502, Co 62175, and Co 957 by gamma radiation. Recently, Dalvi et al. (2021) reported smut disease-resistant mutant clones of CoC 671 variety developed by EMS mutagenesis. Oloriz et al. 2012 observed a higher frequency of rust-resistant mutants from NaN3 treatment in sugarcane variety B4362 Oloriz et al. 2012). Ahmed et al. (2019) have reported somaclonal variants resistant to mosaic virus disease. Concerning resistance to mosaic virus disease, Kona et al. (2019) used chemical-induced mutagenesis on Co 86032 variety to produce yellow leaf disease-resistant clones.

Improvements in agronomic traits such as flowering reduction, increase in cane sugar content, increase in height, and increase in tillering (Sandhu et al. 2019) have also been reported. The first Indian variety developed from CoC 671 by somaclonal variation 'Co 94012 (PhuleSavitri)' was released in 2006 (Jalaja et al. 2006). Co 94012 was selected for early maturity, high sucrose content, and moderate resistance to diseases like red rot and smut. 'Guifu 98–296' developed by mutation breeding was released in 2011. It was selected for high yield and high sugar content (Song et al. 2011). In 2016, VSI 434 variety developed through somaclonal variation was released for early maturity and high sugar content (Tawar et al. 2016). Although various somaclonal variants have been developed with improved agronomic traits, their performance in the field is a benchmark for their commercial release.

Somaclonal variation appears to be a promising tool for genetic improvement of sugarcane because (i) it can be optimized for recalcitrant varieties; (ii) it is simpler, faster, and more cost-effective compared to conventional breeding programs and the genetic transformation technologies; (iii) approval for the release of the improved varieties produced through somaclonal variation or induced mutagenesis is legally acceptable in most of the countries; and (iv) both somaclonal variation and induced mutagenesis can be applied to change specific characters in promising variety without altering its other characteristics.

3.3 GENETIC TRANSFORMATION

The development of efficient regeneration protocols and the particle bombardment method enabled the transformation of sugarcane using the embryogenic callus and geneticin selection system (Bower and Birch 1992). Later, transgenic sugarcane plants were also reported by *Agrobacterium*-mediated transformation (Arencibia et al. 1998; Enríquez-Obregón et al. 1998). Optimization of transformation protocols using direct and indirect embryogenesis has been reported (Taparia et al. 2012). Comparison of *Agrobacterium*-mediated and biolistic transformation showed that both methods resulted in similar transformation efficiency, numbers of single-copy events, and transgene expression level, but the transgenic lines obtained from biolistic transformation showed stable and better field performance (Jackson et al. 2013; Joyce et al. 2014; Wu et al. 2015). This result could be attributed to the fact that *Agrobacterium*-mediated transformation involves indirect somatic embryogenesis protocol and a larger duration in tissue culture, leading to somaclonal variations that can affect the field performance of the plants.

Selection systems play an important role in reducing transgenic escapes and the development of nonchimeric transgenic plants. The most frequently used selection systems used for the development of sugarcane transgenics are neomycin phosphotransferase (*nptII*, Bower and Birch 1992), phosphinothricin acetyl transferase (*bar/pat*, Gallo-Meagher and Irvine 1996), phosphor mannose isomerase (*manA*, Jain et al. 2007), and hygromycin phosphor transferase (*hpt*, Arencibia et al. 1998).

Considerable progress has been made in the direction of introducing agronomically important traits in sugarcane (Table 3.1), such as resistance to diseases and pests; increasing sucrose accumulation (van der Merwe et al. 2010; Hamerli and Birch 2011), and tolerance to abiotic factors such as drought, salinity, and low and high temperatures (Devarumath et al. 2019; Budeguer et al. 2021). As sugarcane production is largely affected by drought, the development of more water-deficit tolerant cultivars is imperative to all sugarcane-producing countries (Ferreira et al. 2017). Glycine betaine is an osmo-regulator that helps the cell maintain the water potential as well as the biological functions during dehydration arising from drought stress. Choline dehydrogenase is an enzyme involved in the synthesis of glycine betaine (Sugiharto 2018). In sugarcane, bacterial *betA*

TABLE 3.1
A Few Examples of Transgenic Sugarcane Developed for Different Stress Responses

Gene	Method	Trait	Reference
CP	Microprojectile	SCYLVdisease resistance	Zhu et al. 2011
CP	*Agrobacterium*	SrMVdisease resistance	Guo et al. 2015
b-1,3-glucanase	*Agrobacterium*	Red rot disease resistance	Nayyar et al. 2017
CP	*Agrobacterium*	CMV disease resistance	Apriasti et al. 2018
CP	Microprojectile	CMV disease resistance	Aslam et al. 2018
Chitinase Class-II	Microprojectile	Red rot disease resistance	Tariq et al. 2018
cry1Ac	Microprojectile	Pest resistance	Gao et al. 2016
cry1Ab	*Agrobacterium*	Pest resistance	Islam et al. 2016
CaneCPI-1	Microprojectile	Pest resistance	Schneider et al. 2017
cry1Aa3	*Agrobacterium*	Pest resistance	Kalunke et al. 2009
cry1Ab	*Agrobacterium*	Pest resistance	Wang et al. 2017
cry1Ac	Microprojectile	Pest resistance	Zhou et al. 2018
cry1Ab and *cry2Ab*	*Agrobacterium*	Pest resistance	Cristofoletti et al. 2018
cry2A	Microprojectile	Pest resistance	Gao et al. 2018
cry1Ab	*Agrobacterium*	Pest resistance	Cheavegatti-gianotto et al. 2018
cry1ac	*Agrobacterium*	Pest resistance	Gianotto et al. 2019
cryIAb-cryIAc	*Agrobacterium*	Pest resistance	Koerniati et al. 2020
cry1Ac	*Agrobacterium*	Pest resistance	Dessoky et al. 2020
Vip3A	*Agrobacterium*	Pest resistance	Riaz et al. 2020
Epsps	Microprojectile	Herbicide tolerance	Racedo et al. 2019
AVP1	*Agrobacterium*	Salt stress tolerance	Kumar et al. 2014
H+-Ppase*AVP1*	Microprojectile	Drought stress tolerance	Raza et al. 2016
P5CS	Microprojectile	Salt stress tolerance	Guerzoni et al. 2014
RmBetA	*Agrobacterium*	Drought stress tolerance	Waltz 2014
OsglyII	Microprojectile	Methylglyoxal and salt stress tolerance	Rani et al. 2012
EaHSP70	*Agrobacterium*	Drought stress tolerance	Augustine et al. 2015a
EaHSP70	*Agrobacterium*	Drought and salinity tolerance	Augustine et al. 2015b
AtDREB2A	*Agrobacterium*	Drought stress tolerance	Reis et al. 2014
EaDREB2 and*PDH45*	*Agrobacterium*	Drought and salinity tolerance	Augustine et al. 2014
Isopentenyltransferase (IPT)	Microprojectile	Cold stress tolerance	Belintani et al. 2012
AtBI-1	Microprojectile	Drought stress tolerance	Ramiro et al. 2016
Expansin (*EXPA1*)	*Agrobacterium*	Drought stress tolerance	Narayan et al. 2021

gene encoding for choline dehydrogenase (CDH) has successfully been introduced and resulted in the transgenic drought-tolerance sugarcane. The over-expression of the *betA* gene increased GB contents that act as an osmoprotectant and help sugarcane acclimate in water deficit conditions. Transgenic sugarcane genotype expressing a bacterial gene encoding a choline dehydrogenase developed by Persero (PT Perkebunan Nusantara XI) also became the first commercially released drought-tolerant transgenic sugarcane in the world (Parisi et al. 2016).

As a potential biofuel crop with high biomass content, sugarcane is an ideal candidate for the production of lignocellulosic ethanol. In this regard, genetically engineered sugarcane that can facilitate improved saccharification efficiency has been developed (Jung et al. 2013). The production of recombinant cellulolytic enzymes *in planta* can substantially reduce the saccharification costs for the production of lignocellulosic bioethanol. Achieving high yields of recombinant cellulolytic enzymes is one of the key challenges when the enzymes are expressed in plants to facilitate biomass conversion in situ. Vacuolar-targeted *in planta* production of cellulases is a novel strategy to bring down the cost of cellulosic ethanol by producing a large quantity of enzymes in juice without affecting the normal plant metabolism. It has been successfully demonstrated in sugarcane (Harrison et al. 2011; 2014).

Also, the carbon-partitioning balance has been altered to produce triacylglycerols in the vegetative tissues of sugarcane for biodiesel production (Zale et al. 2016). Production of bioplastics like PHAs (polyhydroxy alkanoates) and PHBs (polyhydroxy butrate) have been attempted in sugarcane (Petrasovits et al. 2007; 2012). Due to its high biomass and fast growth, the use of sugarcane may be used as a platform for the production of high-value compounds or pharmaceuticals in the future (Palaniswamy et al. 2016; Appunu et al. 2017).

3.4 GENOME EDITING

Genome-editing technologies hold the potential for crop improvement, especially for improving the agronomic traits such as improvement of yield during drought stress, disease resistance, herbicide resistance, and improved nutritive value, as well as better shelf life. An important application of genome editing in polyploids is that it opens up the possibility to study the gene function, especially in the crops that have not been fully sequenced as yet due to their huge genomes. Genome editing has been reported in several polyploid crops (as reviewed by Schaart et al. 2021). Although transgenic technology has been demonstrated in sugarcane, the random insertion of transgene in the huge polyploid nuclear genome of sugarcane often results in transgene silencing (Altpeter and Karan 2018). Genome editing would allow targeted genome modification in the complex sugarcane genome, avoiding these position effects and resulting in efficient expression of the gene of interest. Thus, precise genome modifications provide a cost-effective platform for product development because they reduce the undesirable side effects, resulting in a predictable performance of the transgene (Ainley et al. 2013). Genome editing also offers relatively simpler methods for the characterization of transgenic events. Sequential transgene stacking has been demonstrated using genome editing, which could become a potential tool to improve the crop yield, a polygenic trait.

Genome-editing technology uses sequence-specific nucleases (SSNs) that allow precise and targeted modifications of the complex genome by site-specific insertion, deletion, and substitution of the DNA. SSNs are programmed to recognize the preselected genomic sites, and they use cellular DSB repair mechanisms (Puchta et al. 1993) such as non-homologous end joining (NHEJ) or homology-directed repair (HDR). NHEJ disrupts a gene by random insertion or deletion of DNA at the cut site before the free DNA ends are reattached (Puchta 2005). HDR involves the addition of a donor DNA of choice, which is homologous to the site of the break used by the cell to repair the DNA (Haber 2000). Using the HDR pathway, scientists can introduce a new gene with a vital function or correct a mutation by replacing the mutated sequence with a healthy sequence (Symington and Gautier 2011). SSNs have been classified into four major classes viz. engineered homing endonucleases or mega nucleases (MNs), zinc finger nucleases (ZFNs), transcription activator-like effector nucleases (TALENs), and clustered regularly-interspaced short palindromic repeats (CRISPR) and CRISPR-associated (Cas) nuclease 9 (CRISPR/Cas9). Each class of SSNs differs in its ability to recognize the target DNA. Meganucleases, ZFNs, and TALENs recognize their sequence targets through protein/DNA interactions. CRISPR/Cas9 uses a guide RNA (sgRNA), which has a specific sequence that can target the genomic sequence of interest and form an RNA/DNA duplex, which is then cleaved by the Cas9 nuclease. This creates a DSB, which is repaired by NHEJ, resulting in indels in the protein-coding regions and causing frameshift or knock-down of the desired genes (Basso et al. 2020). CRISPR/Cas9 method of genome editing was quickly adopted in many crops owing to its simplicity of DNA targeting through base pairing.

Genome editing in sugarcane has been reported using TALEN-mediated (Jung et al. 2016; Kannan et al. 2018) and CRISPR/Cas9 (Zhao et al. 2021; Eid et al. 2021; Oz et al. 2021) approaches. The first report by Jung et al. (2016) used TALEN-mediated approach to target the *COMT* (Caffeic acid O-methyl transferase) gene to reduce the lignin content in sugarcane cultivar CP88–1762. Two years later, the follow-up study reported the agronomic performance of genome-edited *COMT* mutant lines of sugarcane with improved saccharification efficiency 39%–44% [much higher compared to RNAi suppression of *COMT* in sugarcane (19%–32%, Jung et al. 2013)] and reduced lignin content. The polyploid and highly redundant genome of sugarcane demands co-editing at a large number of target sites. This study reported extensive co-editing of 107 of the 109 identified *COMT* copies/alleles of the lignin biosynthetic gene caffeic acid O-methyltransferase (*COMT*), without compromising the agronomic performance under the field conditions. Zhao et al. (2021) reported a short homology arm (~30 nucleotides) mediated error-free genome editing in sugarcane by HDR, but this method resulted in significantly lower efficiency compared to site-specific recombination by Cre/lox and necessitated the improvement of gene-targeting methods in sugarcane. CRISPR/Cas9 approach has been successfully demonstrated in sugarcane for efficient multi-allelic editing of magnesium chelatase subunit I (*MgCh*) (Eid et al. 2021). This study reported co-editing of 49 (out of 59) copies/alleles of magnesium chelatase in sugarcane, which resulted in reduced chlorophyll contents conferring scorable phenotype to the *in vitro* regenerated plants. CRISPR-Cas9 mediated genome editing for herbicide tolerance in sugarcane has also been reported by the gene targeting of the *ALS* gene (Oz et al. 2021). This study reports efficient and reproducible multi-allelic GT in sugarcane for the first time. The disruption of *ALS* gene leading to herbicide tolerance in sugarcane can be a potential selectable marker system; in combination with multiplexed GT, it can replace transgenes as selectable markers (Altpeter and Karan 2018).

Genome editing, also referred to as a new breeding technique, introduces small changes in the genomes of plants that could be found in nature. These modifications do not involve the introduction of foreign DNA into the plants, unlike the older approaches for genome modification; genome editing allows researchers to make more targeted changes in the genome. Hence, genome editing and derived products are more likely to get regulatory approval, although international discussions are seeking legal clarity on this issue (Lassoued et al. 2021). The soybean variety-producing oil with a longer shelf life was launched as the first commercialized genome-edited food product in 2019 and the gene-edited tomato with higher amounts of γ-aminobutyric acid (GABA) in 2021; since then, the chances of regulatory approval, as well as public acceptance of other genome-edited crops, have broadened. Genome-editing tools hold the promise for future food security, and the regulatory approval of genome-edited crops will play a crucial role in this regard.

3.5 APPLICATION OF MOLECULAR MARKERS AND MARKER-ASSISTED SELECTION

Molecular markers play a crucial role in the development of new sugarcane varieties. Identification of molecular markers associated with important agronomic traits assists the breeding program through selection of best parents, which accelerates the rate of genetic gain and significantly reduces the duration and the costs needed to develop new sugarcane varieties (Racedo et al. 2016). Marker systems have been successfully used in the last few decades for studying the genetics of sugarcane, especially for analysis of genetic diversity, genetic mapping, quantitative trait locus (QTL) mapping, identification of syntenic genes, and estimation of co-ancestry (Garcia et al. 2013). Several types of molecular markers have been used to construct genetic maps in sugarcane, for example, RFLPs, AFLPs, TRAP (Garcia et al. 2006, Andru et al. 2011), EST-SSRs (Oliveira et al. 2007; Parida et al. 2010; Farsangi et al. 2018b), and DARTs (Manners et al. 2011). RFLP (Lu et al. 1994), ITS or rDNA (Glaszmann et al. 1990; Farsangi et al. 2018c), RAPD (Kawar et al. 2009, Vaze et al. 2010), ISSR (Kalwade et al. 2012; Farsangi et al. 2017), SSR (Cordeiro et al. 2003; Devarumath et al. 2012), SSR-SSCP (Kalwade and Devarumath, 2014), AFLP (Selvi et al. 2005; Creste et al. 2010), and TRAP (Devarumath et al. 2013) have been able to successfully identify and differentiate the members of the *Saccharum* complex (Suprasanna et al. 2011). At the organellar level, chloroplast and mitochondrial markers have been used to differentiate *Saccharum* and its wild relatives (D'Hont et al. 2003; Viola et al. 2011; Nerkar et al. 2015; Lekshmi et al. 2017, Aristya et al. 2020). Farsangi et al. (2018a) have also employed TRAP markers for the identification of potential salt tolerance candidate genes and their validation on selected tolerant and susceptible sugarcane genotypes.

Sugarcane is predominantly a vegetatively propagated crop, and its breeding programs have involved a small number of cycles of intercrossing, suggesting that linkage disequilibrium is extensive in this crop (Raboin et al. 2008; Wei et al. 2010). But, due to low-density markers and nonrefined statistical methods, the association studies are still in their infancy. Also, the application of marker-assisted selection (MAS) in sugarcane-breeding programs appears to be a challenging task. In sugarcane, to date, MAS has been reported for the Bru-1 and Bru-2 haplotypes, which have potential use in the identification of rust-resistant gene in sugarcane germplasm (Le Cunff et al. 2008). This is not surprising because the two most important traits, like sucrose content and yield, are governed by multiple quantitative trait loci, while each QTL contributes to only a small proportion of the overall phenotypic effect (Aitken et al. 2008).

3.6 GENOMICS AND TRANSCRIPTOMICS APPROACHES FOR SUGARCANE IMPROVEMENT

The bioinformatics and computational analysis of the genomes and transcriptomes provide a better understanding of the connections between genes, proteins, and metabolites (Mustafa et al. 2018). The omics approaches have been used to obtain enormous information on the molecular mechanisms of sugarcane resistance and tolerance to herbicides, cold, drought, and salinity stress, as well as plant development (Kaur et al. 2017). Genome-wide association studies (GWAS) may be used to identify marker-trait associations (MTAs) and then assist breeders in better managing crosses and selecting superior genotypes in breeding programs.

The marker-trait associations that help the breeders in selecting superior genotypes for breeding programs can be identified using genome-wide association studies. A few examples include the identification of resistance loci to orange rust and yellow leaf virus diseases (Yang et al. 2019) and MTAs related to soluble solid content, stalk height, stalk number, stalk weight, and cane yield traits in sugarcane (Barreto et al. 2019). Using the transcriptomics approach, genes associated with important agronomic traits, such as cellulose and lignin biosynthesis (Kasirajan et al. 2018), leaf abscission (Li et al. 2016), ripening (Cunha et al. 2017), and photosynthesis (Mattiello et al. 2015), have been identified. With a recent availability of an allele-defined genome of the autopolyploid sugarcane *S. spontaneum*, future research on sugarcane improvement would be significantly improved (Zhang et al. 2018).

3.7 CONCLUSION

Biotechnological tools have contributed significantly to the genetic improvement of sugarcane. The last three decades have witnessed tremendous progress in the areas of mutation breeding, genetic engineering, molecular markers, genomics, and transcriptomics. Sugarcane research has come a long way from identifying genes related to important agronomic traits to their genetic transformation and generation of stable transgenic lines with improved agronomic performance in the field. The recent emergence of the new area of genome editing has opened up the possibility of creating targeted modifications in the sugarcane genome without leaving a transgenic footprint, which has been the long-standing goal of the sugarcane breeders across the world. The co-editing of multiple alleles in the polyploid genome of sugarcane shall provide a unique opportunity of improving the expression of the gene of interest offering a better-performing genome-edited sugarcane crop. The regulatory approval of genome-edited products in a few countries holds a promising future for genome-edited crops that can meet the food and fuel requirements of the global population under the changing climatic conditions. Promoting dialogue between biotechnologists on one hand and farmers and breeding companies on the other can help gain more public acceptance for the genome-edited sugarcane in the future.

REFERENCES

Ahmed M.S., Gardezi D.A., Batley J., Iqbal M.Z., and Javid A. (2019). Somaclonal variations for red rot and sugarcane mosaic virus resistance and candidate genes integrity assessment in somaclones of selected sugarcane varieties (*Saccharum officinarum* L). *Pakistan Journal of Agricultural Sciences*, 56(1): 15–27.

Ainley W.M., Sastry-DentL., Welter M.E., et al. (2013). Trait stacking via targeted genome editing. *Plant Biotechnology Journal*, 11: 1126–1134.

Aitken K.S., Hermann S., Karno K., Bonnett G.D., McIntyre L.C., and Jackson P.A. (2008). Genetic control of yield related stalk traits in sugarcane. *TheorAppl Genet*, 117: 1191–1203.

Altpeter F., and Oraby H. (2010). Sugarcane. In: Kempken F. and Jung C. (Eds.) *Genetic Modification of Plants*, Springer, Berlin, pp. 453–472.

Altpeter F., and Karan R. (2018). Genetic improvement of sugarcane by transgenic, intragenic and genome editing technologies In: Rott P. (Ed.) *Achieving Sustainable Cultivation of Sugarcane*, Vol. 2, Burleigh Dodds Science Publishing, Cambridge, pp. 133–154.

Andru S., Pan Y.B., Thongthawee S., Burner D.M., and Kimbeng C.A. (2011). Genetic analysis of the sugarcane (*Saccharum* spp.) cultivar 'LCP 85-384'. I. Linkage mapping using AFLP, SSR, and TRAP markers. *TheorAppl Genet*, 123: 77–93.

Appunu C., Ram B., and Subramonian N. (2017). Sugarcane: an efficient platform for molecular farming In: Mohan C. (Ed.) *Sugarcane Biotechnology: Challenges and Prospects*, Springer, Cham.

Apriasti R., Widyaningrum S., Hidayati W.N., Sawitri W.D., Darsono N., Hase T., et al. (2018). Full sequence of the coat protein gene is required for the induction of pathogen-derived resistance against sugarcane mosaic virus in transgenic sugarcane. *Mol. Biol. Rep.*, 45: 2749–2758.

Arencibia A.D., Carmona E.R., Téllez P., Chan M.T., Yu S.M., Trujillo L.E., and Oramas P. (1998). An efficient protocol for sugarcane (*Saccharum* spp. L.) transformation mediated by *Agrobacterium tumefaciens*. *Transgenic Research*, 7(3): 213–222.

Aristya G.R., Putri F., Kasiamdari R.S., and Musthofa A. (2020). DNA Barcoding and Phylogenetic Analysis of Sugarcane (*Saccharumofficinarum* L.) based on *matK* (maturase K) gene. *Key Engineering Materials*, 840: 162–170.

Aslam U., Tabassum B., Nasir I.A., Khan A., and Husnain T. (2018). A virus-derived short hairpin RNA confers resistance against sugarcane mosaic virus in transgenic sugarcane. *Transgenic Res*, 27: 203–210.

Augustine S.M., Cherian A.V., Syamaladevi D.P., and Subramonian N. (2015a). *Erianthusarundinaceus* HSP70 (EaHSP70) acts as a key regulator in the formation of anisotropic interdigitation in sugarcane (Saccharum spp. hybrid) in response to drought stress. *Plant Cell Physiol*, 56(12): 2368–2380.

Augustine S.M., Narayan J.A., Syamaladevi D.P., Appunu C., Chakravarthi M., Ravichandran V., and Subramonian N. (2015b). *Erianthusarundinaceus* HSP70 (EaHSP70) overexpression increases drought and salinity tolerance in sugarcane (*Saccharum* spp. hybrid). *Plant Sci*, 232: 23–34.

Augustine S.M., Narayan J.A., Syamaladevi D.P., Appunu C., Chakravarthi M., Ravichandran V., Tuteja N., and Subramonian N. (2014). Overexpression of EaDREB2 and pyramiding of EaDREB2 with the pea DNA helicase gene (PDH45) enhance drought and salinity tolerance in sugarcane (*Saccharum* spp. hybrid). *Plant Cell Rep*, 34(2): 247–263.

Barreto F.Z., Rosa J.R.B.F., Balsalobre T.W.A., Pastina M.M., Silva R.R., Hoffmann H.P., de Souza A.P., Garcia A.A.F., and Carneiro M.S. (2019).A genome-wide association study identified loci for yield component traits in sugarcane (*Saccharum* spp.). *PLoS ONE*, 14: e0219843.

Belintani N.G., Guerzoni J.T.S., Moreira R.M.P., and Vieira L.G.E. (2012). Improving low-temperature tolerance in sugarcane by expressing the ipt gene under a cold inducible promoter. *Biol Plant*, 56(1): 71–77.

Bower R., and Birch R.G., (1992). Transgenic sugarcane plants via microprojectile bombardment. *Plant Journal*, 2(3): 409–416.

Budeguer F., Enrique R., Perera M.F., Racedo J., Castagnaro A.P., Noguera A.S., and Welin B. (2021). Genetic Transformation of Sugarcane, Current Status and Future Prospects. *Front. Plant Sci.*, 12: 768609.

Cha-um S., Chantawong S., Siriwatana C., Ashraf M., and Kirdmanee C. (2013). Field screening of sugarcane (*Saccharum* spp.) mutant and commercial genotypes for salt tolerance. *Not Bot HortiAgrobo*, 41(1): 286–293.

Cheavegatti-gianotto A., Gentile A., Oldemburgo D.A., and Cheavegattigianotto A. (2018). Lack of detection of Bt sugarcane Cry1Ab and NptII DNA and proteins in sugarcane processing products including raw sugar. *Front. Bioeng.Biotechnol.*, 6: 24. doi: 10.3389/fbioe.2018.00024

Cordeiro G.M., Pan, Y.B., and Henry, R.J. (2003). Sugarcanemicrosatellites for the assessment of genetic diversity insugarcane germplasm. *Plant Science*, 165: 181–189.

Creste S., Sansoli D.M., Tardiani A.C.S. et al. (2010). Comparison of AFLP, TRAP and SSRs in the estimation of genetic relationships in sugarcane. *Sugar Tech*, 12: 150–154.

Cristofoletti P.T., Kemper E.L., Capella A.N., Carmago S.R., Cazoto J.L., Ferrari F., et al. (2018). Development of transgenic sugarcane resistant to sugarcane borer. *Trop. Plant Biol.*, 11: 17–30.

Cunha C.P., Roberto G.G., Vicentini R., Lembke C.G., Souza G.M., Ribeiro R.V., Machado E.C., Lagôa A.M., and Menossi M. (2017). Ethylene-induced transcriptional and hormonal responses at the onset of sugarcane ripening. *Sci. Rep.*, 7: 43364.

Dalvi S.G., Tawar P.N., Suprasanna P., Dixit G.B., and Prasad D.T. (2021). EMS-Based In Vitro Mutagenesis and Mutant Screening for Smut Resistance with Agronomic Traits in Sugarcane. *Sugar Tech.*, 23(4): 854–864.

Daniels J., and Roach B.T. (1987). Taxonomy and evolution. In: Heinz D.J. (Ed.) *Sugarcane Improvement through Breeding*, vol. 11, Elsevier, Amsterdam, the Netherlands, pp. 7–84.

Dermawan H., Karan R., Jung J.H., Zhao Y., Parajuli S., Sanahuja G., and Altpeter F., (2016). Development of an intragenic gene transfer and selection protocol for sugarcane resulting in resistance to acetolactate synthase-inhibiting herbicide. *Plant Cell Tissue and Organ Culture*, 126(3): 459–468.

Dessoky E.S., Ismail R.M., Elarabi N.I., Abdelhadi A.A., and Naglaa A. (2020). Improvement of sugarcane for borer resistance using Agrobacterium mediated transformation of cry1Ac gene. *GM Crops Food*, 12: 47–56.

Devarumath R.M., Mirajkar S.J., Thorat A.S., Farsangi F.J., and Suprasanna P. (2019). Genomic Landscapes of Abiotic Stress Responses in Sugarcane. In: Rajpal V., Sehgal D., Kumar A., Raina S. (Eds.) *Genomics Assisted Breeding of Crops for Abiotic Stress Tolerance, Vol. II. Sustainable Development and Biodiversity*, vol 21. Springer, Cham.

Devarumath R.M., Doule R.B., Kawar P.G., Naikebawane S.B., and Nerkar Y.S. (2007). Field performance and RAPD analysis to evaluate genetic fidelity of tissue culture raised plants viv-à-vis conventional setts derived plants of sugarcane. *Sugar Tech*, 9: 17–22.

Devarumath R.M., Kalwade S.B., Bundock P., Eliott F.G., and Henry R. (2013). Independent target region amplification polymorphism and single-nucleotide polymorphism marker utility in genetic evaluation of sugarcane genotypes. *Plant Breeding*, 132: 736–747.

Devarumath R.M., Kalwade S.B., Kawar P.G., and Sushir K.V. (2012).Assessment of Genetic Diversity in Sugarcane Germplasm Using ISSR and SSR Markers. *Sugar Tech*, 14: 334–344.

D'Hont, A. , Grivet, L. , Feldmann, P., et al. (1996). Characterisation of the double genome structure of modern sugarcane cultivars (Saccharum spp.) by molecular cytogenetics. *Molecular Genetics & Genomic Medicine*, 250: 405–413.

Eid A., Mohan C., Sanchez S., Wang D., and Altpeter F. (2021). Multiallelic, Targeted Mutagenesis of Magnesium ChelataseWith CRISPR/Cas9 Provides a Rapidly Scorable Phenotype in Highly Polyploid Sugarcane. *Frontiers in Genome Editing*, 3: 654996.

Enríquez-Obregón G.A., Vázquez-Padrón R.I., Prieto-Samsonov D.L., De la Riva G.A., and Selman-Housein G. (1998). Herbicide-resistant sugarcane (Saccharumofficinarum L.) plants by *Agrobacterium*-mediated transformation. *Planta*, 206(1): 20–27.

Farsangi F., Thorat A., and Devarumath R. (2017). Analysis of Molecular Assortment in Sugarcane Varieties using RAPD and ISSR markers. *Research Journal of Biotechnology*, 20: 20–28

Farsangi F., Thorat A., and Devarumath R. (2018a) Molecular characterization of sugarcane genotypes for their salinity and susceptibility using trap markers. *International Journal of Current Research*, 10: 68947–68951.

Farsangi F., Thorat A., and Devarumath R. (2018b) Assessment of the Utility of TRAP and EST-SSRs Markers for Genetic Diversity Analysis of Sugarcane Genotypes. *Cytology and Genetics*, 52: 467–477.

Farsangi F., Thorat A., and Devarumath R. (2018c) Assessment of genetic diversity among different sugarcane genotypes using internal transcribed spacer (ITS) region of the ribosomal DNA (rDNA). *GSC Biological and Pharmaceutical Sciences*, 05(02): 017–025.

Ferreira T.H.S., Tsunada M.S., Bassi D., Araújo P., Mattiello L., Guidelli G.V., et al. (2017). Sugarcane water stress tolerance mechanisms and its implications on developing biotechnology solutions. *Front. Plant Sci.*, 8: 1077.

Gadakh S.S., Patel D.U., and Patil A.B. (2015). Evaluation of sugarcane (Saccharum spp. complex) mutants for yield, yield contributing traits and quality parameters. *Int J AdvBiol Res.*, 5(3): 220–228.

Gallo-Meagher M., and Irvine J.E. (1996).Herbicide resistant transgenic sugarcane plants containing the bar gene. *Crop Science*, 36(5): 1367–1374.

Gao S., Yang Y., Wang C., Guo J., Zhou D., Wu Q., et al. (2016). Transgenic sugarcane with a cry1Ac gene exhibited better phenotypic traits and enhanced resistance against sugarcane borer. *PLoS One*, 11: 0153929.

Gao S., Yang Y., Xu L., Guo J., Su Y., Wu Q., et al. (2018). Particle bombardment of the cry2A gene cassette induces stem borer resistance in sugarcane. *Int. J. Mol. Sci.*, 19: 1692.

Garcia A.A., Kido E.A., Meza A.N., Souza H.M., Pinto L.R., Pastina M.M., Leite C.S., Silva J.A., Ulian E.C., Figueira A. et al. (2006). Development of an integrated genetic map of a sugarcane (*Saccharum* spp.) commercial cross, based on a maximum-likelihood approach for estimation of linkage and linkage phases. *TheorAppl Genet*, 112: 298–314.

Garcia A., Mollinari M., Marconi T. et al. (2013).SNP genotyping allows an in-depth characterisation of the genome of sugarcane and other complex autopolyploids. *Sci Rep*, 3, 3399.

Ghosh A., and Gadgil V.N. (1979). Shift in ploidy level of callus tissue: a function of growth substances. *Indian Journal of Experimental Biology*, 17: 562–564.

Gianotto A.C., Rocha M.S., Cutri L., Lopes F.C., Dal'Acqua W., Hjelle J.J., et al. (2019). The insect-protected CTC91087-6 sugarcane event expresses Cry1Ac protein preferentially in leaves and presents compositional equivalence to conventional sugarcane. *GM Crops Food*, 10, 208–219.

Glaszmann J.C., Lu Y.H., and Lanaud C. (1990). Variation of nuclear ribosomal DNA in sugarcane. *Journal of Genetics and Breeding*, 44: 191–198

Guerzoni J.T.S., Belintani N.G., Moreira R.M.P., Hoshino A.A., Domingues D.S., Filho J.C.B., et al. (2014). Stress-induced 11-pyrroline-5-carboxylate synthetase (P5CS) gene confers tolerance to salt stress in transgenic sugarcane. *Acta Physiol. Plant.*, 36: 2309–2319.

Guo J., Gao S., Lin Q., Wang H., Que Y., and Xu L. (2015). Transgenic sugarcane resistant to Sorghum mosaic virus based on coat protein gene silencing by RNA interference. *Biomed Res. Int.*, 2015: 861907.

Haber J.E. (2000). Partners and pathways: repairing a double-strand break. *Trends in Genetics*, 16: 259–264.

Hamerli D., and Birch R.G. (2011). Transgenic expression of trehalulose synthase results in high concentrations of the sucrose isomer trehalulose in mature stems of field-grown sugarcane. *Plant Biotechnol. J.*, 9, 32–37.

Harrison M.D., Geijskes J., Coleman H.D., Shand K., Kinkema M., Palupe A., Hassall R., Sainz M., Lloyd R., Miles S., and Dale J.L. (2011). Accumulation of recombinant cellobiohydrolase and endoglucanase in the leaves of mature transgenic sugar cane. *Plant Biotechnology Journal*, 9: 884–896.

Harrison M.D., Zhang Z., Shand K. et al. (2014). The combination of plant-expressed cellobiohydrolase and low dosages of cellulases for the hydrolysis of sugar cane bagasse. *Biotechnol Biofuels*, 7: 131.

Heinz D.J., and Mee W.P.. (1969). Differentiation from callus tissue of Saccharum species. *Crop Sci.*, 9: 346–348.

Islam N., Laksana C., and Chanprame S. (2016). Agrobacterium-mediated transformation and expression of bt gene in transgenic sugarcane. *J. ISSAAS Int. Soc. Southeast Asian Agric. Sci.*, 22: 84–95.

Jackson M.A., Anderson D.J., and Birch R.G. (2013). Comparison of Agrobacterium and particle bombardment using whole plasmid or minimal cassette for production of high-expressing, lowcopy transgenic plants. *Transgenic Research*, 22(1): 143–151.

Jain M., Chengalrayan K., Abouzid A., and Gallo M. (2007). Prospecting the utility of a PMI/mannose selection system for the recovery of transgenic sugarcane (*Saccharum* spp. hybrid) plants. *Plant Cell Reports*, 26(5): 581–590.

Jalaja N.C., Sreenivasan T.V., Pawar S.M. et al (2006). Co 94012 – a new sugarcane variety through somaclonal variation. *Sugar Tech*, 8: 132–136

Joyce P., Hermann S., O'Connell A., Dinh Q., Shumbe L., and Lakshmanan P. (2014). Field performance of transgenic sugarcane produced using Agrobacterium and biolistics methods. *Plant Biotechnology Journal*, 12(4): 411–424.

Jung J.H., and F. Altpeter. (2016). TALEN mediated targeted mutagenesis of the caffeic acid O-methyltransferase in highly polyploidy sugarcane improves cell wall composition for production of bioethanol. *Plant Molecular Biology*, 92: 131–142.

Jung J.H., Vermerris W., Gallo M., Fedenko J.R., Erickson J.E., and Altpeter F. (2013). RNA interference suppression of lignin biosynthesis increases fermentable sugar yields for biofuel production from field-grown sugarcane. *Plant Biotechnology Journal*, 11: 709–716.

Kalunke R.M., Kolge A.M., Babu K.H., and Prasad D.T. (2009). *Agrobacterium*-mediated transformation of sugarcane for borer resistance using *cry 1Aa3* gene and one-step regeneration of transgenic plants. *Sugar Tech*, 11: 355–359.

Kalwade S.B., and Devarumath R.M. (2014). Single strand conformation polymorphism of genomic and EST-SSRs marker and its utility in genetic evaluation of sugarcane. *PhysiolMolBiol Plants*, 20(3): 313–321.

Kalwade S.B., Devarumath R.M., Kawar P.G., and Sushir K.V. (2012). Genetic profiling of sugarcane genotypes using Inter Simple Sequence Repeat (ISSR) markers. *Electronic Journal of Plant Breeding*, 3(1): 621–628.

Kannan B., Jung J.H., Moxley G.W., Lee S.-M., and Altpeter F. (2018). TALEN-mediated targeted mutagenesis of more than 100 COMT copies/alleles in highly polyploid sugarcane improves saccharification efficiency without compromising biomass yield. *Plant Biotechnology Journal*, 16: 856–866.

Kasirajan L., Hoang N.V., Furtado A., Botha F.C., and Henry R.J. (2018). Transcriptome analysis highlights key differentially expressed genes involved in cellulose and lignin biosynthesis of sugarcane genotypes varying in fiber content. *Sci. Rep.*, 8: 11612.

Kaur L., Dharshini S., Ram B., and Appunu C. (2017). Sugarcane Genomics and Transcriptomics. In *Sugarcane Biotechnology: Challenges and Prospects*. Springer, Cham, Switzerland, pp. 13–32.

Kawar P.G., Devarumath R.M., and Nerkar Y.S. (2009). Use of RAPD markers for assessment of genetic diversity in sugarcane cultivars. *Indian Journal of Biotechnology*, 8: 67–71.

Koerniati S., Sukmadjaja D., and Samudra I.M. (2020). C synthetic gene of CryIAb-CryIAc fusion to generate resistant sugarcane to shoot or stem borer. *IOP Conf. Ser. Earth Environ. Sci.*, 418: 012069.

Kona P., Kumar M.H., Reddy K., et al. (2019). Regeneration and evaluation of somaclones of sugarcane variety Co86032 for yellow leaf disease resistance and yield traits. *J Biosci.*, 44(29).

Kumar T., Uzma, Khan M.R., Abbas Z., and Ali G.M. (2014). Genetic improvement of sugarcane for drought and salinity stress tolerance using Arabidopsis vacuolar pyrophosphatase (AVP1) gene. *Mol. Biotechnol.*, 56: 199–209.

Kwon-Ndung E.H., and Ifenkwe O.P. (2000). Mutational studies on sugarcane using gamma irradiation. *Sugar Tech.*, 2: 29–35.

Lakshmanan P., Geijskes R.J., Aitken K.S., Grof C.L.P., Bonnett G.D., and Smith G.R., (2005). Sugarcane biotechnology: The challenges and opportunities. *In Vitro Cellular and Developmental Biology - Plant*, 41(4): 345–363.

Lassoued R., Phillips P.W.B., Macall D.M., Hesseln H., and Smyth S.J. (2021). Expert opinions on the regulation of plant genome editing. *Plant Biotechnology Journal*, 19(6): 1104–1109.

Le Cunff L., Garsmeur O., Raboin L.M., Pauquet J., Telismart H., Selvi A., Grivet L., Philippe R., Begum D., Deu M. et al. (2008). Diploid/ polyploidsyntenic shuttle mapping and haplotype-specific chromosome walking toward a rust resistance gene (*Bru1*) in highly polyploid sugarcane (2n approximately 12x approximately 115). *Genetics*, 180: 649–660.

Lekshmi Maya, Pazhany A.S., Sobhakumari V.P., and Premachandran M.N. (2017). Nuclear and cytoplasmic contributions from *Erianthusarundinaceus* (Retz.) Jeswiet in a sugarcane hybrid clone confirmed through genomic in situ hybridization and cytoplasmic DNA polymorphism. *Genet Resour Crop Evol*, 64: 1553–1560.

Li M., Liang Z., Zeng Y., Jing Y., Wu K., Liang J., He S., Wang G., Mo Z., and Tan F. (2016). De novo analysis of transcriptome reveals genes associated with leaf abscission in sugarcane (*Saccharumofficinarum* L.). *BMC Genom.*, 17: 195.

Lu Y.H., A. D'Hont, D.I.T. Walker, P.S. Rao, P. Feldmann, and Glaszmann J.C. (1994). Relationships among ancestral species of sugarcane revealed with RFLP using single copy maize nuclear probes. *Euphytica*, 78: 7–18.

Manners J.M., and Casu R.E. (2011).Transcriptome analysis and functional genomics of sugarcane. *Trop Plant Biol*, 4: 9–21.

Masoabi M., Lloyd J., Kossmann J., and van der Vyver C. (2018). Ethyl Methane sulfonate mutagenesis and In Vitro polyethylene glycol selection for drought tolerance in sugarcane (*Saccharum* spp.). *Sugar Tech.*, 20: 50–59.

Mattiello L., Riaño-Pachón D.M., Martins M.C.M., da Cruz L.P., Bassi D., Marchiori P.E.R., Ribeiro R.V., Labate M.T.V., Labate C.A., and Menossi M. (2015). Physiological and transcriptional analyses of developmental stages along sugarcane leaf. *BMC Plant Biol.*, 15: 300.

Ming R., Moore P.H., Wu K., D' Hont A., Glaszmann J.C., Tew T.L., Mirkow T.E., Silva J.D., Jifon J., Rai M., Schnell R.J., Brumbley S.M., Lakshmanan P., Comstock J.C., and Paterson A.H. (2006). Sugarcane improvement through breeding and biotechnology. *Plant Breeding Review*, 27: 15–118

Mustafa G., Joyia F.A., Anwar S., Parvaiz A., and Khan M.S. (2018). Biotechnological interventions for the improvement of sugarcane crop and sugar production. In *Sugarcane-Technology and Research*; IntechOpen: London, UK, pp. 113–138.

Narayan A., Chakravarthi M., Gauri Nerkar, Manoj V.M., Dharshini S., Subramonian N., Premachandran M.N., Arun Kumar R., Krishna Sunder K., Hemaprabha G., Ram Bakshi, and Appunu C. (2021). Overexpression of expansin EaExpa1, a cell wall loosening protein, enhances drought tolerance in sugarcane. *Industrial Crops and Products*, 159: 113035.

Nayyar S., Sharma B.K., Kaur A., Kalia A., Sanghera G.S., Thind K.S., et al. (2017). Red rot resistant transgenic sugarcane developed through expression of b-1,3-glucanase gene. *PLoS One*, 12: e0179723.

Nerkar G., Thorat A., Sheelavantmath S., Babu K.H., and Devarumath R. (2018). Genetic Transformation of Sugarcane and Field Performance of Transgenic Sugarcane. In: Gosal S.S., and Wani S.H. (Eds.) *Biotechnologies of Crop Improvement*, vol. 2. Springer Nature, Switzerland AG.

Nerkar G., Farsangi F., and Devarumath R. (2015). Organellar Genome Diversity in Saccharum and Erianthus spp. revealed by PCR-RFLP. *Molecular Plant Breeding*, 6: 1–11

Nikam A.A., Devarumath R.M., Ahuja A., Babu K.H., Shitole M.G., Suprasanna P. (2015). Radiation-Induced in Vitro Mutagenesis System for Salt Tolerance and Other Agronomic Characters in Sugarcane (Saccharumofficinarum L.). *The Crop Journal*, 3: 46–56.

Nikam A.A., Devarumath R.M., Shitole M.G., Ghole V.S., Tawar P.N., and Suprasanna P. (2014). Gamma radiation, in vitro selection for salt (NaCl) tolerance, and character- ization of mutants in sugarcane (*Saccharumofficinarum* L.). *In Vitro Cell DevBiol Plant*, 50: 766–776.

Oliveira K.M., Pinto L.R., Marconi T.G., Margarido G.R.A., Pastina M.M., Teixeira L.H.M., Figueira A.V., Ulian E.C., Garcia A.A.F., and Souza A.P. (2007). Functional integrated genetic linkage map based on ESTmarkers for a sugarcane (Saccharum spp.) commercial cross. *Mol Breed*, 20: 189–208.

Oloriz M.I., Gil V., Rojas L., Veitía N., Höfte M., and Jiméne E. (2012). Selection and characterisation of sugarcane mutants with improved resistance to brown rust obtained by induced mutation. *Crop Pasture Sci.*, 62: 1037–1044

Oz M.T., Altpeter A., Karan R., Merotto A., and Altpeter F. (2021). CRISPR/Cas9-Mediated Multi-Allelic Gene Targeting in Sugarcane Confers Herbicide Tolerance. *Frontiers in Genome Editing*, 3: 673566.

Palaniswamy H., Syamaladevi D.P., Mohan C., Philip A., Petchiyappan A., and Narayanan S. (2016). Vacuolar targeting of r-proteins in sugarcane leads to higher levels of purifiable commercially equivalent recombinant proteins in cane juice. *Plant Biotechnol J.*, 14(2): 791–807.

Parida S.K., Pandit A., Gaikwad K., Sharma T.R., Srivastava P.S., Singh N.K., Mohapatra T. (2010). Functionally relevant microsatellites in sugarcane unigenes. *BMC Plant Biol*, 10: 251.

Parisi C., Tillie P., Rodríguez-Cerezo E. (2016). The global pipeline of GM crops out 30 to 2020. *Nat Biotechnol*, 34: 32–36.

Patade V.Y., Suprasanna P. (2008). Radiation Induced in Vitro Mutagenesis for Sugarcane Improvement. *Sugar Tech.*, 10: 14–19. doi: 10.1007/s12355-008-0002-4.

Petrasovits L.A., Purnell M.P., Nielsen L.K., and Brumbley S.M. (2007). Production of polyhydroxybutyrate in sugarcane. *Plant Biotechnol.*, 5, 162–172.

Petrasovits L.A., Zhao L., McQualter R.B., Snell K.D., Somleva M.N., Patterson N.A., et al. (2012). Enhanced polyhydroxybutyrate production in transgenic sugarcane. *Plant Biotechnol. J.*, 10, 569–578. doi: 10.1111/j.1467-7652.2012.00686.x

Prasad V., and Naik G.R. (2000). In vitro strategies for selection of eye-spot resistant sugarcane lines using toxins of Helminthosporiumsacchari. *Indian Journal of Experimental Biology*, 38: 69–73.

Puchta H. (2005). The repair of double-strand breaks in plants: mechanisms and consequences for genome evolution. *Journal of Experimental Botany*, 56: 1–14.

Raboin L.M., Pauquet J., Butterfield M., D'Hont A., and Glaszmann J.C. (2008). Analysis of genome-wide linkage disequilibrium in the highly polyploid sugarcane. *TheorAppl Genet*, 116: 701–714.

Racedo J., Budeguer F., Perera M.F., SoriaFemeías M.J., Ovejero S.N., Cuenya M.I., et al. (2019). Development of a new transgenic sugarcane variety resistant to glyphosate herbicide. *Proc. ISSCT*, 30, 515–521.

Racedo J., Gutiérrez L., Perera M.F. *et al.* (2016). Genome-wide association mapping of quantitative traits in a breeding population of sugarcane. *BMC Plant Biol*, 16: 142.

Ramiro D.A., Melotto-Passarin D.M., de Almeida Barbosa M., Santos F.D., Gomez S.G.P., MassolaJúnior N.S., Lam E., and Carrer H. (2016). Expression of Arabidopsis Bax Inhibitor-1 in transgenic sugarcane confers drought tolerance. *Plant Biotechnol J*, 14: 1826–1837

Rani K., Sandhu S.K., and Gosal S.S. (2012). Genetic augmentation of sugarcane through direct gene transformation with Osgly II gene construct. *Sugar Tech*, 14(3): 229–236

Rao S., Jabeen F.T.Z. (2013). In vitro selection and characterization of polyethylene glycol (PEG) tolerant callus lines and regeneration of plantlets from the selected callus lines in sugarcane (Saccharumofficinarum L.). *Physiology and Molecular Biology of Plants*, 19: 261–268.

Raza G., Ali K., Ashraf M.Y., Mansoor S., Javid M., and Asad S. (2016). Overexpression of an H+-PPase gene from Arabidopsis in sugarcane improves drought tolerance, plant growth, and photosynthetic responses. *Turk. J. Biol.*, 40, 109–119.

Reis R.R., Barbara A.D.B.C., Polyana K.M., Maria T.B.M., Jean C.A., Antonio C.J., Alan C.A., Ana P.R., Feng Q., Junya M., Yamaguchi-Shinozaki K., Kazuo N., Josirley F.C.C., Carlos A.F.S., Alexandre L.N., Adilson K.K., and Hugo B.C.M. (2014). Induced over-expression of AtDREB2A CA improves drought tolerance in sugarcane. *Plant Sci*, 221–222: 59–68

Riaz S., Nasir I.A., Bhatti M.U., Adeyinka O.S., Toufiq N., Yousaf I., et al. (2020). Resistance to Chiloinfuscatellus (Lepidoptera: pyraloidea) in transgenic lines of sugarcane expressing Bacillus thuringiensis derived Vip3A protein. *Mol. Biol. Rep.*, 47, 2649–2658.

Rutherford R.S., Snyman S.J., and Watt M.P. (2014). In vitro studies on somaclonal variation and induced mutagenesis: progress and prospects in sugarcane (*Saccharum* spp.) – a review, *The Journal of Horticultural Science and Biotechnology*, 89(1): 1–16.

Schaart J.G., van de Wiel C.C.M., and Smulders M.J.M. (2021). Genome editing of polyploid crops: prospects, achievements and bottlenecks. *Transgenic Research*, 30: 337–351.

Schneider V.K., Soares-Costa A., Chakravarthi M., Ribeiro C., Chabregas S.M., Falco M.C., et al. (2017). Transgenic sugarcane overexpressing CaneCPI-1 negatively affects the growth and development of the sugarcane weevil Sphenophoruslevis. *Plant Cell Rep.*, 36, 193–201.

Selvi A., Nair N.V., Noyer J.L., Singh N.K., Balasundaram N., Bansal K.C., Koundal K.R., and Mohapatra T.. 2005. Genomic constitution and genetic relationship among the tropical and sub-tropical Indian sugarcane cultivars revealed by AFLP. *Crop Science*, 45: 1750–1757.

Shamansori S., Dehkordi M.K., and Shomeili M. (2018). Development of in vitro somaclonal drought tolerant variants of Sugarcane (Saccharumofficinarum L.) using Iranian commercial varieties. *Environmental Stresses in Crop Sciences*, 11(3) 481–489.

Song L.I., Jian-Hua Y., Kun-Xing Y.U., Hong-Jian L., Jian W.E.I., Min L.L., Ming D., Fing T., and Man-Man L.. (2011). Mutation Breeding of New Sugarcane Variety GUIFU 98-296. *Journal of Nuclear Agriculture and Science*, 24: 1177–1181.

Sugiharto B. (2018). Biotechnology of drought-tolerant sugarcane. In: de Oliveira A.B. (Ed.) *Sugarcane – Technology and Research*, InTech, London.

Suhesti S., Khumaida N., Wattimena G.A., Syukur M., Husni A., and Hadipoentyanti E. (2015). Gamma Irradiation and in Vitro Selection Could Increase Drought Tolerance in Sugarcane. *International Journal of Sciences: Basic and Applied Research*, 23 (2): 370–380.

Suprasanna P., Patade V.Y., Desai N.S. et al. (2011). Biotechnological Developments in Sugarcane Improvement: An Overview. *Sugar Tech*, 13, 322–335.

Symington L.S., and Gautier J. (2011). Double-Strand Break End Resection and Repair Pathway Choice. *Annual Review of Genetics*, 45(1): 247–271.

Taparia Y., Gallo M., and Altpeter F. (2012). Comparison of direct and indirect embryogenesis protocols, biolistic gene transfer and selection parameters for efficient genetic transformation of sugarcane. *Plant Cell Tissue and Organ Culture*, 111(2): 131–141.

Tariq M., Khan A., Tabassum B., Toufiq N., Bhatti M.U., Riaz S., et al. (2018). Antifungal activity of chitinase II against Colletotrichumfalcatum Went. causing red rot disease in transgenic sugarcane. *Turk. J. Biol.*, 42: 45

Tawar P.N., Sawant R.A., Sushir K.V., Devarumath R.M., Hapase R.S., Meti N.T. (2016). VSI 434: New Sugarcane Variety Obtained Through Somaclonal Variation. *Agric Res.*, 5(2): 127–136.

van der Merwe M.J., Groenewald J.H., Stitt M., Kossmann J., and Botha F.C. (2010). Downregulation of pyrophosphate: D-fructose-6- phosphate 1-phosphotransferase activity in sugarcane culms enhances sucrose accumulation due to elevated hexosephosphate levels. *Planta*, 231: 595–608.

Vaze A., Nerkar G., Pagariya M., Devarumath R.M., and Theertha Prasad D. (2010). Isolation and PCR amplification of genomic DNA from dry leaf samples of sugarcane. *International Journal of Pharma and Bio Sciences*, 1(2): 1–3.

Viola V.R., Lekshmi M., and Premachandran M.N. (2011). Differentiation of cytoplasm of Saccharum and Erianthus species by mitochondrial DNA polymorphism. *Ind J Genet*, 71: 1–3

Wagih M.E., Ala A., and Musa Y. (2004). Regeneration and evaluation of sugarcane somaclonal variants for drought tolerance. *Sugar Tech*, 6(1&2): 35–40.

Waltz E. (2014). Beating the heat. *Nat Biotechnol*, 32(7): 610–613

Wang W.Z., Yang B.P., Feng X.Y., Cao Z.Y., Feng C.L., Wang J.G., et al. (2017). Development and characterization of transgenic sugarcane with insect resistance and herbicide tolerance. *Front. Plant Sci.*, 28: 1535.

Wei X., Jackson P.A., Hermann S., Kilian A., Heller-Uszynska K., and Deomano E. (2010).Simultaneously accounting for population structure, genotype by environment interaction, and spatial variation in marker–trait associations in sugarcane. *Genome*, 53: 973–981.

Wu H., Awan F.S., Vilarinho A., Zeng Q., Kannan B., Phipps T., McCuiston J., Wang W., Caffall K., and Altpeter F. (2015). Transgene integration complexity and expression stability following biolistic or Agrobacterium-mediated transformation of sugarcane. *In Vitro Cellular and Developmental Biology – Plant*, 51(6), 603–611.

Yang X., Sood S., Luo Z., Todd J., and Wang J. (2019). Genome-wide association studies identified resistance Loci to orange rust and yellow leaf virus diseases in sugarcane (*Saccharum* spp.). *Phytopathology*, 109, 623–631.

Zale J., Jung J.H., Kim J.Y., Pathak B., Karan R., Liu H., et al. (2016). Metabolic engineering of sugarcane to accumulate energy-dense triacylglycerols in vegetative biomass. *Plant Biotechnol. J.*, 14, 661–669.

Zhang J., Zhang X., Tang H., Zhang Q., Hua X., Ma X., Zhu F., Jones T., Zhu X., and Bowers J. (2018). Allele-defined genome of the autopolyploid sugarcane *Saccharumspontaneum* L. *Nat. Genet.*, 50: 1565.

Zhao Y., Karan R., and Altpeter F. (2021). Error-free recombination in sugarcane mediated by only 30 nucleotides of homology and CRISPR/Cas9 induced DNA breaks or Cre-recombinase. *Biotechnology Journal*, 16: 2000650.

Zhou D., Liu X., Gao S., Guo J., Su Y., Ling H., et al. (2018). Foreign cry1Ac gene integration and endogenous borer stress-related genes synergistically improve insect resistance in sugarcane. *BMC Plant Biol.*, 18: 342.

4 Sugarcane Biotechnology: Developments and Opportunities

Sarita Kumari and Vimanth S
Department of Agricultural Biotechnology and Molecular Biology, CBS&H, RPCAU-Pusa, Samastipur, Bihar, India

Sumeet Kumar Singh
Department of Seed Science and Technology, PGCA, RPCAU – Pusa, Samastipur, Bihar, India

CONTENTS

4.1 INTRODUCTION

Sugarcane is a perennial grass grown in tropical and subtropical regions. This important cash crop for industries acts as treasure for sugar, biofibre, bioplastic, biofuel, and waxes. Sugarcane originated in New Guinea and has grown over 35° N and S of equator. As of 2019, more than 100 countries have produced over 26.8 million ha with annual production of 1949 million tonnes. Brazil leads countries with an annual production of 752 million tonnes; India follow with 405 million tonnes. Sugarcane is the major and prime source for sugar production. Around 179.66 million metric tons of sugar is produced globally, out of which 171.8 million metric tons are being consumed. Global sugar consumption increased from 123.454 million metric tonnes to 171.8 million metric tonnes between 2001 and 2021 (https://knoema.com/atlas/topics/Agriculture/Crops-Production-Area-Harvested/Sugar-cane-area-harvested 2019). On the other hand, the production is also being constrained by biotic and abiotic stresses (Nasir et al., 2000; Khaliq et al., 2005). Reports suggest that drought reduces sugarcane yield up to 17%, while salinity stress tremendously affects the sugarcane production since it is glycophyte (Suprasann, 2010).

Sugarcane is the one of the most important crops among industries. Use of the sugarcane crop can be grouped into two categories: a) direct application and b) indirect applications (Figure 4.1). Direct application of sugarcane is being used for cane juice, syrup, jaggery, etc., while whole plants in immature form can be used for fodder. On the other hand, indirect application includes

DOI: 10.1201/9781003292425-4

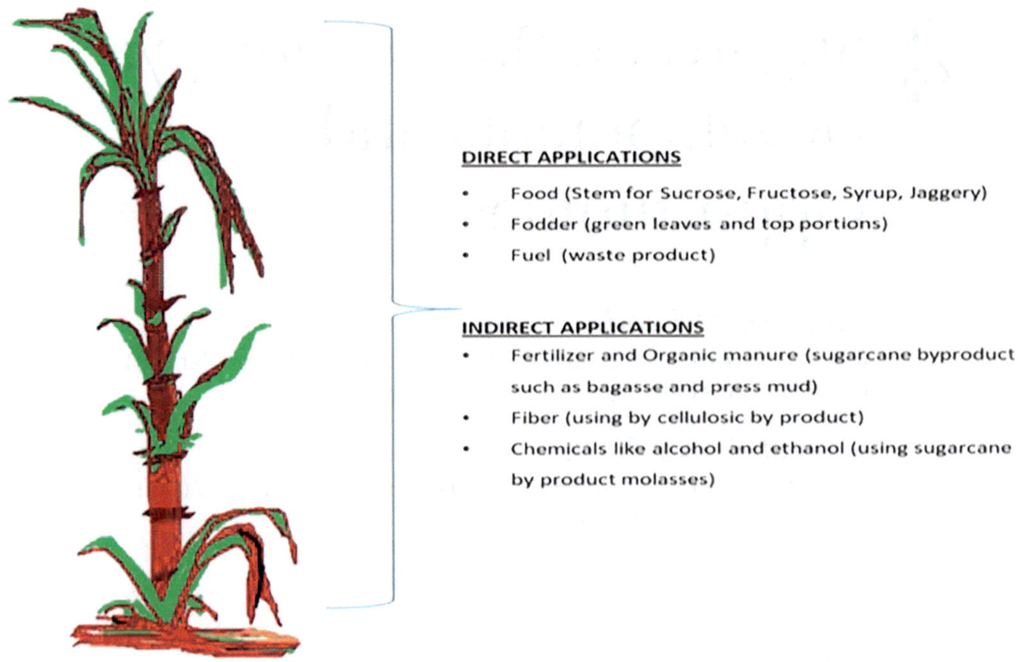

FIGURE 4.1 Direct and Indirect application of sugarcane plant.

fertilisers, fibre, chemicals, etc., which are generated using agricultural and industrial waste, such as press mud, bagasse, molasses, and spent wash. The sugar industry is the second-most important leading industry, after cotton, that produces a huge amount of by-product, such as field waste, waste separation at cleaning station, bagasse, press mud, molasses, etc. (Almazan et al., 1998; Dotaniya et al., 2016). The waste produced by sugar industries is summarised in Figure 4.2. The waste product after harvest can also be used as traditional fuel after drying. Bagasse and press mud are being used to improve biological, physical, and chemical properties of soil in the form of organic manure and fertilizers. However, the by-product of sugar industries is diverted to organic manures, fibre, etc., but their management is a cumbersome process. Researchers suggested different possible solutions for using the waste. One major and direct utilisation is the incorporation of these wastages into the soil for organic decomposition using microbial stain for remobilisation of the nutrient from wastages. The varieties of organic acids released during the microbial decomposition may enhance the mobilisation of insoluble phosphorus (P) (Bhattacharyya et al., 2003). Different industrial application, increasing the rate of sugarcane consumption and high production of sugarcane wastage, is urging to the field of science for quantity and quality enhancement of sugarcane and by-products.

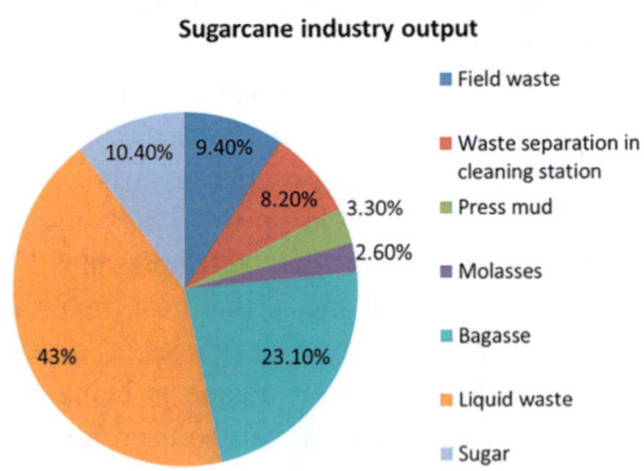

FIGURE 4.2 Different waste product from sugarcane industry.

4.2 INTERVENTION OF BIOTECHNOLOGY TOOLS FOR SUGARCANE IMPROVEMENT

The genetic improvement in sugarcane is essential for the maintenance of sufficient sucrose levels, increased agricultural productivity with upright architecture, higher fibre content, excellent sprouting, and greater resistance to biotic and abiotic stresses. Conventional breeding programmes include: parent selection, hybridisation, and selection of superior genotypes. Selection of superior genotypes involves three to four phases. The release of a new variety for sugarcane takes 12–13 years, after being tested for genotypic and phenotypic performance (Gazaffi et al., 2010; Landell, 2003; Landell et al., 2010). The success rate of the breeding programme depends upon the parent selection, trait heritability, reliable phenotypic methodology, and application of a suitable statistical model (Gazaffi et al., 2010). The major challenge for sugarcane breeding programmes in most of the country is the flowering under subtropical conditions that limits the seed production (Scortecci et al., 2012). Moreover, complex genome with high polyploidy/aneuploidy, narrow gene pool, poor fertility, and long breeding/selection cycle are the major constraints for conventional breeding programmes focused on the genetic improvement in sugarcane (Lakshmanan et al., 2006). Therefore, the use of biotechnology offers new windows in sugarcane improvement. Biotechnology is the field of science that provides different tools: molecular markers, genomics, transcriptomics, tissue culture, genetic engineering, and genome editing *etc.*

4.3 MOLECULAR MARKERS IN SUGARCANE

A molecular marker is a DNA stretch with known locations in the genome. It is an important tool, widely being used to study the germplasm for genetic diversity, genotype identification, and fingerprinting; the population's structure; and trait mapping and introgressions (Kesawat et al., 2009; Nadeem et al., 2018). Molecular typing can be easily performed using a small amount of genomic DNA with PCR machines. Results produced by molecular markers are not affected by the genotypic locations, stages of growth, and stress conditions. More than 20 types and subtypes of markers have been reported to date, with different groups. They have been classified broadly based on their dominant/codominant relationship, method of detection, i.e., PCR/Non-PCR based (Nadeem et al., 2018).

The complex poly-aneuploid genome of sugarcane limits the commercial cultivation of popular varieties using a conventional breeding programme. Molecular markers – SSR, ISSR, TRAP, AFLP, RFLP, SNP, DArT, and RAPD – are widely used for sugarcane. Diversity analysis and germplasm characterisation using SSR markers and TRAP markers are widely used for sugarcane (Ali et al., 2019; You et al., 2016; Singh et al., 2017; Medeiros et al., 2020). Genetic linkage map for sugarcane can be easily performed using RFLP, AFLP, and retro transposon-based molecular markers (Alwala et al., 2008; Palhares et al., 2012). Apart from these, a number of marker studies have been reported for the screening of yield, sucrose content, and biotic & abiotic stress tolerance traits in sugarcane (Table 4.1). However, SNP discovery and genotyping is a cumbersome process in sugarcane; scientists have reported few successful examples of SNP genotyping in association with the results of other markers, like SSR for leaf-scald disease resistance in sugarcane (Gutierrez et al., 2018; Fickett et al., 2019). Despite so many hurdles with genome sequencing in sugarcane, an axiom array platform for 100K SNP genotyping in sugarcane has been developed (You et al., 2019).

4.4 GENOMICS IN SUGARCANE

Agronomically important genes are being transferred from one species to other species, such as interspecific transfer between *Saccharum officinarum* and *Sacharum spontaneum* species through hybridisation programmes; however, transfer of some important genes is a cumbersome process for aneuploidy species (Cheavegatti-Gianotto et al., 2011; Waclawovsky et al., 2010). Moreover, the average duration of sugarcane breeding programmes is 10–15 years (Morais et al., 2015). Despite

TABLE 4.1
Different Molecular Markers and Their Uses for Sugarcane

Studies	Molecular Markers	References
Genetic diversity, Germplasm characterisation and population structure	Target region amplification polymorphism (TRAP)	Singh et al., 2017; Medeiros et al., 2020
	SSR (Simple Sequence Repeats)	Ali et al., 2019; You et al., 2016
Genetic linkage map constructions	RFLP, AFLP and retro transposon based molecular markers	Alwala et al., 2008; Palhares et al., 2012
For drought resistance in sugarcane	SSR (Simple Sequence Repeats)	Prasitsom et al., 2019
	SCAR	Srivastava et al., 2012
	ISSR	Khaled et al., 2017
	RAPD	Gadakh et al., 2017
QTLs mapping for sucrose content, fibre and yield in sugarcane	RFLP derived markers	Pinto et al., 2010
	SSR, AFLP and TRAP	Liu et al., 2016
GWAS for disease resistance traits in sugarcane.	AFLP, DArT and SSR markers	Gouy et al., 2015
Yellow leaf viruses in Sugarcane	RFLP, AFLP, DArT and SSR markers	Costet et al., 2012
spotted stem borer in Sugarcane	SSR, RFLP	Nibouche et al., 2012
For screening of salt and drought tolerant lines of sugarcane	RAPD	Gadakh et al., 2017
For selection of salinity tolerant genotypes in sugarcane	TRAP	Farsangi et al., 2018

that, a lot of varieties have been released for different biotic and abiotic stress tolerance in sugarcane (Cursi et al., 2021). Genomic region is the target platform for genetic gain, either through hybridisation or transgenic approaches; it is essential to unveil them at full length. However, a number of studies have reported on sugarcane genome sequences, but they are still an enigma because of their complexity, lots of interaction between cytosolic genetic materials, and dynamic nature of genetic ploidy level. The ploidy level of sugarcane genome is 10 or more with genome size around 10 GB, which is much more than maize (5500 Mb, 2n = 20), sorghum (1600 Mb, 2n = 20), or rice (860 Mb, 2n = 24) (D'Hont et al., 2005).

The interspecific, polyploid, and aneuploid characters of sugarcane are hindering their genome assembly. The monoploid genomes of *S. officinarum* and *S. spontaneum* are 930 Mb (megabase pairs) and 750 Mb, respectively, more than twice the size of the *Oryza sativa* genome (390 Mb) (Ali et al., 2019). Haploid of *S. spontaneum* (AP85–441) has been used for the development of autopolyploid, allowing the connection of 32 pseudo-chromosomes comprising eight homologous groups of four members each, including 35,525 genes (Zhang et al., 2018). The fission of two ancestral chromosomes and subsequent translocation to four different chromosomes decreased the number of essential chromosomes from 10 to eight in *S. spontaneum.* Surprisingly, 80% of the genes related to disease resistance were found in four rearranged chromosomes with 51% in rearranged areas (Zhang et al., 2018). Geneticists are attempting to make out the correlations between the complex genomes of sugarcane and other crops/plants. A homologous species, such as sorghum genome, has been used as a reference to assemble the BAC-based data for sugarcane monoploid genome (Thirugnanasambandam et al., 2018). Based on the genome-wide analysis, the minimal tiling path of the 4,660 BAC, which covered the gene-rich region of sorghum genomic portion, was chosen, sequenced, and assembled into a 382 Mb of a high-quality sequence that yields a total of 25,316 protein-coding gene models, of which 17% showed no collinearity with their sorghum orthologs (Garsmeur et al., 2018). Most of the studies showed the transposable elements (TEs) and chromosomal rearrangements of two species; *S. officinarum* and *S. spontaneum* explain the different genomic size and basic chromosome count of current sugarcane cultivars (De Setta et al., 2014). Plants of the poaceae family have genomes ranging from diploid to polyploidy. The gene sequence is responsible for the maintenance and genesis of gene function. In grasses, the genome expansion significantly supports the TE, especially between coding genes. The LTR (long terminal repeat) retrotransposon is also the most abundant reverse transcription element in plants. A transpose protein creates an insertion–deletion of the base pair that causes variation in the genic and intergenic regions of the genome. The retrotransposon's movement is controlled by the active site of transcription, which is reinserted into the genome after each breeding cycle to increase the copy number (Shingote et al., 2019). TEs may be triggered for unravelling the role of different candidate genes in sugarcane using the functional genomics method. Different sugarcane characteristics, such as sucrose accumulation, fibre content, and pathogen-resistance proteins, can be revealed.

Mutant-like transposons are the most descriptive transposon transcripts in the sugarcane transcriptome, according to recent research. In monocots/true dicots, mutant-like transposons offer at least four (I–IV) groupings of evidence, according to sugarcane reverse genetics and transcriptome research by Rossi M (2004). Despite sugarcane's genetic complexity, genome-wide association studies (GWAS) may be utilised to discover marker-trait associations (MTAs), which can subsequently help the breeders better manage crosses and choose superior genotypes in breeding programmes. GWAS using a multi-locus mixed-model revealed resistance loci to orange rust and yellow leaf virus infections in sugarcane; 23 MTAs linked to soluble solid content, stalk height, stalk number, stalk weight, and cane yield characteristics in sugarcane were discovered (*Saccharum spp.*) by (Barreto et al., 2019). However, an allele-defined genome of *S. spontaneum* is available, allowing for the identification and phylogeny of different gene families in sugarcane on a genome-wide scale. Furthermore, RNA-Seq data may be used to identify the expression patterns of these gene families (Santiago et al., 2018).

4.5 SUGARCANE TRANSCRIPTOMICS

Genomic size and number of genes are still enigma for sugarcane. A famous report has shown that the genome of sugarcane is around 10 GB with 3 lakhs 75000 genes (Souza et al., 2019). The presence of the genome's aneuploidy/polyploidy nature is a big hurdle for functional genomics. Transcriptomics is the study of total RNA transcript with respect to their location, level of expression, and functions. Techniques involved in transcript study may be grouped as low throughput and high throughput technologies. Northern blot, reverse transcriptase PCR (RT-PCR) are the technologies used for the gene-expression analysis of a few genes, and hence, may be regarded as low throughput technologies (Alwine et al., 1977; Becker-André and Hahlbrock, 1989; Morozova et al., 2009). However, few technologies such as serial analysis of gene expression (SAGE) (Velculescu et al., 1995), DNA microarrays (Lockhart et al., 1996), and massively parallel signature sequencing (MPSS) (Brenner et al., 2000) can study more than 20000 genes at a time. Few limitations are associated with these technologies; they are technologically demanding, extensive data analysis is required, and poorly expressed genes may be less represented. RNA-Seq technology is another high throughput technology superior over SAGE and MPSS for studying the complete transcriptome. It is a single high-throughput sequencing assay for both the discovery and quantification of transcripts. However, a next-generation-based sequencing (NGS) platform for RNA sequencing has some inherent limitations. It is not estimating accurately multiple copies of full-length transcript assembled through short read, isoform, and RNA variants; that is the major problem for eukaryotic systems due to alternate splicing. Scientists today use third-generation sequencing technologies with longer read length (Wang et al., 2019). Longer reads have improved the sugarcane gene model, where complex genome was cloudy for gene estimation and annotation (Hoang et al., 2017).

Transcriptomics technologies act as a complementary tool for functional genomics to unravel the gene function with respect to their expression level at certain conditions (Lowe et al., 2017). Transcriptome data may be used for identification and isolation of differentially expressed genes for agronomically important traits, e.g., RNA-Seq technology has been used to identify around 2000 differentially expressed genes for smut disease (Wu et al., 2013) and several genes for cellulose and lignin biosynthesis to improve fiber in sugarcane (Kasirajan et al., 2018). A comprehensive transcriptome analysis was done with 30 genotypes of sugarcane using cDNA-microarray and RT-PCR to unveil the role of genes for hormone signalling (auxin, ethylene, jasmonates), protein kinase and phosphatase signalling, calcium signalling, abiotic stress responses (drought, cold, oxidative), cell wall metabolism, sugar metabolism, and transcription factors (Papini-Terzi et al., 2005). Differential expression of genes identifies and studies different sugarcane tissues, such as leaf-stem culm and root using cDNA microarray technology (Manners, 2011).

Transcriptome data may be used for marker development in genic regions; it is highly important for the development of functional markers. EST markers are widely being developed through transcriptome data that is effective for construction of a genetic linkage map. The first EST database was developed for leaf role in sugarcane in 2000 (Carson et al., 2000). The largest EST database for sugarcane is the Brazalian SUCEST project (Vettore et al., 2001). Genic SSR (8630) & SNP (442,152) have been identified using contrasting genotypes of sugarcane for leaf abscission, tolerance to pokkah boeng disease, and drought stress (Xu et al., 2018). Moreover, transcriptome data may be used to isolate functional RNA, especially microRNA. Micro RNAs, such as miR5671, miR5054, miR5783, miR5221, and miR6478, have been identified in sugarcane for their role in pathogenesis for smut disease (Su et al., 2017). This confers that transcriptomics study is essential for any crops, especially for sugarcane where misinterpretation of genomic data can be complemented with transcriptome data.

4.6 PROTEOMICS FOR SUGARCANE

The study of a cell's total protein is called proteomics. It gives the overall comparative study of differentially expressed genes between two specimens with respect to their location and level of expression, post-translational modification, interactions, etc. Functional annotation of genes at the protein level deduced more direct and reliable functions to decode the role of candidate genes. ELISA, western blotting, and chromatography-based techniques are conventional techniques that give primitive data for gene-expression study. While sodium dodecyl sulphate polyacrylamide gel electrophoresis (SDS-PAGE), two-dimensional gel electrophoresis (2-DGE), and two-dimensional differential gel electrophoresis (2D-DIGE) are the gel-based advanced techniques but it gives only relative estimation of gene-expression analysis. Nowadays, isotope-coded affinity tag (ICAT), stable isotope labelling with amino acids in cell culture (SILAC), isobaric tag for relative and absolute quantitation (iTRAQ), and protein array are more advanced high throughput techniques available for proteome analysis (Aslam et al., 2017). Protein-based techniques are important for characterising the genes; they code for an enzyme involved in metabolism. The expression of genes is also regulated at the RNA level because of posttranscriptional modification, lack of ribosomal binding sites, and codon optimisation; the transcript is not entirely translated into protein. Moreover, the protein-level expression is regulated by the posttranslational modification, protein folding, and protein targeting to correct cellular location. Thus, proteomics provide the more direct and reliable estimation of gene characterisation. Sucrose metabolism is the major target for quality enhancement and juice improvement in sugarcane (Ostengo et al., 2021). Comparative proteome study based on iTRAQ labelling and MS/MS analysis identified 25 genes for sucrose metabolism in sugarcane, out of which key proteins are UDP-glucose 6-dehydrogenase, fructokinase, β-D-glucosidase, and α-glucan phosphorylase (Qin et al., 2021). Similarly, based on LC-MS/MS analysis, 37 transcription factors have been identified for drought stress tolerance in sugarcane (Salvato et al., 2019). Thirty proteins based on 2D- PAGE and MALDI-TOF were identified in sugarcane that was related to smut disease (Singh et al., 2019). Seven candidate genes: UDP-glycosyltransferase, a nonspecific lipid-transfer protein, a cytochrome P450 protein, a cytosolic glyceroldehyde-3-phosphate dehydrogenase, an argonaute family protein, the ubiquitin-activating enzyme E1, and the photosystem I P700 apo-protein A1 (chloroplast) were identified for leaf-scald disease in sugarcane using iTRQ and MS/MS proteomics technology (Meng et al., 2020).

4.7 TISSUE CULTURE IN SUGARCANE

Tissue culture is a technique for growing cell/tissue or any part of plant as an explant on an artificial media under controlled conditions. It provides disease-free, mass multiplication of both the flowering and nonflowering species across the season. It preserves the heterozygousity, provides an important system for long-duration germplasm conservation, and produces a true-to-type mother plant (Khan et al., 2009). The possibility of disease transfer via seed cane from generation to generation, low replication rates, and labour-intensive and time-consuming protocol limit the seed propagation and restrict the effectiveness of traditional sugarcane propagation. Commercial propagation and breeding programmes face a significant bottleneck because of the lengthy time for its propagation. As a result, traditional propagation techniques were unable to meet the increasing demand for newly released types (Lakshmanan et al., 2006). The use of plant tissue culture methods offers an alternate approach for sugarcane multiplication and development (Sandhu et al., 2009). Plant tissue culture, which uses micro propagation of sugarcane to produce high-quality, phytosanitary planting material at a quicker pace is the most effective technique. The tissue culture has the ability to enhance proliferation potential by 20–35 times. Plantation over one hectare requires around 18520 plants that can be generated from a single shoot via micro propagation even though 8.8 tonnes of cane seed is needed in traditional techniques. As a result, the multiplication ratio of tissue culture plants is 100–150 times that of traditional cane sets, resulting in a significant decrease in seed cane requirements. Moreover, in comparison to the traditional method, plants may be disease indexed, with healthy material reproduced within a short time. As a result, micro propagation has been used to create more effective regeneration protocols for propagation. Nowadays, tissue culture is the only practical way to achieve fast, large-scale production of disease-free seed canes of newly discovered sugarcane varieties to accelerate the sugarcane breeding and commercialization process. From single shoot tips of seed cane, it is predicted that 260,000 shoots can be produced in four months. Furthermore, in the field, micro propagated plants (CV. Co 83) had up to 44.96% more canes/plots and 22.9% higher cane yield/plot than plants propagated from three-budded sets. It is also claimed that the gross yield gained from *in vitro* multiplication-derived seed cane plants is more than the source plants (Biradar et al., 2009). According to Benisheikh et al. (2012), the gross yield achieved from *in vitro* multiplication derived seed cane plants is more than the source plants. Tissue culture responses have been recorded for different explant and hormone compositions with basal media in sugarcane (Table 4.2).

4.8 GENETIC ENGINEERING APPROACHES IN SUGARCANE

The unusual genetic complexity in the sugarcane genome is creating a challenging task for genetic manipulation in sugarcane. The analysis of transgenic sugarcane is cumbersome due to the number of copies and the location of transformations. However, genetic manipulation in this area is tedious; scientists have made it easy, though, using different transformation methodology with different traits in sugarcane (Singh et al., 2013). Both direct and indirect methods of transformation have been used for transgenic generation in sugarcane. *Agrobacterium*-mediated transformation is widely used for indirect transformations (Enriquez-Obregon et al., 1998; Manickavasagam et al., 2004). The micro-projectile bombardment method is the method of choice for direct transformation in sugarcane (Bower et al., 1992; Frank et al., 1991). Other methods, such as electroporation, have also been used to transfer cryIA(b) gene in Saccharum officinarum L. against stem borer (Arencibia et al., 1997).

TABLE 4.2
Different Explant, Hormone and Tissue Culture Responses for Sugarcane

Explant	Hormone	Response	References
Young leaf disc	MS + BAP (1.5 mg/L) + NAA (+0.5 mg/L) + casein hydrolysate (500 mg/L)	Direct regenerations	Ali et al., 2012
Young leaf segment	MS + NAA (5 mg/L) + Kinetin (0.5 mg/L)	Direct regenerations	Gill et al., 2006
Shoot tip	MS + BAP (2 mg/L)	Direct regenerations	Biradar et al., 2009
Young leaf	MS + 2, 4-D (3 mg/l) + 15% coconut water.	High percentage of somatic embryogenic callus responses	Khamrit et al., 2012
Shoot tip	MS + 2, 4-D (2 mg/l) + Kinetin (0.5 mg/l)	Rapid callus growth in all sugarcane genotypes	Srivong et al., 2015
Callus	MS + Paclobutrazol (40 mg/l)	Increases the production of plantlets and reduce the number of dead leaves and height of shoots	Panti, 2016
Callus	MS + BAP (2–2.5 mg/l)	High percentage of shoot formation and number of shoots per clump of callus	Hapsoro et al., 2012
Apical stem	MS + BAP (0.5 mg/L) + Activated charcoal (2.0 g/L)	High multiplication rate	Mittal et al., 2016
Callus	MS + BAP (1.0 mg/L) + GA3 (0.25 mg/L) + sucrose (20 g/L) + Agar (7 g/L)	Microtillering	Patel et al., 2015
Callus	MS or ½ MS + NAA (2.5 to 4 mg/L) or MS or ½ MS + NAA (0.5 mg/L) + IBA (2.5 to 3 mg/L)	High root number with good Root length	Satpal et al., 2011; Godheja et al., 2014; Tuan et al., 2015
Callus	Half-MS medium + 3 – 5 mg/l NAA + 50 g/l sucrose	Profuse rooting	Yadav et al., 2012; Tesfa et al., 2016
Young leaf	MS + 2,4-D (3 mg/L) reported	Highest Embryogenic calli	Ijaz et al., 2012

TABLE 4.3
Genetic Improvement in Sugarcane Using Transgenic Technology with Gene for Different Agronomically Important Trait

Trait	Genes	References
Herbicide tolerance	bar gene and nptII gene	Manickavasagam et al., 2004
	bar gene and neo gene	Falco et al., 2000
Insect / Pest resistance (cry genes)	cry1Aa3 gene for borer resistance	Kalunke et al., 2009
	cry1Ab gene for controlling Diatraea saccharalis (Lepidoptera: Cram- bidae).	Braga et al., 2003
	Aprotinin gene for stem borer	Christy et al., 2009
	Chitinase gene for targeting Insect/Pest	Khamrit et al., 2012
Disease resistance	Viral CP genes for Sugarcane mosaic potyvirus	Joyce et al., 2010
Drought stress tolerance	*Grifola* Trehalose synthase gene	Zhang et al., 2006
	Arabidopsis bax inhibitor-1 gene	Ramiro et al., 2016
	EaDREB2 transcription factor and pea DNA helicase gene (PDH45)	Augustine et al., 2015
Osmotic stress tolerance	Trehalose synthase gene	Wang et al., 2005
Salinity stress tolerance	D1-pyrroline-5-carboxylate synthetase (P5CS) gene	Guerzoni et al., 2014
Quality improvement	Invertase genes for sucrose metabolism	Wang et al., 2017

Establishment of tissue culture conditions make genetic transformation for different agronomic traits in sugarcane (Table 4.3).

Bacillus thuringiensis (Bt) produces insecticidal crystalline proteins generated during sporulation of the bacterium cause toxicity. These cry toxins have been shown to have specific insecticidal properties against the orders *Lepidoptera* (butterflies and moths), *Coleoptera* (beetles), *Hymenoptera* (ants, bees, sawflies, wasps), *Diptera* (flies and mosquitoes), and nematodes. When an insect consumes endo-toxin crystals, the extremely alkaline digestive gut denatures the insoluble (non-toxic) crystals, rendering them soluble (toxic). The cry toxins enter the insect gut's cell membrane and paralyse the digestive system. In certain cases, the microbe that colonises the targeted insect is genetically modified, rendering the pest incapable of infecting the crops. For example, to regulate *D. saccharalis*, scientists genetically modified the endophytic bacteria *Pantoea agglomerans* (considered a good sugarcane growth stimulator). *P. agglomerans* colonised the pest in the sugarcane stalks and killed it by preventing larval development and stunting growth. Sugarcane converted with modified forms of insecticidal genes like cry1Ac has previously been shown to be resistant to the stalk borer, *E. lignosellus*. Following these successes, Arencibia et al. (1997) used intact cells to electroporate a shortened cryIA(b) gene (encoding the Bt-endotoxin) into sugarcane. Braga and his colleagues (2003) created transgenic sugarcane plants by using the cry1A(b) gene to confer resistance to *D. saccharalis* without affecting other characteristics of the transgenic plants. Micro-projectile bombardment was used to convert sugarcane embryogenic calli with the synthetic cry1Ac gene. As a result, stem borer was completely killed by the sugarcane plants expressing the synthetic cry1Ac protein (Bonatelli et al., 2017). Sugarcane has been successfully converted using the cry1Aa3 gene for stem-borer resistance (Babu et al., 2021). The particle bombardment has been used to transform sugarcane with the cry2A gene to enhance resistance to *D. saccharalis* (F) without alteration in sugar production (Gao et al., 2016). In August 2018, the United States authorised the sugarcane event CTC175-A (IR) for food use exclusively. Other sugarcane events, CTC93209-4 and CTC91087-6, carrying the cry1Ac gene produced by CTC, were authorised for food, feed, and cultivation in Brazil in 2019. The noncrystalline poisons produced by *B. thuringiensis* during vegetative development are known as vegetative insecticidal proteins (VIP) (Iqbal et al., 2021). These VIPs have been proven to be effective against a variety of economically significant insect pests. The VIP toxin family is divided into four sub-families based on amino acid similarity, which also influences VIP naming. Vip1, Vip2, Vip3, and Vip4 proteins have been identified with vegetative insecticidal protein properties. Vip1 and Vip2 poisons have been shown to be effective against the bug orders, *Coleoptera* and *Hemiptera* respectively. Vip3 is the protein shown to have anti-lepidopteran action. Vip3A, the first vip in this family identified, has the ability to target receptors in the midgut of insects that are different from those targeted by cry proteins, making them a suitable combination for effective resistance management. Transgenic plants expressing both the Vip3Aa and cry genes have been produced, and Vip3Aa-expressing transgenic maize and cotton have been marketed (Lemes et al., 2017). Cry1Ac and Vip3Aa were shown to be the most promising gene combination for transgenic sugarcane targeting secondary pests. Moreover, transgenic sugarcane expressing vip3A toxin has found resistance against stem borer (*C. infuscatellus*) (Riaz et al., 2020). Their results also indicated that the vip3A gene may be used in gene pyramiding with the Bt toxin to achieve efficient and long-lasting resistance. Use of proteinase inhibitors is another important approach for insect pest resistance in sugarcane, e.g., aprotinine gene for stem borer in sugarcane (Christy et al., 2009). The genes bar, neo, and nptII have been targeted for herbicidal tolerance in sugarcane (Falco et al., 2000; Manickavasagam et al., 2004). However, genes like DREB, PDH45 helicase genes, and trehalose synthase are the important transcription factors for drought-stress tolerance in sugarcane (Zhang et al., 2006; Augustine et al., 2015). However, invertase gene is the target for sugarcane yield improvement (Wang et al., 2017). Different agronomic traits modified in sugarcane using transgenic technology has been summarised in Table 4.3.

4.9 CRISPR-CAS SYSTEM AS A GENOME EDITING TOOL FOR SUGARCANE

CRISPR/Cas [CRISPR (clustered regularly interspaced short palindromic repeat)-Cas (CRISPR-associated)] system is a powerful genome editing tool that uses Cas nuclease to create a double-stranded break in a sequence-specific area of the genome, targeted using RNA-DNA hybrids. It has been evolved as a genome-editing tool from microbial adaptive immune systems (Kumari et al., 2021).

The first successful genome targeting using CRISPR/Cas technology was used for rice and wheat (Shan et al., 2014). CRISPR/Cas is a site-directed mutagenesis tool that edits specific loci under *in vivo* conditions. It acts as a novel tool that can target different genes of same species and all the homeoalleles of polyploid species simultaneously. It has been used to mutate the GW2, GW5, and TGW6 of rice to increase the seed size (Xu et al., 2016). Oil seed crop, *Brassica napus,* has been mutated for higher number of seeds per siliqua, increased grain weight by targeting the CLVTA3 gene using CRISPR/Cas9, and increased grain weight and length in wheat by targeting the gene TaGW2 (Wang et al., 2018).

Moreover, it has been used to target the several genes related to abiotic stress and biotic stress to improve the agronomic performances of crops (Haque et al., 2018). CRISPR/Cas technology can be used against multiple targets simultaneously; therefore, it is the most potent technique for genetic improvement in sugarcane for different agronomically important traits. Large genome size, polyploidy, low transformation efficiency, transgene silencing, and lack of high throughput screening techniques are certainly great challenges for genome editing in sugarcane (Mohan et al., 2016). Despite these challenges, the technology has been used to target genes for biotic and abiotic stress tolerances. Herbicide-tolerant sugarcane has been developed for herbicides: sulfonylurea, imidazolinones, triazolopyrimidines, pyrimidinyloxybenzoates, and sulfanilamide-carbonyl-thiazolidinones, by targeting ALS homeoalleles using HDR (homology-directed repair) mechanism of the CRISPR/Cas system (Oz et al., 2021). The magnesium chelatase gene responsible for chlorophyll synthesis has been modified to elevate the canopy photosynthetic efficiency in sugarcane (Eid et al., 2021). However, there are different obstructions for targeted mutagenesis in sugarcane. Even so, the opportunities to edit sugarcane genome using CRISPR/Cas must be explored, as various reports convey (Mohan et al., 2022).

4.10 CONCLUSIONS AND PERSPECTIVES

Sugarcane is the world's highest producing crop, especially for Brazil, India, and China. Increasing consumption and industrial applications encourage the scientific community to implement their tools for genetic improvement in sugarcane. Poor flowering, low fertility, perennial nature, long breeding cycle, and large and complex genomes are the impediments for genetic improvement using sugarcane-breeding programmes. Intervention of biotechnological tools, such as omics approaches, molecular-marker technology, tissue culture, and transgenic approaches unravel the genome size, gene estimation, genetic linkage map construction, gene annotation and identification, and functional characterisation of agronomically important gene for sugarcane.

REFERENCES

Ali A., Pan Y.B., Wang Q.N., Wang J.D., Chen J.L., and Gao S.J. (2019). Genetic diversity and population structure analysis of Saccharum and Erianthus genera using microsatellite (SSR) markers. *Scientific Reports*, 9: 1–10.

Ali S., Khan M.S. and Iqbal. J. (2012). In vitro direct plant regeneration from cultured young leaf segments of sugarcane (*Saccharum officinarum* L.). *Journal of Animal and Plant Sciences*, 22(4): 1107–1112.

Almazan O., Gonzalez L., and Galvez L. (1998). *The sugar cane, its by-products and co-products. Maurice paturau memorial lecture keynote address.* AMAS, Food and Agricultural Research Council, Réduit, Mauritius.

Alwala S., Kimbeng C.A., Veremis J.C., and Gravois K.A. (2008). Linkage mapping and genome analysis in Saccharum interspecific cross using AFLP, SRAP and TRAP markers. *Euphytica*, 164: 37–51.

Alwine J.C., Kemp D.J., and Stark G.R. (1977). Method for detection of specific RNAs in agarose gels by transfer to diazobenzyloxymethyl-paper and hybridization with DNA probes. *PNAS*, 74: 5350–5354.

Arencibia A., Vazquez R.I., Prieto D., Tellez P., and Carmona E.R. (1997). Transgenic sugarcane plants resistant to stem borer attack. *Molecular Breeding*, 3: 247–255.

Aslam B., Basit M., Nisar M.A., Khurshid M., and Rasool M.H. (2017). Proteomics: Technologies and Their Applications. *Journal of Chromatographic Science*, 55(2): 182–196.

Augustine S.M., Narayan J.A., Syamaladevi D.P., Appunu C., Chakravarthi M., Ravichandran V., Tuteja, N., and Subra-monian N. (2015). Over-expression of EaDREB2 and pyramiding of EaDREB2 with the pea DNA helicase gene (PDH45) enhance drought and salinity tolerance in sugarcane (*Saccharum spp.* hybrid). *Plant Cell Report*, 34: 247–263.

Babu K.H., Devarumath R.M., Thorat A.S., Nalavade V.M., Saindane M., Appunu C., and Suprasanna P. (2021). Sugarcane transgenics: Developments and opportunities. In *Genetically modified crops* (pp. 241–265). Springer, Singapore.

Barreto F.Z., Rosa J.R.B.F., Balsalobre T.W.A., Pastina M.M., Silva R.R., Hoffmann H.P., de Souza A.P., Garcia A.A.F., and Carneiro M.S. (2019). A genome-wide association study identified loci for yield component traits in sugarcane (*Saccharum spp.*). *PLoS ONE*, 14: e0219843.

Becker-André M., and Hahlbrock K. (1989). Absolute mRNA quantification using the polymerase chain reaction (PCR). A novel approach by a PCR aided transcript titration assay (PATTY). *Nucleic Acids Research*, 17: 9437–9446.

Benisheikh A., Mala M., Ahmadu U., Fatima B., Zainab M., et al. (2012). Virus free plantlets production of sugar (*Saccharum officinarum* L.) through in vitro micropropagation of shoot tip culture. *Continental Journal of Biological Sciences*, 5(1): 42–46.

Bhattacharyya P., Datta S.C., and Dureja P. (2003). Interrelationship of pH, organic acids and phosphorus concentration in soil solution of rhizosphere and non-rhizosphere of wheat and rice crops. *Communications in Soil Science and Plant Analysis*, 34(1–2): 231–245.

Biradar S., Biradar D., Patil V.C., Patil S., Kambar N., et al. (2009). In vitro plant regeneration using shoot tip culture in commercial cultivar of sugarcane. *Karnataka Journal of Agriculture and Science* 22(1): 21–24.

Bonatelli M.L., Quecine M.C., Silva M.S., and Labate C.A. (2017). Characterization of the contaminant bacterial communities in sugarcane first-generation industrial ethanol production. *FEMS microbiology letters*, 364(17).

Bower R., and Birch R.G. (1992). Transgenic sugarcane plants via microprojectile bombardment. *Plant Journal*, 2: 409–416.

Braga D.P.V., Arrigoni E.D.B., Silva-Filho M.C., and Ulian E.C. (2003). Expression of the Cry 1 Ab protein in genetically modified sugarcane for the control of *Diatraea saccharalis* (Lepidoptera: *Crambidae*). *Journal of New Seeds*, 5: 209–221.

Brenner S., Johnson M., Bridgham J., Golda G., Lloyd D.H., Johnson, D. et al. (2000). Gene expression analysis by massively parallel signature sequencing (MPSS) on microbead arrays. *Nature Biotechnology*, 18: 630–634.

Carson D.L., and Botha F.C. (2000). Preliminary analysis of expressed sequence tags for sugarcane. *Crop Science*, 40: 1769–1779.

Cheavegatti-Gianotto A., De Abreu H.M.C., Arruda P. et al. (2011). Sugarcane (*Saccharum officinarum*): A Reference Study for the Regulation of Genetically Modified Cultivars in Brazil. *Tropical Plant Biology*, 4: 62–89.

Christy L.A., Aravith S., Saravanakumar M., Kanchana M., Mukhunthan N. et al. (2009). Engineering sugarcane cultivars with bovine pacncreatic trypsin inhibitor (aprotinin) gene for protection against to borer (*Scripophaga excerptalis* Walker). *Plant Cell Report*, 28: 175–184.

Costet L., Raboin L., Payet M., D'Hont A., and Nibouche S. (2012). A major quantitative trait allele for resistance to the Sugarcane yellow leaf virus (Luteoviridae). *Plant Breeding*, 131: 637–640.

Cursi D.E., Hoffmann H.P., Barbosa G.V.S., Bressiani J.A., Gazaffi R., Chapola R.G. et al. (2021). History and Current Status of Sugarcane Breeding, Germplasm Development and Molecular Genetics in Brazil. *Sugar Tech.* 10.1007/s12355-021-00951-1

D'Hont A. (2005). Unraveling the genome structure of polyploids using FISH and GISH; examples of sugarcane and banana. *Cytogenetic and Genome Research*, 109: 27–33. doi: 10.1159/000082378

De Setta N., Monteiro-Vitorello C.B., Metcalfe C.J., Cruz G.M.Q., Del Bem L.E., Vicentini R., Nogueira F.T.S., Campos R.A., Nunes S.L., and Turrini P.C.G. (2014). Building the sugarcane genome for biotechnology and identifying evolutionary trends. *BMC Genomics*, 15: 540.

Dotaniya M.L., Datta S.C., Biswas D.R., Dotaniya C.K., Meena B.L., Rajendiran S., Regar K.L., and Lata M. (2016). Use of sugarcane industrial by-products for improving sugarcane productivity and soil health. *International Journal of Recycling of Organic Waste in Agriculture*, 5: 185–194.

Eid A., Mohan C., Sanchez S., Wang D., and Altpeter F. (2021). Multiallelic, Targeted Mutagenesis of Magnesium Chelatase With CRISPR/Cas9 Provides a Rapidly Scorable Phenotype in Highly Polyploid Sugarcane. *Frontiers in Genome Editing*, 3: 654996.

Enriquez-Obregon G.A., Vazquez-Padron R.I., Pieto-Samsonov D.L., De la Riva G.A., and Selman-Housein G. (1998). Herbicide-resistant sugarcane (*Saccharum officinarum* L.) plants by Agrobacterium-mediated transformation. *Planta*, 206: 20–27.

Falco M.C., Neto A.T., and Ulian E.C. (2000). Transformation and expression of a gene for herbicide resistance in a Brazilian sugarcane. *Plant Cell Report*, 19: 1188–1194.

Farsangi F.J., Thorat A.S., and Devarumath R.M. (2018). Molecular characterization of sugarcane genotypes for their salinity and susceptibility using TRAP markers. *International Journal of Current Research*, 10: 68947–68951.

Fickett N., Gutierrez A., Verma M., Pontif M., Hale A., Kimbeng C., and Baisakh N. (2019). Genome-wide association mapping identifies markers associated with cane yield components and sucrose traits in the Louisiana sugarcane core collection. *Genomics*, 111: 1794–1801.

Franks T., and Birch R.G. (1991). Gene transfer into intact sugarcane cells using microprojectile bombardment. *Australian Journal of Plant Physiology*, 18: 471–480.

Gadakh S., Patel D., and Singh D. (2017). Use of RAPD markers to characterize salt and drought lines of sugarcane. *International Journal of Advanced Research In Biological Sciences*, 4: 50–57.

Gao S., Yang Y., Wang C., Guo J., Zhou D., and Wu Q. (2016). Transgenic sugarcane with acry1ac gene exhibited better phenotypic traits and enhanced resistance against sugarcane Borer. *PLoS ONE*, 11(4): e0153929.

Garsmeur O., Droc G., Antonise R., Grimwood J., Potier B., Aitken K., Jenkins J., Martin G., Charron C., and Hervouet C. (2018). A mosaic monoploid reference sequence for the highly complex genome of sugarcane. *Nature Communication*, 9: 2638.

Gazaffi R., Oliveira K.M., Souza A.P., and Garcia A.A.F. (2010). The Importance of the Germoplasm in Developing Agro-Energetic Profile Sugarcane Cultivars. In: *Sugar Cane Bioethanol: R&D for productivity and sustainability*, L.A.B. Cortez (Ed.), Blucher, ISBN: 9788521205302, São Paulo, pp. 333–343.

Gill R., Malhotra P.K., and Gosal S.S. (2006). Direct plant regeneration from cultured young leaf segments of sugarcane. *Plant Cell Tissue Organ Culture*, 84: 227–231.

Godheja J., Shekhar S.K., and Modi D.R. (2014). The standardization of protocol for large scale production of sugarcane (co-86032) through micropropagation. *International Journal of Plant Animal and Environmental Sciences*, 4(4): 135–143.

Gouy M., Rousselle Y., Chane A.T., Anglade A., Royaert S., Nibouche S., and Costet L. (2015). Genome wide association mapping of agro—Morphological and disease resistance traits in sugarcane. *Euphytica*, 202: 269–284.

Guerzoni J.T.S., Belintani N.G., Moreira R.M.P., Hoshino A.A., Domingues D.S., Filho J.C.B., and Vieira L.G.E. (2014). Stress-induced D1-pyrroline-5-carboxylate synthetase (P5CS) gene confers tolerance to salt stress in transgenic sugarcane. *Acta Physiologiae Plantarum*, 36: 2309–2319.

Gutierrez A., Hoy J., Kimbeng C., and Baisakh N. (2018). Identification of genomic regions controlling leaf scald resistance in sugarcane using a bi—Parental mapping population and selective genotyping by sequencing. *Frontiers in Plant Science*, 9: 877.

Hapsoro D., Astrina Putri Febrianie A.P., and Yusnita (2012). In vitroshoot formation on sugarcane (*Saccharum officinarum* L.) callus as affected by Benzyladenine concentrations. *Jurnal Agronomi Indonesia*, 40(1): 56–61.

Haque E., Taniguchi H., Hassan M.M., Bhowmik P., Karim M.R., Smiech M., Zhao K., Rahman M., and Islam T. (2018). Application of CRISPR/Cas9 Genome Editing Technology for the Improvement of Crops Cultivated in Tropical Climates: Recent Progress, Prospects, and Challenges. *Frontiers in Plant Science*, 9: 617.

Hoang N.V., Furtado A., Mason P.J., Marquardt A., Kasirajan L., Thirugnanasambandam P.P., Botha F.C., and Henry R.J. (2017). A survey of the complex transcriptome from the highly polyploid sugarcane genome using full-length isoform sequencing and de novo assembly from short read sequencing. *BMC Genomics*, 18: 395.

Ijaz S., Rana I.A., Khan I.A., Saleem M. (2012). Establishment of an in vitro regeneration system for genetic transformation of selected sugarcane genotypes. *Genetics and Molecular Research*, 11(1): 512–530.

Iqbal A., Khan R.S., Khan M.A., Gul K., Jalil F., Shah D.A. et al. (2021). Genetic Engineering Approaches for Enhanced Insect Pest Resistance in Sugarcane. *Molecular Biotechnology*, 63(7): 557–568.

Joyce P., Kuwahata M., Turner N., and Lakshmanan P. (2010). Selection system and co-cultivation medium are important determinants of *Agrobacterium*-mediated transformation of sugarcane. *Plant Cell Report*, 29(2): 173–183.

Kalunke R.M., Kolge A.M., Babu K.H., and Prasad D.T. (2009). Agrobacterium mediated transformation of sugarcane for borer resistance using cry 1Aa3 gene and one step regeneration of transgenic plants. *Sugar Tech*, 11: 355–359.

Kasirajan L., Hoang N.V., Furtado A., Botha F.C., and Henry J.R. (2018). Transcriptome analysis highlights key differentially expressed genes involved in cellulose and lignin biosynthesis of sugarcane genotypes varying in fiber content. *Scientific Report*, 8: 11612.

Kesawat M.S., and Das Kumar B. (2009). Molecular markers: It's application in crop improvement. *Journal of Crop Science and Biotechnology* 12: 169–181.

Khaled K., El-Sherbeny S., and Abdelhadi A. (2017). R-ISSR as marker assisted selection for drought tolerance in sugarcane. *Journal of Agricultural Chemistry and Biotechnology*, 8: 91–97.

Khaliq A., Ashfaq M., Akram W., Choi J.K., and Lee J. (2005). Effect of plant factors, sugar contents, and control methods on the Top Borer (*Scirpophaga nivella* F.) infestation in selected varieties of sugarcane. *Entomological Research*, 35: 153–160.

Khamrit R., Jaisil P., and Bunnag P. (2012). Callus induction, regeneration and transformation of sugarcane (*Saccharum officinarum* L.) with chitinase gene using particle bombardment. *African Journal of Biotechnology*, 11: 6612–6618.

Khan I., Dahot M., Seema N., Yasmeen A., Naqvi M., et al. (2009). Direct regeneration of sugarcane plantlets: a tool to unravel genetic heterogeneity. *Pakistan Journal Botany*, 41(2): 797–814.

Kumari S., Singh S.K., Sharma V.K., Kumari R., Mathur M., Upadhyay T.K., and Prajapat R.K. (2021). CRISPR-Cas: A continuously evolving technology. *The Indian Journal of Agricultural Sciences*, 91(9): 1274–1279.

Lakshmanan P., Geijskes R.J., Wang L., Elliott A., Grof C.P., Berding N., and Smith G.R. (2006). Developmental and hormonal regulation of direct shoot organogenesis and somatic embryogenesis in sugarcane (*Saccharum spp*. Interspecific hybrids) leaf culture. *Plant Cell Report*, 25(10): 1007–1015.

Landell M. (2003). ProCana – O Programa Cana-de-açúcar do Instituto Agronômico. *O Agronomico*, 55(1) 5–8 ISSN 0365-2726.

Landell M.G.A., Creste S., Pinto L.R., Xavier M.A., and Bressiani J.A. (2010). The Strategy For Regional Selection For The Development of Energy Sugarcane Cultivar Profile. In: *Sugar Cane Bioethanol: R&D for productivity and sustainability*, L.A.B. Cortez (Ed.), Blucher, ISBN: 9788521205302, São Paulo, pp. 345–352.

Lemes A.R.N., Figueiredo C.S., Sebastião I., Marques da Silva L., da Costa Alves R., de Siqueira H.Á.A., Lemos M.V.F., Fernandes O.A., and Desidério J.A. (2017). Cry1Ac and Vip3Aa proteins from Bacillus thuringiensis targeting Cry toxin resistance in *Diatraea flavipennella* and *Elasmopalpus lignosellus* from sugarcane. *Peer Journal*, 5:e2866.

Liu P., Chandra A., Que Y., Chen P.H., Grisham M.P., White W.H., Dalley C.D., Tew T.L., and Pan Y.B. (2016). Identification of quantitative trait loci controlling sucrose content based on an enriched genetic linkage map of sugarcane (*Saccharum spp.* hybrids) cultivar 'LCP 85–384'. *Euphytica*, 207: 527–549.

Lockhart D.J., Dong H., Byrne M.C., Follettie M.T., Gallo M.V., Chee M.S. et al. (1996). Expression monitoring by hybridization to high-density oligonucleotide arrays. *Nature Biotechnology*, 14: 1675–1680.

Lowe R., Shirley N., Bleackley M., Dolan S., and Shafee T. (2017). Transcriptomics technologies. *PLOS Computational Biology*, 13(5): e1005457.

Manickavasagam M., Ganapathi A., Anbazhagan V.R., Sudhakar B., Selvaraj N., Vasudevan A., and Kasthurirengan S. (2004). Agrobacterium-mediated genetic transformation and development of herbicide-resistant sugarcane (*Saccharum spp.* hybrids) using axillary buds. *Plant Cell Report*, 23: 134–143.

Manners J.M., and Casu R.E. (2011). Transcriptome analysis and functional genomics of sugarcane. *Tropical Plant Biology*, 4(1): 9–21.

Medeiros C., Balsalobre T.W.A., and Carneiro M.S. (2020). Molecular diversity and genetic structure of *Saccharum* complex accessions. *PLoS ONE*, 15(5): e0233211.

Meng J.Y., Ntambo M.S., Rott P.C., Fu H.Y., Huang M.T., Zhang H.L., and Gao S.J. (2020). Identification of differentially expressed proteins in sugarcane in response to infection by *Xanthomonas albilineans* using iTRAQ quantitative proteomics. *Microorganisms*, 8(1): 76. PMCID: PMC7023244

Mittal P., Devi R., and Gosal S.S. (2016). Effect of genotypes and activated charcoal on high frequency in vitro plant regeneration in sugarcane. *Indian Journal of Biotechnology*, 15: 261–265.

Mohan C. (2016). Genome editing in sugarcane: challenges ahead. *Frontiers in plant science*, 7: 1542.

Mohan C., Easterling M., and Yau Y.Y. (2022). Gene Editing Technologies for Sugarcane Improvement: Opportunities and Limitations. *Sugar Tech*, 24: 369–385. 10.1007/s12355-021-01045-8

Morais L.K., Aguiar M.S., Silva P.A., Câmara T.M.M., Cursi D.E., Fernandes Júnior A.R., Chapola R.G., Carneiro M.S., and Bespalhok Filho J.C. (2015). Breeding of sugarcane. In *Industrial crops: breeding for bioenergy and bioproducts*, V.M.V. Cruz, and D.A. Dierig (Eds), Editora Springer, New York, pp. 29–42.

Morozova O., Hirst M., and Marra M.A. (2009). Applications of new sequencing technologies for transcriptome analysis. *Annual Review of Genomics and Human Genetics*, 10: 135–151.

Nadeem M.A., Nawaz M.A., Shahid M.Q., Doğan Y., Comertpay G., Yıldız M., Hatipoğlu R., Ahmad F., Alsaleh A., Labhane N., Özkan H., Chung G., and Baloch F.S. (2018). DNA molecular markers in plant breeding: Current status and recent advancements in genomic selection and genome editing. *Biotechnology & Biotechnological Equipment*, 32(2): 261–285.

Nasir N.M., Qureshi R.H., and Aslam M. (2000). Effect of salinity on emergence of sugarcane lines. *Pakistan Sugar Journal*, 15: 12–14.

Nibouche S., Raboin L.M., Hoarau J.Y., D'Hont A., and Costet, L. (2012). Quantitative trait loci for sugarcane resistance to the spotted stem borer *Chilo sacchariphagus*. *Molecular Breeding*, 29: 129–135.

Ostengo S., Rueda Calderón M.A., Bruno C., Cuenya M.I., and Balzarini M. (2021). A protocol for identifying characteristic sucrose accumulation curves of sugarcane genotypes (*Saccharum spp.*). *Sugar Tech*, doi: 10.1007/s12355-020-00926-8

Oz M.T., Altpeter A., Karan R., Merotto A., and Altpeter F. (2021). CRISPR/Cas9-Mediated Multi-Allelic Gene Targeting in Sugarcane Confers Herbicide Tolerance. *Fronteirs in Genome Editing*, 3: 673566.

Palhares A.C., Rodrigues-Morais T.B., Van Sluys M.A., Domingues D.S., Maccheroni W., Jordão H., Souza A.P., Marconi T.G., Mollinari M., Gazaffi, R. et al. (2012). A novel linkage map of sugarcane with evidence for clustering of retrotransposon-based markers. *BMC Genetics*, 13: 51.

Panti N.C. (2016). Paclobutrazol (PBZ)-Cultar regulator in sugarcane: The effect of different PBZ-Cultar concentrations on sugarcane (*Saccharum*) of variety CPCL99-4455, A Thesis Submitted to the University of Belize in Fulfillment of BIOL 4992 -Independent Research.

Papini-Terzi F.S., Rocha F.R., Nicoliello Vêncio R.Z., Oliveira, K.C., De Maria Felix et al. (2005). Transcription profiling of signal transduction-related genes in sugarcane tissues. *DNA research*, 12(1): 27–38.

Patel V.S., Mehta R., Naik K.H., Singh D., Patel D.U., and Mali S.C. (2015). Callus induction & whole plant regeneration in sugarcane (*Saccharum* spp. complex) variety Co 86032. *Green Farming*, 6 (5): 935–939.

Pinto L.R., Garcia A.A.F., Pastina M.M., Teixeira L.H.M., Bressiani J.A., Ulian E.C., Bidoia M.A.P., and Souza A.P. (2010). Analysis of genomic and functional RFLP derived markers associated with sucrose content, fiber and yield QTLs in a sugarcane (*Saccharum spp.*) commercial cross. *Euphytica*, 172: 313–327.

Prasitsom C., Jubsab N., Klomsa-ard P., Sriroth K., and Keawsompong S. (2019). Selection of SSR Markers for drought resistant sugarcane in Thailand. *International Journal of Agricultural Technology*, 15(6): 997–1010.

Qin C.X., Chen Z.L., Wang M., Li A.M., Liao F., Li Y.R., Wang M.Q., Long M.H., Lakshmanan P., and Huang D.L. (2021). Identification of proteins and metabolic networks associated with sucrose accumulation in sugarcane (*Saccharum spp.* interspecific hybrids). *Journal of Plant Interactions*, 16(1): 166–178.

Ramiro D.A., Melotto P.D.M., Barbosa M.D.A., Santos F.D., Gomez S.G.P., Junior N.S.M., Lam E., and Carrer H. (2016). Expression of Arabidopsis bax inhibitor-1 in transgenic sugarcane confers drought tolerance. *Plant Biotechnology Journal*, 14: 1826–1837.

Riaz S., Nasir I.A., Bhatti M.U., Adeyinka O.S., Toufiq N., Yousaf I., and Tabassum B. (2020). Resistance to *Chilo infuscatellus* (Lepidoptera: Pyraloidea) in transgenic lines of sugarcane expressing *Bacillus thuringiensis* derived Vip3A protein. *Molecular biology reports*, 47(4): 2649–2658.

Rossi M., Araujo P., De Jesus E., Varani A., and Van Sluys M.A. (2004). Comparative analysis of Mutator-like transposases in sugarcane. *Molecular Genetics and Genomics*, 272: 194–203.

Salvato F., Loziuk P., Kiyota E., Daneluzzi G.S., Araújo P., Muddiman D.C., and Mazzafera P. (2019). Label-free quantitative proteomics of enriched nuclei from sugarcane (*Saccharum spp.*) stems in response to drought stress. *Proteomics*, 19(14): e1900004.

Sandhu S., Gosal S., Singh K., Cheema G., Meeta M., Singh K., and Cheema G.S. (2009). Field performance of micro propagated plants and potential of seed cane for stalk yield and quality in sugarcane. *Sugar Technology* 11(1): 34–38.

Santiago T.R., Pereira V.M., de Souza W.R., Steindorff A.S., Cunha B.A., Gaspar M., Favaro L.C., Formighieri E.F., Kobayashi A.K., and Molinari H.B. (2018). Genome-wide identification, characterization and expression profile analysis of expansins gene family in sugarcane (Saccharum spp.). *PLoS ONE*, 13:e0191081

Satpal S.B., Routray A.K., and Mishra R. (2011). Rapid *in vitro*propagation technique for sugarcane variety 018. *International Journal of Pharma and Bio Sciences*, 2(4): B242–B279.

Scortecci K.C., Creste S., Calsa T., Xavier M.A., Landell M.G.A., Figueira A., and Benedito V.A. (2012). Challenges, Opportunities and Recent Advances in Sugarcane Breeding, Plant Breeding, Ibrokhim Y. Abdurakhmonov, Intech Open, DOI: 10.5772/28606.

Shan Q., Wang Y., Li J., and Gao C. (2014). Genome editing in rice and wheat using the CRISPR/Cas system. *Nature Protocol*, 9: 2395–2410.

Shingote P.R., Mithra S.A., Sharma P., Devanna N.B., Arora K., Holkar S.K., Khan S., Singh J., Kumar S., and Sharma, T. (2019). LTR retrotransposons and highly informative ISSRs in combination are potential markers for genetic fidelity testing of tissue culture-raised plants in sugarcane. *Molecular Breeding*, 39: 25.

Singh P., Song Q.Q., Singh R.K., Li H.B., Solanki M.K., Malviya M.K., Verma K.K., Yang L.T., and Li Y.R. (2019). Proteomic analysis of the resistance mechanisms in sugarcane during *Sporisorium scitamineum* infection. *International Journal of Molecular Sciences*, 20(3): 569.

Singh R.B., Singh B., Singh R.K. (2017). Study of genetic diversity of sugarcane (*Saccharum*) species and commercial varieties through TRAP molecular markers. *Indian Journal of Plant Physiology*, 22: 332–338.

Singh R.K., Kumar P., Tiwari N.N., Rastogi J., and Singh S.P. (2013). Current Status of Sugarcane Transgenic: an Overview. *Advancements in Genetic Engineering*, 2: 112.

Souza G.M., Marie-Anne V.S., Carolina G.L., Hayan L., Gabriel R.A.M., Carlos T.H., Jonas W.G., Augusto L.D., Mauro de Medeiros O., Sávio de S.F., Milton Y.N., Felipe T.C. et al. (2019). Assembly of the 373k gene space of the polyploid sugarcane genome reveals reservoirs of functional diversity in the world's leading biomass crop. *Giga Science* 8: giz129.

Srivastava M.K., Li C.N., and Li Y.R. (2012). Development of sequence characterized amplified region (SCAR) marker for identifying drought tolerant sugarcane genotypes. *Australian Journal of Crop Science*, 6: 763.

Srivong T., Zhu Y.J., Pongdontri P., Pliansinchai U., Sakuanrungsirikul S., Borthakur D., Nagai C., and Kosittrakun M. (2015). Optimization of callus induction and plant regeneration in sugarcane (*Saccharum* spp.) for a study of sucrose accumulation in relation to soluble acid invertase expression. *Chiang Mai Journal of Science*, 42(4): 797–805.

Su Y., Zhang Y., Huang N., Liu F., Su W., Xu L., Ahmad W., Wu Q., Guo J., and Que Y. (2017). Small RNA sequencing reveals a role for sugarcane miRNAs and their targets in response to *Sporisorium scitamineum* infection. *BMC Genomics*, 18: 325.

Suprasann P. (2010). Biotechnological interventions in sugarcane improvement: Strategies, methods and progress. *Nuclear Agric. & Biotech. Div., BARC Newsletter* 316: 47–53.

Tesfa M., Admassu B., and Bantte K. (2016). *In vitro* rooting and acclimatization of micropropagated elite sugarcane (*Saccharum officinarum* L.) genotypes-N52 and N53. *Journal of Tissue Science and Engineering*, 7: 164.

Thirugnanasambandam P.P., Hoang N.V., and Henry R.J. (2018). The Challenge of Analyzing the Sugarcane Genome. *Fronteirs in Plant Science*, 9: 616. doi: 10.3389/fpls.2018.00616.

Tuan V.A., Hanh H.H., Phuong P.T.T., Thuy P.T.T., Thuy P.T., Vinh D.N., and Khanh T.D. (2015). Rapid *in viro* multiplication of some sugarcane cultivars (*Saccharum officnarum* L.) via embryogenic callus culture of young leaf tissues. *International Journal of Development Research*, 5(12): 6139–6146.

Velculescu V.E., Zhang L., Vogelstein B., and Kinzler K.W. (1995). Serial analysis of gene expression. *Science*, 270: 484–487.

Vettore A.L., Silva F.R.D., Kemper E.L., and Arruda P. (2001). The libraries that made SUCEST. *Genetics and Molecular Biology*, 24: 1–7.

Waclawovsky A.J., Sato P.M., Lembke C.G., Moore P.H., and Souza G.M. (2010). Sugarcane for bioenergy production: an assessment of yield and regulation of sucrose content. *Plant Biotechnology Journal*, 8(3): 263–276.

Wang B., Kumar V., Olson A., and Ware D. (2019). Reviving the transcriptome studies: an insight into the emergence of single-molecule transcriptome sequencing. *Frontiers in Genetics*, 10: 384.

Wang B., Regulski M., Tseng E., Olson A., Goodwin S., McCombie W.R., et al. (2018). A comparative transcriptional landscape of maize and sorghum obtained by single-molecule sequencing. *Genome Research*, 28: 921–932.

Wang L., Zheng Y., Ding S., Zhang Q., Chen Y., and Zhang J. (2017). Molecular cloning, structure, phylogeny and expression analysis of the invertase gene family in sugarcane. *BMC Plant Biology*, 17: 109.

Wang Z.Z., Zhang S.Z., Yang B.P., and Ruil Y. (2005). Trehalose synthase gene transfer mediated by *Agrobacterium tumefaciens* enhances resistance to osmotic stress in sugarcane. *Sugar Technology*, 7(1): 49–54.

Wu Q., Xu L., Guo J., Su Y., and Que Y. (2013).Transcriptome Profile Analysis of Sugarcane Responses to *Sporisorium scitaminea* Infection Using Solexa Sequencing Technology. *BioMed Research International*, 2013: 9.298920.

Xu R., Yang Y., Qin R., Li H., Qiu C., Li L., et al. (2016). Rapid improvement of grain weight via highly efficient CRISPR/Cas9-mediated multiplex genome editing in rice. *Journal of Genetics and Genomics*, 43: 529–532. doi: 10.1016/j.jgg.2016.07.003.

Xu S., Wang J., Shang H., Huang Y., Yao W., Chen B., and Zhang M. (2018). Transcriptomic characterization and potential marker development of contrasting sugarcane cultivars. *Scientific Report*, 8: 1683.

Yadav S., Ahmad A., and Lal M. (2012). Effect of different auxins and cytokinins on in vitro Multiplication and rooting of shoot cultures in Sugarcane. *International Journal of Bio-Pharma Research*, 3(6): 814–818.

You Q., Pan Y.B., Xu L.P., Gao S.W., Wang Q.N., Su Y.C., Yang Y.Q., Wu Q.B., Zhou D.G., and Que Y.X. (2016). Genetic diversity analysis of sugarcane germplasm based on fluorescence—Labelled simple sequence repeat markers and a capillary electrophoresis Based genotyping platform. *Sugar Technology*, 18: 380–390.

You Q., Yang X., Peng Z., Islam M.S., Sood S., Luo Z., Comstock J., Xu L., and Wang J. (2019). Development of an Axiom Sugarcane 100K SNP array for genetic map construction and QTL identification. *Theoretical and Applied Genetics*, 132: 2829–2845.

Zhang J., Zhang X., Tang H., Zhang Q. Hua X., Ma X., Zhu F., Jones T., Zhu X., and Bowers J. (2018). Allele-defined genome of the autopolyploid sugarcane *Saccharum spontaneum* L. *Nature Genetics*, 50: 1565.

Zhang S.Z., Yang B.P., Feng C.L., Chen R.K., Luo J.P., Cai W.W., et al. (2006). Expression of the *Grifola frondosa* trehalose synthase gene and improvement of drought-tolerance in sugarcane (*Saccharum officinarum* L.). *Journal of Integrative Plant Biology*, 48: 453–459.

5 Augmenting Sugar Accumulation Potential in Sugarcane: Conventional and Molecular Approaches

M. Swapna
Division of Crop Improvement, ICAR-Indin Institute of Sugarcane Research, Lucknow, Uttar Pradesh, India

CONTENTS

5.1 INTRODUCTION

Sugarcane, an excellent source of sugar and various diversified products, has an important role in the global economy in terms of productivity, profitability, and sustainability. The crop is being grown in more than 100 countries worldwide under varying climatic conditions; it is the harbinger of a gradual shift toward a bio-based economy in many of the sugarcane growing countries. Literally, each and every part of the sugarcane plant can be effectively utilized in one form or the other in the pre-harvest or post-harvest form. This versatility makes the crop an immensely attractive option for enhancing the profitability and sustainability of the sweetener industry. Even with the present-day demands of diversification and value addition, sugar, the primary product from sugarcane, has retained its relevance to the full extent. Therefore, the R&D programmes in the sector still emphasize the improvement of the sugar accumulation and final sugar content in the newly developed sugarcane varieties, along with other traits.

As it has been well-documented, the sugarcane plant is genetically complex with its large genome size, heterozygosity, deviation from the normal with respect to chromosome transmission, etc. (Sreenivasan et al., 1987). Even though great strides had been made in varietal improvement through conventional hybridization and selection, leading to many promising clones, the hybridization programmes still more or less remain a hit-and-miss affair, with the many nuances of

DOI: 10.1201/9781003292425-5

the crop's reproductive behaviour yet to be elucidated. There appears to be a general narrowing of genetic base in the newly developed commercial sugarcane clones. Such challenges, along with the complex genetic nature of the crop, have led to a plateauing of the economically important traits, like cane yield and sugar recovery (Jackson, 2019, Ram et al., 2022). This reduction has necessitated the use of nonconventional/molecular tools also, to supplement the ongoing conventional sugarcane improvement programmes.

From the initial attempts of nobilization and interspecific hybridization in sugarcane, leading to the first hybrid variety Co 205, the global sugarcane varietal breeding scenario has passed through many exciting developments; these advancements mainly involve improved location-specific sugarcane varieties, parental material, improved population for specific traits, and the application of molecular biology, including the application of genomic tools. In India, too, the impact of sugarcane improvement programmes is visible through the more than 60 improved sugarcane varieties notified for commercial cultivation over approximately 50 years (Sinha et al., 2016). An overall improvement in the cane yield, sugar production, and sugar recovery has occurred during the past decades in the country, even though the increments in sugar recovery have not been substantial.

5.2 SUGARCANE IMPROVEMENT: THE INITIAL PHASE

The first attempts of nobilization and development of improved varieties like PoJ series and Co varieties resulted in varieties with better cane yield and sugar, improved stress resistance/tolerance, and other traits. These varieties/clones became the base material for use in commercial breeding programmes worldwide (Hemaprabha et al., 2022). In India, sugarcane cultivation has been reported to date back to 5000 BC, with the genetic improvement through conventional strategies being initiated during the 19th century. The application of interspecific hybridization resulting in the first commercial sugarcane hybrid Co 205 at ICAR-Sugarcane Breeding Institute, Coimbatore in India, gave a major impetus to the subsequent sugarcane improvement programmes all over the world, including India. The introduction of various species accessions of *Saccharum*, like *S. sinense*, *S. officinarum*, in the country during the 18th and 19th centuries, along with the other improved types of canes, were important milestones in the history of sugarcane improvement (Ram et al., 2022). The early 20th century reported many of the local varieties, selections, and accessions introduced from the other sugarcane-growing countries to India, like Java and Mauritius. In India, with the establishment of sugarcane research centres at Coimbatore, Shahjahanpur, and at other places, coupled with the efforts of earlier sugarcane experts like George Clarke and Dr. C.A. Barber, targeted sugarcane improvement was initiated in a systematic manner. This was followed by the establishment of a network of sugarcane research stations. Subsequent efforts on natural hybridization and seedling production, leading to the development of popular varieties like Co 210, Co 213, Co 214, etc., laid a strong foundation for the future sugarcane improvement through conventional hybridization and selection, not only in India, but also in the other sugarcane-growing countries.

The sugarcane improvement programmes have been concentrating mostly on improving the cane and sugar yield, sugar content, and resistance to stresses, with diversification being added as a component in the recent years. The benefits from an improvement in sugar content has been reported to be 1.8 times higher than that from a proportional improvement in cane yield; thus, improving sugar content per se is more economical than a corresponding improvement in sugar yield through enhanced cane yield (Jackson, 2005; Hogarth et al., 1981, Kapur and Swapna, 2014). The breeding programmes in the major sugarcane-growing countries have resulted in a quantum jump in sugarcane and sugar productivity, as well as in sugar recovery in the sugar factories during the past few decades. In 1950 and 1951, India produced 69.2 million tons of sugarcane with a mean yield of 40.5t/ha. Sugar production was approximately 11.0 million tons with a sugar recovery of 10.03% (Anonymous, 2022). There has been a substantial improvement with respect to all the parameters, with the cane productivity and sugar recovery increasing by 87.7% and 3.3%, respectively, from 1950–1951 to 2020–2021. The sugar production and sugar recovery peaked at

33.16 million tons and 11.01%, respectively, during the year 2018–2019. The world sugar production also has recorded an increase of more than 200% at 171.0 million tons of sugar being produced during 2020–2021 compared to that in the early 1960s. This improvement in cane and sugar production has been possible, to a great extent, due to the robust sugarcane improvement programmes in the different countries using the conventional and nonconventional tools, coupled with appropriate innovative and sustainable management technologies (Lakshmanan et al., 2022).

5.3 IMPROVING SUGAR CONTENT: CONVENTIONAL AND MODERN TOOLS

The conventional strategies of germplasm collection and utilization, introgression breeding for base broadening, population improvement, pre-breeding for development of superior parental clones, varietal breeding involving hybridization and selection, etc., have been the backbone of sugarcane improvement programmes, with the biotechnological tools yet to make a significant impact in terms of improved genotypes and varieties. Some of the interesting developments in these areas are mentioned in the subsequent sections.

5.3.1 Genetic Resources and Their Utilization

The vast genetic resources available in this crop in the form of species clones and related genera, cultivated varieties, and parental stocks, apart from the world germplasm collections at Miami, Florida, and Kannur, India, have greatly contributed to the improvement of cane and sugar productivity (Nayak et al., 2014, Govindaraj and Mahadevaswamy, 2021a). Each country has its own set of local germplasm collections, also being utilized in the development of improved high-sugar varieties (Hemaprabha et al., 2022). Germplasm collection expeditions are regularly conducted by researchers in all countries. In India, the Spontaneum Expedition Scheme (SES) and the other continuing efforts have helped in bringing together a large number of *Saccharum* complex accessions, and these are characterized for further utilization (Nair et al., 2007, Manjunatha et al., 2018, Govindaraj and Mahadevaswamy, 2021a,b, Appunu et al., 2020, Sobhakumari, 2020). Evolution and taxonomy of sugarcane and its species and related genera have been studied by many researchers, but still many unresolved concerns exist in this area (Kellog, 2013; Welker et al., 2015). The complex network resulting from the interspecific and intergeneric matings within the groups led to conflicting conclusions, with a recent investigation pointing to the involvement of three ancestral genomes in the origin of modern sugarcane (Pompidor et al., 2021). Several research groups have studied in detail, the relationships among the *Saccharum* complex (Mukherjee, 1957; Clayton and Renvoize, 1986; Daniels and Roach, 1987; Hodkinson et al., 2002; Welker et al., 2015). The general acceptance among the sugarcane workers still seems to be toward recognizing the six species within *Saccharum* spp. and grouping *Erianthus* and others under related genera (Hemaprabha et al., 2022). The conventional and molecular studies carried out to date, starting from the early investigators (Brandes, 1956, Mukherjee, 1957, Daniels and Roach, 1987, Sreenivasan et al., 1987, Irvine, 1999, D'Hont and Glaszmann, 2001, Selvi et al., 2005, 2006, Nair et al., 2002, 2012, Welker et al., 2015, Zhang et al., 2019, Pompidor et al., 2021), leave some scope for evolutionary analyses in this crop. Suffice to say, the information from all these investigations has facilitated base broadening as well as a substantial improvement with respect to cane yield and sugar content in the newly developed sugarcane genotypes.

The early introgression studies in the late 1800s carried out in India and Indonesia led to the development of promising varieties of POJ series, EK series, and Co series of sugarcane clones. POJ 2878, one among the historically important varieties, is a typical example. Co 205 and other Co varieties from India have also played a major role in sugarcane improvement worldwide in terms of cane and sugar yield. Interspecific, intraspecific, as well as intergeneric hybridizations have been attempted on a large scale in the different countries, with the aim of broadening the genetic base leading to an improvement in economically important traits, including sugar content (detailed in later sections). *Saccharum officinarum,* is the major source of sugar and nobilization through repeated crossing with S. *officinarum* counterpart, has resulted in an improvement in sugar content. The 2n+n chromosome transmission observed in the interspecific crosses resulting in an increase in the proportion of *S. officinarum* genome is the major factor leading to the better progeny performance. Apart from *S. officinarum,* the accessions in the other species, as well as genera, have had an important role in broadening of genetic base as well as overcoming the abiotic and biotic stresses, thereby indirectly resulting in improvement in cane and sugar yield. There are many reports of successful utilization of *S. spontaneum* clones, resulting in progenies with commercial success, from several sugarcane-growing countries (Wang et al., 2008, Sookasthan et al., 1992, Heinz, 1980).

5.3.2 Varietal Breeding for Improving Sugar Content

The breeding programmes for economically important traits have cane and sugar yield as the centre point, with other traits like stress tolerance being an integral part of the programmes. As mentioned before, the lion-share of the improvement in sugar yield had come from an improvement in cane yield, with no compromise in the sugar content of the genotypes. Thus, the increment in sugar content *per se* in the newly developed genotypes seems to have not been very substantial in many of the programmes, especially in the newly developed genotypes (Jackson, 2005, 2019, Kapur et al., 2015). This is true specifically with respect to the subtropical sugarcane growing zone in India. Introgression of new set of genes for base broadening, use of the existing variability to identify the ideal combination of genes leading to high sugar accumulation, hybridization and selection and application of nonconventional tools are the usual methods adopted for improving the sugar content in these genotypes. Biparental crosses, general collections, polycrosses, and selfed populations have given rise to many promising varieties in the various sugarcane improvement efforts. The historical varieties like Co 205, Co 419, PoJ 2878, and other POJ varieties, EK series, NCo series, etc., have spearheaded the development of promising varieties in the sugarcane-growing countries all over the world. Several reports mention the Co varieties from India and other sugarcane varieties being the progenitors of varieties developed at later stages (Amalraj and Balasundaram, 2006, Ram et al., 2022, Santchurn et al., 2022, Kennedy, 2022, Dumont et al., 2022). The cross Co 421 × Co 312 made at Sugarcane Breeding Institute, Coimbatore in 1938 and the fluff imported to South Africa, led to the development of the world-famous varieties NCo 310 and NCo 376 (Nuss and Brett, 1995, Nuss 2001). Introductions from other countries have mostly been the mainstay of the Vietnam sugar industry until recent years (Cao and Doan, 2022). Efforts to use varieties from other countries, especially for base broadening, is going on in many sugarcane-growing countries, like Thailand, Vietnam and Fiji. Private sector organizations in some of the countries also initiate specialized programmes to address the slow improvement in sucrose content, using the varieties from other countries. One such programme in Florida initiated in 2015 used the populations from Carribean and Barbados and is a typical example (Hale et al., 2022). The contributions of other private initiatives, like the erstwhile COPERSUCAR program in Brazil (Cursi et al., 2022), sugarcane R&D in Ecuador, are also noteworthy (Castillo and Cifuentes, 2022, Dumont et al., 2022).

The long-term utilization of improved varieties, starting from the initial attempts of nobilization, have invariably led to an improvement in sugar accumulation potential in the newly developed genotypes all over the world in recent years (Ram et al., 2022, Widyasari et al., 2022, Santchurn et al., 2022, Zhou 2022, Hale et al., 2022). In India, a comparative study of the commercial clones developed from Coimbatore before 1960, and those developed during 1991–2000, showed a steady improvement in sucrose % in juice. Many of the recent Co canes have been reported to have a sucrose content as high as 22–23% (Ram et al., 2022, Hemaprabha et al., 2015). The sugarcane breeding programmes in Lousiana reported an improvement of around 300% for average sugar per hectare and more than 100% for sugar recovery from the 1880s through the 2010s (Hale et al., 2022).

5.3.2.1 Varietal Breeding for Sugar Content: R&D Set up in India

In India, the sugarcane-improvement programme had its origin at the Sugarcane Breeding Institute at Coimbatore, which is considered the "cradle of sugarcane breeding" worldwide. A plaque dedicated to the first interspecific hybrid, Co 205, at the ICAR-Sugarcane Breeding Institute, Coimbatore, India speaks volumes about the impact the variety made on the global sugarcane-improvement scenario. Consequent to the establishment of six sugarcane research stations in the Pre-British period, several other institutions were set up to tackle the various problems, like susceptibility to diseases and pests, low yield, and sugar recovery.

The present sugarcane research network in India consists of research institutes under National Agricultural Research System (NARS) comprising of the Indian Council of Agricultural Research (ICAR), New Delhi, and state agricultural universities (SAUs). These are complemented by the state sugarcane departments and also the private organizations. The ICAR-Sugarcane Breeding Institute (ICAR-SBI), Coimbatore, and ICAR-Indian Institute of Sugarcane Research (ICAR-IISR), Lucknow, work in association with the sugarcane research stations of the state agricultural universities, state research centres like the Uttar Pradesh Council of Sugarcane Research (UPCSR), Shahjahanpur in the state of Uttar Pradesh, state agricultural departments, and private R&D organizations mentored by sugar mills, e.g., Vasantdada Sugar Institute (VSI), Pune, Maharashtra, KJ Somaiyya Institute of Applies Agricultural Sciences, Karnataka, M/S EID Parry Ltd. etc. All these institutes are engaged in sugarcane varietal development and related programmes for improving the cane yield and sugar recovery (Table 5.1).

Apart from these organizations, the All India Co-ordinated Research Project on Sugarcane (AICRP(Sugarcane)), with its headquarters at ICAR-IISR, Lucknow, facilitates the pooling of resources and sharing of the improved varieties and agro-technologies across the whole country, thus addressing the problems of national and regional importance (Sinha et al., 2016). The varietal development programmes under the AICRP (Sugarcane) is the mechanism through which promising varieties are tested across the locations and finally identified for release for commercial cultivation. Until now, the system has facilitated the identification of approximately 60 high-yielding high-sugar sugarcane varieties for commercial cultivation across the country (Sinha et al., 2016, Shukla et al., 2018).

This widespread network of sugarcane research stations engaged in varietal development makes use of the varieties and other clones in the National Hybridization Garden at Coimbatore and in the Distant Hybridization Facility at Agali, Kerala for making crosses. The improvement in cane and sugar yield during the last few years is evident from the increase in sugar yield by more than 180% during the period from 1950 to 2021. The sugar recovery has gone up by almost 9% during the period, with the highest recovery of 11.0% being recorded during the 2018–2019 (Figure 5.1). The gradual plateauing of the gains for traits like sugar recovery is true in many of the sugarcane-growing countries (Kapur et al., 2015, Jackson et al., 2005). Some of the molecular studies have, indeed, pointed to the existence of ample genetic diversity in the available genetic pool (Parida et al., 2010), and this leaves much scope for exploiting the genetic variability in the gene pool, along with the introgression of new gene combinations into the existing genetic background. The biparental crosses with proven progeny performance constitute the main component of the hybridization programme in the sugarcane-improvement programmes, with new combinations involving new parents being attempted, which later evolve into proven crosses depending on progeny performance.

Utilization of species-level clones and related genera directly in the crossing programmes has its own limitations due to the irregular flowering, low pollen fertility, seed set, and seed viability. Intraspecific/intergeneric improvement of the species clones has been the norm with these improved clones being used as parents in commercial crosses (Nair et al., 2006, Nair et al., 2017). Utilization of improved *Saccharum robustum* clones, *S. officinarum* × *S. robustum* clones etc., have been

TABLE 5.1
R&D Network for Sugarcane and Sugar Technology in India

Sl. No	Institute/ Organization/ Departments	Affiliation	Link
1.	ICAR-Sugarcane Breeding Institute, Coimbatore, Tamil Nadu	Indian Council of Agricultural Research (ICAR), New Delhi	https://sugarcane.icar.gov.in
2.	ICAR-Indian Institute of Sugarcane Research, Lucknow, Uttar Pradesh	Indian Council of Agricultural Research (ICAR), New Delhi	https://iisr.icar.gov.in
3.	Research Stations of Agricultural Universities	Respective State Agricultural Universities (SAUs)	Web-site of the respective SAUs
4.	UP Council of Sugarcane Research, Shahjahanpur, Uttar Pradesh	UP Council of Sugarcane Research (UPCSR), Shahjahanpur, Uttar Pradesh	http://www.upcsr.org
5.	Vasantdada Sugar Institute (VSI), Pune, Maharashtra	Co-operative sugar factories of Maharashtra state	https://www.vsisugar.com
6.	KJ Somaiyya Institute of Applied Agricultural Research, Sameerwadi, Karnataka	Godavari Biorefineries	https://godavaribiorefineries.com/agriculture_research
7.	M/s EID Parry, Tamil Nadu	EID Parry, Tamil Nadu	https://www.eidparry.com
8.	State cane departments	Respective state governments	Web-site of respective state cane departments/state governments
9.*	National Sugar institute (NSI), Kanpur, Uttar Pradesh (For sugar technology related research)	Department of Food & Public Distribution Ministry of Consumer Affairs, Food & Public Distribution, Government of India	https://www.nsi.kanpur
10.*	S. Nijalingappa, Sugar Institute, Belagavi, Karnataka	Government of Karnataka	https://www.nijalingappasugar.com

Notes
* Institutions carrying out sugar technology related research

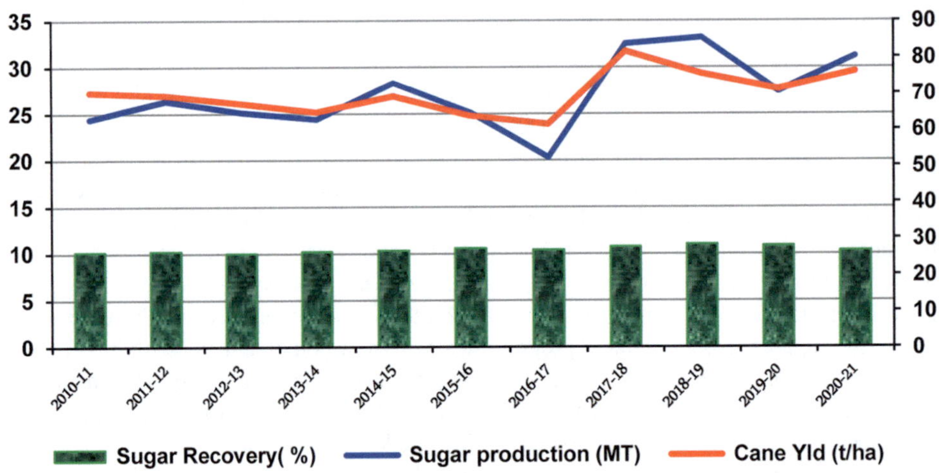

FIGURE 5.1 Sugarcane and sugar scenario in India in the last decade.

documented in India, too, with an increase in sucrose content up to 26% in these improved clones (Karuppaiyyan et al., 2020, Alarmelu et al., 2014, 2017). These can be directly used in commercial crossing programmes. Some of the Co canes developed in India have *S. robustum* and other species clones in their parentage. The interspecific hybridization programmes in India also have proved their utility in developing improved sugarcane varieties like Co 97017, Co 98003, Co 99007, Co 99016, and Co 02005 (Nair et al., 2007), and elite high-sugar accumulating clones CoLk 97147, CoLk 97169, and CoLk 99270 (Kapur and Swapna, 2014). Such successful attempts of utilizing the species clones for bettering the sugar content in the newly developed genotypes are well documented from other sugarcane growing countries also (Heinz, 1967). Using the ISH series of clones developed in India (Ram and Hemaprabha, 1990, 1992, 1995, 1998), F1s, back-crossed hybrids, and intercrossed hybrids were developed (Kandasamy and Thiravium, 1987). It was demonstrated that back-crossing strategy was favourable for improving the sugar content, whereas intercrossing of the F1 hybrids mostly favours a better cane yield (Ram and Hemaprabha, 1993).

The Australian sugarcane improvement program has also identified commercially improved sugarcane clones through interspecific and intergeneric hybridization involving *S.spontaneum and Erianthus* (Wei et al., 2022). Pre-breeding for sucrose accumulation and other traits at Mauritius emphasized introgression involving interspecific hybridization to broaden the genetic base. The use of parental clones with high sugar accumulation, along with larger stalk diameter, was an important strategy, as mentioned in some of the reports (Badaloo and Ramdoyal, 2004). The contribution of wild relatives in the Mauritius sugarcane-improvement programmes has been amply highlighted (Santchurn et al., 2019a,b). The utilization of *S. spontaneum* clones for improvement of sugar production was an important activity in the Caribean countries too. This was taken up in early 1900s (McIntosh, 1944, Stevenson, 1945, 1960). The very high sucrose-accumulating populations generated through recurrent selection (Brix values ranging from 27.0 to 31.2) are the derivatives of crosses involving species clones and related genera (Kennedy, 2022). The Lousiana sugar industry boasts of continuous genetic gains in economically important traits, mainly due to the introgression programs that had led not only to an increased tolerance to the abiotic and biotic stresses in the newly developed sugarcane genotypes, but also to an improvement in sugar yield. Many of the upcoming commercial varieties have the wild accessions like *S. spontaneum* as their progenitors, and this holds true for the sugarcane genotypes of the United States sugar industry as a whole (Hale et al., 2022).

Intervarietal crosses have been extensively carried out for enhancing the sugar content in the recently developed sugarcane varieties/promising genotypes. An analysis of the pattern of crosses carried out in the varietal improvement programme in India from 2001 to 2016 revealed that biparental crosses using specific parents constitute almost 60% of the total hybridizations. Around 30% comprise general collections (GCs). The polycrosses and selfs form approximately 4% and 2%, respectively, of the total crosses carried out (Durai et al., 2017). This pattern holds true for traits like sugar content also, where the specific combining ability of the desirable parents are exploited to the maximum extent to obtain promising varieties with high sugar accumulation potential. The polycrosses and general crosses involve large number of parents, and the populations exhibit wide variability. Also, these are easier to obtain and are less labour intensive. The proportion of crosses in each category varies across the countries and programmes, depending on the strategy of the programme (Cursi et al., 2022). The high level of heterozygosity in the genome of sugarcane is a boon to the breeders in this context since there can be any number of permutations and combinations of the genomic regions, even in selfed progenies, with the possibility of obtaining transgressive segregants. Once realized, this variability can be fixed for the later stages through the clonal propagation in this crop. The superiority of the biparental crosses has also been documented in a detailed study of the parental value of Co canes developed over a century (1918–2017). Some of the proven parents and cross combinations involving the Co canes for the tropical sugarcane-growing region, specifically for the sugar-related traits, were identified (Table 5.2). The study identified a novel parent specifically for the sugar-related traits, OH-44, an interspecific hybrid, which is a new representation in the genetic pool (Hemaprabha et al., 2020).

TABLE 5.2
Some Proven Parental Combinations for Sugar Related Traits

Sl. No.	Trait	Best Performing Parental Combinations
1.	Brix	Co 8371 × Co 86011, Co 86032 × Co 86011, Co 86032 × Co 05001, Co 7201 × Co 775, Co 85002 × Co 86011, Co 86002 × Co 775, Co 86011 × CoT 8201, Co 88002 × Co 62174, CoC 671 × CoT 8201, CoC 671 × Co 6806
2.	Sucrose % in juice	Co 8371 × Co 86011, Co 86032 × Co 86011, Co 8371 × Co 85002, Co 86002 × Co 775, Co 86032 × Co 05001, Co 86011 × CoT 8201, Co 85002 × Co 86011, Co 85002 × OH-44
3.	CCS %	Co 86032 × Co 86011, Co 86002 × Co 775, Co 8371 × Co 86011, Co 8371 × Co 85002, Co 86011 × CoT 8201, Co 86032 × Co 05001, CoC 671 × CoT 8201, CoC 90063 × Co 88013, Co 85002 × Co 86011, Co 85002 × OH-44

Source: Hemaprabha et al., 2020

In India, the main crossing programme takes place from October to December at the National Hybridization Garden and Distant Hybridization Facility situated at Sugarcane Breeding Institute. VSI, Pune also has its own hybridization facility at Amboli, Maharashtra. The researchers from all over India attempt the desired matings, depending on the synchronization of flowering. The seeds are sent to the respective stations, and evaluations are carried out at the respective research stations. After a series of evaluations, initially in the seedling stage and in the subsequent clonal stages, the promising clones are identified. For improving the sugar accumulation, family selection has been widely used in the early selection stages, followed by individual clonal selections within the superior families (Shanthi et al., 2008, McRae et al., 1993). The seedling generation is screened for HR Brix and general morphology. The promising clumps are selected and carried forward, and the sucrose content in juice and commercial sugar are evaluated in the clonal stages. Each station has its own mechanism to identify the promising clones to be evaluated in the national multi-location trials under All India Co-ordinated Research Project on Sugarcane. After an initial varietal trial (IVT) and advanced varietal trials (AVT) in plant and ratoon crops for two years at the designated locations, promising varieties are identified for commercial cultivation based on their performance. The varieties are finally notified for cultivation in the specific sugarcane-growing zones by the Central Sub-Committee on Crop Standards, Notification and Release of Agricultural Crops. The individual states have their own varietal release mechanisms for varieties that do not perform well in a particular zone but are promising in the respective states; the varieties get identified by the state varietal release committee. The notification is finally done by the Central Sub-Committee.

The continuous selection for high sugar accumulation has resulted in several promising clones with sucrose% juice values more than 20–21% (Annual Report, SBI, Annual Report IISR). Approximately 30 varieties specific to each zone, with improved sugar content and sugar recovery have been notified for commercial cultivation all over the country in the last decade.

5.3.3 Pre-Breeding for Sugar Content

Availability of superior parental clones for use in hybridization programmes is of utmost importance in sugarcane improvement. All the sugarcane-growing countries have a population improvement and pre-breeding programme that goes hand in hand with the mainstream varietal development (Lakshmanan et al., 2022). As mentioned before, the increment in sugar content *per se* has been plateauing in most of the programmes, even though an improvement in cane yield has mainly contributed to the improvement in sugar yield. A substantial increase in sugar content *per se* had been difficult, especially in the subtropical sugarcane-growing regions, and this might be due to the relatively low frequency with which the best possible genome combinations that favour a high sugar content have materialized in the newly developed varieties (Jackson 2005, 2019). Introgression of new genes/gene combinations into the existing gene pool and base broadening to enhance the variability are important to overcome these challenges. This will facilitate the development of new sugarcane genotypes with improvement in sugar accumulation potential. Such improved genotypes will serve as excellent parental candidates, provided the high sugar-accumulating trait is transferred to the progenies. Development of high sugar-accumulating sugarcane clones have been reported from several countries from the pre-breeding programmes (Kapur et al., 2015, Swapna et al., 2020, Santchurn et al., 2022, Ram et al., 2022). Recurrent selection cycles involving convergent breeding have invariably been the method of choice, with an improvement in sucrose content documented in the later cycles. In India, recurrent selection for improving the sugar accumulation in sugarcane have reported an improvement in sugar content of more than 10% over the base populations, with sufficient variability being retained in the different selection cycles (Shanthi and Alarmelu, 2012, Alarmelu et al., 2015). Long-term pre-breeding efforts initiated since the early 1990s have helped identify more than 60 high-sugar genotypes as parental clones, and some of these with good progeny performance have been registered as high-sugar genetic stocks (Kapur et al., 2015, Kapur and Swapna, 2014). A preliminary evaluation of progenies from advanced recurrent cycles indicated an improvement of 8%–10% in sucrose % in juice over that in the base populations (Swapna et al., 2020). These results substantiate the earlier findings regarding the suitability of recurrent selection as a tool for pre-breeding for high sugar content in this crop. Similar efforts have been reported from other sugarcane-growing countries also, with pre-breeding for sugar content forming an inherent part of the sugarcane improvement programmes (Ramdoyal and Badaloo, 2002, Santchurn et al., 2022, Hale et al., 2022, Wei et al., 2022, Luzaran et al., 2022).

5.4 BIOTECHNOLOGICAL TOOLS FOR IMPROVING SUGAR CONTENT

The sugarcane-improvement programmes all over the world relied to a large extent on the conventional tools of plant breeding, like introduction of new germplasm, hybridization involving elite clones, and selecting the promising genotypes. A quantum jump in the cane and sugar yield has been the result of this strategy, and this has been successfully put to use in the development of energy canes, too. The genetic peculiarities of the crop viz. the large genome size, polyploidy, and chromosomal segregation pattern have been impediments in understanding the genetic inheritance of important traits in this crop. The slow pace of improvement in the traits like cane and sugar yield in the recent years has also emerged as a challenge; this, to some extent, can be attributed to the genetic complexities in this crop. Thus, nonconventional tools like molecular markers and functional genomics need to be an integral part of the improvement programmes to facilitate continuous progress in the development of high-sugar sugarcane genotypes catering to our need. The molecular studies in this crop had lagged behind, compared to those in the other diploid crops, until the beginning of this decade. Ample classical genetic and cytogenetic studies using tools like transposon tagging, addition-deletion-substitution lines, monosomics, nullisomics etc. had provided some advantage for other crops like rice, wheat, and maize in which such tools were available (Swapna and Srivastava, 2012). Yet, the molecular studies in sugarcane had gained pace at later stages, and, at present, there is seldom any novel molecular strategy that has not been applied in sugarcane, too, even though several challenges still exist (Henry and Kole, 2010, Thirugnanasambandam, 2018). Perhaps the most exciting development in sugarcane molecular biology and genomics in the recent years has been the international initiative and other attempts to sequence the sugarcane genome (Garsmeur et al., 2018, Zhang et al., 2018, Souza et al., 2019 Trujillo-Montenegro, 2021). These draft reference genome sequences that have been reported can facilitate other applications, including genetic and genomic studies in this crop.

5.4.1 Molecular Marker Applications for Sugar Improvement

Both molecular cytogenetic and DNA-based markers have been used on a large scale to study the sugarcane genome. Even though molecular cytogenetic studies specific to sugar content have seldom been reported in this crop, the investigations using genomic in situ hybridization (GISH) and fluorescent in situ hybridization (FISH) have provided valuable information on the genome composition of sugarcane hybrids, including genome size estimation, basic chromosome number of species clones, ploidy level, evolutionary relationships among the species clones and within *Saccharum* complex, etc. (D'Hont et al., 1994, 1995, 1996, Ming et al., 2006, Piperidis et al., 2010).

DNA-based markers have been extensively used to study the diversity of sugarcane with respect to their sugar content. In fact, the very early investigations to study the enzymes involved (Hatch et al.,1963, Hatch and Glaszious, 1963, Sacher et al., 1963) can be considered as the initial step towards the nucleic-acid based marker studies for sugar accumulation. The application of DNA-based markers in sugarcane started with the marker systems like RFLP, PCR-based systems like RAPD, microsatellites, and AFLP (Ming et al., 2006, Swapna and Srivastava, 2012). The sequences related to the genes involved in sugar metabolism have been used for understanding the underlying variations (Grivet et al., 2001, Sugiharto et al., 1997, Lingle and Dyer, 2004, da Silva and Bressiani, 2005, Alwala et al., 2006a, 2006b, Hemaprabha et al., 2006, Singh et al., 2008, Swapna et al., 2011, Parida et al., 2010, Medeiros et al., 2020). Several allele-specific differences with respect to the sugar-gene sequences were identified in these studies. The results can facilitate pyramiding of different sucrose-specific loci/alleles, leading to improved populations and genetic stocks with better sucrose-accumulation potential. Such studies have also thrown up a lot of information about the extent of diversity and also the transferability of these marker systems. This, in turn, supports the role of Indian sugarcane varieties in the lineage of elite clones in other countries (Pan, 2006).

Efforts to construct well-saturated linkage maps and to map loci linked to important traits can be traced back to early 1990 (Wu et al., 1992, Ming et al., 2001, 2002, 2006). The early attempts to map loci linked to sugar content used interspecific segregating populations (Ming et al., 2001, 2002, Aitken et al., 2006, 2008, Alwala et al., 2009) and selfed populations (Hoarau et al., 2002). Populations from cultivated varieties have also been utilized (Singh et al., 2008, Pinto et al., 2009, Banerjee et al., 2015). Several sugar content linked QTLs/QTA are documented in these studies, along with the gene action, inheritance of the trait, etc. (reviewed in Ming et al., 2006, Swapna and Srivastava, 2012). Association mapping using diverse panels of sugarcane clones helps to do away with the development of biparental mapping populations for molecular mapping. This strategy has also led to identification of sugar content related marker-trait associations (Banerjee et al., 2015, Singh et al., 2016, Racedo et al., 2016, Banerjee et al., 2020) in this crop.

As in other crops, the advent of high-throughput next-generation sequencing (NGS) technologies like genotyping-by-sequencing (GBS), genome-wide association studies (GWAS) in large, representative diversity panels has given a huge impetus to the molecular mapping studies. Traits like sugar content, yield-related traits and stress resistance/tolerance have been widely studied (Balsalobre et al., 2017, Barreto et al., 2019, Yang et al., 2017, Banerjee et al., 2020, Ram et al., 2022). Marker-trait associations (MTAs) have been identified that can facilitate the effective utilization of these markers (Fickett et al., 2019, Racedo et al., 2016, Ram et al., 2022). Genomic selection is an upcoming area where precise identification of favourable genomic regions across an entire genome facilitates a precise prediction of favourable parental/gene combinations, thereby enhancing the genetic gains. The strategy has been found to be highly useful for traits like cane and sugar yield where the genetic gains through phenotypic selection alone is very low (Jackson, 2005, Yadav et al., 2020, 2021, Mahadevaiah et al., 2021). Genomic selection can be applied to the characterization of field gene banks, core collections, prediction of suitable parental combinations for higher genetic gains, progeny evaluation and clonal selections, recurrent selection for pre-breeding, base broadening, introgression of genes for sucrose content, yield and stress related traits (Gouy et al., 2013, Mahadevaiah et al., 2021, Yadav et al., 2021). This makes use of genome-wide high throughout markers like SNPs that are capable of capturing the complete genetic variability present in the population/germplasm for precise selection and introgression of favourable alleles. The approach appears to be highly promising in this crop and is being put to use in the sugarcane improvement programmes (Gouy et al., 2013, Deomano et al., 2020, Hayes et al., 2021, Voss-Fels et al., 2021, Yadav et al., 2021) where multi-location testing of the promising clones demands a reliable and precise evaluation of their performance in all the locations. As in other countries, for the sugarcane breeding programmes in India too, the superiority of the genetic gains through genomic selection compared to the phenotypic selection alone, not only in terms of the varietal performance but also in terms of net returns, unit cost, time, and resource allocation have been ample incentives for adopting the genomic selection strategy in the present-day sugarcane breeding programmes (Mahadevaiah et al., 2021).

5.4.2 Functional Genomics

Functional genomics in sugarcane have targeted many traits, like sucrose accumulation, reaction to abiotic and biotic stresses, and traits related to bioenergy and other diversified products, thereby facilitating the pathways related to the traits studied. The regulatory mechanisms involving microRNAs, transcriptomes, and other regulatory factors have been investigated through these studies (Zanca et al., 2010, Silva et al., 2021, reviewed in Swapna and Kumar, 2017), even though a clearcut blueprint regarding the mechanism of regulation of these traits or the regulatory mechanisms of the genes or factors speculated to be involved, might not have been prepared until now. The role of micro RNAs in regulation of sucrose metabolism has been speculated in other crops, like switchgrass, sweet sorghum and grain sorghum, that might be applicable to sugarcane also (Xie et al., 2014. Yu et al., 2015). Trancsriptomic studies have also targeted several sugar metabolism-related genes/ factors, confirming the suitability of functional genomics studies for understanding the regulatory mechanisms of sugar accumulation and regulation of sugar metabolic pathway in this crop (Thirugnanasambadam, 2018, 2019, Chen et al., 2019, Correr et al., 2020, Dantas et al., 2020, Ma et al., 2020, Calderan-Rodrigues, 2021). Reverse genetics approaches involving transgenics also has helped the manipulation of sugar-metabolic pathways, even though at a preliminary level (Wu and Birch, 2007, van der Merwe et al., 2010, Rossouw et al., 2010). Genome-editing technologies have been applied in sugarcane (Mohan, 2016, Eid et al., 2021, Oz et al., 2021), and these will be of help in manipulating the strategic control points in the sugar-metabolic pathway.

5.4.3 Bioinformatics

The genome sequencing and functional genomics, along with the other molecular biology tools, have necessitated the use of bioinformatics to handle the voluminous data generated out of these studies. The tools that are applicable to the normal diploid plants may not be sufficient for handling the sugarcane data, but the latest developments in the field of sugarcane genomics prove that sugarcane bioinformatics also has come of age (Casu, 2010). Development of new marker systems and their utilization, increased coverage and saturation of molecular maps, developments in sugarcane genome-sequencing comparative genomics and functional genomics, genome-editing strategies, etc., would not have been possible without a very fast-evolving field of bioinformatics in sugarcane research.

5.5 FUTURE RESEARCH THRUSTS

Sugarcane-improvement programmes have efficiently made use of the conventional and cutting-edge technologies to bring about a quantum jump in the cane and sugar productivity. The complexity of the genome necessitates the application of precision breeding and molecular biology tools that can help in a better understanding of the various metabolic pathway, which can ultimately pave the way for the development of wonder varieties. Precision breeding for promising varieties and parental clones; robust molecular markers for marker-assisted breeding, thereby improving the breeding efficiency; highly saturated genetic maps; genome-sequencing strategies with better coverage; genome editing for manipulation of complex traits like sugar accumulation, yield, stress response, etc., are some of the targets for the future sugarcane-improvement programmes. These strategies should complement the sustainable and innovative production practices with the ultimate aim of catering to the Sustainable Development Goals.

REFERENCES

Aitken K.S., Jackson P.A., and McIntyre C.L. (2006). Quantitative trait loci identified for sugar related traits in sugarcane (*Saccharum* spp.) cultivar x *Saccharum officinarum* population. *Theor Appl Genet* 112: 1306-1

Aitken K.S., Hermann S., Karno K., Bonnett G.D., McIntyre L.C., and Jackson P.A. (2008). Genetic control of yield related stalk traits in sugarcane. *Theor Appl Genet* 117: 1191–1203

Alarmelu S., Nagarajan R., Shanthi R.M., Hemaprabha G., and Nair N.V. (2014). Development and evaluation of backcross progenies of improved *Saccharum* spp. for yield and quality traits. *J. Sugarcane Res.* 4(1): 19–32.

Alarmelu S., Hemaprabha G., and Shanthi R.M. (2015). Potential hybrids through recurrent selection for yield and quality in sugarcane. *Proceedings Joint Convention of the Sugar Technologists Association of India and the Deccan Sugar Technologists Association* 11: 248–258.

Alarmelu S., Pazhany A.S., Jayabose C., and Manjunatha T. (2017). Pre-breeding utilizing the improved intra specific clones of *Saccharum* spp. In:Hemaprabha G., Viswanathan R., Ramasubramaniam T., Bhaskaran A., Mohanraj K., Bakshi Ram (eds.). *Proc. International Symp. Sugarcane Research since Co 205: 100 years and beyond (SucroSym 2017).* ICAR-Sugarcane Breeding Institute, Coimbatore. pp. 7–10.

Alwala S., Suman A., Arro J.A., Veremis J.C., and Kimberg C.A. (2006a) Target Region Amplification Polymorphism (TRAP) for assessing genetic diversity in sugarcane germplasm collections. *Crop Sci* 46: 448–455.

Alwala S., Kimberg C.A., Gravois K.A., and Bischoff K.P. (2006b) Trap, a new tool for sugarcane breeding: Comparison with AFLP and co-efficient of parentage. *Sugarcane International* 24(6): 11–21.

Alwala S., Collins A., Kimbeng J., Veremis C., and Gravois K.A. (2009). Identification of molecular markers associated with sugar-related traits in a *Saccharum* interspecific cross *Euphytica.* 167: 127–142.

Amalraj V.A., and Balasundaram N. 2006. Status of sugarcane genetic resources in India. *PGR Newsletter FAO-Bioversity* 148: 26–31.

Anonymous. (2022). Indian Sugar Vol. LX XIII (3): 48–58.

Appunu C., Narayan A.J., Mahadevaswamy H.K., Karthigeyan S., Valarmathi R., Mahadevaiah C., Kumar R., Meena M.R., and Bakshi Ram. (2020). Variability and molecular diversity of wild sugarcane germplasm collected from low temperature regions Lohit and Changlang of Arunachal Pradesh. *Indian J. Biotechnol.* 19: 159–168.

Badaloo M.H.G., and Ramdoyal K. (2004). Variation and inheritance of quantitative traits in Commercial x *Saccharum spontaneum* L. Crosses. In. (ed.)Lalouette J.A., Bachraz D.Y., and Bheenick K.J. *Proc. 6th AnnualMeeting of Agricultural Scientists Reduit: Food and Agricultural Council.* 169–178.

Balsalobre T.W.A., Pereira G.S., Margarido G.R.A. et al. (2017). GBS-based single dosage markers for linkage and QTL mapping allow gene mining for yield-related traits in sugarcane. *BMC Genomics* 18: 72. DOI 10.1186/s12864-016-3383-x

Banerjee N., Siraree A., Yadav S., Kumar S., Singh J., Kumar S., Pandey D.K., and Singh R.K. (2015). Marker-trait associations for sucrose and yield contributing traits in sugarcane (*Saccharum* spp. Hybrid). *Euphytica.* 205: 185–201.

Banerjee N., Khan M.S., Swapna M., Singh R.K., and Kumar S. (2020). Progress and Prospects of Association Mapping in Sugarcane (*Saccharum* Species Hybrid), a Complex Polyploid Crop. *Sugar Tech.* 22 (6): 939–953. 10.1007/s12355-020-00852-9

Barreto F.Z., Rosa J.R.B.F., Balsalobre T.W.A. et al. (2019). A genome-wide association study identified loci for yield component traits in sugarcane (*Saccharum* spp.). *PLoS ONE* 14(7): e0219843. 10.1371/journal.pone.0219843

Brandes E.W. (1956). Origin, dispersal and use in breeding of the Melanesian garden sugarcanes and their derivative, *Saccharum officinarum. L. Proc. Int. Soc. Sugar Cane Technol.* 9: 709–760.

Calderan-Rodrigues M.J., de Barros Dantas L.L., Cheavegatti Gianotto A., and Caldana C. (2021). Applying Molecular Phenotyping Tools to Explore Sugarcane Carbon Potential. *Front. Plant Sci.* 12: 637166. doi: 10.3389/fpls.2021.637166709-760

Cao A.D. and Doan L.T. (2022). Sugarcane breeding, germplasm development and supporting genetic research in Vietnam. *Sugar Tech* 14 (1): 272–278. 10.1007/s12355-021-01068-1

Castillo R.O. and Cifuentes E.S. (2022). Sugarcane breeding and supporting genetic research in Ecuador. *Sugar Tech* 24 (1): 222–231. 10.1007/s12355-021-01057-4

Casu R.E. (2010). Role of bioinformatics as a tool for sugarcane research. In: Henry R.J. and Kole C. (Eds.). *Genetics, Genomics and Breeding of Sugarcane*. Science Publishers, USA. 229–248.

Chen Z., Qin C., Wang M., Liao F., Liao Q., Liu X., et al. (2019). Ethylene mediated improvement in sucrose accumulation in ripening sugarcane involves increased sink strength. *BMC Plant Biol.* 19: 285. doi: 10.1186/s12870-019-1882-z

Clayton W.D. and Renvoize S.A. (1986). *Genera graminum: grasses of the world.* Her Majesty's Stationery Office, London. UK.

Correr F.H., Hosaka G.K., Barreto F.Z., Valadão I.B., Balsalobre T.W.A., Furtado A., et al. (2020). Differential expression in leaves of Saccharumgenotypes contrasting in biomass production provides evidence of genesinvolved in carbon partitioning. *BMC Genomics* 21: 673. doi: 10.1186/s12864-020-07091-y

Cursi D.E., Hoffman H.P., Barbosa G.V.S., Bressiana J.A. et al. (2022). History and current status of sugarcane breeding, germplasm development and molecular genetics in Brazil. *Sugar Tech* 24(1): 112–233. 10.1007/s12355-021-00951-1

D'Hont A., Lu Y.H., Le'on D.G.D., Grivet L., Feldmann P., Panaud E., and Glaszmann J.C. (1994). A molecular approach to unravelling the genetics of sugarcane, a complex polyploid of Andropogonaea tribe. *Genome* 37: 222–230.

D'Hont A., Grivet L., Feldmann P., Rao P.S., Berding N., and Glaszmann J.C. (1995). Identification and characterization of sugarcane intergeneric hybrids, *Saccharum officinarum* x *Erianthus arundinaceus* with molecular markers and DNA *in situ* hybridization. *Theor Appl Genet* 91: 320–326.

D'Hont A., Grivet L., Feldmann P., Rao P.S., Berding N., and Glaszmann J.C. (1996). Characterisation of the double genome structure of modern sugarcane cultivars (*Saccharum* spp.) by molecular cytogenetics. *Mol Gen Genet* 250: 405–413.

D'Hont A. and Glaszmann J.C. (2001). Sugarcane genome analysis with molecular markers, a first decade of research. *Proc. Int. Soc. Sugar Cane Technol.* 24: 556–559.

da Silva J.A. and Bressiani J.A. (2005). Sucrose synthase molecular marker associated with sugar content in elite sugarcane progeny. *Genet Mol Biol*: 28: 294–298.

Daniels J. and Roach B.T. (1987). *Taxonomy and evolution in Sugarcane Improvement through Breeding* In: Heinz D.J. (eds.), Elsevier, Amsterdam, pp. 77–84.

Dantas L.L.B., Almeida-Jesus F.M., de Lima N.O., Alves-Lima C., Nishiyama Jr M.Y., Carneiro M.S., et al. (2020). Rhythms of transcription in field-grown sugarcane are highly organ specific. *Sci. Rep.* 10: 6565. doi: 10.1038/s41598-020-63440-7

Deomano E., Jackson P., Wei X., Aitken K., Kota R., and Pérez-Rodríguez P. (2020). Genomic prediction of sugar content and cane yield in sugar cane clones in different stages of selection in a breeding program, with and without pedigree information. *Mol. Breed.* 40: 4. doi: 10.1007/s11032-020-01120-0

Dumont T., Barau L., Thong-Chane A., Dijoux J., Mallin M., Daugrois J., and Hoarau J.-Y. (2022). Sugarcane breeding in Reunion. Challenges, achievements and future prospects. *Sugar Tech* 24 (1): 181–192. 10.1007/s12355-021-00998-0

Durai A.A., Pazhany A.S., Hemaprabha G., and Ram B. (2017). Appraisal of different crossing methods adopted by fluff receiving centres of AICRP(S) for sugarcane improvement programme in India. In: Hemaprabha G., Viswanathan R., Ramasubramaniam T., Bhaskaran A., Mohanraj K., Bakshi Ram (eds.). *Proc. International Symp. Sugarcane Research since Co 205: 100 years and beyond (SucroSym 2017).* ICAR- Sugarcane Breeding Institute, Coimbatore, pp. 114–116.

Eid A., Mohan C., Sanchez S., Wang D., and Altpeter F. (2021). Multiallelic, targeted mutagenesis of magnesium chelatase with CRISPR-Cas9 provides a rapidly scorable phenotype in highly polyploid sugarcane. *Front. Genome Ed.* 3: 654996. doi: 10.3389/fgeed.2021.654996

Fickette N., Gutierrez A., Verma M., Pontif M., Hale A., Kimbeng C. and Baisakh N. (2019). Genome-wide association mapping identifies markers associated with cane yield components and sucrose traits in the Louisiana sugarcane core collection. *Genomics* 111 (2019): 1794–1801. 10.1016/j.ygeno.2018.12.002

Garsmeur O., Droc G., Antonaise R. et al. (2018). A mosaic monoploid reference sequence for the highly complex genome of sugarcane. *Nat. Comm.* 9: 2638. DOI: 10.1038/s41467-018-05051-5

Gouy M., Rousselle Y., Bastianelli D., Lecomte P., Bonnal L., Roques D., et al. (2013). Experimental assessment of the accuracy of genomic selection in sugarcane. *Theor. Appl. Genet.* 126, 2575–2586. doi: 10.1007/s00122-013-2156-z

Govindaraj. P. and Mahadevaswamy H.K. (2021a). Collection, Characterization and Diversity Analysis of New Wild Sugarcane Germplasm Collected from Western Ghats: A Rich Biodiversity Spot in India. *Sugar Tech.* 23 (1): DOI:10.1007/s12355-020-00933-9

Govindaraj P. and Mahadevaswamy H.K. (2021b). Expedition for the Collection and Conservation of Saline and Waterlogging Tolerant Sugarcane Wild Germplasm from West Bengal and Assam. *Sugar Tech* 23 (16): DOI:10.1007/s12355-021-00982-8

Grivet L., Glaszmann J.C., and Arruda P. (2001). Sequence polymorphism from EST data in sugarcane: a fine analysis of 6-phosphogluconate dehydrogenase. *Genet Mol Biol* 24 (1-4): 161–167

Hale A.L., Todd J.R., Gravios K.A., Mollov D. et al. (2022). Sugarcane breeding programs in USA. *Sugar Tech* 24 (1): 97–111. 10.1007/s12355-021-01018-x

Hatch M.D., Sacher J.A., and Glasziou K.T. (1963). Sugar accumulation cycle in sugarcane I. Studies on enzymes of the cycle. *Plant Physiol* 38: 338–343.

Hatch M.D. and Glasziou K.T. (1963). Sugar accumulation cycle in sugarcane. II. *Relationship of invertase activity to sugar content & growth rate in storage tissue of plants grown in controlled environments Plant Physiol* 38: 344–348.

Hayes B.J., Wei X., Joyce P., Atkin F., Deomano E., Yue J., et al. (2021). Accuracy of genomic prediction of complex traits in sugarcane. *Theor. Appl.Genet.* 134, 1455– 1462. doi: 10.1007/s00122-021-03782-6

Heinz D.J. (1967). Wild *Saccharum* species for breeding in Hawaii. *Proc. ISSCT*, 12, 1037–1043

Heinz D.J. (1980). Thailand *S. spontaneum* hybrid progeny as a new germplasm source in Hawaii. *Proc. ISSCT* 17: 1347–1356.

Hemaprabha G., Natarajan U.S., Balasundaram N., and Singh N.K. (2006). STMS based genetic divergence among common evolution in sugarcane (*Saccharum* sp.). *Sugarcane International* 24(6): 22–27.

Hemaprabha G., Alarmelu S., and Shanthi R.M. (2015). Genetic improvement of sugarcane varieties developed over decades. In: Proceedings of 11th Joint Convention of STAI and DSTA, 219–227.

Hemaprabha G., Mohanraj K., Alarmelu S., and Ram B. (2020). Assessment of breeding methods and parental value of Co canes developed during 1918 – 2017. *J. Sugarcane Res.* 10: 24–31. 10.37580/JSR.2020.1.10.24-31

Hemaprabha G., Mohanraj K., Jackson P.A., Lakshmanan P., Ali G.S., Li A.M., Huang D.L., and Ram B. (2022). Sugarcane genetic diversity and major germplasm collections. *Sugar Tech* 24(1): 279–297.

Henry R.J. and Kole C. (2010). *Genetics, Genomics and Breeding of Sugarcane*. Science Publishers, USA. pp. 268

Hodkinson T.R., Chase M.W., Liedo N., Salamin, and Renvoize S.A. (2002). Phylogenetics of *Miscanthus, Saccharum* and related genera (*Saccharinae, Andropogoneae, Poaceae*) based on DNA sequences from ITS nuclear ribosomal DNA and plastid trnL intron and trnL-F intergenic spacers. *J. Plant Res.* 115: 381–392.

Hogarth D.M., Wu K.K., and Heinz D.J. (1981). Estimating genetic varuiance in sugarcane using a factorial cross design. *Crop Sci.* 21: 21–25.

Hoarau J.Y., Grivet L., Offmann B., Raboin L.M., Diorflar J.P., Payet J., Hellmann M., D'Hont A., and Glaszmann J.C. (2002). Genetic dissection of a modern sugarcane cultivar (*Saccharum* spp.). II. Detection of QTLs for yield components. *Theor Appl Genet* 105: 1027–1037.

Irvine J.E. (1999). *Saccharum* species as horticultural classes. *Theor. Appl. Genet.* 98: 186-194.

Jackson P.A. (2005). Breeding for improved sugar content in sugarcane. *Field Crops Res* 92: 277–290.

Jackson P.A. (2019). Why are yields of sugarcane not increasing as much as of sugar beet (or other crops)? *Proc. International Society of Sugar Cane Technologists*. 30: 128–137.

Kandasamy P.A. and Thiravium A. (1987). Utilization of *Saccharum robustum* in breeding. *Sugar Cane*. 3: 3–7.

Kapur R. and Swapna M. (2014). Sugarcane. In: Chopra V.L. (Ed.) *Breeding Field Crops II: Advances*. Studium Press, LLc, USA, pp. 111–129.

Kapur R., Duttamajumder S.K., and Ram kumar (2015). Identifying parents conferring high sugar accumulation potential to the progeny. Proc 11th Joint Conv STAI & DSTA: 163–169.

Karuppaiyyan R., Nair N.V., Ram B., Mohanraj K., Durai A.A., and Amudha P. (2020). Evaluation of intra-specific hybrids of Saccharum robustum for yield, quality and yellow leaf disease incidence. *Indian J. Sugar Technol.* 35(1): 28–32.

Kellog E.A. (2013). Phylogenetic relationship of *Saccharinae* and *Sorghinae*. *Genomics of Saccharinae*, Springer, New York. 3–21.

Kennedy A.J. (2022). Sugarcane variety development in the Caribbean. *Sugar Tech* 24 (1): 64–72. 10.1007/s12355-020-00936-6

Lakshmanan P., Jackson P.A., Hemaprabha G. and Li Y.R. (Eds). (2022). Special Issue: History of Sugarcane Breeding, Germplasm Development and Molecular Biology. *Sugar Tech* 24 (1): 385.

Lingle S. and Dyer J.M. (2004). Polymorphism in the promoter region of the sucrose synthase-2 gene of *Saccharum* genotypes. *J Am Soc Sugar Cane Technol* 24: 241–249.

Luzaran R.T., Engle L.M., Villariez H.P., and Oquias G.B. (2022). Sugarcane breeding and germplasm development in Philippines. *Sugar Tech* 24 (1): 210–221. 10.1007/s12355-021-00979-3

Ma P., Zhang X., Chen L., Zhao Q., Zhang Q., Hua X., et al. (2020). Comparative analysis of sucrose phosphate synthase (SPS) gene family between *Saccharum officinarum* and *Saccharum spontaneum. BMC Plant Biol.* 20: 422. doi: 10.1186/s12870-020-02599-7

Mahadevaiah C., Appunu C., Aitken K., Suresha G.S., Vignesh P., MahadevaSwamy H.K., Valarmathi R., Hemaprabha G., Alagarasan G., and Ram B. (2021). Genomic Selection in Sugarcane: Current Status and Future Prospects. *Front. Plant Sci.* 12: 708233. doi: 10.3389/fpls.2021.708233

Manjunatha T., Chandran K., Mahesh P., Mohanraj K., and Appinu C. (2018). Characterization of *Saccharum spontaneum* accessions from North-Eastern India. *Electronic Journal of Plant Breeding*, 9(1): 38–50. DOI: 10.5958/0975-928X.2018.00005.4

McIntosh A.E.S. (1944). Nobilisation in cane breeding at the British West Indies Central Sugar Cane Breeding Station and its practical results to date. In: Proceedings of British West Indies Sugar Technology, 1944, 26–40.

Mc Rae T.A., Hogarth D.M., Foreman J.F., and Braithwaite M. (1993). Selection of sugarcane families in the Burdekin district. In: Imrie B.C. and Hacker J.B. (eds.) *Focused Plant Improvement: Towards responsible and sustainable agriculture, Proc. 10th Aust. Plant breeding conference*, Vol 1, Organizing Committee, Australian Convention and Travel Service, Canberra, pp. 77–82.

Medeiros C., Balsalobre T.W.A., and Carneiro M.S. (2020). Molecular diversity and genetic structure of *Saccharum* complex accessions. *PLOS ONE* 10.1371/journal.pone.0233211

Ming R., Liu S.C., Moore P.H., Irvine J.E., and Paterson A.H. (2001). QTL analysis in a complex autopolyploid: genetic control of sugar content in sugarcane. *Genomic Res* 11: 2075–2084.

Ming R., Wang Y.W., Draye X., Moore P.H., Irvine J.E., and Paterson A.H. (2002). Molecular dissection of complex traits in autopolyploids: mapping QTLs affecting sugar yield and related traits in sugarcane. *Theor Appl Genet* 105: 332–345.

Ming R., Moore P.H., Wu K.K., D'Hont A., Glaszmann J.C., and Tew T.L. (2006). Sugarcane Improvement through breeding and biotechnology. *Plant Breed Rev* 27: 15–118.

Mohan C. (2016). Genome Editing in Sugarcane: Challenges Ahead. *Front. Plant Sci.* 7: 1542. doi: 10.3389/fpls.2016.01542

Mukherjee S.K. (1957). Origin and distribution of *Saccharum. Bot Gaz* 119: 55–61.

Nair N.V., Selvi A., Sreenivasan T.V., and Pushpalatha K.N. (2002). Molecular diversity in Indian sugarcane cultivars as revealed by Random Amplified DNA Polymorphisms. *Euphytica* 127: 219–225.

Nair N.V., Selvi A., Sreenivasan T.V., Pushpalatha K.N., and Mary S. (2006). Characterization of intergeneric hybrids of *Saccharum* using molecular markers. *Genet Resour. Crop Evol.* 53: 163–169.

Nair N.V. (2007). Sugarcane genetic resources: an Indian perspective. In: Singh S.B., Rao G.P., Solomon S. and Gopalasundaram P. (eds.) *Sugarcane Crop Production and Improvement.* Studium Press LLC, Texas, USA, 2007, pp. 1–20.

Nair N.V. (2012). Sugarcane genetic resources: Status, potential and role in sugarcane improvement. In:Hemaprabha G., Viswanathan R., Mohanraj K., and Bakshi Ram (eds.) *Proc. International Symp. New Paradigms in Sugarcane Res. (ISNPSR-2012)*, ICAR-Sugarcane Breeding Institute, Coimbatore. pp. 1–8.

Nair N.V., Mohanraj K., Sunadaravelpandian K., Suganya A., Selvi A., and Appunu C. (2017). Characterization of an intergeneric hybrid of *Erianthus procerus x Saccharum officinarum* and its backcross progenies. *Euphytica.* 213: 267. DOI 10.1007/s10681-017-2053-7

Nayak S.N., Song J., Villa A., Pathak B., Ayala-Silva T., Yang X., Todd J., Glynn N.C., Kuhn D.N., Glaz B., Gilbert R.A., Comstock J.C., and Wang J. (2014). Promoting utilization of *Saccharum* spp. genetic resources through genetic diversity analysis and core collection construction. *PLOS ONE* 9(10): e110856 doi: 10.1371/journal.pone.0110856.t005

Nuss K.J. and Brett P.G.C. (1995). The release of variety NCo 310 in 1945 and its impact on the sugar industry. *Proc. South African Sugar Technol. Assoc.* 69: 1–8.

Nuss K.J. (2001). The contribution of variety NCo376 to sugar production in South Africa from 1995 to 2000 and its value as a parent in the breeding programme. *Proc. South African Sugar Technol. Assoc.* 75: 154–158.

Oz M.T., Altpeter A., Karan R., Merotto A., and Altpeter F. (2021). CRISPR/Cas9-Mediated Multi-Allelic Gene Targeting in Sugarcane Confers Herbicide Tolerance. *Front. Genome Ed.* 3: 673566. doi: 10.3389/fgeed.2021.673566

Pan Y.B. (2006). Highly polymorphic microsatellite DNA markers for sugarcane germplasm evaluation and variety identity testing. *Sugar Tech* 8(4): 246–2456.

Parida S.K., Pandit A., Gaikwad K., Sharma T.R., Srivastava P.S., Singh N.K., and Mohapatra T. (2010). Functionally relevant microsatellites in sugarcane unigenes. *BMC Plant Biol* 10: 251.

Pinto L. R., Garcia A. A. F., Pastina M. M., Teixeira L. H. M., Bressiani J. A., Ulian E. C., Bidoia M. A. P. and Souza A. P. (2009). Analysis of genomic and functional RFLP derived markers associated with sucrose content, fiber and yield QTLs in a sugarcane (*Saccharum* spp.) commercial cross. *Euphytica* 172 : 313–327 doi:10.1007/s10681-009-9988-2

Piperidis G., Piperidis N., and D'Hont A. (2010). Molecular cytogenetic investigation of chromosome composition and transmission in sugarcane. *Mol Genet Genomics* 284: 65–73.

Pompidor N.C., Charron C., Hervouet S. et al. (2021). Three founding ancestral genomes involved in the origin of sugarcane. *Annals of Botany*. XX: 1–14.

Racedo J., Gutiérrez L., Perera M.F., Ostengo S., Pardo E.M., Cuenya M.I., Welin B., and Castagnaro A.P. (2016). Genome-wide association mapping of quantitative traits in a breeding population of sugarcane. *BMC Plant Biol.* Jun 24;16(1): 142. doi: 10.1186/s12870-016-0829-x

Ram B. and Hemaprabha G. (1990). Variability pattern in cultivar x species progenies in sugarcane. *Indian J. Genet.* 50: 400–406.

Ram B. and Hemaprabha G. (1992). Genetic variability in interspecific progenies in sugar cane (*Saccharum* spp.). *Indian J. Genet.*, 52, 192–198.

Ram B. and Hemaprabha G. (1993). Genetic divergence of stalk yield and juice quality and their components in flowering clones of *Saccharum officinarum. Indian J. Genet.*, 53, 28–33.

Ram B. and Hemaprabha G. (1995). Influence of noble and commercial hybrid clones on economic traits in nobilization of *Saccharum* species. *Indian J. Genet.*, 55,166–169.

Ram B. and Hemaprabha G. (1998). Nature and pattern of genetic divergence of sugar yield and its components in the progenies of S*accharum barberi. Indian J. Genet.* 58: 193–199.

Ram B., Hemaprabha G., Singh B.D., and Appunu C. (2022). History and current sugarcane breeding, germplasm development and molecular biology in India. *Sugar Tech* 24(1): 4–29.

Ramdoyal K. and Badaloo Goolam M.H. (2002). Pre-breeding in sugarcane with an emphasis on the programme of the Mauritius sugar industry research institute. In: Engel J.M.M., Rao V.R., Brown A.H.D., and Jackson M.T. (eds.) *Managing plant genetic diversity*, CABI, Rome. 307–322. 10.1079/9780851995229.0000

Rossouw D., Kossmann J., Botha F.C. and Groenewald J.H. (2010). Reduced neutral invertase activity in the culm tissues of transgenic sugarcane plants results in a decrease in respiration and sucrose cycling and an increase in the sucrose to hexose ratio. *Funct. Plant Biol.* 37: 22-31 doi:10.1071/fp08210

Sacher J.A., Hatch M.D., and Glasziou K.T. (1963). The sugar accumulation cycle in sugarcane. III. Physical & metabolic aspects of cycle in immature storage tissue. *Plant Physiol* 38: 348–354.

Santchurn D., Badaloo Goolam M.H., Zhou M., and Labuschagne M.T. (2019a). Contribution of sugarcane crop wild relatives in the creation of improved varieties in Mauritius. *Cambridge Core: Plant Genetic Resources (PGRCU)* 17: 151–163.

Santchurn D., Badaloo Goolam M.H., Zhou M., and Labuschagne M.T. (2019b). Genetic studies on sugar and fibre accumulation patterns among different types of energy canes in Mauritius. *Sugar Tech* 21: 879–890.

Santchurn D., Badaloo M.G.H., Koonjah S., and Dookun-Saumtally A. (2022). Sugarcane breeding and supporting genetic research in Mauritius. *Sugar Tech* 24 (1): 48–65. 10.1007/s12355-021-00960-0

Selvi A., Nair N.V., Noyer J.L., Singh N.K., Balasundaram N., Bansal K.C., Koundal K.R., and Mohapatra T. (2005). Genomic constitution and genetic relationship among the tropical and subtropical Indian sugarcane cultivars revealed by AFLP. *Crop Sci.* 45: 1750–1757.

Selvi A., Nair N.V., Noyer J.L., Singh N.K., Balasundaram N., Bansal K.C., Koundal K.R., and Mohapatra T. (2006). AFLP analysis of the phenetic organization and genetic diversity in the sugarcane complex. *Saccharum* and *Erainthus. Genetic Resour. Crop Evol.* 68: 831–842.

Shanthi R.M., Bhagyalakshmi K.V., Hemaprabha G., Alarmelu S., and Nagarajan R. (2008). Relative performance of the sugarcane families in early selection stages. *Sugar Tech* 10(12): 114–118.

Shanthi R.M. and Alarmelu S. (2012). Genetic stocks for high sugar content generated over two selection cycles in sugarcane. *Indian J Genet* 72 (3): 336–342.

Shukla S.K., Zubair A., Awasthi S.K., and Pathak A.D. (2018). *Sugarcane varieties identified by AICRP (S) in India. AICRP Technical Publication-II.* ICAR-All India Co-ordinated Research Project on Sugarcane ICAR-IISR Lucknow. pp. 11.

Silva J. O. L., da Silva R.G., Nogueira, L. F.and Zingaretti S. M. (2021). MicroRNAs regulate tolerance mechanisms in sugarcane (*Saccharum* spp.) under aluminum stress. *Crop Breed. Appl. Biotechnol.*, 21(i):e34442115 doi:10.1590/1984-70332021v21n1a5

Singh R.K., Srivastava S., Singh S.P., Sharma M.L., Mohapatra T., Singh N.K., and Singh S.B. (2008). Identification of new microsatellite DNA markers for sugar and related traits in sugarcane. *Sugar Tech* 10 (4): 156–162.

Singh R.K., Banerjee N., Khan M.S., Yadav S., Kumar S., Duttamajumder S.K., Lal R.J., Patel J.D., Guo H., Zhang D., and Paterson A.H. (2016). Identification of putative candidate genes for red rot resistance in sugarcane (*Saccharum* species hybrid) using LD-based association mapping. *Mol Genet. Genomics*; 291: 1363–1377. DOI 10.1007/s00438-016-1190-3

Sinha O.K. Kumar, R., Chaudhary, S.K., and Zubair, A. (2016). *Forty Five years of AICRP on Sugarcane*. All India Co-ordinated Research Project on Sugarcane, ICAR-Indian Institute of Sugarcane Research, Lucknow. India p77.

Sobhakumari V.P. (2020). Evolutionary dynamics of chromosomal diversity in *Saccharum spontaneum* population from Punjab and Haryana, India. *Cytologia* 85: 313–318.

Sookasthan K., Nagatomi S., Naritoom K., and Sadakorn J. (1992). *Saccharum* germplasm collected in Thailand and its significance. *Proc. ISSCT* 21: 324–336.

Souza G.M., Sluys M.V., Lembke C.G. et al. (2019). Assembly of the 373k gene space of the polyploid sugarcane genome reveals reservoirs of functional diversity in the world's leading biomass crop. *Gigascience*. 8(12): giz129. 10.1093/gigascience/giz129

Sreenivasan T.V., Ahloowalia B.S., and Heinz D.J. (1987). Cytogenetics. In: Heinz D.J. (ed.) *Sugarcane Improvement through Breeding*. Elseveir Science Publishers, the Netherlands, pp. 211–254.

Stevenson G.C. (1945). Sugar cane varieties in Barbados: an historical review. B.W.I. Central Sugar Cane Breeding Station Bulletin No 37.

Stevenson G.C. (1960). *Sugar cane varieties in Barbados: An historical review*. Barbados: B.W.I. Central Sugar Cane Breeding Station Bulletin No 39, 29.

Sugiharto B., Sakakibara H., Sumadi, and Sugiyami T. (1997). Differential expression of two genes for Sucrose-Posphate Synthase in sugarcane: Molecular cloning of the cDNAs and comparative analysis of gene expression. *Plant Cell Physiol* 38(8): 961–965.

Swapna M., Sivaraju K., Sharma R.K., Singh N.K., and Mohapatra T. (2011). Single-strand conformational polymorphism of EST-SSRs: a potential tool for diversity analysis and varietal identification in sugarcane. *Plant Mol Biol Rep*. 29: 505–513.

Swapna M. and Srivastava S. (2012). *Molecular Marker Applications for Improving Sugar Content in Sugarcane*. Springer Brief, Springer, DOI: 10.1007/978-1-4614-2257-0_1

Swapna M. and Kumar S. (2017). MicroRNAs and their regulatory role in sugarcane. *Front. Plant Sci.*, 8:997. doi:10.3389/fpls.2017.00997

Swapna M., Pandey D.K., and Kapur R. (2020). Recurrent selection cycles for pre-breeding in sugarcane: enhancing the sugar-accumulation potential in selection cycles. *Int Sugar J* 122 (1459): 496–502

Thirugnanasambadam P.P., Hoang N.V., and Henry R.J. (2018). The Challenge of analyzing the sugarcane genome. *Frontiers in Plant Sci*. 9: 616. doi: 10.3389/fpls.2018.00616

Thirugnanasambandam P.P., Mason P.J., Hoang N.V., Furtado A., Botha F.C., and Henry R.J. (2019). Analysis of the diversity and tissue specificity of sucrose synthase genes in the long read transcriptome of sugarcane. *BMC Plant Biol*. 19: 160. doi: 10.1186/s12870-019-1733-y

Trujillo-Montenegro J.H., Rodríguez Cubillos M.J., Loaiza C.D., Quintero M., Espitia-Navarro H.F., Salazar Villareal F.A., Viveros Valens C.A., González Barrios A.F., De Vega J., Duitama J., and Riascos J.J. (2021). Unraveling the Genome of a High Yielding Colombian Sugarcane Hybrid. *Front. Plant Sci.*12: 694859. doi: 10.3389/fpls.2021.694859

van der Merwe, M.J. Groenewald J.H. Stitt M., Kossmann J. and Botha F.C. (2010). Downregulation of pyrophosphate: d-fructose-6-phosphate 1-phosphotransferase activity in sugarcane culms enhances sucrose accumulation due to elevated hexose-phosphate levels. *Planta*. 231: 595–608 doi:10.1007/s00425-009-1069-1

Voss-Fels K.P., Wei X., Ross E.M., Frisch M., Aitken K.S., Cooper M., et al. (2021). Strategies and considerations for implementing genomic selection to improve traits with additive and non-additive genetic architectures in sugarcane breeding. *Theor. Appl. Genet.* 134, 1493–1511. doi: 10.1007/s00122-021-03785-3

Wang L., Jackson P.A., Lu X., Fan Y. et al. (2008). Evaluation of Sugarcane × *Saccharum spontaneum* Progeny for Biomass Composition and Yield Components. *Crop Sci.*, 48(3): 951–961. 10.2135/cropsci2007.10.0555

Wei X., Eglinton J., Piperidis G., Atkin F., Morgan T., Parfitt, and Hu F. (2022). Sugarcane breeding in Australia. *Sugar Tech* 24 (1): 151–165. 10.1007/s12355-021-00969-5

Welker C.A.D., Souza-Chies T.T., Longhi-Wagner H.M., Peichoto M.C., McKain M.R., and Kellog E.A. (2015). Phylogenetic analysis of *Saccharum* s.1 (Poaceae; Andropogoneae), with emphasis on the circumscription of the South American species. *Am. J. Bot.* 102 248–263.

Widyasari W.B., Putra L.K., Ranomahera M.R.R., and Puspitasari A.R. (2022). Historical notes, germplasm development and molecular approaches to support sugarcane breeding program in Indonesia. *Sugar Tech* 24 (1): 30–47. 10.1007/s12355-021-01069-0

Wu L. and Birch R. G. (2007). Doubled sugar content in sugarcane plants modified to produce a sucrose isomer. *Plant Biotechnol. J.* 5: 109–117 doi:10.1111/j.1467-7652.2006.00224.x

Wu K.K., Burnquist W., Sorrells M.E., Tew T.L., Moore P.H., and Tanksley S.D. (1992). The detection and estimation of linkage in polyploids using single-dose restriction fragments. *Theor Appl Genet* 83: 294–300.

Xie F., Stewart C. N., Taki F. A., He Q., Liu H. and Zhang B. (2014). High-throughput deep sequencing shows that microRNAs play important roles in switchgrass responses to drought and salinity stress. *Plant Biotechnol. J.* 12: 354–366 doi: 10.1111/pbi.12142

Yadav S., Jackson P., Wei X., Ross E.M., Aitken K., Deomano E., et al. (2020). Accelerating genetic gain in sugarcane breeding using genomic selection. *Agronomy* 10, 1–21. doi: 10.3390/agronomy10040585

Yadav S., Wei X., Joyce P., Atkin F., Deomano E., Sun Y., et al. (2021). Improved genomic prediction of clonal performance in sugarcane by exploiting non- additive genetic effects. *Theor. Appl. Genet.* 134, 2235–2252. doi: 10.1007/s00122-021-03822-1

Yang X., Luo Z., Todd J., Sood S., and Wang J. (2017). Genome-wide association study of multiple yield traits in a diversity panel of polyploid sugarcane (*Saccharum* spp.) *The Plant Genome*. 2020;13: e20006. 10.1002/tpg2.20006

Yu H., Cong L., Zhu Z., Wang C., Zou J., Tao C., Shi Z. and Lu X. (2015). Identification of differentially expressed microRNA in the stems and leaves during sugar accumulation in sweet sorghum. *Gene.* 571 (2) : 221–230 doi:10.1016/j.gene.2015.06.056.

Zanca, A.S., Vicentini, R., Ortiz-Morea, F. A., Del Bem, L.E.V., da Silva, M.J., Vincentz, M. and Nogueira, F.T.S. (2010). Identification and expression analysis of microRNAs and targets in the biofuel crop sugarcane. *BMC Plant Biol.*, 10: 260 doi:10.1186/1471-2229-10-260.

Zhang J., Zhang Q., Li L., Tang H., and Ming R. (2019). Recent polyploidization events in three *Saccharum* founding species. *Plant Biotechnol. J.* 17: 264–274.

Zhang J., Zhang H., Tang H. et al. (2018). Allele-defined genome of the autopolyploid sugarcane *Saccharum spontaneum* L. *Nat. Genet.* 50: 1565–1573. 10.1038/s41588-018-0237-2

Zhou M. (2022). History and current status of sugarcane breeding, germplasm development and supporting molecular research in South Africa. *Sugar Tech.* 24 (1): 86–96. 10.1007/s12355-021-00961-z

6 Omics Applications to Decipher Host-Pathogen Interactions and to Understand Pathogens in Sugarcane

R. Viswanathan, A. Ramesh Sundar, P. Malathi, and K. Nithya
ICAR-Sugarcane Breeding Institute, Coimbatore, India

CONTENTS

DOI: 10.1201/9781003292425-6

6.1 INTRODUCTION

Sugarcane is a highly complex autopolyploid and aneuploid crop. The present-day cultivars are mostly the product of crosses between two autopolyploid species, *Saccharum officinarum,* a high sugar but stress-susceptible and *S. spontaneum,* a low-sugar but stress tolerant. The noble cane *S. officinarum* (2n = 80, x = 10) contributed 70%–80% of the genome, and the wild cane *S. spontaneum* (2n = 40–128, x = 8) contributed 10%–20% of the genome to the commercial-cultivated hybrids (Garsmeur et al. 2018, Ming et al. 2006). It has a polploidy genome of ~10 Gb, and a complete genome details for the modern sugarcane cultivars are yet to come (Zhang et al. 2018; Garsmeur et al. 2018). It is an important field crop grown for its high sucrose content and for bio ethanol and bio energy production in tropics and subtropics across the world.

Biotic stress factors like insect pests and diseases have constantly placed crop plants under challenging situations. The stress-responsiveness of plants can be studied by analyzing changes in physiological and metabolic perturbations. Recent advances in metabolomics, genomics, and transcriptomic and metabolic modelling offer new opportunities to address this question and generate a system-level understanding of metabolic interactions at the host-pathogen interface (Rico et al. 2011). Different domains of "omics," namely genomics, transcriptomics, proteomics, and metabolomics collectively contribute to decipher the functional genomics component in any plant-pathogen interactions. Unwinding of whole genome sequences, followed by RNA and protein expression profiling, added strength to comprehensively determine how genes can contribute to complex phenotypes.

The last two decades had applied genomic tools to understand sugarcane and its responses to various environmental factors; hence, the need for proteomic tools to complement datasets accumulating from genomic and metabolomic studies are becoming emphatic. Establishing protocols to understand alteration in sugarcane proteome under normal and stressed conditions marks the "initiation phase" of sugarcane proteomics. Proteomics find potential applications in understanding complex plant-pathogen interactions at the molecular level. The possible outcome would be identification of proteins that confer disease resistance and pathogenicity determinants from the host and pathogen sides, respectively. With the accomplishments in elucidating sugarcane ESTs, which was ably supported to unlock the secrets of pathogenomics and pathoproteomics by exploiting the next-generation sequencing platforms, it is now possible to improve our understanding of disease resistance in sugarcane. Hence, the integration of the tools of "*omics*" approaches would enable researchers to reconstruct the whole cascade of cellular events, leading to rapid responses and adaptation to the various important abiotic/biotic stress stimuli. Sugarcane proteomics, though, at its infancy, is progressing at a steady-slow pace when compared to other monocots, like rice, wherein well-established protocols for proteomic studies on several tissues, organs, and organelles are available with several 2-D reference gels. Size and complexity of its genome further hampers the progress; so, specifically, proteomics has emerged as an indispensable tool in understanding the crop. Disease resistance in sugarcane has been extensively reviewed, concluding that an integration of the tools of "omics," namely genomics, proteomics, metabolomics, etc., would further strengthen deciphering plant-pathogen interactions at the molecular level (Sundar et al. 2015). Recently, progress in understanding of *Colletotrichum falcatum,* the fungal pathogen causing red rot in sugarcane, has been reviewed comprehensively (Viswanathan 2021a,b, Viswanathan et al. 2018).

Integrating the tools of "omics" is essential for making a comprehensive analysis of the complex interaction. Recent years have witnessed metabolomics-assisted breeding approaches successfully being employed in many crop-improvement programs. Omics as an integrated tool can be successfully used to study sugarcane behaviour in response to biotic and abiotic stresses. Robust data generated from different facets of omics tools would help in understanding physiology, function, and collinearity of sugarcane with other related crops/wild relatives. Multiple omics analysis is required to integrate omics data generated from comparable platforms, which will enhance our understanding of plant-pathogen interaction in a holistic manner. Data integration on a horizontal axis helps to map finer details obtained from different experiments so that a meaningful interpretation of complex data can be made. In spite of the recent advances made in big data management, it still remains a daunting task to analyze results obtained from different platforms. A comprehensive analysis from a holistic perspective will lead to deciphering the interaction between various components in plant-pathogen interaction, wherein "interactomics" come into play.

Disease resistance in crop plants is an enigma to be unravelled, in spite of advances made in plant biology. Understanding plant-pathogen interaction precisely is a fascinating area. It forms a basis to develop disease-resistant varieties in agriculture. The science of plant disease resistance has undergone a paradigm shift in understanding, starting with the gene-for-gene concept to the present age of guard-decoy model to decipher disease-resistant genes in crop plants. This paradigm shift in the understanding of plant disease resistance is because of the advent of robust molecular tools. An integration of the tools of "omics," namely genomics, proteomics, metabolomics, etc., has strengthened in deciphering plant-pathogen interactions at the molecular level. More specifically, genomics research with the phenomenal upsurge to the next-generation sequencing platform has proven to be robust enough to decipher complete genetic information coding for useful traits of interest. Proteomics – a complementary tool to genomics – is gaining substantial progress in many crop plants like rice, wheat, maize, sorghum, etc., besides having been well established in the model plant – *Arabidopsis*. There is an unprecedented acceleration in the pace with which this tool of "*omics*" is carried progressively to address many unresolved issues in agricultural research. It is evident that proteomics is a powerful tool in the post-genomic era and plays a key role in empowering systems biology in plants.

In sugarcane, contribution of multiple alleles to complex traits such as disease resistance is a basic question strangling the breeding efforts, which could only be addressed by the development novel/new strategies that integrate various tools of biotechnology. However, sugarcane genome sequencing is lagging way behind when compared to other related monocots, like rice, sorghum, and maize, due to its high level of polyploidy, large genome size, and complexity of chromosomes. SUCEST – an initiative project of Brazilian researchers to index large-scale expressed sequence tags (i.e >2,38,000 ESTs) in sugarcane – has revolutionized sugarcane genomics research (Vettore et al. 2001, 2003). With this momentum, functional genomics of sugarcane has progressed a lot (through gene-expression profiling) and is leading to the definition of gene networks through various tools/platforms, viz., cDNA-AFLP, SSH, SAGE, microarray, and next-generation sequencing. As far as the status of molecular basis of sugarcane disease resistance is concerned, the progress is not quite encouraging due to the hurdles stated above. So far, only one R gene, Bru1 for resistance to brown rust caused by *Puccinia melanocephala,* has been identified and functionally validated in sugarcane (Costet et al. 2012).

The current review summarizes the significant breakthroughs achieved in the major fungal pathogens of sugarcane, namely *Colletotrichum falcatum* and *Sporisorium scitaminuem,* causing red rot and smut, respectively, in sugarcane and viruses infecting sugarcane. Also, attempts were made to highlight major breakthroughs in other pathogens and phytoplasma infecting sugarcane. This is a compilation of the results achieved so far by employing various tools of omics in understanding resistance mechanism against different diseases. Most of the available information on the subject is exploitation of various genomics-based strategies, as multiple tools are utilised by sugarcane researchers globally in understanding host resistance and pathogen dynamics (Manners 2011).

6.2 SUGARCANE-*COLLETOTRICHUM FALCATUM* INTERACTION

6.2.1 Accomplishments Made in Pre-Genomics Era

By virtue of the genomic complexity, and with no clear understanding of the inheritance pattern of red rot resistance, achieving durable disease resistance in sugarcane is not yet fully accomplished. However, concerted efforts by the researchers at ICAR- Sugarcane Breeding Institute, Coimbatore have laid down a solid foundation in understanding red rot resistance in sugarcane. Most of the published information on red rot resistance in sugarcane is from the ICAR-SBI group and forms a major part of the available global literature. Substantial work progress has been made in deciphering red rot resistance in sugarcane by employing various tools of omics. Considering the significance of defense-related PR-proteins in plant-disease resistance, a focussed search was made to identify specific PR-proteins in sugarcane varieties differing in red rot resistance on a temporal level. Viswanathan et al. (2005) reported differential induction of chitinases and thaumatin-like proteins (TLPs) in sugarcane varieties differing in resistance to *Colletotrichum falcatum* and causing red rot, which indicated a possible role of these PR proteins in conferring red rot resistance in sugarcane. Sundar et al. (2008) reported accumulation of potential PR- proteins like chitinase, β-1,3-glucanase, and TLP mediated by *C. falcatum* elicitor induction in suspension-cultured cells of the red rot resistant cultivar. Apart from PR-proteins, specific accumulation of 3-deoxyanthocyanidin phytoalexins at the site of pathogen colonization in incompatible interaction has been found to restrict pathogen progress, whereas in compatible interactions, either the production of phytoalexins is delayed or the pathogen is able to degrade the chemicals (Viswanathan et al. 1996, Kumar et al. 2015, Nandakumar et al. 2021a). This was also further validated with histological assays including GFP-tagged *C. falcatum* in sugarcane varieties varying in red rot resistance. In the assays, the fungal pathogen makes unrestricted progress in its spread over the inoculated host tissues, whereas, in cases of resistant interaction, the pathogen could not progress beyond the inoculated site or with limited progress in mycelial growth (Ashwin et al. 2020a, Nandakumar et al. 2020).

6.2.2 Early Phase of Omics Era

From elucidating biochemical indices on red rot resistance, further studies on using different molecular approaches to uncover the resistance mechanism in sugarcane against *C. falcatum* were undertaken. Viswanathan et al. (2009) identified differential transcripts representing candidate genes *viz.* resistance gene analogues (RGA's), defense-related genes, transcription factors, and signaling pathway genes induced by utilising differential display (DD)-RT-PCR protocol. By employing DD-RT-PCR, differentially expressed genes in response to *C. falcatum* infection in sugarcane stalk tissue were functionally categorized and a set of candidate genes homologous to genes known to be actively involved in various pathogenesis-related functions in plant species were identified (Prathima et al. 2013). Rahul et al. (2016) standardized an *in vitro* system of using sugarcane suspension cells and a crude elicitor of *C.falcatum* for transcriptome analysis; they observed the regulation of many potential defense-related transcripts like putative chitinase, glycine-rich protein, 14-3-3-like protein, xylanase inhibitor protein, calmodulin-related protein, etc. Full-length gene sequences of 14-3-3-like protein and xylanase inhibitor were identified by adopting RACE-PCR, and the genes were characterized to domain level. The results demonstrated that the *in vitro* system of sugarcane suspension cells and Cf-elicitor mimics that of the cane tissue challenged with red rot pathogen and useful to identify defense-related traits in sugarcane against *C. falcatum*. Differential regulation of transcription factors (TFs) viz. WRKY, bZIP, MYB, NAC, and TLP upon *C. falcatum* challenge and systemic-acquired resistance (SAR) priming in sugarcane was established, and the differentially expressed TFs are found to be involved in triggering or coordinating defense against pathogen invasion (Muthiah et al. 2013). Further, molecular profiling of SAR-responsive transcripts in sugarcane challenged with *C. falcatum* revealed differential expression of 12 transcripts out of 22 putative defense-related genes. The transcripts represented predominantly defense-related phenylpropanoid pathway, resistant gene analogues (RGAs), and transcripts coding for phytoalexin synthesis, and these upregulated genes presumably play a potential role in SAR induction and might contribute to defense against C. *falcatum* (Selvaraj et al. 2014).

Viswanathan et al. (2014) indicated differential accumulation of transcripts of the flavanoid biosynthetic pathway-like coumarate-4-hydroxylase, chalcone synthase, chalcone reductase, flavanoid 3′-5′ hydroxylase, and flavanoid glycosyl transferase. This transcript analysis further confirmed the role of sugarcane phytoalexins in red rot resistance. Sathyabhama et al. (2015) employed suppression-subtractive hybridization (SSH) to study the probable existence and integration of the signals: ROS, Ca^{2+}, JA, ET, PI, and MAPKs in sugarcane and *C. falcatum* interaction. A huge set of plasma-membrane localized transporters, the secretory pathway associated transcripts and defense genes/enzymes were upregulated in this study, besides signalling. The set of ESTs generated from this study would contribute as a prime factor to understand the complex interaction between sugarcane and *C. falcatum* in more detail.

Ashwin et al. (2020b) critically reviewed the status of plant-pathogen interaction, involving a systematic and well-orchestrated event of interactions between two distinct, dynamic, and rapidly evolving biological entities. Several hypothesis/models viz. zig-zag model, refined zig-zag using bio-models, iceberg models, etc., have been proposed to explain the quantitative and qualitative outcome of molecular events of plant-pathogen interactions. Ashwin et al. (2020a) profiled the expression pattern of transcripts coding for major defense-signaling pathway-regulatory genes during compatible and incompatible interactions and in response to defense priming using qRT-PCR. Defense priming with synthetic elicitor BTH and *C. falcatum* PAMPs, CfEPL1 and CfPDIP1, showed considerable upregulation of the SA pathway-associated genes. Overall results indicated that the SA-mediated defense pathway is the most predominant pathway reprogrammed during priming with BTH, CfEPL1, and CfPDIP1 and substantiated the earlier findings that these agents indeed induce systemic resistance against *C. falcatum* in sugarcane.

6.2.3 Application of NGS Tools

With the advent of next-generation sequencing (NGS) technology, it is possible to undertake whole-genome and whole-transcriptome sequencing in a high-throughput manner. NGS was judiciously utilised to gain a better understanding of sugarcane-pathogen interaction. Recently, Nandakumar et al. (2021b) validated 20 candidate sugarcane miRNA genes involved in sugarcane defense response during their interaction with *C. falcatum*. The results indicated that there is a variation in ranking among the algorithms. This is a pioneering study on the sugarcane miRNAs as reference genes for normalizing RT-qPCR expression data involving miRNAs during sugarcane × *C. falcatum* interaction and is the first of its kind. Subsequently, the same group studied sugarcane microRNAs and the regulation of the gene targets specifically involved during compatible and incompatible interactions with red rot pathogen *C. falcatum* by adopting the NGS platform. qRT-PCR analyses of eight differentially regulated miRNAs; 26 gene transcript targets expression indicated that these miRNAs have a regulatory effect on the expression of respective target genes. The results revealed that certain miRNAs regulate many target genes that are involved in inciting early responses to the pathogen infection, signaling pathways, endoplasmic reticulum stress, and resistance gene activation through feedback response from various cellular processes during the compatible and incompatible interaction with the red rot pathogen *C. falcatum* (Nandakumar et al. 2021c). Further studies of Nandakumar et al. (2021d) on the temporal expression of selected red rot resistance-associated candidate gene transcripts in sugarcane-*C. falcatum* interaction revealed more efficient defense responses as early recognition in the resistant variety as compared to the susceptible variety. Further, the selected resistance-associated gene transcripts have the potential to be explored for use in screening sugarcane varieties for red rot resistance traits. To circumvent interference by the host varietal background in transcripts expression during the host-pathogen interaction, Viswanathan et al. (2016a) exploited differential behaviour of sugarcane cv Co 7805 to two different pathotypes. The variety behaved susceptible to the virulent pathotype Cf94012 and resistant to the pathotype Cf87012. Hence, in the SSH study, the former was used as driver and the latter as a tester. Subsequently, cDNA population after SSH was subjected to NGS analysis to study the entire resistance and susceptibility associated transcriptomes in the interaction. They have identified 7,849 and 2,899 high-quality reads with BLASTx hits for resistant and susceptible interactions, respectively. Further gene expression assays revealed that the transcripts homologous to CEBiP receptor and transcripts involved in the signals ROS, Ca^{2+}, BR, JA, and ABA were exhibited in both the responses. In addition, MAPK, ET, PI signals, and JA amino conjugation-related transcripts were found only in resistant interaction. Another interesting finding was that many pathogen-specific transcripts found only in the reverse SSH population (susceptible interaction), indicating initiation of *C. falcatum* pathogenesis during the early intervals in the interaction (Sathyabhama et al. 2022).

Kumar et al. (2020) profiled the differentially expressed proteins of red rot resistant and susceptible sugarcane cultivars through two-dimensional gel electrophoresis (2-DE) and mass spectrometry. Based on specific expression of proteins in resistant and susceptible interactions, it was hypothesised that the PAMPs/HDMPs generated during the attempted invasion of the fungus were recognized by PRRs/R-genes in the host. Upon recognition by the receptors, downstream signalling processes, including activation of calcium influx channels, induction of transcription factors (Myb and WRKY), activation of H_2O_2 production, and induction of R-gene expression, are activated. Influx of Ca^{2+} ions activates CDPK and in the presence of 14-3-3 protein leads to the generation of ROS. The ROS production is further supported by the H_2O_2 generated by SOD/NADPH-complex and rapid generation of ROS further supports the scale of downstream signalling processes. The collective action of all the above components probably results in a defense signal of magnitudes well above the threshold level of hypersensitive response, and ultimately, cell death occurs. The proteome also serves as a reference defense proteome for furthering research on sugarcane defense proteomics.

6.2.4 *C. falcatum* Genome Characterization

C. falcatum has caused historic failures of elite varieties in India starting from Co 205, the first man-made hybrid sugarcane variety to the recent epiphytotics on Co 0238. The new epiphytotics are attributed to the emergence of new variants of *C. falcatum,* with greater virulence attacks on the new varieties, which were hitherto resistant to the pathogen (Viswanathan 2021a,b, Viswanathan et al. 2021). During the last few decades, new *C. falcatum* pathotypes CF03, CF07, CF08, CF09, and CF11 from subtropical region and CF12 from the tropical states were characterized and designated (Viswanathan 2017, 2021b). Recently, the isolates caused breakdown effect on the predominant variety nicknamed as wonder cane 'Co 0238' in the subtropical states were characterized and designated as 'CF13 pathotype' (Viswanathan et al. 2022). Characterizing *C. falcatum* isolates based on pathogenicity on a set of host-differentials has been successful in the past. However, molecular variation among the highly variable isolates for pathogenicity could not be clearly established. Initially, RAPD and ITS markers were used to group *C. falcatum* isolates in India (Mohanraj et al. 2002, Suman et al. 2005) and Thailand (Sangdit et al. 2014), respectively. Grouping of the *C. falcatum* pathotypes-based vegetative compatibility grouping (VCG), pathogenicity, and ITS showed a similar pattern (Malathi et al. 2010, Viswanathan et al. 2003). However, recently, virulent strain specific simple sequence repeat (SSR) marker from the genome of *C. falcatum* was identified utilizing the NGS data (Prasanth et al. 2021). To understand pathogenicity of *C. falcatum,* detailed genomics and proteomics studies were carried out mostly at ICAR-SBI, Coimbatore.

Understanding pathogen biology comprehensively is an important step toward deciphering pathogen dynamics and variability. Given the changes in weather and pathogen profile, frequent cases of sugarcane varietal breakdown to red rot are encountered. This is attributed to the emergence of new highly virulent pathotypes of *C. falcatum.* With the advancements in molecular biology tools like next-generation sequencing platforms, it is possible to undertake whole genome and whole transcriptome sequencing of any pathogen of interest. This would help us to identify possible mutations in the pathogen genome/transcriptome, which might account for the change in pathotype. Pathogen characterization at the gene level will lead to barcoding of fungal DNA, which will assist in the choice of specific pathotype to be used for screening breeding clones for disease resistance. Pioneering work on *C. falcatum* pathogenomics by Viswanathan et al. (2016b) resulted in decoding a draft genome of *C. falcatum*, which is about 48.16 Mb in size with 12,270 genes and the results established considerable synteny to *C. graminicola* and *C. sublineola* -the causal organisms of anthracnose in maize and sorghum, respectively. The study has brought a new revelation to explore the lifestyle of this unique pathogen, which is specialized to infect sugarcane stalk tissues in detail. Subsequently, RNA-Seq technology to establish transcriptome of *C. falcatum*. The genes for virulence have been classified and grouped into candidate effectors, transition-specific and secondary metabolites, proteases, transporters, and peptidases, which revealed that *C. falcatum* transcripts encode a large number of secondary metabolites and membrane transporters. Phylogenomics analysis indicated that *C. falcatum* is closely related to *C. graminicola* and *C. sublineola* infecting related host plants, maize and sorghum, respectively (Prasanth et al. 2017). This study provided a comprehensive understanding of *C. falcatum* pathobiology and has identified many candidate genes/putative functions possibly required for its pathogenesis. We sequenced *In planta* transcriptome sequencing of *C. falcatum* under Illumina Hi-Seq platform and expanded search toward finding the small secretory proteins (SSPs) expressed during the host-pathogen interactions. These SSPs were further identified as crucial key pathogenic determinants from *in planta* transcriptome analysis. The repertoire of *C. falcatum* effectors prediction from *in vitro* and *in planta* transcriptome identified several putative genes, which are involved in biotrophy-necrotrophy in functionally diverse patterns. This will facilitate further analysis of stage specific genes, fungal pathogenicity determinants, and characterization of the expression of SSPs in *in planta* (Prasanth et al. 2019). Recently, by performing long read sequencing on PacBio RSII system 56.06 Mb assemblies with 238 contigs of *C. falcatum,* pathotype CF08 were reported by Chandra et al. (2021). They have predicted a total of 18,635 protein-coding genes, probably due to near complete coverage of the genome. About 298 glycoside hydrolases were predominant among 617 CAZymes predicted, and 77 genes having effector functions were identified from 7264 genes associated with pathogenicity/virulence. These results proved to be an important tool for further understanding on the evolutionary and functional genomics in *C. falcatum.*

6.2.5 Characterization of Pathogenicity Gene Homologs/Proteins in *C. falcatum*

To gain more insights on the genes responsible for pathogen virulence, ~28 pathogenicity gene homologs involved in various functions of pathogenesis reported from other *Colletotrichum* spp were studied in *C. falcatum* with specific primers. Phylogenetically differentiated two *C. falcatum* isolates viz. *Cf*671 and *Cf*92020 varying in virulence were employed to understand intraspecific variation. Important gene homolog's amplified include PKS1, SCD1, THR1, Cap5, Cap20, MAF1, Pg1, PG2, Pel1, Pel2, SSD1, APH1, ArpA, ICL1, RGT2, RAC1, CreA, HXT1, and HXT3. Sequence information of the selected pathogenicity genes of *C. falcatum* isolates has shown both intra- and interspecific variation. The expression pattern of the pathogenicity gene homologs has clearly exhibited a differential expression between the two isolates during host–pathogen interaction. This preliminary study confirmed the role of some pathogenicity genes and their expression in *C. falcatum* pathogenesis (Scindiya et al. 2017). In proteomic approach, differentially expressed proteins from mycelia, secreted proteins, and proteins expressed *in-planta* during host–pathogen interaction were documented by 2-D gel electrophoresis in the virulent (*Cf*671) and the least virulent (*Cf*92020) isolates. SSH libraries from the same set of isolates identified differentially expressed genes from the virulent isolate. Differentially expressed transcripts from SSH libraries matched to 27 proteins from mycelium, eight proteins from secretome, and 12 proteins that are characterized as unique proteins in the virulent isolate. Besides, 10 identified proteins were found to be common to mycelium and host-pathogen interaction and four proteins *viz.,* Bys1 family protein, two methyltransferases, and 3-isopropylmalate dehydrogenase were present in all the mycelial, secretome and host-pathogen interaction analyses. Furthermore, the candidate genes differentiated the pathogen virulence *in vitro* and *in planta* (Kaverinathan et al. 2018).

6.2.6 Functional Analysis of Pathogenicity-related Genes

Maruthachalam et al. (2008) standardized *Agrobacterium tumefaciens*-mediated transformation, an effective insertional mutagenesis molecular tool to identify and characterize pathogenicity-related genes in *C. falcatum*. Earlier, a gene knock-out approach for functional validation of chitin synthase gene CHS-1A in *C. falcatum* was demonstrated and proved its role in pathogenicity (Malathi and Viswanathan 2012). Through a reverse genetics approach, the function of a gene can be identified for understanding the molecular mechanisms involved in fungal/oomycete development *viz.*, sporulation, spore germination, appressorium or infection structure formation, appressorium morphogenesis and penetration; fungal nutrition (i.e., the uptake of nutrients such as iron, phosphorous from the host), and compatible or incompatible interactions between fungi/oomycetes and their host plants. Multinucleate nature of filamentous fungal pathogens involves isolation of homokaryotic transformants, which is a challenge for knock-out and insertional mutagenesis for studying the loss of function of a gene. Hence, RNAi, which disrupts the expression of genes by targeting their mRNA, could be useful for determining the expression of gene in the pathogens. RNAi not only elucidates the function of unknown genes but also helps to identify genes that are necessary for growth and pathogenesis of the pathogen. The advantage of this method is, it does not alter the genomic structure and also requires less sequence data (Cheng et al. 2005), making gene knockdown a more convenient and reliable approach. RNAi-mediated silencing for the functional analysis of pathogenicity-related genes in *C. falcatum* was standardized using the fungal transformation vector pSilent-1, which facilitates generation of hair-pin constructs and suppresses the expression of a gene in a sequence-specific manner at the post-transcriptional level. By this approach, functional analyses of acyl transferase (AT) and conidial pigment (CP) domains of pathogenicity-related gene polyketide synthase (PKS1) responsible for conidial pigment polyketide synthase by the knockdown method. The knockdown mutant failed to produce conidia *in vitro,* and in pathogenicity assays, the disease symptoms of both the AT and CP mutants were limited within the inoculated internode until the 15th and 5th day after inoculation, respectively (Scindiya et al. 2019). Subsequently, glucose transporter (GT) and sucrose non-fermenting 1 (SNF1) involved in carbon-source metabolism were carried out; that showed differential expression in response to virulence, and their potential role in pathogenicity in *C. falcatum* was established. Knockdown of the pathogenicity related genes GT and SNF1 in the fungal pathogen by RNAi approach validated their possible role in causing the disease in sugarcane (Scindiya et al. 2021).

Besides, to further confirm and identify roles of candidate-pathogenicity genes in *C. falcatum*, antagonistic *Trichoderma harzianum* strain T20 due to its highly mycoparasitic mechanism of antagonism. The genes' coding for cell-wall degrading enzymes, such as chitinase (Chit33), endochitinase (endo42), β-1, 3-endoglucanase (glu), exochitinase1 (exc), and exochitinase2 (exc2) of *T. harzianum* T20 expressed at a higher level during and after contact with *C. falcatum;* the protease-related genes like alkaline proteinase (prb1), trypsin-like protease (pra1), and subtilisin-like serine protease (ssp) were upregulated during contact. This study identified genes specifically expressed during mycoparasitism in the antagonistic fungus, down regulation of certain pathogenicity-associated genes in *C. falcatum* under *in vitro* and on detached cane leaves (Elamathi et al. 2018). Further, genomic and proteomic analysis of stalk samples during tritrophic interaction with sugarcane, *C. falcatum,* and *Trichoderma* showed that the differentially expressed proteins were mainly host-defense related, and simultaneously, the pathogenicity / virulence–related proteins *viz*., cytochrome p450, putative phosphopyruvate hydratase, Hsp20-like protein, copper / zinc superoxide dismutase, and hypothetical proteins reported from *Colletotrichum* spp were also obtained during the interaction of sugarcane and *C. falcatum.* Molecular analysis on tritrophic interaction among *C. falcatum, T. harzianum,* and sugarcane, the pathogenicity / virulence–related proteins *viz*., cytochrome p450 and hsp20-like proteins, hypothetical proteins of *C. falcatum* were down regulated during the interactions both *in vitro* and *in vivo* conditions (Elamathi et al. 2017, 2019).

Ashwin et al. (2017) conducted a detailed study on comparative secretome analysis of *C. falcatum*, which identified a cerato-platanin protein (EPL1) as a potential pathogen-associated molecular pattern (PAMP) inducing systemic resistance in sugarcane. Further, transcriptional expression of candidate high abundant proteins during host-pathogen interaction demonstrated a potential role of EPL1 in inducing PAMP-triggered immunity (PTI) in sugarcane. Later, Ashwin et al. (2018) identified a novel secreted protein of *C. falcatum* and characterized it as CfPDIP1, which elicited defense responses in sugarcane and triggered hypersensitive response in tobacco; the expression level of CfPDIP1 was greater in incompatible interaction than the susceptible interaction.

6.2.7 Candidate Genes for Fungicidal Targets

Apart from applying genomic tools to identify pathogenicity genes in *C. falcatum*, the omic tools were applied to identify potential targets of fungicidal action in the fungal pathogen *C. falcatum*. Among the various pathogenicity gene homologs established, the role of melanin in pathogenicity was studied in detail in *C. falcatum*, and the genes involved in melanin biosynthesis were characterized at the molecular level. The polyketide synthase (PKS1) gene involved in melanin synthesis was characterized at the molecular level. Also, the role of melanin in pathogenicity was established using melanin inhibitor tricyclazole, which suppressed conidial germination, appressorium formation, melanin synthesis, and, correspondingly, symptom production (Kaverinathan et al. 2017). Among the various pathogenicity-related proteins identified in *C. falcatum*, melanin-synthesizing enzymes and cytochrome p450 are found as potential target sites of azole fungicides. Cytochrome P450 has been identified as a potential target site for fungicidal management using azole fungicide as an inhibitor of the designated protein. Azole fungicide Nativo® (a combination of tebuconazole and strobilurin) was tested against the fungus *in vitro* and *in planta,* and qRT-PCR assays confirmed the pathogenicity gene Cytochrome p450 as the potential target for fungicide action based on the expression of genes. Furthermore, the same pathogenicity protein is suppressed during tritrophic interaction of *C. falcatum* and the antagonist *T. harzianum* in sugarcane.

6.3 SUGARCANE *S. SCITAMINEUM* INTERACTION

Unlike red rot, smut in sugarcane is a disease of global prevalence. During the past two decades, moderate to severe incidences of smut have been reported in various sugarcane-growing regions of India. With the experience gained in exploiting the tools of omics in understanding other fungal diseases of sugarcane, several researchers attempted to use genomics and proteomics tools to unravel the smut-sugarcane interaction, which is enigmatic, by virtue of the biotrophic nature of the sugarcane smut fungus. Molecular basis of smut resistance in sugarcane has been reported by many researchers worldwide. Initially, a differential-expression analysis by cDNA-AFLP of *Saccharum* spp. in response to inoculation with *S. scitamineum* was reported; a majority of the differential transcript-derived fragments (TDFs) were up-regulated in the smut-resistant cv M31/45. The studies suggested a key role for genes involved in the oxidative burst and the lignin pathways in the initial sugarcane defense against the *S. scitamineum* infection (Lao et al. 2008). Recently, functional analysis of oxidative burst in sugarcane smut-resistant and -susceptible genotypes revealed an early smut pathogen-induced modulation of the production and scavenging of ROS during defence responses in resistant sugarcane that coincided with the developmental stages of fungal growth (Peters et al. 2017). At the same time, Schaker et al. (2016) examined whether the transcriptomic changes already described during *S. scitamineum* infection in sugarcane impact changes at the metabolomic level, and the results found that energetic pathways and amino acid pools are affected throughout the interaction. Comprehensively, smut-sugarcane compatible interaction consists of a very complex mechanism involving various steps of metabolic reprogramming in response to the fungal development in plant.

Small RNA sequencing revealed a role for sugarcane miRNAs and their targets in response to *S. scitamineum* infection (Su et al. 2017). The study showed that certain predicted target genes of miR5671, miR5054, miR5783, miR5221, and miR6478 played roles in the mitogen-activated protein kinase (MAPK) signaling pathway, plant-hormone signal transduction, and plant-pathogen interaction. The findings served as an evidence for the association of miRNA expression with the molecular mechanism underlying the pathogenesis of sugarcane smut, particularly on the significance of miRNA levels in relation to the cultivation of smut-resistant sugarcane varieties. Manzoni et al. (2018) presented an overview of data gathered over the past few years using 'omics' techniques (genomics, transcriptomics, proteomics, and metabolomics) in combination with functional studies performed to identify candidate genes involved in the interaction. The study employed omics techniques and highlighted the functioning of molecular cross-talk between *S. scitamineum* and sugarcane genotypes (susceptible and resistant), which require functional validation.

Smut pathogenomics has been extensively studied elsewhere during the past decade. Initially, high-throughput tag-sequencing (tag-seq) analysis of *S. scitamineum*–infected sugarcane was performed by Solexa technology (Wu et al. 2013). Subsequently, genome sequencing of *S. scitamineum* provided insights into its pathogenesis. The genome sequence of *S. scitamineum* was assembled *de novo* with a contig N50 of 41 kb, a scaffold N50 of 884 kb, and genome size 19.8 Mb, containing an estimated 6,636 genes. From the genome of *S. scitamineum,* 192 virulence-associated genes and 68 candidate-secreted effector proteins (CSEPs) were identified, which could possibly be involved in infection and/or triggering defense responses mechanisms (Que et al. 2014). Taniguti et al. (2015) reported complete genome sequencing of individual chromosomes of *S. scitamineum* and its transcriptome profiling during biotrophic interaction with sugarcane. A considerable proportion of genes were found to be differentially expressed during a real-time interaction involving sugarcane and *S. scitamineum* at distinct developmental stages in the life cycle of the smut pathogen. Few of them are plant cell wall-degrading enzymes, proteases, lipases, chitin modification and lignin degradation enzymes, sugar transporters, and transcriptional factors. The smut pathogen also modulated transcription of genes related to surviving against reactive oxygen species and other toxic metabolites produced by the plant. Besides reported effectors, few novel genes, possibly coding for smut pathogenicity, were also proposed in the study.

Comparative transcriptome analysis of *S. scitamineum* SsΔMAT-1b, a *b* locus mutant defective in mating and pathogenicity involving the mutant strain and the wild strain, revealed critical difference in environmental signals for fungal sexual mating and filamentous growth. Experimental validation of the obtained results indicated that the bE/bW heterodimeric transcriptional factor, encoded by the b-locus, could possibly regulate *S. scitamineum* sexual mating and/or filamentous growth via modulating glucose metabolism and Hog1-mediating oxidative response (Yan et al. 2016).

In sugarcane, major pathways in response to *S. scitamineum* infection were identified by transcriptome analyses. Genes coding for hormone responsive signalling viz. auxin, abscisic acid, salicylic acid, and ethylene were induced differentially due to *S. scitamineum* inoculation. It was concluded that hypersensitive response, protein metabolism, polyamine synthesis, and cell wall formation might play an important role in host defense (Huang et al. 2018). With RNA-seq and microscopy assays, McNeil et al. (2018) analysed the resistance mechanisms in sugarcane during *S. scitamineum* infection and observed that the sugarcane response to the pathogen infection is complex, with an early response dominated by external mechanism followed by switching on the internal mechanisms. The results highlighted early induction of genes involved in the phenylpropanoid pathway in sugarcane cultivar CP74–2005 (a cultivar that possibly possesses both internal and external defence mechanisms). Genome survey for resistance-gene analogues (RGAs) in sugarcane indicated that smut resistant and susceptible genotypes recorded a distinct pattern of differential expression of the RGAs. Further, it was speculated that most of the smut-responsive RGAs in commercial varieties have their origin from chromosome 5 of the wild species *S. spontaneum*. Transmembrane domains followed by LRR family (TM-LRR) represented the most responsive up-regulated transcripts during the early phase of host-pathogen interaction, thus suggesting the relevance of an innate immune system (Rody et al. 2019).

Temporal expression of 24 potential pathogenicity-associated genes of cell-wall degrading enzymes (PCWDEs) and candidates secreted effector protein-coding (CSEPs) genes; they varied significantly between the two *S. scitamineum* isolates in their virulence (Nalayeni et al. 2021). A clear transition of distinct stages of *in vitro* life cycle and *in planta* developmental stages of *S. scitamineum* by generating stable GFP transformants was reported. Molecular tagging of GFP transformed *S. scitamineum* in sugarcane tissues during initial stages of pathogenesis uncovered the tricky life cycle of the pathogen, which consists of three distinct growth stages: diploid teliospores, haploid sporidia, and dikaryotic mycelia. The life style transition from a non-pathogenic teliospore to a pathogenic dikaryotic mycelium was clearly established by employing genomics tools. Overall, for the first time, an efficient protoplast-based transformation method was employed to depict clear-cut developmental stages *in vitro* and *in planta* using GFP-tagged strains for better understanding of *S. scitamineum* life cycle development (Agisha et al. 2021).

Proteomics, though at its infancy, provided an additional leverage to unwind the enigmatic knots in sugarcane-pathogen interactions, given the absence of genome information of sugarcane and the smut pathogen during the previous decades. Barnabas et al. (2015) critically reviewed the status of sugarcane proteomics and provided a comprehensive update on the current status, challenges, and future prospects. It was observed that with the surge in genomics research accomplished by NGS platforms, sugarcane proteomics has gained momentum. The recent decade has witnessed a paradigm shift in the pace with which sugarcane proteomics is progressing, as evident by the number of research publications. Despite proteomics of sugarcane being in its infancy, few proteomics-based investigations in sugarcane–pathogen interaction involving sugarcane and smut pathogen have led to the identification of potential defense-related proteins of sugarcane and few putative orthologous effectors of the sugarcane smut pathogen. We investigated the proteome-level alterations occurring in the meristem of a smut-susceptible sugarcane cultivar at the whip emergence stage. Differentially abundant proteins were identified by 2DE, coupled with MALDI-TOF/TOF-MS. Comprehensively, 53 sugarcane proteins identified were related to defence, stress, metabolism, protein folding, energy, and cell division; in addition, a putative effector of *S. scitamineum*, chorismate mutase, was identified. Transcript expression *vis-a-vis* the activity of phenylalanine ammonia lyase was relatively higher in the infected meristem. The abundance of seven candidate proteins in 2D gel profiles correlated with its corresponding transcript-expression levels, as validated by qRT-PCR. The study has opened up new perspectives on the interaction between sugarcane and *S. scitamineum* (Barnabas et al. 2016).

The comparative proteomics of Su et al. (2016) revealed that central metabolism changes are associated with resistance against smut in sugarcane. Their study identified 273 and 341 differentially expressed proteins in sugarcane smut-resistant (Yacheng05–179) and susceptible (ROC22) cultivars after pathogen inoculation by employing an isobaric tag for relative and absolute quantification (iTRAQ). They identified differentially expressed proteins that are closely related to smut resistance, such as β-1,3-glucanase, peroxidase, PR1, endo-1,4- β-xylanase, heat–shock protein, and lectin. Ethylene and gibberellic acid pathways, phenylpropanoid metabolism and PRs, such as PR1, PR2, PR5 and PR14, were more active in Yacheng05–179, which suggested their possible roles in sugarcane smut resistance. The sugarcane smut pathogen relies predominantly on its secretome to successfully colonise its host, in accordance with other related smut fungi. Considering the significance of deciphering its secretome, Barnabas et al. (2017) have examined alterations in the *in vitro* secretome of *S. scitamineum* in response to synthetic and sugarcane meristem tissue amended growth media, to identify host signal responsive secretory proteins and identified differentially abundant and exclusively secreted proteins. The identified proteins were related to cell wall modification, morphogenesis, polysaccharide degradation, and carbohydrate metabolism. Further, *in-planta* gene expression profiling indicated that five *in vitro* secreted proteins were expressed in distinct patterns by

S. scitamineum during different stages of infection, and probably they aid in penetration and colonisation of sugarcane cells.

In vitro and *in planta* secretomics to elucidate the intricate biotrophic interaction between sugarcane and *S. scitaminuem* indicated specific interplay of host-signalling proteins, which possibly decide the outcome of the interaction *viz.* compatible or incompatible. Leveraging proteomics to decipher the interactome of disease resistance in crop plants was initiated by many researchers worldwide. Disease resistance in sugarcane is demonstrated to be heritable and is governed by multifactorial process; however, the exact mechanism of disease resistance and a comprehensive knowledge on factors governing disease resistance is yet to be fully elucidated in sugarcane. Considering the developments in sugarcane proteomics, in spite of the absence of whole-genome information about sugarcane, it is appropriate to state that the potential of proteomics to understand disease resistance is being harnessed in a progressive manner.

6.4 POKKAH BOENG DISEASE

Molecular techniques were applied in different countries to identify and characterize *Fusarium* associated with pokkah boeng (PB). In China, *F. sacchari, F. verticillioides, F. proliferatum, F. oxysporum,* and *F. andiyazi* were reported as the casual organism; however, *F. verticillioides* is the dominant species associated with PB (Lin et al. 2014; Bao et al. 2020; Meng et al. 2020). A species-specific PCR assay was developed to detect and confirm the two species in sugarcane plants (Lin et al. 2014). The majority of 55% *Fusarium* spp associated PB with knife-cut symptoms in Iran were found to be *Fusarium verticillioides* (*F. moniliforme*), followed by *F. proliferatum, F. subglutinans,* and *F. semitectum.* Almost all the isolates were pathogenic except *F. semitectum* isolates, and among pathogenic species, *F. verticillioides* and *F. subglutinans* isolates were more pathogenic than *F. proliferatum* isolates (Khani et al. 2013). The AFLP molecular marker separated *F. proliferatum* isolates infecting maize, rice, sugarcane, and onion into four separate clusters based on their hosts. The genetic variation within *F. proliferatum* population in Iran is reported as the result of vegetative compatibility, parasexual cycle, different host varieties, and high levels of migration to the population, as well as sexual reproduction (Alizadeh et al. 2017). Under Indian conditions, association of *F. sacchari* and *F. proliferatum* was found with the disease. However, the majority of the isolates belonged to the former, and only a few were of the latter. In addition, the same *F. sacchari* was found to cause both wilt and PB in a cane variety (Viswanathan et al. 2017). Before applying molecular tools, correct identity of the pathogen(s) was unclear and by using omic tools we were able to clearly establish epidemiology of wilt and PB. Earlier by applying different molecular tools and pathogenicity assays, we clearly established *F. sacchari* as the wilt-associated pathogen in sugarcane for the first time (Viswanathan et al. 2011). Further grouping of 50 sugarcane wilt-associated *Fusarium* spp using four genomic markers differentiated *F. sacchari* with other non-pathogenic *Fusarium* and further reiterated that *F. sacchari* is the major causative agent of wilt of sugarcane (Poongothai et al. 2015). The molecular mechanisms that regulate the defense responses of sugarcane toward this fungus are not yet fully known. Wang et al. (2019) analysed transcriptomes of *F. verticillioides* inoculated sugarcane cvs YT 94/128 (resistant) and GT 37 (susceptible) and obtained 76,175 unigenes after assembling the sequencing data, and comparisons of unigenes with NR, Swiss-prot, KOG, and KEGG databases confirmed 42,451 unigenes. The analysis of differential-expression genes (DEGs) in each sample revealed 9092 DEGs in the resistant cv YT 94/128,'including 8131 up-regulated DEGs and 961 down-regulated DEGs; the respective DEGs in the susceptible cv GT 37 were 9829, 7552, and 2277. The identified DEGs were mainly involved in catalytic enzyme activity, cell protease, hydrolytic enzymes, peptide enzyme, protein metabolism process of negative regulation, phenylpropanoid metabolism, extracellular region, aldehyde dehydrogenase, endopeptidase, REDOX enzyme, protein kinases, and phosphoric acid transferase categories. Further pathway analysis indicated that the DEGs-related resistance are metabolic pathways of phenylpropanoid biosynthesis, cutin, suberine and wax biosynthesis, nitrogenous metabolism, biosynthesis of secondary metabolites, and plant-pathogen interactions. This study elucidates the mechanism of sugarcane and *F. verticillioides* interactions.

6.5 RUST

Sugarcane is seriously infected by two major rust pathogens: *Puccinia melanocephala,* causing brown rust, and *P. kuehnii,* causing orange rust in different countries (Magarey 2000, Raid and Comstock, 2000). Decades ago, the diseases were considered as minor foliar diseases; however, severe sugarcane rust epidemics in different continents led to research on rust management by conventional and molecular approaches. A major rust resistance gene *Bru*1 to common rust in a progeny derived from selfing a hybrid cv R 570 was identified, and this is the first disease-resistant gene identified from sugarcane (Daugrois et al. 1996). Molecular mechanism of host-pathogen interaction in sugarcane with rust pathogens is important; hence, efforts were made to identify molecular markers for rust resistance in different countries (Glynn et al. 2013; dos Santos et al. 2017; Yang et al. 2018). Quantitative trait loci (QTL) associated markers with the *Bru1* were applied sugarcane research toward breeding for brown rust resistance and to investigate *Bru1* occurrence in panels of sugarcane genotypes (Glynn et al. 2013; Parco et al. 2017. Three QTLs associated with orange rust (OR) resistance were detected by Yang et al. (2018). Further, they also identified putative resistance genes from sugarcane transcripts with orthologs in the sorghum genome. For that, homology searches for leucine-rich-repeat (LRR) and nucleotide binding (NB) sites were performed using five combined databases. There are also reports on identifying a marker associated with orange rust resistance for using it as a probable candidate for marker-assisted selection. Metabolomic approaches to group non-inoculated susceptible and resistant genotypes based on metabolite fingerprinting were reported (Leme et al. 2014). There were also attempts to assess gene-expression pattern after rust pathogen infections in sugarcane. After detailed RNA-Seq analyses, Correr et al. (2020) explained that *P. kuehnii* successfully establishes on sugarcane by effective suppression of the plant immune system or probably plant signaling delay to trigger defense responses in the host.

6.6 SUGARCANE VIRUSES

After applications of NGS technologies in plant virology during 2009 (Hadidi et al. 2016), their applications have been continuing in crop research across the world. It revolutionized the viral genomics mainly on the identification and discovery of novel viruses and redefining the characterization of previously identified viruses (Barba et al. 2014; Hadidi et al. 2016; Blawid et al. 2017; Pecman et al. 2017) from which the viral metagenomics or viromics arrived for the detection of all viruses present in a given sample (Roossinck et al. 2015). Since plant viruses are obligate/biotrophic pathogens, the NGS based omics technologies have led to deciphering the blueprint of the genome, the molecular mechanism of host-virus interaction networks, and functional analysis of the gene pathways/genes, etc. Prior to the arrival of NGS technologies, suppression-subtractive hybridization (SSH), expressed sequence tag (EST), and cDNA-AFLP-sequencing were highly used to revel the genome details (Muthamilarasan et al. 2019). Here, we discuss multi-omics technologies, such as genomics, transcriptomics, and proteomics approaches in sugarcane viruses and phytoplasma and their implications in crop protection and crop improvement.

6.6.1 SUGARCANE BACILLIFORM VIRUS

Sugarcane bacilliform virus (SCBV) causing leaf fleck in sugarcane belonged to the genus *Badnavirus*, family *Caulimovirridae*. Badnaviruses are plant pararetroviruses with 30 × 130 to 150 nm in size and non-enveloped bacilliform particles. It contains a circular dsDNA genome with a size of 7.5 to 8.0 kb but not covalently closed because each strand is interrupted by a site-specific

FIGURE 6.1 Sugarcane leaf fleck affected leaves showing tiny chlorotic yellow strings of fleck symptoms.

discontinuity (Lockhart 1990). The genome contains a tRNAMet-binding site, which serves as a primer for minus-strand synthesis during replication of the genome using RNA-intermediate via a virus-encoded reverse transcriptase (RT) (Bouhida et al. 1993). SCBV was first reported in Cuba in the cv B34104 during the year 1985 and later in Morocco (Autrey et al. 1992). In India, the presence of SCBV was reported in *S. officinarum* genotypes Black Tanna and Listada and a hybrid genotype D1135 during 1992 based on electron microscopy and ELISA assays (Viswanathan et al. 1996). Since then, the virus occurrence has been reported in more than 20 countries viz. Australia, Brazil, Cuba, India, and Madagascar, etc. (Viswanathan et al. 2018). SCBV mainly induce foliar symptoms on the infected plant, which is often variable as a tiny chlorotic yellow string of dots and/or flecks more clearly visible against the sunlight, or chlorotic stripes and mottling in the old leaves (+4 to +5) from the tips downward (Figure 6.1). In severe cases, the affected leaf tip and lamina turned a reddish-orange colour and start showing necrotic drying symptoms (Lockhart et al. 1996; Viswanathan et al. 2019; Balan et al. 2022).

6.6.1.1 SCBV Genomics

The first whole-genome sequence of SCBV was revealed from Sugarcane bacilliform Mor virus (SCBMV) during 1993 in U.S.A. (Bouhida et al. 1993) from a series of plasmid template sequencing by following the Sanger sequencing chemistry of the dideoxy nucleotide chain-termination method (Sanger et al. 1977). Later, an Australian isolate sugarcane bacilliform IM virus (SCBIMV) was completely sequenced by following the same methodology in 2002 (Geijskes et al. 2002). From these two complete genomes, three open reading frames (ORFs) of SCBV like other typical badnaviruses were revealed with 72% nucleotide sequence identity. Based on the whole-genome sequence, a universal badnavirus specific primers Badna FP and Badna RP were designed from the RT/RNaseH genome region (Yang et al. 2003) and is being widely used to date for SCBV diversity studies across many countries. The complete genome revealed a typical *Badnavirus* genome organization, consisting of three major ORFs in SCBV. ORF1 encodes a small protein with unknown function; ORF2 encodes a virion-associated protein (Geering and Hull 2012); ORF3 encodes a large polyprotein that is cleaved by a viral aspartic protease to produce functional proteins encoding an intercellular movement protein, a coat protein, an aspartic protease, a reverse transcriptase (RT) and a ribonuclease H (RNase H) protein (Bouhida et al. 1993; Geijskes et al. 2002; Muller et al. 2011; Karuppaiah et al. 2013).

ICTV has described two different species, SCBV-Mor and SCBV-IM, and named as SCBMV and SCBIMV, respectively (Geijskes et al. 2002; Fauquet et al. 2005). In India, five SCBV genomes were amplified by overlapping primers from the *S. officinarum* and hybrid genotypes viz. BO 91, Black Reunion, Black Tanna, Boetatoe Bilatoe, Iscambine, Ireng Maleng, and Batavia with nucleotide length of 7,553 to 7,884 nt, of which SCBV-BO91 presumed to be widely prevalent in *Saccharum* hybrid genotypes. The Indian SCBV isolates shared nucleotide identities of 69%–85% for the complete genomic sequence within themselves, indicating wide genetic diversity among them, and shared 70%–82% identity with SCBIMV from Australia and SCBMV, as well as 43%–46% identity with a Banana streak virus (BSV) and BSV-related SCBV species from Guadeloupe, indicating that the Indian SCBV isolates are distinct from other SCBV isolates. Besides, it had more than 20% nucleotide variation in RT/RNase H-region along with a variation of more than 30% in other genomic regions, which makes them all viz. SCBBBV, SCBBOV, SCBBRV, SCBIMV, and SCBMV to propose as new SCBV species from India. Around 13 SCBV phylogroups were designated based on analysis of partial RT/RNase H genomic region (SCBV-A to SCBV-M) (Muller et al. 2011; Karuppaiah et al. 2013; Rao et al. 2014), which represent a high diversity of global SCBV isolates.

Recently, the fourth-generation sequencing technology Nanopore MinION platforms (Oxford Nanopore sequencing Technologies-ONT) was used to obtain the whole genome sequence of *S. officinarum* genotype Baragua at ICAR-SBI, Coimbatore. The entire SCBV-Baragua has 7896 nucleotides, which were obtained as a very long single read in the small portable MinION devise in a short time with less cost. However, the error rate was found as high as 10%–20% with N substitution for all unknown nucleic acid residues, but with availability three technical replicates data for the sample, the error rate was rectified. The consensus sequences of NCBI Blastn analysis showed 90.50% identity with SCBIMV and 90.45% identity with SCBBRU genome. In addition, the whole-genome sequencing was performed without any errors by rolling-cycle amplification (RCA) and primer walking (Viswanathan, unpublished).

Our recent studies with partial RT - RNAase H genome broadly classified 105 isolates into 25 genotypes, of which five new genotypes from the study named as SCBV-U, V, W, X and Y and 20 other genotypes reported from all over the world (SCBV A-T). A new SCBV species (SCBV-W) was identified from the variability study based on the species demarcation criterion of >20% variation in the RT-RNase H region of Badnaviruses. Indian SCBV isolates have the most divergent population with 19 genotypes out of 25 reported genotypes all over the world. Five genotypes that were already established in other sugarcane growing countries viz., SCBV-G, SCBV-Q, SCBV-R, SCBV-S, and SCBV-T from China and France were discovered in Indian *Saccharum* populations. SCBV-U, a novel genotype, can be considered as the most frequently occurring genotype in India from *Saccharum* hybrids and interspecific spp. Intra-SCBV recombination was found as a major reason for the probable heterogeneity and exchange of gene fragments between the parental sequences for the emergence of new variants (Janiga et al. 2021).

At present, 16 SCBV whole-genome sequences are available in the NCBI genbank viz. SCBMV from Morocco (Bouhida et al. 1993); SCBIMV from Australia (Geijskes et al. 2002); Sugarcane bacilliform Guadeloupe A virus (SCBGAV) and Sugarcane bacilliform Guadeloupe D virus (SCBGDV) (Muller et al. 2011); SCBV-CHN1, -CHN2, -YG 40, -ROC 22, -WZ G, and -GX YLG from China (Wu et al. 2016, NCBI); SCBV-BO 91, SCBV-BB, SCBV-BT, SCBV-Iscam and SCBV-BRU from India (Karuppaiah et al. 2013). All the complete genome sequences are highly variable in their nucleotide sequences.

6.6.2 Sugarcane Mosaic

Among the sugarcane viral diseases, mosaic has been widely reported in all the sugarcane-growing countries (Grisham et al. 2012; Wu et al. 2012). It is caused by sugarcane mosaic virus (SCMV),

FIGURE 6.2 Leaves in the spindle show typical mosaic symptoms (Left); Sugarcane plants expressing severe mosaic symptoms of yellowing and drying of leaves(Right).

sorghum mosaic virus (*Potyvirus, Potyviridae*), and sugarcane streak mosaic virus (*Poacevirus, Potyviridae*) (Grisham 2000; Viswanathan et al. 2007). In nature, SCMV is transmitted by aphids in a non-persistent manner and also by mechanical inoculation with infected leaf sap (He et al. 2020). The major symptoms of the disease include chlorotic yellow and green stripes, blotchy chlorotic patches of light green to yellow and dark green patches more clearly visible against the sunlight; in severe infection, all the leaves exhibit pale green or yellow canopy with systemic yellowing (Figure 6.2). Among the causative viruses, SCSMV is most predominant in India (Bagyalakshmi et al. 2019a,b; Viswanathan and Karuppaiah 2010). Studies conducted during different periods at ICAR-SBI have clearly shown that mosaic significantly impact all the cane growth, physiological and yield parameters in the popular cvs Co 740, CoC 671 and CoJ 64 (Bagyalakshmi et al. 2019b).

6.6.2.1 Genomics of Mosaic Associated Viruses

SCMV contains a positive-sense single-stranded RNA molecule with a genome size of ~10 kb enveloped by a 700-750-nm long flexuous filamentous virion. SCSMV has flexuous filamentous particles of 890 X 15 nm in size with a single-stranded, positive-sense RNA genome of ~10 kb in size. The whole-genome sequence of SCSMV-IND isolates were obtained from sugarcane cv CoC 671, *Saccharum* sp. clone R-71, *S. spontaneum* SES 369 and *Saccharum* hybrid CP 52–68 from experimental fields of ICAR-SBI, Coimbatore, India (Parameswari et al. 2014; Bagyalakshmi et al. 2019a). The complete genome sequence of SCSMV-IND was determined by the first-generation Sanger-sequencing dideoxy-chain termination method by using consensus regions overlapping primers. The complete genome of SCSMV-IND had 9,786 nts excluding the poly (A) tail at 3' terminus (Table 6.1). The genome consists of 5' untranslated region of 199 nts followed by a single large ORF of 9,396 nts encoding a polyprotein of 3,131 amino acids followed by a 3' UTR of 191 nts. Nine potential cleavage sites for three viral proteinases were identified in the translated amino acid (aa) sequences of SCSMV-IND. Except the P1 and NIa proteins, all other proteins of SCSMV-IND are similar in size to that of the previously reported SCSMV isolates. The P1 protein of SCSMV-IND is a single serine protease with conserved catalytic triad H-8X-D-32X-S and P1 consensus motif FVVL(R) G at aa positions 263–1308 and 332–336, respectively. The role of P1 in the RNA-silencing suppressor has been established (Bagyalaksmi and Viswanathan 2020).

TABLE 6.1
Details of SCSMV Whole Genome Sequences

S. No	Country	SCSMV Isolates	Cultivar	Genbank Accession No
1	China	YN-YZ 211	*Saccharum* sp. hybrid cv. YZ05–211	KJ187047.1
2		HN-YZ49	*Saccharum* sp. hybrid cv. YZ05–49	KJ187048.1
3		JP 1	*Saccharum* hybrid	JF488064.1
4		JP 2	*Saccharum* hybrid	JF488065.1
5	Myanmar	MYA- Formosa	*Saccharum* sp. hybrid cv. Formosa	KJ187049.1
6		ID	*Saccharum* hybrid	JF488066.1
7	India	INDR-71	*Saccharum* sp. clone R-71	KX823356.1
8		IND 369	*Saccharum spontaneum* clone SES 369	KX823355.1
9		IND 5268	*Saccharum* hybrid cultivar CP5268	KX823354.1
10		IND 671	*Saccharum* hybrid cultivar CoC 671	JN941985.1
11		TPT	*Saccharum* hybrid	GQ246187.1
12	Iran	IR-Khuz57	*Saccharum officinarum*	KX430772.1
13		IR- Khuz6	*Saccharum officinarum*	KX430771.1
14	Pakistan	PAK	*Saccharum* hybrid	NC_014037.1
15		PAK	*Saccharum* hybrid	GQ388116.1
16	Thailand	THA-NP3	*Saccharum* hybrid	JN163911.1

The HC-Pro protein (42 kDa) of SCSMV-IND is 358 aa long and has putative protease motif C-71X-H (aa 718–790) at the C-terminal region. It also lacks the conserved motifs associated with aphid transmission at its N-terminal region and suppression of RNA silencing at the central region.

Comparative sequence analyses revealed that the SCSMV-IND shared 81%–93% nt identity with those of other completely sequenced SCSMV isolates. They also showed the highest 93% nt and 97% aa identity with SCSMV-PAK and shared only 81% nt identity with SCSMV from China, Indonesia, and Thailand, which is higher than the species-demarcation threshold value of 75%–76%. Amino acid sequence identities between SCSMV-IND and other SCSMV isolates revealed that P1 is the most variable protein (10%–12% divergence). SCSMV-IND and -Pak isolates shared the highest aa similarity in all ORFs except P1 (10% divergence), suggested the role of P1 in the host specificity (Parameswari et al. 2013).

The whole-genome sequence of SCMV-IND isolate was obtained from three different sugarcane commercial cultivars viz. CoC 671, CoN 05072, CoSi 776 and foreign hybrid CP 52–68, and genotype UB 1 by first-generation Sanger-sequencing dideoxy chain-termination method using seven sets of consensus regions overlapping primers (Bagyalakshmi et al. 2019a). The whole genomes of SCMV-IND671, -IND5268, and -INDUB1 were 9574, 9570, and 9519 nt, respectively, the isolates SCMV-INDN05072 and -INDSi776 were 9573 nt in length, excluding polyA tail (Figure 6.3). The differences in length were due to the deletions in the CP genome. The recently discovered putative protein Pretty Interesting Potyviridae (PIPO) was also found with a frame-shifting motif of G^2A^6 within the P3 protein gene in all the five isolates.

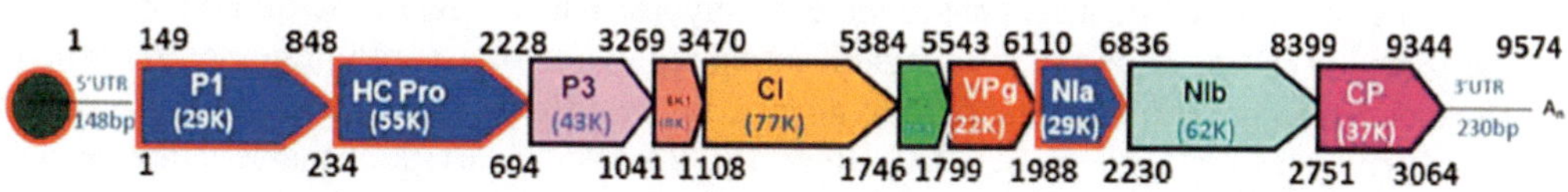

FIGURE 6.3 Schematic representation of SCMV-IND genome organization with nucleotide positions of different open reading frames encoding various proteins and their molecular weight.

Comparative genome analyses of five new isolates with previously reported SCMV full genome sequences of sugarcane, maize, sorghum, and canna by paired-end sequence-identity matrix and phylogenetic analyses have shown that Indian isolates are closely related to sugarcane-infecting isolates reported from Australia, Argentina, China, and Iran, and they diverged as a separate subgroup from other reported maize-infecting isolates from Mexico, China, Ohio, Spain, Germany, Iran, Ethiopia, Kenya, and Ecuador. Among the genomic region, CI gene has exhibited comparatively more recombination hotspots followed by HC-Pro, unlike other reported isolates (Bagyalakshmi et al. 2019a). Other than sugarcane, SCMV has been recently characterized from *Canna* sp, an ornamental crop in China by following NGS-small RNA deep-sequencing methods (Li et al. 2019). SCMV-canna with 9579 nt shared 78% nt and 85% aa sequence identities with other SCMV isolates present in the GenBank. Protein sequence analysis based on the BLASTX identified nine putative cleavage sites for the 10 mature proteins as being like other SCMV isolates.

About 108 SCMV whole-genome sequences are available in NCBI Genbank to date from Australia, Argentina, Brazil, China, Ecuador, Ethiopia, Germany, Kenya, India, Iran, the United States, Rwanda, Spain, Thailand, Mexico, and Nigeria (Cheng et al. 2002; Zhong et al. 2005; Achon et al. 2007; Viswanathan et al. 2009; Gao et al. 2012; Moradi et al. 2016; Xie et al. 2016; Tang et al. 2018; Bagyalakshmi et al. 2019; Yahaya et al. 2019). Similarly, 16 SCSMV whole genome sequences are available in NCBI Genbank till now (Table 6.1).

6.6.2.2 Sugarcane Mosaic Transcriptomics

Whole transcriptome/RNA-seq of sugarcane infected with mosaic-causing viruses (Medeiros et al. 2014; Dong et al. 2017) and differential-expressed gene profiling were used for the identification of genes and pathways related to sugarcane. Viswanathan et al. (2014b) identified and validated miRNAs encoded by SCSMV from the cv Co 86032 for better understanding of the host target genes. Plastidial isopropnoid biosynthesis was reported as metabolically targeted by the SCSMV encoded miRNA. Besides, genes encoding nonspecific lipid transfer protein 1, SNF2 domain-containing helicase, disease-resistant protein, and disease resistant RPP13-like protein 3 were reported as targets and cleavage by SCSMV miR16, by which systemic spread of host-silencing signals and systemic-acquired resistance (SAR) are lost, leading to progress in virus infection in the infected plants. Overall, SCSMV miR16 was reported as the target of many cellular transcripts for the establishment and completion of a virus life cycle and helped to overcome host defence. Later, genetic mechanisms of sugarcane in response to SCSMV infection were revealed in detail by high-throughput, paired-end RNA sequencing in which 38,505 unigenes were annotated with COG, GO, KEGG, KOG, Pfam, Swiss-Prot and Nr. Around 4982 differently expressed genes (DEGs) were screened, of which 3791 were up-regulated and 50 were down-regulated. The DEG of unbiquitin proteolytic system, proteasome, and translational pathways in endoplasmic reticulum (ER) were up-regulated in SCSMV challenged sugarcane plants compared to control (Dong et al. 2017).

Subsequently, da Silva et al. (2020) from Brazil reported SCMV-mediated transcriptomic changes in resistant and susceptible cultivars of IACSP95–5000 and IAC91–1099, respectively, by RNAseq analysis. The study revealed SCMV-mediated changes in epigenetic modifications, such as cytosine methylation pattern in resistant genotypes, and its role in gene expression regulation was analysed by differentially methylated fragments (DMF) and differentially transcribed fragments (DTFs) in both resistant and susceptible genotypes. Further, real-time qRT-PCR study of both DMF and DTF transcripts showed significant up regulation of aspartyl protease, VQ protein, stress responses, epigenetic changes, photosynthesis, lipid transport, and oxidoreductases genes in resistance genotypes (da Silva et al. 2020). Recently, RNA-seq analysis of maize lethal necrosis affected maize infected by maize chlorotic mottle virus (MCMV) and SCMV from Kenya and Ethiopia revealed 23 assembled SCMV sequences ranging from 2,191 bp to 9,632 bp (Braidwood et al. 2019).

FIGURE 6.4 Sugarcane yellow leaf symptoms (left); intense midrib yellowing clearly visible in the abaxial leaf surface (right).

6.6.3 Sugarcane Yellow Leaf Disease

Yellow leaf disease (YLD) is one of the important viral diseases caused by sugarcane yellow leaf virus (ScYLV) (*Polerovirus*, *Solemoviridae*). This disease is reported in 25 major sugarcane-growing countries, including India, Brazil, China, Colombia, Kenya, Martinique, Mauritius, Mexico, Philippines, Reunion of Islands, South Africa, Cuba, and the United States (ElSayed et al. 2015; Holkar et al. 2020; Viswanathan et al., 2009). Recently, widespread occurrence of the disease was reported from Argentina (Bertani et al. 2014), Kenya (Amata et al. 2016) and Philippines (Mendoza et al. 2020). In India, its occurrence is widespread, and the disease attained epidemic status; for that reason, large-scale crop losses were recorded (Viswanathan 2013, 2016, 2018, 2021). The disease is responsible for varietal degeneration on different varieties either alone or with sugarcane mosaic-causing viruses, thereby yield potential of the high-yielding and high-sugar varieties are severely affected (Bagyalakshmi et al. 2019; Viswanathan 2016; Viswanathan et al. 2014a). (Figure 6.4)

6.6.3.1 ScYLV Genomics

Genomic application played a major role in revealing ScYLV genome organization, gene functions, and genotypic variations from all the sugarcane-growing countries. From India, four complete genomes of the virus from the cvs Co 86032 (IND2), CoV 92102 (IND3), and CoC 85061 (IND 4), and one foreign hybrid B 38192 (IND 1) were reported (Chinnaraja et al. 2013). The length of ScYLV genome ranged from 5,612 to 5,899 nts in all the genotypes reported from different countries. Based on phylogenetic analysis of partial or full-length genome sequences of ScYLV and its geographical origin and molecular characterization, 10 different ScYLV genotypes were reported worldwide till date viz. BRA (Brazil), PER (Peru), CUB (Cuba), HAW (Hawaii), COL (Colombia), REU (Reunion Island), IND (India), CHN1, CHN2, and CHN3 (China) (ElSayed et al. 2011; Moonan and Mirkov 2002; Ahmad et al. 2006a,b; Wang and Zhou 2010; Gao et al. 2012; Wang et al. 2012; Chinnaraja et al. 2013; Lin et al. 2014). So far, 36 whole-genome sequences of ScYLV have been characterized and submitted, of which 33 are from sugarcane and three from sorghum. Seven ScYLV whole-genome sequences from France (Reunion) (Ahmad et al. 2006b; Lin et al. 2014), six from India (Chinnaraja et al. 2013; Gaur et al. 2003), three ScYLV from Mauritius (Joomun and Dookun-Saumtally 2010), three from Hawaii (Elsayed et al. 2011), one each representing from Cuba, Colombia and Brazil (Ahmad et al. 2007) and the

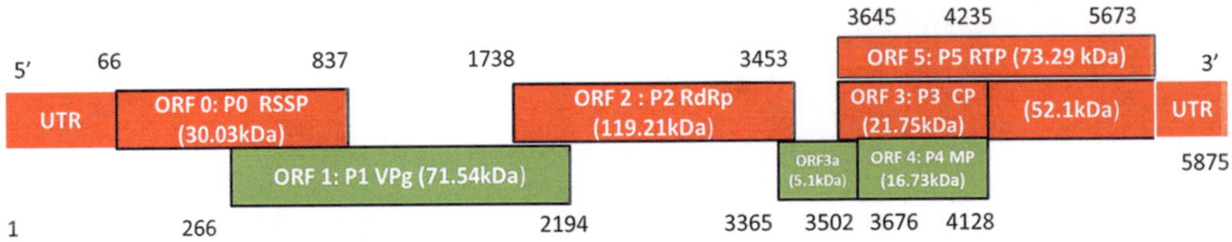

FIGURE 6.5 Schematic representation of ScYLV-IND 2 genome organization with nucleotide positions of different open reading frames encoding various proteins and their molecular weight.

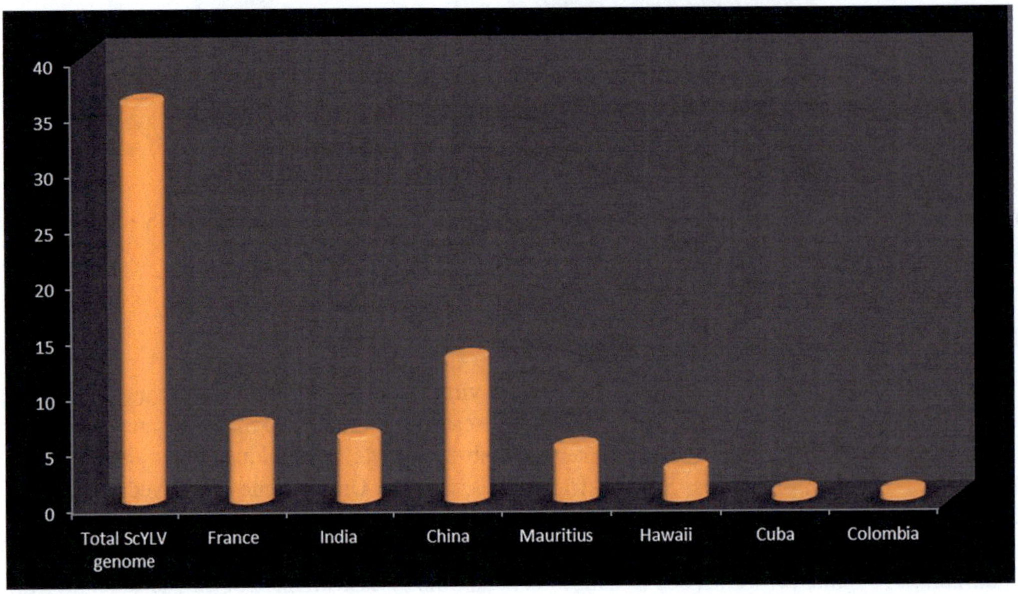

FIGURE 6.6 Graphical representation of available ScYLV genome in public database from different countries.

remaining were from United States (ElSayed et al. 2015; Moonan et al. 2000;) and China (Lin et al. 2014; Wang et al. 2012) (Figure 6.6). Mangwende et al. (2009) experimentally proved suppression of posttranscriptional gene silencing (PTGS) by ScYLV P0 and also the gene-activated cell death phenotype in the model plant.

6.6.4 Sugarcane Viral Metagenomics

The advent of metagenomic approaches led to identification and characterization of many new/ novel viruses infecting plants and animals (Delwart 2007; Roossinck 2012). In plant virology, the viability of metagenomics was taken up during 2009 with the usage of the virus-derived small-interfering RNA (siRNA) based approach (Donaire et al. 2009; Kreuze et al. 2009). Subsequently, it quickly became popular; many studies of plant viruses and viroids have been published using this technique (Wu et al. 2015), along with identification of 86 unknown plant viruses (Roossinck et al. 2015). Plant-viral metagenomics data also helps in analysing virus evolution and diversity studies, viral surveillance, developing new diagnostics and molecular viral epidemiology (Bernardo et al. 2013). Plant viral metagenomics is an effective tool for assessing the virome (the genomes of all the viruses that inhabit a particular organism/ plants of interest), which led to discover unknown/new/novel viruses existence in any plant. Using this approach, prevalence of plant viruses at the agro-ecosystem of France and South Africa were estimated as 25.8% to 35.9% (Bernardo 2014).

Recently, 19 new viruses have been reported to infect sugarcane in South Africa, Egypt, France, Nigeria, China, and USA (Boukari et al. 2020). All these new viruses have been identified/diagnosed in the respective countries using NGS-based metagenomic approaches. In sugarcane, a novel uncharacterized *Mastrevirus* was identified from the two Egyptian sugarcane plants maintained for a number of years at the CIRAD Sugarcane Quarantine Station in Montpellier, France using the siRNA- and VANA-based approaches (Candresse et al. 2014). Later, the same virus was diagnosed from 21.7% of the plant materials originated from Sudan and named as Sugarcane white streak virus (SWSV) based on the symptoms. It has been reported that the novel SWSV had escaped from the routine quarantine diagnostic assays for the last two decades, and now it has been detected in several sugarcane growing countries viz. Sudan, Barbados, Guadeloupe, and Reunion Island. During 2012, VANA-based metagenomics was used to analyse the virome of 300 sugarcane varieties and the ScYLV diversity in the Guadeloupe Island. Among the total VANA-reads, ScYLV-related reads were present in 134 samples with an infection rate of 45% (Debibakas et al. 2014; Filloux et al. 2015). The sugarcane viral metagenome study conducted during the year 2013–2014 from the commercial cultivars and germplasm collections at Florida revealed new sugarcane-infecting mastrevirus, Sugarcane striate virus and two potential new viruses in the genera *Chrysovirus* and *Umbravirus* along with high prevalence of ScYLV among the known viruses and emergence of its new genotypes (Filloux et al. 2018).

Besides, sugarcane chlorotic streak virus in Nigeria (Yahaya et al. 2017), sugarcane striate virus in Florida, Guadeloupe and Reunion of island (Boukari et al. 2017), sugarcane white streak virus and sugarcane striate virus from the germplasm material imported from Sudan at CIRAD, France (Boukari et al. 2020) and sugarcane striate virus from China (Lin et al. 2019), Maize yellow mosaic virus (*Polerovirus, Luteoviridae*) in Nigeria and China (Yahaya et al. 2017; Sun et al. 2019, NCBI), and an an undetermined umbravirus (*Tombusviridae*). Although no umbravirus was known to cause disease in sugarcane, (Tahir et al. 2017) were reported in major sugarcane-growing countries based on NGS metagenome based virome analysis.

Mollov and Malapi-Wight (2016) adopted NGS technologies (Illumina) for the detection of viral pathogens ScYLV, SCSMV, SCMV, SCBV, and SrMV in sugarcane, St. Augustine grass and Columbus grass of foreign and domestic origin maintained at the USDA quarantine PGQP and their study showed that NGS is a reliable diagnostic tool for rapid virus identification that can be applied in sugarcane quarantine, certification, and breeding programs. This highlights the NGS approach as a robust way of detecting viruses, especially in cases where primer sequences may not match viral genomic targets sufficiently to achieve enzymatic amplification.

The main challenge in metagenomic analysis has been reported as an accurate assembly of *de novo* metagenomic reads, which usually consist of a mixture of DNA from different organisms such as viruses, bacteria, fungi, host, and other endophytic organisms (Ghurye et al. 2016; Rose et al. 2016). To assemble the metagenomics dataset, several computational algorithms were developed, such as the greedy method, the overlap-layout-consensus approach, and the de-Bruijn graph approach, the last being widely used since it is based on the relationship between substrings of fixed length k (k-mers) derived from the reads (Ghurye et al. 2016). In addition, several alignment-based approaches have been proposed for identifying and classifying metagenomics reads, such as prediction of coding sequences (CDS) and annotation with domain family homologs, conserved domain database (CDD) (Marchler-Bauer et al. 2015), Pfam which could be scanned using position-specific scoring matrices (PSSMs) or hidden Markov models (HMMs) (Lobb et al. 2015). Likewise, NGS-based technologies applications in plant virology have been progressing all over the world. Several countries, like Australia, Canada, France, New Zealand, and the United States, have adopted the NGS-based metagenomics in quarantine, in addition to the existing legal diagnostic tests, and have formed an internal working group along with multiple agencies to discuss various NGS-based applications in plant protection and quarantine policy operations (Hadidi et al. 2016).

6.7 SUGARCANE GRASSY SHOOT DISEASE

Sugarcane grassy shoot (SCGS) is one of the important destructive diseases in Asia. It is characterized by proliferation of lean, lanky, chlorotic, or green tillers in numerous numbers, which gives typical grassy appearance of shoot with severe stunting. Therefore, the disease was named by its symptoms as grassy shoot disease. It occurs in all growth stages viz. germination, tillering, formative and maturity phase of the crop. However, the symptom expression depends on the phytoplasma titre in the infected plant (Viswanathan and Rao 2011; Nithya et al. 2020a). The SCGS disease is caused by SCGS-phytoplasma (*Candidatus* phytoplasma saccahri), a small pleomorphic, obligate, cell wall-less intracellular bacterial parasites. The phytoplasma was characterized and classified based on sequence analysis of highly conserved 16S RNA, and it belonged to the 16SrXI/ rice yellow dwarf (RYD) group (Nasare et al. 2007; Rao et al. 2008; Viswanathan et al. 2011). Further, based on *in silico* RFLP analysis using iPhyClassifier, which classifies the phytoplasma based on 16S rRNA gene sequences with threshold similarity coefficient of less than 0.85 and 0.97 for new group, and subgroup delineation, respectively based on the specific set of 17 restriction enzymes (Wongkeaw et al. 1997; Lee et al. 1998; Zhao et al. 2009) and actual RFLP analysis of SCGS phytoplasm XI-F was reported in India (Yadav et al. 2017). All the SCGS phytoplasma isolates from India had 99%–100% similarity with previously submitted isolates; 99.6% similarity with SCWL phytoplasma; 97.3% and 97.1% similarity with Bermuda grass white leaf (BGWL) (*Ca* P. cynodontis) and *Brachiaria* grass white leaf (BraWL) phytoplasmas respectively (Rao et al. 2008; Yadav et al. 2017; Nithya et al. 2020b). Of the 742 SCGS phytoplasma sequences, 80% are from ribosomal and the remaining 20% are from non-ribosomal regions.

The NGS technologies have revolutionized the phytoplasma genomics research in all the crops. Around 34 *Ca.* phytoplasma belonging to various groups mostly from fruits and vegetable crops have been sequenced and deposited at draft or complete genome level in the last few years. In sugarcane, SCGS phytoplasma from India was sequenced at whole genome level using NGS Illumina HiSeq platform and Oxford nanopore technology (ONT) platform. Initially, 22,765,620 reads were obtained from Illumina platform and 50,541 reads were resulted from ONT platform after quality filtering (>Q7) using GUPPY. The SCGS genome has 505,173 nt with 19.86% GC, including a putative plasmid of 2976nt and predicted with 404 protein-coding genes, 12 tRNA and two rRNA genes. The 16Sr RNA of this SCGS assembly showed 98.07% similarity with reference sequence of *Ca.* Phytoplasma cynodontis (Kirdat et al. 2020). Like sugarcane viruses, SCGS phytoplasma transcriptomics and proteomics research details are not available.

6.8 CONCLUSION

Given the paradigm shift in the post-genomics era, wherein an array of omics platforms is available to comprehensively understand plant-pathogen interactions, metabolomics seems to have ample and exciting scope in disease management by way of harnessing and integrating the upcoming potential high-throughput, robust tools of systems biology. Future applications of metabolomics may include developing metabolome assisted breeding markers to mitigate biotic/abiotic stress and to enhance crop improvement.

The developments in NGS technologies have greatly advanced our ability to comprehensively investigate the viral genome, molecular mechanism of host-virus interaction networks, identification of defence pathways/genes, prediction of population diversity, new viruses, novel genotypes, etc. *Saccharum* spp whole genome, whole transcriptome, and proteome are not available until now, rather only partial genome sequences are available. In sugarcane virology, omics approaches are in the early stage only despite the sugarcane viral-genome characterization by first-generation sequencing methods. Application of NGS in sugarcane viral transcriptomics and proteomics to elucidate plant-virus interactions are yet to be explored in India. In yellow leaf and leaf fleck diseases, NGS-based omics approaches such as genomics, transcriptomics, and proteomics applications by using second-generation sequencing platforms Illumina Hiseq, Miseq, Novaseq, and third-generation PacBIO SMRT platforms are not reported due to genomic complexity of sugarcane.

Proteomics of sugarcane was in its infancy, especially when dealing with the stalk tissues, since the first sugarcane stalk reference proteome was established at ICAR-SBI, Coimbatore (Sundar et al. 2010). A systematic proteome analysis of stalk tissue was investigated in sugarcane, wherein the stalk tissue is well known for its rigidity, fibrous nature, and the presence of oxidative enzymes, phenolic compounds, and extreme levels of carbohydrates, thus making the protein extraction complicated. A total of 36 nonredundant proteins were identified, and this is the very first basic study on sugarcane stalk-proteome analysis and laid a solid foundation for furthering sugarcane proteome research. Big data management generated from multiple omics platforms will facilitate biotechnological intervention for engineering durable disease resistance in sugarcane (Ali et al. 2019).

The robust omics technologies have been recently boomed up in the last 5–6 years due to less cost of per sample analysis. Of course, the genomics has helped to reveal the blueprint of the viral genome, thus helping the plant pathologists and biotechnologists design management strategies by *in silico* modelling using various bio informatics tools and implementing with RNAi and CRISPR Cas9 mediated delivery methods. Hopefully, more outputs will come from viral transcriptome, proteome, and metabolome analysis in the future for better understanding of the host viral pathogen interaction networks in all the *Saccharum* and related genera.

REFERENCES

Agisha, V.N., N.M.R. Ashwin, T. Vinodhini, et al. 2021. Protoplast-mediated transformation in *Sporisorium scitamineum* facilitates visualization of in planta developmental stages in sugarcane. *Molecular Biology Reports* 48: 7921–7932. 10.1007/s11033-021-06823-x.

Achon, M.A., L. Serrano, N. Alonso-Duenas and C. Porta. 2007. Complete genome sequences of Maize dwarf mosaic and Sugarcane mosaic virus isolates co infecting maize in Spain. *Archives of Virology* 152: 2073–2078.

Ahmad, Y.A., L. Costet, J.H. Daugrois et al. 2007. Variation in infection capacity and in virulence exists between genotypes of *Sugarcane yellow leaf virus*. *Plant Disease* 91: 253–259.

Ahmad, Y.A., M. Royer, J.H. Daugrois et al. 2006b. Geographical distribution of four *sugarcane yellow leaf virus* genotypes. *Plant Disease* 90: 1156–1160.

Ahmad, Y.A., L. Rassaby, M. Royer et al. 2006a. Yellow leaf of sugarcane is caused by at least three different genotypes of *sugarcane yellow leaf virus*, one of which predominates on the Island of Reunion. *Archives of Virology* 151: 1355–1371.

Ali, A., M. Khan, R. Sharif, et al. 2019. Sugarcane omics: An update on the current status of research and crop improvement. *Plants* 8(9): 344.

Alizadeh, A., M. Javan-Nikkhah, G.S. Jouzani et al. 2017. AFLP, pathogenicity and mating type analysis of Iranian *Fusarium proliferatum* isolates recovered from maize, rice, sugarcane and onion. Mycologia Iranica 4(1): 13–28, 2017, DOI: 10.22043/mi.2017.115180

Amata, R.L., E. Fernandez, D. Filloux et al. 2016. Prevalence of *Sugarcane yellow leaf virus* in sugarcane-producing regions in Kenya revealed by reverse-transcription loop-mediated isothermal amplification method. *Plant Disease* 100: 260–268.

Ashwin, N.M.R., K.V. Lakshana, D. Amalamol, et al. 2020b. Tête-à-Tête during plant-pathogen interactions: Intricacies involved and beyond. *Plant Disease Research* 35(2): 89–96.

Ashwin, N.M.R., L. Barnabas, A.R. Sundar, et al. 2017. Comparative secretome analysis of *Colletotrichum falcatum* identifies a cerato-platanin protein (EPL1) as a potential pathogen-associated molecular pattern (PAMP) inducing systemic resistance in sugarcane. *Journal of Proteomics* 169: 2–20.

Ashwin, N.M.R., L. Barnabas, A.R. Sundar, et al. 2018. CfPDIP1, a novel secreted protein of *Colletotrichum falcatum*, elicits defense responses in sugarcane and triggers hypersensitive response in tobacco. *Applied Microbiology and Biotechnology* 102: 6001–6021.

Ashwin, N.M.R., L. Barnabas, D. Amalamol, et al. 2020a. Transcriptional reprogramming of major defense-signaling pathways during defense priming and sugarcane-*Colletotrichum falcatum* interaction. *Molecular Biology Reports* 47: 8911–8923.

Autrey, L.J.C., S. Boolell, B.E.L. Lockhart et al. 1992. *Proceedings of International Society of Sugar Cane Technologists* 21: 527–541.

Bagyalakshmi, K. and R. Viswanathan. 2020. Identification of the RNA silencing suppressor activity of sugarcane streak mosaic virus P1 gene. *Virus Disease* 31(3): 333–340. 10.1007/s13337-020-00618-7

Bagyalakshmi, K., B. Parameswari, and R. Viswanathan. 2019a. Phylogenetic analysis and signature of recombination hotspots in sugarcane mosaic virus infecting sugarcane in India. *Phytoparasitica* 47: 275–291.

Bagyalakshmi, K., R. Viswanathan, and V. Ravichandran. 2019b. Impact of the viruses associated with mosaic and yellow leaf disease on varietal degeneration in sugarcane. *Phytoparasitica* 47: 591–604. 10.1007/s12600-019-00747-w

Balan, S., K. Nithya, K.A. Cherian, and R. Viswanathan. 2022. True Seed Transmission of Sugarcane bacilliform virus (SCBV) in Sugarcane. *Sugar Tech* 24: 513–521. 10.1007/s12355-021-01031-0

Bao, Y.X., Z. Huang, T.M. Li, Z.T. Yao, C.A. Powell, and M.Q. Zhang 2020. First report of *Fusarium andiyazi* causing pokkah boeng disease of sugarcane in China. *Plant Dis.* 104: 286–286.

Barba, M., H. Czosnek, and A. Hadidi. 2014. Historical Perspective, Development and Applications of Next-Generation Sequencing in Plant Virology. *Viruses* 6(1): 106–136. 10.3390/v6010106

Barnabas, L., N.M.R. Ashwin, and A.R. Sundar. 2015. Sugarcane proteomics: An update on current status, challenges, and future prospects. *Proteomics* 15: 1658–1670. 10.1002/pmic.201400463

Barnabas, L., N.M.R. Ashwin, K. Kaverinathan, et al. 2016. Proteomic analysis of a compatible interaction between sugarcane and *Sporisorium scitamineum. Proteomics* 16: 1111–1122.

Barnabas, L., N.M.R. Ashwin, K. Kaverinathan, et al. 2017. *In vitro* secretomic analysis identifies putative pathogenicity-related proteins of *Sporisorium scitamineum*–The sugarcane smut fungus. *Fungal Biology* 121: 199–211.

Bernardo, P., M. Golden, M. Akram et al. 2013. Identification and characterisation of a highly divergent geminivirus: Evolutionary and taxonomic implications. *Virus Research* 177: 35–45.

Bernardo, P. 2014. Ecology, diversity, and discovery of phytoviruses at the scale of two agro-ecosystems in a spatio-temporal framework using geometagenomics. p. 278, Doctoral thesis: Microbiology and parasitology, University Montpellier 2.

Bertani, P.R., M.F. Perera, M.E. Arias et al. 2014. A study of the sugarcane yellow leaf disease in Argentina. *Plant Disease* 98: 1036–1042.

Blawid, R., J.M.F. Silva, and T. Nagata. 2017. Discovering and sequencing new plant viral genomes by next-generation sequencing: description of a practical pipeline. *Annals of Applied Biology* 170: 301–314. 10.1111/aab.12345

Bouhida, M., B.E.L. Lockhart, and N.E. Olszewski. 1993. An analysis of the complete sequence of a sugarcane bacilliform virus genome infectious to banana and rice. *Journal of General Virology* 74(1): 15–22.

Boukari, W., D. Filloux, H.J. Daugrois et al. 2020. Molecular detection of sugarcane striate virus and sugarcane white streak virus and their prevalence in the Miami World Collection of sugarcane and related grasses. *Plant Pathology* 69: 1060–1069. 10.1111/ppa.13192

Boukari, W., R.I. Alcala-Briseno, S. Kraberger et al. 2017. Occurrence of a novel mastrevirus in sugarcane germplasm collections in Florida, Guadeloupe and Reunion. *BMC Virology Journal* 14: 146

Braidwood, L., Y. Sebastian, Muller, and D. Baulcombe. 2019. Extensive recombination challenges the utility of *Sugarcane mosaic virus* phylogeny and strain typing. *Scientific Reports* 9: 20067. 10.1038/s41598-019-56227-y

Candresse, T., D. Filloux, B. Muhire et al. 2014. Appearances can be deceptive: Revealing a hidden viral infection with deep sequencing in a plant quarantine context. *PLoS One* 9: e102945.

Chandra, A., D. Singh, D. Joshi, et al. 2021. A highly contiguous reference genome assembly for *Colletotrichum falcatum* pathotype Cf08 causing red rot disease in sugarcane. *3 Biotech* 11: 1–7.

Cheng, Y., J. Chen, and J. Chen. 2002. The complete sequence of a sugarcane mosaic virus isolate causing maize dwarf mosaic disease in China. *Science in China Series C-Life Sciences* 45: 322–330.

Cheng, P., Q. He, Q. He, et al. 2005. Regulation of the *Neurospora* circadian clock by an RNA helicase. *Genes & Development* 19: 234–241.

Chinnaraja, C., R. Viswanathan, R. Karuppaiah et al. 2013. Complete genome characterization of *sugarcane yellow leaf virus* from India: Evidence for RNA recombination. *European Journal of Plant Pathology* 135: 335–349.

Correr, F.H., G.K. Hosaka, S.G.P. Gómez, et al. 2020. Time-series expression profiling of sugarcane leaves infected with *Puccinia kuehnii* reveals an ineffective defense system leading to susceptibility. *Plant Cell Reports* 39: 873–889. 10.1007/s00299-020-02536-w

Costet, L., L.L. Cunff, S. Royaert, et al. 2012. Haplotype structure around Bru1 reveals a narrow genetic basis for brown rust resistance in modern sugarcane cultivars. *Theoretical and Applied Genetics* 125: 825–836.

da Silva, M.F., M.C. Goncalves, M.D.S. Brito et al. 2020. Sugarcane mosaic virus mediated changes in cytosine methylation pattern and differentially transcribed fragments in resistance-contrasting sugarcane genotypes. *PLoS ONE* 15(11): e0241493. 10.1371/journal.pone.0241493

Daugrois, J.H., L. Grivet, D. Roques, et al. 1996. A putative major gene for rust resistance linked with an RFLP marker in sugarcane cultivar R570. *Theoretical and Applied Genetics* 92: 1059–1064.

Debibakas, S., S. Rocher, O. Garsmeur et al. 2014. Prospecting sugarcane resistance to Sugarcane yellow leaf virus by genome-wide association. *Theoretical Applied Genetics* 127: 1719–1732.

Delwart, E.L. 2007. Viral Metagenomics. *Reviews in Medical Virology* 17/2: 115–131.

Donaire, L., Y. Wang, D. Gonzalez-Ibeas et al. 2009. Deep sequencing of plant viral small RNAs reveals effective and widespread targeting of viral genomes. *Virology* 392: 203–214.

Dong, M., G. Cheng, L. Peng et al. 2017. Transcriptome Analysis of Sugarcane Response to the Infection by *Sugarcane Streak Mosaic Virus* (SCSMV). *Tropical Plant Biology* 10: 45. 10.1007/s12042-016-9183-2

dos Santos, F.R., M.I. Zucchi, J.W. Park, et al. 2017. New sugarcane microsatellites and target region amplification polymorphism primers designed from candidate genes related to disease resistance. *Sugar Tech* 19: 219–224. 10.1007/s12355-016-0457-7

Elamathi, E., P. Malathi, R. Viswanathan, and A.R. Sundar. 2017. Molecular analysis of the suppression of *Colletotrichum falcatum* by *Trichoderma harzianum* in sugarcane. *Journal of Plant Pathology* 99: 211–218.

Elamathi, E., P. Malathi, R. Viswanathan, and A.R. Sundar. 2018. Expression analysis on mycoparasitism related genes during antagonism of *Trichoderma* with *Colletotrichum falcatum* causing red rot in sugarcane. *Journal of Plant Biochemistry and Biotechnology* 27: 351–361.

Elamathi, E., P. Malathi, R. Viswanathan, and A.R. Sundar. 2019. Identification and characterization of differentially expressed proteins from *Trichoderma harzianum* during interaction with *Colletotrichum falcatum* causing red rot in sugarcane. *Sugar Tech* 21: 765–772.

ElSayed, A.I., E. Komor, M. Boulila et al. 2015. Biology and management of *Sugarcane yellow leaf virus*: an historical overview. *Archives of Virology* 160: 2921- 1934 doi: 10.1007/s00705-015-2618-5

ElSayed, A.I., A.R..Weig, and E. Komor. 2011. Molecular characterization of Hawaiian *sugarcane yellow leaf virus* genotypes and their phylogenetic relationship to strains from other sugarcane growing countries. *European Journal of Plant Pathology* 129: 399–412.

Fauquet, C., M. Mayo, J. Maniloff et al. 2005. Virus Taxonomy: Eighth Report of the International Committee on Taxonomy of Viruses. Elsevier 318 Academic Press.

Filloux, D., E. Fernanadez, J.C. Comstock et al. 2018. Viral metagenomic-based screening of sugarcane from Florida reveals occurrence of six sugarcane-infecting viruses and high prevalence of *Sugarcane yellow leaf virus*. *Plant Disease* 102: 2317–2323. 10.1094/PDIS-04-18-0581-RE

Filloux, D., S. Dallot, A. Delaunay et al. 2015. Metagenomics approaches based on virion-associated nucleic acids (VANA): An innovative tool for assessing without a priori viral diversity of plants. *Methods in Molecular Biology* 1302: 249–257.

Gao, S.J., Y. H. Lin, Y.B Pan et al. 2012. Molecular characterization and phylogenetic analysis of Sugarcane yellow leaf virus isolates from China. *Virus Genes* 45: 340–349.

Garsmeur, O., G. Droc, R. Antonise et al. 2018. A mosaic monoploid reference sequence for the highly complex genome of sugarcane. *Nature Communications* 9(1): 2638. 10.1038/s41467-018-05051-5

Gaur, R.K., A.P. Singh, M. Singh et al. 2003. Reliability of serological identification of sugarcane mosaic potyvirus (SCMV) and sugarcane yellow leaf luteovirus (SCYLV) from cane stalk juice. *Sugar Cane International* 5: 18–21.

Geijskes, R.J., K.S. Braithwaite, J.L. Dale et al. 2002. Sequence analysis of an Australian isolate of Sugarcane bacilliform badnavirus. *Archives of Virology* 147(12): 2393–2404.

Geering, A. D. W. and Hull, R. 2012. Family *Caulimoviridae*. In Virus taxonomy classification and nomenclature of viruses ninth report of the international committee on taxonomy of viruses, Eds, King A. M. Q., Adams M. J., Carestens E. B., & Lefkowitz E. J., (pp. 424–443), Elsevier, San Diego, USA.

Ghurye, J.S., V. Cepeda-Espinoza, and M. Pop. 2016. Metagenomic assembly: Overview, challenges and applications. *Yale Journal of Biology and Medicine* 89: 353–362

Glynn, N.C., C. Laborde, R.W. Davidson, et al. 2013. Utilization of a major brown rust resistance gene in sugarcane breeding. *Molecular Breeding* 31: 323–331. 10.1007/s11032-012-9792-x

Grisham, M.P. 2000. Mosaic. In *A Guide to* Sugarcane *Diseases*, eds. P. Rott, R.A. Bailey, J.C. Comstock, and B.J. Croft, 249–254. CIRAD Publication Services: Montpellier, France.

Grisham, M. P., Maroon-Lango, C. J., and Hale, A. L. 2012. First report of *Sorghum mosaic* virus causing mosaic in *Miscanthus sinensis*. Plant Disease, 96:150. 10.1094/pdis-07-11-0617

Hadidi, A., R. Flores, T. Candresse, and M. Barba. 2016. Next-Generation Sequencing and Genome Editing in Plant Virology. *Frontiers in Microbiology* 7: 1325. 10.3389/fmicb.2016.01325

He, Z., Z. Donga, and H. Gan. 2020. Genetic changes and host adaptability in sugarcane mosaic virus based on complete genome sequences. *Molecular Phylogenetics and Evolution* 149: 106848.

Holkar, S.K., B. Parameswari, A. Kumar, K. Nithya et al. 2020. Present status and future management strategies for sugarcane yellow leaf virus: A major constraint to the global sugarcane production: An overview. *Plant Pathology Journal* 36(6): 1–22.

Huang, N., H. Ling, Y. Su, et al. 2018. Transcriptional analysis identifies major pathways as response components to *Sporisorium scitamineum* stress in sugarcane. *Gene* 678: 207–218. 10.1016/j.gene.2018.08.043

Janiga, P.K., K. Nithya, and R. Viswanathan. 2021. Emergence of novel sugarcane bacilliform virus genotypes with evidences of recombination. Proceedings of International Conference on Sugarcane Research: Sugarcane for Sugar and Beyond (CaneCon2021) p.589, Coimbatore, India.

Joomun, N. and A. Dookun-Saumtally. 2010. Occurrence of three genotypes of Sugarcane yellow leaf virus in a variety collection in Mauritius. Sugar Tech 12: 312–316.

Karuppaiah, R., R. Viswanathan, and V.G. Kumar. 2013. Genetic diversity of sugarcane bacilliform virus isolates infecting *Saccharum* spp. in India. *Virus Genes* 46: 505–516.

Kaverinathan K., P. Malathi, R. Viswanathan, and A.R. Sundar. 2018. Identification and characterization of genes/proteins related to *Colletotrichum falcatum* pathogenesis in sugarcane. ISSCT XII Pathology Workshop, Sep 03-07, 2018 – Coimbatore, India, pp. 77–78.

Kaverinathan, K., M. Scindiya, P. Malathi, et al. 2017. Role of melanin in *Colletotrichum falcatum* pathogenesis causing sugarcane red rot. *Sugar Tech* 19: 584–591.

Khani K.T., A. Alizadeh, R.F. Nejad, and A.S. Tehrani. 2013. Pathogenicity of *Fusarium proliferatum*, a new causal agent of pokkah boeng in sugarcane. Proc Intern Soc Sugar Cane Technol 28: 1–5.

Kirdat, K., B. Tiwarekar, V. Thorat et al. 2020. Draft Genome Sequences of Two Phytoplasma Strains associated with Sugarcane Grassy Shoot (SCGS) and Bermuda Grass White Leaf (BGWL) diseases. *Molecular Plant-Microbe Interaction* 33(5): 715–717 doi: 10.1094/MPMI-01-20-0005-A

Kreuze, J.F., A. Perez, M. Untiveros et al. 2009. Complete viral genome sequence and discovery of novel viruses by deep sequencing of small RNAs: A generic method for diagnosis, discovery and sequencing of viruses. *Virology* 388: 1–7.

Kumar, V.G., R. Viswanathan, P. Malathi, et al. 2015. Differential induction of 3-deoxyanthocyanidin phytoalexins in relation to *Colletotrichum falcatum* resistance in sugarcane. *Sugar Tech* 17: 314–321.

Kumar, V.G., R. Viswanathan, P. Malathi, et al. 2020. Identification of differential expressed proteins and establishing a defense proteome of sugarcane in response to *Colletotrichum falcatum* infection. *Journal of Plant Pathology* 102: 685–702.

Lao, M., A.D. Arencibia, E.R. Carmona, et al. 2008. Differential expression analysis by cDNA-AFLP of Saccharum spp. after inoculation with the host pathogen *Sporisorium scitamineum. Plant Cell Reports* 27: 1103–1111.

Leme, G.M., I.D. Coutinho, S. Creste, et al. 2014. HPLC-DAD method for metabolic fingerprinting of the phenotyping of sugarcane genotypes. *Analytical Methods* 6(19): 7781–7788.

Lin Z., S. Xu, Y. Que, J. Wang, J.C. Comstock, et al. 2014. Species-specific detection and identification of Fusarium species complex, the causal agent of sugarcane pokkah boeng in China. PLoS ONE 9(8): e104195. doi:10.1371/journal.pone.0104195

Lee, I.M., D.E. Gundersen, R.E. Davis, and I. Bartoszyk. 1998. Revised classification scheme of phytoplasmas based on RFLP analysis of 16S rRNA and ribosomal protein gene sequences. *International Journal of Systematic and Evolutionary Microbiology* 48: 1153–1169.

Li, Y., F. Xia, Y. Wang et al. 2019. Characterization of a highly divergent Sugarcane mosaic virus from *Canna indica* L. by deep sequencing. *BMC Microbiology* 19: 260. 10.1186/s12866-019-1636-y

Lin, Y.H., S.J. Gao, M.B. Damaj et al. 2014. Genome characterization of *Sugarcane yellow leaf virus* from China reveals a novel recombinant genotype. *Archives of Virology* 159: 1421–1429.

Lin, Y., N. Ali, L. Zhou et al. 2019. Identification of sugar cane striate virus and its incidence rate in Guangxi China, NCBI submissions of complete genome, Sugarcane striate virus isolate SCStV-NN-j (MH286123.1)

Lobb, B., D.A. Kurtz, G. Moreno-Hagelsieb, and A.C. Doxey. 2015. Remote homology and the functions of metagenomic dark matter. *Frontiers in Genetics* 6: 234.

Lockhart, B. E. L., Irey, M. S., and Comstock, J. C. 1996. Sugarcane bacilliform virus, Sugarcane mild mosaic virus, and Sugarcane yellow leaf syndrome. In: Sugarcane Germplasm Conservation and Exchange, Eds, Croft B.J., Piggin C.T., Wallis E.S., and Hogarth D.M., (pp. 108–112), Australian Centre for International Agricultural Research (ACIAR), Canberra.

Lockhart, B. E. L. 1990. Evidence for a double-stranded circular genome in a second group of plant viruses. *Phytopathology* 80: 127–131.

Mangwende, T., L.W. Ming, W. Borth et al. 2009. The P0 gene of *Sugarcane yellow leaf virus* encodes an RNA silencing suppressor with unique activities. *Virology* 384: 38–50

Magarey, R.C. 2000. Orange rust. In *A guide to sugarcane diseases*, eds. P. Rott, R.A. Bailey, J.C. Comstock, B.J. Croft, and A.S. Saumtally, 121–125, Montpellier: CIRAD, France.

Malathi, P. and R. Viswanathan. 2012. Identification of pathogenicity determinants in *Colletotrichum falcatum* using wild and mutant cultures. *Sugar Tech* 14: 383–390.

Malathi, P., R. Viswanathan, A.R. Sundar, et al. 2010. Variability among *Colletotrichum falcatum* pathotypes used for screening red rot resistance in sugarcane. *Sugar Cane International* 28(2): 47–52.

Manners, J.M. 2011. Functional genomics of sugarcane. *Advances in Botanical Research* 60: 89–168.

Manzoni, C., D.A., Kia, J. Vandrovcova, et al. 2018. Genome, transcriptome and proteome: the rise of omics data and their integration in biomedical sciences. Briefings in bioinformatics 19: 286–302. doi:10.1093/bib/bbw114

Marchler-Bauer, A., M.K. Derbyshire, N.R. Gonzales et al. 2015. CDD: NCBI's conserved domain database. *Nucleic Acids Research* 43: D222–D226.

Medeiros, C.N.F., M.C. Goncalves, R. Harakava et al. 2014. Sugarcane transcript profiling assessed by cDNA-AFLP analysis during the interaction with *Sugarcane mosaic virus*. *Advances in Microbiology* 4: 511–520. 10.4236/aim.2014.49057

Mendoza, J.V.S., M.S. Pinili, and M.D.C. Fe. 2020. Incidence and detection of sugarcane yellow leaf disease in the Philippines. *Australasian Plant Disease Notes* 15: 42. 10.1007/s13314-020-00412-w

Ming, R., P.H. Moore, K.K. Wu et al. 2006. Sugarcane improvement through breeding and biotechnology. *Plant Breed Reviews* 27: 15–118.

Ming, R., Moore, P. H., Wu, K.-K., D'hont, A., Glaszmann, J. C., Tew, T. L., Mirkov, T. E., da Silva, J., Jifon, J., Rai, M., Schnell, R. J., Brumbley, S.M., Lakshmanan, P., Comstock, Jack C., and Paterson, A.H.2010. Sugarcane improvement through breeding and biotechnology, Plant Breeding Reviews 27: 15–11810.1002/9780470650349.ch2.

Mollov, D. and M. Malapi-Wight. 2016. Next Generation Sequencing: a useful tool for detection of sugarcane viruses in quarantine programs. *Proceedings of International Society of Sugar Cane Technologists* 29: 631–1635

Moonan, F. and T.E. Mirkov. 2002. Analyses of genotypic diversity among North, South, and Central American isolates of *Sugarcane yellow leaf virus*: evidence for Colombian origins and for intraspecific spatial phylogenetic variation. *Journal of Virology* 76: 1339–1348.

Moradi, Z., M. Mehrvar, E. Nazifi, and M. Zakiaghl. 2016. The complete genome sequences of two naturally occurring recombinant isolates of Sugarcane mosaic virus from Iran. *Virus Genes* 52: 270–280.

Moonan, F., J. Molina, and T.E. Mirkov. 2000. *Sugarcane yellow leaf virus*: an emerging virus that has evolved by recombination between luteoviral and poleroviral ancestors. *Virology* 269: 156–171.

Muller, E., V. Dupuy, L. Blondin et al. 2011. High molecular variability of Sugarcane bacilliform viruses in Guadeloupe implying the existence of at least three new species. *Virus Research* 160: 414–419.

Muthamilarasan, M., N.K. Singh, and M. Prasad. 2019. Multi-omics approaches for strategic improvement of stress tolerance in underutilized crop species: a climate change perspective. *Advances in Genetics* 103: 1–38. doi: 10.1016/bs.adgen.2019.01.001

Maruthachalam, K., N.V. Nair, H.-S. Rho, et al. 2008. *Agrobacterium tumefaciens*-mediated transformation in *Colletotrichum falcatum* and *C. acutatum*. *Journal of Microbiology and Biotechnology* 18: 234–241.

McNeil, M.D., S.A. Bhuiyan, P.J. Berkman, et al. 2018. Analysis of the resistance mechanisms in sugarcane during *Sporisorium scitamineum* infection using RNA-seq and microscopy. *PLoS One* 13: e0197840.

Meng J.R., H.J. Huang, Y.X. Li, Y.J. Li, J.Q. Li, B.S. Chen. 2020. First report of *Fusarium sacchari* causing sugarcane pokkah boeng in China. Plant Dis. 10.1094/PDIS-05-19-0906-PDN

Mohanraj, D., S. Kumaresan, and T.V. Sreenivasan. 2002. Molecular characterisation of isolates of the sugarcane red rot pathogen. *Indian Phytopathology* 55: 147–151.

Muthiah, M., N.M.R. Ashwin, A.R. Sundar, et al. 2013. Expression profiling of transcription factors (TFs) in sugarcane X *Colletotrichum falcatum* interaction. *Journal of Plant Biochemistry and Biotechnology* 22: 286–294.

Nasare, K., A. Yadav, A.K. Singh et al. 2007. Molecular and symptom analysis reveal the presence of new phytoplasmas associated with sugarcane grassy shoot disease in India. *Plant Disease* 91: 1413–1418.

Nalayeni, K., N.M.R. Ashwin, L. Barnabas, et al. 2021. Comparative expression analysis of potential pathogenicity-associated genes of high- and low-virulent *Sporisorium scitamineum* isolates during interaction with sugarcane. *3 Biotech* 11(7): 1–14.

Nandakumar, M., P. Malathi, A.R. Sundar, and R. Viswanathan. 2020. Use of green fluorescent protein expressing *Colletotrichum falcatum*, the red rot pathogen for precise host–pathogen interaction studies in sugarcane. *Sugar Tech* 22: 112–121.

Nandakumar, M., P. Malathi, A.R. Sundar, and R. Viswanathan. 2021a. Host-pathogen interaction in sugarcane and red rot pathogen: exploring expression of phytoalexin biosynthesis pathway genes. *Indian Phytopathology* 74: 529–535.

Nandakumar, M., R. Viswanathan, P. Malathi, and A.R. Sundar. 2021b. Selection of reference genes for normalization of microRNA expression in sugarcane stalks during its interaction with *Colletotrichum falcatum. 3 Biotech* 11: 1–9.

Nandakumar, M., P. Malathi, A.R. Sundar, et al. 2021c. Role of miRNAs in the host–pathogen interaction between sugarcane and *Colletotrichum falcatum*, the red rot pathogen. *Plant Cell Reports* 40: 851–870.

Nandakumar M., P. Malathi, A.R. Sundar, and R. Viswanathan. 2021d. Expression analyses of resistance-associated candidate genes during sugarcane-*Colletotrichum falcatum* Went interaction. *Sugar Tech* 23: 1056–1063.

Nithya, K., B. Parameswari, A. Bertaccini et al. 2020a. Grassy shoot: the destructive disease of sugarcane. *Phytopathogenic Mollicutes* 10 (1): 10–24

Nithya, K., B. Parameswari, and R. Viswanathan. 2020b. Mixed infection of *Sugarcane yellow leaf virus* and Grassy Shoot Phytoplasma in yellow leaf affected Indian sugarcane cultivars. *Plant Pathology Journal* 36(4): 364–377.

Parameswari, B., K. Bagyalakshmi, C. Chinnaraja, and R. Viswanathan. 2014. Molecular characterization of Indian Sugarcane streak mosaic virus isolates reveals recombination and negative selection in the P1 gene. *Gene* 552: 199–203.

Parco, S.A., A.L. Hale, M.C. Avellaneda, et al. 2017. Distribution and frequency of Bru1, a major brown rust resistance gene, in the sugarcane world collection. *Plant Breeding* 136: 637–6511–15. 10.1111/pbr.12508.

Parameswari, B., K. Bagyalakshmi, R. Viswanathan, and C. Chinnaraja. 2013. Molecular characterization of Indian sugarcane streak mosaic virus isolates. *Virus Genes* 46(1): 186–189. doi: 10.1007/s11262-012-0827-5

Pecman, A., D. Kutnjak, I. Gutierrez-Aguirre et al. 2017. Next Generation Sequencing for Detection and Discovery of Plant Viruses and Viroids: Comparison of Two Approaches. *Frontiers in Microbiology*. 10.3389/fmicb.2017.01998

Peters, L.P., G. Carvalho, M.B. Vilhena, et al. 2017. Functional analysis of oxidative burst in sugarcane smut-resistant and-susceptible genotypes. *Planta* 245: 749–764.

Poongothai, M., R. Viswanathan, P. Malathi, A. Ramesh Sundar, C. Naveen Prasanth and C.G. Balaji. 2015. Genetic relatedness among Fusarium populations associated with sugarcane wilt in India: bridging molecular variability and phylogenetic diversity. *Journal of Sugarcane Research* 5(1): 33–48.

Prasanth, C.N., R. Viswanathan, N. Krishna, et al. 2017. Unraveling the genetic complexities in gene set of sugarcane red rot pathogen *Colletotrichum falcatum* through transcriptomic approach. *Sugar Tech* 19: 604–615.

Prasanth, C.N., R. Viswanathan, P. Malathi, and A.R. Sundar. 2019. Comparative transcriptome analysis of candidate secretory effector proteins from *Colletotrichum falcatum* infecting sugarcane. *Agri Gene* 13: 100089.

Prasanth, C.N., R. Viswanathan, P. Malathi, and A.R. Sundar. 2021. Development and characterization of genomic SSR marker for virulent strain-specific *Colletotrichum falcatum* infecting sugarcane. *3 Biotech* 11: 1–10.

Prathima, P.T., M. Raveendran, K.K. Kumar, et al. 2013. Differential regulation of defense-related gene expression in response to red rot pathogen *Colletotrichum falcatum* infection in sugarcane. *Applied Biochemistry and Biotechnology* 171: 488–503.

Que, Y., L. Xu, Q. Wu, et al. 2014. Genome sequencing of *Sporisorium scitamineum* provides insights into the pathogenic mechanisms of sugarcane smut. *BMC genomics* 15: 1–20.

Rahul, P.R., V.G. Kumar, R. Viswanathan, et al. 2016. Defense transcriptome analysis of sugarcane and *Colletotrichum falcatum* interaction using host suspension cells and pathogen elicitor. *Sugar Tech* 18: 16–28.

Raid, R.N. and J.C. Comstock. 2000. Common rust In: A guide to sugarcane diseases, eds, Rott P, Bailey RA, Comstock J.C, Croft B J and Saumtally AS, pp 85–88, CIRAD and ISSCT CIRAD publications, Montpellier, France.

Rao, G.P., S.K. Sharma, D. Singh et al. 2014. Genetically diverse variants of sugarcane bacilliform virus infecting sugarcane in India and evidence of a novel recombinant badnavirus variant. *Journal of Phytopathology* 162(11-12): 779–787.

Rao, G.P., S. Srivastava, and P.S. Gupta. 2008. Detection of sugarcane grassy shoot phytoplasma infecting sugarcane in India and its phylogenetic relationships to closely related phytoplasmas. *Sugar Tech* 10: 74–80.

Rico, A., S.L. McCraw, and G.M. Preston. 2011. The metabolic interface between *Pseudomonas syringae* and plant cells. *Current Opinion in Microbiology* 14: 31–38.

Rody, H.V.S., R.G.H. Bombardelli, S. Creste, et al. 2019. Genome survey of resistance gene analogs in sugarcane: genomic features and differential expression of the innate immune system from a smut-resistant genotype. *BMC Genomics* 20: 1–17.

Roossinck, M.J. 2012. Plant virus metagenomics: Biodiversity and ecology. *Annual Review of Genetics* 46: 359–369.

Roossinck, M.J., D.P. Martin, and P. Roumagnac. 2015. Plant virus metagenomics: Advances in virus discovery. *Phytopathology* 105: 716–727.

Rose, R., B. Constantinides, and A. Tapinos et al. 2016. Challenges in the analysis of viral metagenomes. *Virus Evolution* 2: vew022.

Sangdit, P., C. Leksomboon, and R. Lertsrutaiyotin. 2014. Cultural, morphological and pathological characterization of *Colletotrichum falcatum* causing red rot disease of sugarcane in Thailand. *Agriculture and Natural Resources* 48: 880–892.

Sanger, F., S. Nicklen and A.R. Coulson. 1977. DNA sequencing with chain-terminating inhibitors. Proceedings of the National Academy of Sciences of the United States of America. 74: 5463–5467 HYPERLINK "https://en.wikipedia.org/wiki/Doi_(identifier)"doi:10.1073/pnas.74.12.5463.

Sanju B., K. Nithya, K. Anita Cherian, and R. Viswanathan. 2022. True seed transmission of Sugarcane bacilliform virus (SCBV) in sugarcane. Sugar Tech 24: 513–521 10.1007/s12355-021-01031-0

Sathyabhama, M., R. Viswanathan, C.N. Prasanth, et al. 2022. Differential host responses of sugarcane to Colletotrichum falcatum reveal activation of probable effector triggered immunity (ETI) in defence responses. *Plant Cell Reports* 41:1461–1476 DOI: 10.1007/s00299-022-02870-1

Sathyabhama, M., R. Viswanathan, M. Nandakumar, et al. 2015. Understanding sugarcane defence responses during the initial phase of *Colletotrichum falcatum* pathogenesis by suppression subtractive hybridization (SSH). *Physiological and Molecular Plant Pathology* 91: 131–140.

Schaker, P.D.C., A.C. Palhares, L.M. Taniguti, et al. 2016. RNA-seq transcriptional profiling following whip development in sugarcane smut disease. *PLoS One* 11: e0162237.

Scindiya, M., P. Malathi, K. Kaverinathan, et al. 2017. Molecular characterization of pathogenicity gene homologs in *Colletotrichum falcatum* causing red rot in sugarcane. *Sugar Tech* 19: 563–572.

Scindiya, M., P. Malathi, K. Kaverinathan, et al. 2019. RNA-mediated silencing of PKS1 gene in Colletotrichum falcatum causing red rot in sugarcane. *European Journal of Plant Pathology* 153: 371–384.

Scindiya, M., P. Malathi, K. Kaverinathan, et al. 2021. Knock-down of glucose transporter and sucrose non-fermenting gene in the hemibiotrophic fungus *Colletotrichum falcatum* causing sugarcane red rot. *Molecular Biology Reports* 48: 2053–2061.

Selvaraj, N., R. Ashwin, A.R. Sundar, et al. 2014. Molecular profiling of systemic acquired resistance (SAR)-responsive transcripts in sugarcane challenged with *Colletotrichum falcatum*. *Applied Biochemistry and Biotechnology* 174: 2839–2850.

Su, Y., L. Xu, Z. Wang, et al. 2016. Comparative proteomics reveals that central metabolism changes are associated with resistance against *Sporisorium scitamineum* in sugarcane. *BMC Genomics* 17: 1–21.

Su, Y., Y. Zhang, N. Huang, et al. 2017. Small RNA sequencing reveals a role for sugarcane miRNAs and their targets in response to *Sporisorium scitamineum* infection. *BMC Genomics* 18: 1–19.

Suman, A., S. Lal, A.K. Shasany, A. Gaur, and P. Singh. 2005. Molecular assessment of diversity among pathotypes of *Colletotrichum falcatum* prevalent in sub-tropical Indian sugarcane. *World Journal of Microbiology and Biotechnology* 21: 1135–1140.

Sun, S.R., J.S. Chen, J. Yang, et al. 2019. First report of *Maize yellow mosaic virus* infecting sugarcane in China. *Plant Disease* 103: 2482.

Sundar, A.R., N.M.R. Ashwin, E.L. Barnabas, P. Malathi, and R. Viswanathan. 2015. Disease resistance in sugarcane–An overview. *Scientia Agraria Paranaensis* 14: 200–212.

Sundar, A.R., R. Velazhahan, S. Nagarathinam, and P. Vidhyasekaran. 2008. Induction of pathogenesis-related proteins in sugarcane leaves and cell-cultures by a glycoprotein elicitor isolated from *Colletotrichum falcatum*. *Biologia Plantarum* 52: 321–328.

Sundar, A.R., S. Nagarathinam, V.G. Kumar, et al. 2010. Sugarcane proteomics: Establishment of a protein extraction method for 2-DE in stalk tissues and initiation of sugarcane proteome reference map. *Electrophoresis* 31: 1959–1974.

Tang, W., Z. Y. Yan, T. S. Zhu et al. 2018. The complete genomic sequence of Sugarcane mosaic virus from Canna spp. in China. *Virology Journal* 15:1–4.

Tahir, M.N., S. Grinstead, and D.S. Mollov. 2017. New virus from the family *Tombusviridae* infecting sugarcane (Abstr.). *Phytopathology* 107:S5.98

Taniguti, L.M., P.D.C. Schaker, J. Benevenuto, et al. 2015. Complete genome sequence of *Sporisorium scitamineum* and biotrophic interaction transcriptome with sugarcane. *PloS One* 10: e0129318.

Vettore, A.L., F.R. da Silva, E.L. Kemper, and P. Arruda. 2001. The libraries that made SUCEST. *Genetics and Molecular Biology* 24: 1–7.

Vettore, A.L., F.R. da Silva, E.L. Kemper, et al. 2003. Analysis and functional annotation of an expressed sequence tag collection for tropical crop sugarcane. *Genome Research* 13: 2725–2735.

Viswanathan, C., J. Anburaj, and G. Prabu. 2014a. Identification and validation of sugarcane streak mosaic virus encoded microRNAs and their targets in sugarcane. *Plant Cell Reports* 33: 265–276. DOI 10.1 007/s00299-013-1527-x

Viswanathan, R. 2013. Sustainable eco-friendly disease management systems in sugarcane production under the changing climate a review. *Journal of Mycology and Plant Pathology* 43: 12–27

Viswanathan, R. 2016. Varietal degeneration in sugarcane and its management in India. *Sugar Tech* 18: 1–7.

Viswanathan, R. 2017. Pathogen virulence in sugarcane red rot pathogen versus varieties in cultivation: Classical case of loss in virulence in the pathotype CF06 (Cf671). *Sugar Tech* 19: 293–299

Viswanathan, R. 2018. Changing scenario of sugarcane diseases in India since introduction of hybrid cane varieties: path travelled for a century. *Journal of Sugarcane Research* 8(1): 1–35

Viswanathan, R. 2021a. Red rot of sugarcane (*Colletotrichum falcatum* Went). *CAB Reviews* 16, No. 023, doi: 10.1079/PAVSNNR202116023

Viswanathan, R. 2021b. Sustainable sugarcane cultivation in India through threats of red rot by varietal management. *Sugar Tech* 23: 239–253 DOI: 10.1007/s12355-020-00882-3

Viswanathan, R. 2021c. Impact of yellow leaf disease in sugarcane and its successful disease management to sustain crop production. *Indian Phytopathology* 74: 573–586. 10.1007/s42360-021-00391-7

Viswanathan, R., K.C. Alexander, and I.D. Garg. 1996. Detection of sugarcane bacilliform virus in sugarcane germplasm. *Acta Virologica* 40: 5–8

Viswanathan, R., M. Balamuralikrishnan, and R. Karuppaiah. 2007. Sugarcane mosaic in India: A cause of combined infection of Sugarcane mosaic virus and Sugarcane streak mosaic virus. *Sugar Cane International* 25: 6–14.

Viswanathan, R., C. Chinnaraja, P. Malathi et al. 2014b. Impact of *sugarcane yellow leaf virus* (ScYLV) infection on physiological efficiency and growth parameters in sugarcane in India. *Acta Physiologiae Plantarum* 36: 1805–1822.

Viswanathan, R., C. Chinnaraja, R. Karuppaiah et al. 2011. Genetic diversity of sugarcane grassy shoot (SCGS)-phytoplasmas causing grassy shoot disease in India. *Sugar Tech* 13: 220–228.

Viswanathan, R. and R. Karuppaiah. 2010. Distribution pattern of RNA viruses causing mosaic symptoms and yellow leaf in Indian sugarcane varieties. *Sugar Cane International* 28(5): 202–205.

Viswanathan, R., R. Karuppaiah, K. Bagyalakshmi, et al. 2019. Emergence of leaf fleck caused by *Sugarcane bacilliform virus* in sugarcane as a serious disease under field conditions in India. International Sugar Journal 121: 146–153.

Viswanathan, R., P. Malathi, and P. Padmanaban. 2003. Variation in sugarcane red rot pathogen *Colletotrichum falcatum* Went. In *Frontiers of Fungal Diversity in India*. 639–667, eds. G.P. Rao, C. Manoharachari, D.J. Bhat, R.C. Rajak and T.N. Lakhanpal, International Book Distributing Co: Lucknow, India.

Viswanathan, R., D. Mohanraj, P. Padmanaban, and K.C. Alexander. 1996. Synthesis of phytoalexins in sugarcane in response to infection by *Colletotrichum falcatum* Went. *Acta Phytopathologica et Entomologica Hungarica* 31: 229–237.

Viswanathan, R., B. Parameswari, and K. Nithya. 2018. Molecular characterization of sugarcane viruses and their diagnostics. In *Crop Improvement through Microbial Biotechnology*, eds. R. Prasad, S.S. Gill and N. Tuteja, 175–193, Elsevier: Amsterdam.

Viswanathan, R., C.N. Prasanth, P. Malathi, and A.R. Sundar. 2016b. Draft genome sequence of *Colletotrichum falcatum*-A prelude on screening of red rot pathogen in sugarcane. *Journal of Genomics* 4: 1–3. doi: 10.7150/jgen.13585

Viswanathan, R., and G.P. Rao. 2011. Disease scenario and management of major sugarcane diseases in India. *Sugar Tech* 13: 336–353.

Viswanathan, R., M. Sathyabhama, P. Malathi, and A.R. Sundar. 2016a. Transcriptome analysis of host-pathogen interaction between sugarcane and *Colletotrichum falcatum* by Suppression Subtractive Hybridization and Illumina sequencing. *Proceedings of International Society of Sugar Cane Technologists* 29: 1639–1644.

Viswanathan, R., S.P. Singh, R. Selvakumar, et al. 2022. Varietal break down to red rot in the sugarcane variety Co 0238 mimics Vertifolia effect: characterizing new *Colletotrichum falcatum* pathotype CF13. *Sugar Tech* 24: 1479–1496 10.1007/s12355-021-01070-7

Viswanathan, R., A.R. Sundar, P. Malathi, et al. 2005. Differential induction of chitinases and thaumatin-like proteins in sugarcane in response to infection by *Colletotrichum falcatum* causing red rot disease. *Journal of Plant Diseases and Protection* 112: 417–425.

Viswanathan, R., A.R. Sundar, P. Malathi, et al. 2009. Interaction between sugarcane and *Colletotrichum falcatum* causing red rot: Understanding disease resistance at transcription level. *Sugar Tech* 11: 44–50.

Viswanathan, R., R. Karuppaiah, and M. Balamuralikrishnan. 2009. Identification of new variants of SCMV causing sugarcane mosaic in India and assessing their genetic diversity in relation to SCMV type strains. *Virus Genes* 39: 375–386.

Viswanathan, R., A.R. Sundar, P. Malathi, et al. 2014. Molecular understanding of red rot resistance in sugarcane through genomics and proteomics. *Annual Review of Plant Pathology* 6: 189–213.

Viswanathan, R., A.R. Sundar, R. Selvakumar, and P. Malathi. 2018. Progress in understanding fungal diseases affecting sugarcane: red rot. In *Achieving sustainable cultivation of sugarcane Volume 2: Breeding, pests and diseases*, ed.Rott, P., 201–220, Burleigh Dodds Science Publishing: Cambridge, UK.

Wang Z., Y. Li, C. Li, X. Song, Y. Lei, Y. Gao, and Q. Liang. 2019. Comparative transcriptome profiling of resistant and susceptible sugarcane genotypes in response to the airborne pathogen *Fusarium verticillioides*. Molecular Biology Reports. 10.1007/s11033-019-04820-9

Wu, Q., L. Xu, J. Guo, Y. Su, and Y. Que. 2013. Transcriptome profile analysis of sugarcane responses to *Sporisorium scitaminea* infection using Solexa sequencing technology. *BioMed Research international*. Article ID 298920 10.1155/2013/298920

Wang, M.Q., D.L. Xu, R. Li, and G.H. Zhou. 2012. Genotype identification and genetic diversity of *Sugarcane yellow leaf virus* in China. *Plant Pathology* 61: 986–993.

Wang, M.Q. and G.H. Zhou. 2010. A near-complete genome sequence of a distinct isolate of *Sugarcane yellow leaf virus* from China, representing a sixth new genotype. *Virus Genes* 41: 268–272.

Wongkeaw, P., Y. Hanboonsong, C. Sirithorn et al. 1997. Differentiation of phytoplasmas associated with sugarcane and gramineous weed white leaf disease and sugarcane grassy shoot disease by RELP and sequencing. *Theoretical and Applied Genetics* 95: 660–663.

Wu, L., X. Zu, S. Wang, and Y. Chen. 2012. Sugarcane mosaic virus - Long history but still a threat to industry. *Crop Protection* 42: 74–78.

Wu, Q., S.W. Ding, Y. Zhang, and S. Zhu. 2015. Identification of viruses and viroids by next-generation sequencing and homology-dependent and homology-independent algorithms. *Annual Review of Phytopathology* 53: 425–444.

Wu, X.B., O.J. Alabi, M.B. Damaj, et al. 2016. Prevalence and RT/RNase H genealogy of sugarcane bacilliform virus isolates from China. *Journal of Phytopathology* 164: 595–607.

Xie, X., W. Chen, Q. Fu et al. 2016. Molecular variability and distribution of Sugarcane mosaic virus in Shanxi. China. *PLoS One* 11: e0151549.

Yadav, A., V. Thorat, S. Deokule et al. 2017. New subgroup 16SrXI-F phytoplasma strain associated with sugarcane grassy shoot (SCGS) disease in India. *International Journal of Systematic and Evolutionary Microbiology* 67: 374–378.

Yahaya, A., D.B. Dangora, M.D. Alegbejo et al. 2017. Identification and molecular characterization of a novel sugarcane streak mastrevirus and an isolate of the A-strain of maize streak virus from sugarcane in Nigeria. *Archives of Virology* 162: 597–602.

Yahaya, A., D.B. Dangora, P.L. Kumar et al. 2019. Prevalence and genome characterization of field isolates of Sugarcane mosaic virus (SCMV) in Nigeria. *Plant Disease* 103: 818–824.

Yan, M., W. Dai, E. Cai, et al. 2016. Transcriptome analysis of *Sporisorium scitamineum* reveals critical environmental signals for fungal sexual mating and filamentous growth. *BMC Genomics* 17: 1–11.

Yang, I.C., G.J. Hafner, P.A. Revill et al. 2003. Sequence diversity of South Pacific isolates of Taro bacilliform virus and the development of a PCR-based diagnostic test. *Archives of Virology* 148: 1957–1968.

Yang, X., J. Song, Q. You et al. 2017. Mining sequence variations in representative polyploid sugarcane germplasm accessions. *BMC Genomics* 18: 594.

Yang, X., M.S. Islam, S. Sood, et al. 2018. Identifying quantitative trait loci (QTLs) and developing diagnostic markers linked to orange rust resistance in sugarcane (*Saccharum* spp). *Frontiers in Plant Sciences* 9: 1–10. 10.3389/fpls.2018.00350

Yangm, X., S. Sood, N. Glynn. 2017. Constructing high-density genetic maps for polyploid sugarcane (*Saccharum* spp.) and identifying quantitative trait loci controlling brown rust resistance. *Molecular Breeding* 37(10). 10.1007/s11032-017-0716-7

Zhang, J., X. Zhang, H. Tang et al. 2018. Allele-defined genome of the autopolyploid sugarcane *Saccharum spontaneum* L. *Nature Genetics* 50(11): 1565–1573. 10.1038/s41588-018-0237-2

Zhong, Y., A. Guo, C. Li et al. 2005. Identification of a naturally occurring recombinant isolate of Sugarcane mosaic virus causing maize dwarf mosaic disease. *Virus Genes* 30: 75–83.

Zhao, Y., W. Wei, I.M. Lee et al. 2009. Construction of an interactive online phytoplasma classification tool, iPhyClassifier, and its application in analysis of the peach X-disease phytoplasma group (16SrIII). *International Journal of Systematic and Evolutionary Microbiology* 59: 2582–2593.

7 Genomics-Assisted Precision Breeding for Drought Tolerance in Sugarcane

Swapan K. Tripathy
Department of Agricultural Biotechnology, Odisha University of Agriculture and Technology, College of Agriculture, Siripur Sq., Bhubaneswar, Odisha, India

CONTENTS

7.1 INTRODUCTION

Sugarcane (*Saccharum officinarum* L., Family-Poaceae, Tribe-Andropogoneae) is an important cash crop, and it accounts for 70% of the world's sugar. The crop, being a C4 type, fixes 2–3% of the solar radiation and produces 100 tonnes of green matter every year, which is more than twice the yield of most other commercial crops. Therefore, it also has enormous potential for biofuel (ethanol) production. Besides, the green top of sugarcane is *in vogue,* used as fodder and cattle feed. Dried filter cake is used as an animal feed supplement, fertilizer, and source of sugarcane wax. The filter mud resulting from sugar processing is utilized as manure. Bagasse is used as a bio-sorbent for wastewater purification and also for the manufacture of paper, paper board products, and hardboard, as well as fuel to run boilers for boiling the juice. Among the changing climate scenarios, drought is the single most important factor for reducing the sugar yield, biomass, and juice quality (Khaled et al. 2018). Cane yield loss can be as high as 49% due to drought

DOI: 10.1201/9781003292425-7

(Mahadevaiah et al. 2021). Sugarcane is a highly water-consumptive crop. Annual water requirements for sugarcane production are estimated to be 1600–2500 mm (Inman-Bamber 2004), which is equivalent to 16–20 irrigations to obtain high yield. The use of supplementary irrigation results in an increase in production costs, which threaten the sustainability of the sugar industry.

Plants respond and adapt to the abiotic stresses with a variety of physiological and biochemical adjustments referred as tolerance. Avoidance mechanisms rely on the capacity of the plant to escape from the severity of drought stress. Tolerance to drought is regulated by a variety of transcriptional factors (TFs) and overexpression of stress-inducible genes. Several biochemical pathways under drought, salinity, extreme temperatures, and oxidative stress often happen to cross-talk, inflicting similar cellular damage (disruption of homeostasis and ion disruption, denaturation of functional and structural proteins, etc). Exploring the mechanisms by which the plant recognizes environmental signals and their transmission to cellular machinery to trigger adaptive responses is crucial for the development of breeding strategies.

Conventional breeding meant slow progress in developing drought-tolerant elite sugarcane varieties due to narrow gene pool, higher ploidy (2n = 100–120), rare flowering, high seed sterility, large genome size, long breeding cycle, and high environmental interactions. However, genomic-assisted breeding and molecular manipulation via *in vitro* culture may accelerate the breeding process. In this context, production of transgenic sugarcane can be a better alternative that can pave a way forward to integrate desired gene (s) into its genome; the transgenic plant can be multiplied extensively using micro-propagation techniques. Such a technique avoids the problem of linkage drag normally associated with conventional breeding, besides shortcutting the period of breeding in sugarcane. Therefore, the present study highlights the potential of genomics for precision breeding to improve drought tolerance in the crop.

7.2 ORIGIN, TAXONOMY, AND CULTIVATION OF SUGARCANE

Sugarcane is native to tropical South Asia with the centre of diversity thought to be in Papua New Guinea (Daniels and Roach 1987), as evidenced by amplified fragment length polymorphism (AFLP) marker analysis (Aitken et al. 2006). The centre of origin is also believed to be in India and Indo-Burma (Kumar and Singh 2012). Sugarcane is classified into Kingdom, Plantae; Phylum, Magnoliophyta; Order, Poales; Family, Poaceae; Genus, *Saccharum* and is composed of two wild species, e.g., *S. spontaneum* and *S. robustum*; and four cultivated species, e.g., *S. officinarum* (Linn.), S. edule (Hassk), *S. barberi* (Jeswiet), and *S. sinense* (Roxb.) (D'Hont et al. 1998). Besides, the taxonomy and phylogeny of sugarcane is complicated as the four allied genera *Erianthus*, *Sclerostachya*, *Narenga*, and *Miscanthus* together with *Sachharum* share common characteristics and form a closely related interbreeding group known as the "*Saccharum* complex." It is proposed that *S. officinarum* evolved from the selection of sweet forms of S. robustum. The use of *S. sinense* and *S. barberi* in breeding has been limited due to their shy flowering and low fertility. *S. edule* is morphologically similar to *S. robustum* in growth habit. The present-day sugarcane cultivars are founded on interspecific hybrids between *S. officinarum* (2n = 8x = 80, female parent) and *S. spontaneum* (2n = 8x = 64 or 80, male parent) resulting progenies with 2n + n (100 to 130) (Sreenivasan et al. 1987), which were subjected to repeated backcrosses to *S. officinarum* to improve sugar content, along with tolerance to a number of abiotic stresses through rigorous selective breeding. The transmission of 2n chromosomes from *S. officinarum* is thought to be either through endoreduplication or fusion of two nuclei during meiosis (Bremer 1923). This facilitated to quickly achieve "*officinarum*" background in the hybrids after fewer rounds of backcrossing so as to breed sucrose cultivars; the phenomenon is termed as Nobilization (Sreenivasan et al. 1987). Commercial varieties in use today are typically generated by crosses between other commercial or pre-commercial hybrids. *S. officinarum* accumulates very high levels of sucrose in the stem but is highly susceptible to diseases (Lakshmanan et al. 2005), whereas *S. spontaneum* is characterized by low sucrose content, thinner stalks, high rationing ability, higher fibre content, and tolerance to biotic and abiotic stresses including drought stress (Jackson 2005). In fact, the nobilized sugarcane cultivars share about 80% genome derived from *S. officinarum* and 10%–15% from *S. spontaneum* and the remainder being recombinant chromosomes from the two parental species as confirmed by both genomic *in situ* hybridisation (GISH) and AFLP markers (Hoarau et al. 2001; Cuadrado et al. 2004). Hybridisation between *S. officinarum* and *S. spontaneum* culminated with the release of a cultivar called POJ2878 ("Java Wondercane") in 1921 in Java (Indonesia), which became an important cultivar, allowing for a 35% increase in sugar production over the previous best cultivars (Cox 2000). This cultivar has provided the genetic heritage for many modern cultivars.

The ICAR Sugarcane Breeding Institute (SBI), Coimbatore and the U.S. Department of Agriculture (USDA-ARS), in Miami, Florida, are two key repositories of the world sugarcane genetic resources. Many modern "Co" canes (bred SBI, India) have been successful in 28 other countries, either as commercially cultivated varieties or as parents for developing location-specific varieties. Among *Sachharum* species, *S. spontaneum* clones are highly variable. Nowadays, a lot of genetic erosion happened and only dwarf, early flowering, weedy forms of *S. spontaneum* are still seen in paddy fields.

Sugarcane is grown in both tropical and subtropical regions. At present, sugarcane is grown as a commercial crop primarily in South America (e.g., Argentina, Brazil and Colombia), North/Central America (e.g., Guatemala, Mexico, and the United States), Asia (e.g., China, India, and Thailand), Africa (e.g., Egypt, Kenya, and South Africa), Australia, and the Pacific Islands. It covers about 26.0 million hectare over 90 tropical and subtropical countries, with a worldwide harvest of 1.91 billion tons in 2018 (FAO 2019) and sugar recovery of 179.66 million tons. India is the second (376.9 mill metric tons) largest producer of sugarcane next to Brazil (746.8 million metric tons) during 2019, but the average productivity is just 60% of Peru with high cane yield potential (121 metric tons ha^{-1}). The productivity of sugarcane in the tropical belt is higher than in the subtropical belt. As a cash crop, it contributes nearly 1.1% to the National GDP (Solomon 2016). This can be raised as a ratoon crop after harvest of the main crop, thus saving cost of production (Anonymous 2011).

7.3 PHYSIOLOGICAL AND BIOCHEMICAL RESPONSE TO DROUGHT STRESS

Water is the vital component of the cell. It maintains turgor and serves as a carrier of nutrients, a intercellular reaction medium, and a cooling agent. Besides, plants use water and CO_2 to form sugars and complex carbohydrates. Sugarcane is highly sensitive to water stress during tillering and the stem-elongation stage (Machado et al. 2009). Under moisture stress, plants manifest osmotic and oxidative stress, which drastically affects its growth and sucrose content (Hussain et al. 2018) due to reduced nutrient uptake from soil (Vasantha et al. 2017), decreased relative water content (RWC), low osmotic potential, build-up of reactive oxygen species (ROS), hormonal imbalance (Khonghintaisong et al. 2018), disruption of ion homeostasis (Boutraa et al. 2010), altered molecular dynamics (Farooq et al. 2008), denaturation of functional and structural proteins, reduced synthesis of nucleic acid synthesis, reduced CO_2 assimilation, and decreased cell division and elongation (Ferreira et al. 2017) in drought-sensitive genotypes. However, tolerant varieties adopt a number of physiological and biochemical mechanisms to survive and maintain growth under moisture stress, although they are not able to achieve the same yield as in control. These include deep root system (Tardieu 2012) "stay-green (delayed senescence" trait (Blum 2005), leaf rolling (Sahoo et al. 2013), stomatal control (Snyman et al. 2016), maintenance of turgor pressure, and osmotic potential (Sharma et al. 2012), biosynthesis of osmolytes (proline, trehalose, glycine betaine, etc.), antioxidant enzymes and epicuticular wax; stabilize some protein complexes, cytosolic calcium ion build-up (for signal transduction), cell membrane permeability, programmed cell death, and lipid peroxidation. Drought tolerance was also shown improved using propagules from mother plants grown under water deficit conditions (Marcos et al. 2018).

Water deficit causes disruption of chlorophyll and the thylakoid membranes (Ashraf and Harris 2013). Besides, the rate photosynthesis is drastically affected by a decrease in activity of ribulose- 1,

5-biphosphate carboxylase (RuBisCO), phosphoenolpyruvate carboxylase (PEPcase), malic enzyme (NADP-ME), fructose-1,6-bisphosphatase (FBPase), and pyruvate orthophosphate dikinase (PPDK) (Lakshmanan and Robinson 2014). Sales et al. (2012) reported an increase in starch hydrolysis in drought-tolerant varieties, leading to higher levels of soluble sugars that helped sustain carbon supply, even in a reduced CO_2 fixation due to stomatal closure, facilitating growth recovery after stress.

Under moisture stress, the first signal released by a plant is abscisic acid (ABA), which is synthesized in roots (Xu et al. 2018) and transported through the xylem to the shoots (Peleg et al. 2011). In the leaves, ABA rapidly alters the osmotic potential of the stomatal guard cells, causing a drop in turgor pressure and the stomata to close. It has also role in cellular signal transduction. Some known responses include improving deeper root system growth for water uptake and maintenance of water status, decreasing transpiration rate through inhibiting shoot growth, preventing dehydration through promoting stomatal closure, and facilitation of leaf senescence (Parent et al. 2009). Reactive oxygen species (ROS) (O^{-2}, OH^- and H_2O_2) particularly, H_2O_2 is over produced, which damages the cell components (chlorophyll, lipids, proteins, and nucleic acids) (Ashraf 2009) and cell membrane by generating mutagenic toxic molecule, malondialdehyde (MDA) through lipid peroxidation (Anjum et al. 2011). In addition to antioxidant enzymes, some non-enzymatic compartible molecules (tocopherols, flavanones, carotenoids, anthocyanins, etc.) maintain cellular ion homeostasis by scavenging ROS (Sharma et al. 2012). Proline, another compartible osmolyte, is also reported as a scavenger of hydroxide ion (OH) (Das and Roychoudhury 2014), in addition to its role in combating osmotic stress.

7.4 VARIETAL SCREENING FOR DROUGHT TOLERANCE

Morphological, cytological, anatomical, physiological as well as biochemical markers are now being used for screening drought-tolerant varieties. Morphological markers are based on the traditional botanical descriptions of visible characters and were the first markers to be utilized. They are of limited value because in sugarcane, they are not inherited in a simple Mendelian manner (Hogarth 1987). Cytogenetic markers observed from mitotic and meiotic chromosomes provide additional information, but the small and variable size of sugarcane chromosomes and their abnormal pairing behavior make cytogenetic observations difficult (Sreenivasan et al. 1987). However, anatomical markers can be used as important markers for drought-stress assessment in sugarcane germplasm against moisture stress. The Indonesian *Saccharum* cv. 'UT12' demonstrated higher sensitivity toward anatomical characteristics, e.g., cell wall and cuticle thickness, vascular bundle size, stomatal size, and density than *Saccharum* cv. 'UT13' (Taratima et al. 2020). Among allied species, *S. spontaneum* in general and particularly its clones, e.g., BOT-53, BOT- 54 and BOT-62 were most productive and drought-tolerant (Table 7.1 and Table 7.2).

Plants immediately respond to drought stress in terms of altered physiological and biochemical characteristics. Leaf chlorophyll content (SPADindex), leaf and canopy temperature, photosynthesis rate, stomatal conductance, canopy conductance, leaf fluorescence, and transpiration rate are also used as an indirect selection criteria for sugarcane genotypes tolerant to water stress. Co 85019 is reported to be the most stable drought-tolerant variety using AMMI biplot analysis (Mahadevaiah et al. 2021) (Table 7.3). The clones Co 12009 and Co 12012 survived better under moisture stress, with a higher level of tolerance based on shoot population, plant height, leaf area index, SPAD chlorophyll index, and photochemical efficiency (Fv/Fm), in addition to maintaining better cane yield and juice quality, and these were at par with the resistant standards Co 86032 and Co 99004 (Gomathi et al. 2020). Varieties, e.g., K93–219, KPS01–12, UT12, MPT10-52, and MPT03–320, maintaining higher photosynthetic efficiency (Fv/Fm) and chlorophyll content (SPAD index) showed the potential to achieve greater growth under water-stress conditions (Prasitsom et al. 2019). Cha-um et al. (2012) identified four sugarcane varieties (K84–200, K95-84, K97-32, and LK92-11) as drought tolerant based on proline content, photosynthetic capacity, chlorophyll fluorescence, and growth

TABLE 7.1
Characteristic Features of Different Saccharum Spp. and Related Genera Under Moisture Stress

Genera/Species	Characteristic Features	Sugar Content	Reference
S. officinarum L	High cane yield, high nutrient use efficiency, but highly sensitive to moisture deficit	High	Shrivastava et al. (2001)
S. spontaneum	Tolerance to drought, cold, water logging, low nutrient requirement	Very low	Sreenivasan et al. (2001)
S. barberi	Tolerance to cold at high altitudes	low	Sreenivasan et al. (2001)
S. *sinense*	Moderately tolerant to moisture stress	low	Priji and Hemaprabha (2014)
S. robustum	Tolerance to water logging	Very low	Daniels and Roach(1987)
S. edule Hassk	Compact inflorescence, used as vegetable	Low	Daniels and Roach(1987)
Erianthus spp.	Robust growth, tolerance to drought and salinity, low nutrient requirement	Low	Sreenivasan et al. (2001)
Miscanthus spp	Tolerance to cold at high altitudes	Low	Sreenivasan et al. (2001)
Narenga spp.	Drought tolerance	low	Sreenivasan et al. (2001)

TABLE 7.2
List of Drought Tolerant Clones Belonging to Wild Spp. of *Saccharum* and *Erianthus* Species

S. spontaneum	*S. barberi*	*S. sinense*	*S. robustum*	*Erianthus*	Reference
SES 106 B, SES 168, SES 561, SES600, SES51517, Iritty 2, IND90–813, BOT-53, BOT-54 and BOT-62	Kewali, Khatuia, Pathri, Saretha	Chuckche, Ikhri, Uba white	IJ 76 336, IJ 76 337, NG 77–59, NG 77–122	IK 76–48, IK 76–62, IK 76–91, IK 76–99, IND 84–363	Priji and Hemaprabha (2014); Munawarti et al. (2013)

TABLE 7.3
Sugarcane (*S. officinarum*) Hybrids and Varieties Adapted to Moisture Deficit Dry Lands

Variety	Features for Abiotic Stress Tolerance	Reference
ISH-007, ISH-135 (All hybrids)	Drought tolerance	Srivastava and Srivastava (2016)
Co 740, ISH 100, NS 83/247, Co 85019, Co 997, Co 99008 (All hybrids)	Drought tolerance	Hemaprabha et al. (2012)
Co 85019	Highly stable under drought stress	Mahadevaiah et al. (2021)
Co 12009 and Co 12012, Co 86032, Co 99004	Drought tolerance	Gomathi et al. (2020)
K84–200, K95-84, K97-32 and LK92-11	Higher proline content, photosynthetic efficiency and chlorophyll fluorescence under moisture stress	Cha-um et al. (2012)
PR 980, Co 312	Drought tolerance	Machado Jr and Burnquist (1986)
Co. 997	Drought tolerance	Khaled et al. (2018)
PoJ 2878, NCo 310	Wide adaptation to varied climates	Machado Jr and Burnquist(1986)
K86–161	Drought tolerance	Jangpromma et al. 2007

(Continued)

TABLE 7.3 (Continued)
Sugarcane (*S. officinarum*) Hybrids and Varieties Adapted to Moisture Deficit Dry Lands

Variety	Features for Abiotic Stress Tolerance	Reference
Katha	Widely adaptable to hot and dry climate	Sreenivasan 2004
Agoul, Hemja	Adapted to early drought and late water logging	Sreenivasan 2004
POJ 2727	Adapted to dry, rocky lands	Srivastava and Srivastava (2016)
F141, Co396, Co285, B63340 and KD-01	Drought tolerance	Ishaq et al. (2012)
Khari, Ikhri, Khagri	Drought and water logging	Sreenivasan 2004
SP81–3250	Drought tolerance	Belesini et al. (2017)
Batjan	Vigorous growth, high cane yield, adapted to poor marginal lands	Earle 1928
Sp8032-80, Co997 and BOT41	Drought tolerance	Khaled et al. (2017)
Cavengiri and Co-210	Adapted to dry lands	Srivastava and Srivastava (2016)
EI 8–129 and EI 24-3	Drought tolerance	Fawaz et al. (2013)
Co210	Adapted to hard and dry lands	Earle 1928
RB867515	Drought tolerance	Vantini et al. (2015)
Cavengi	Good yield on poor dry soils	Earle 1928
Kalkya, Khadya, Bansi, Nannal	Drought tolerance	Sreenivasan 2004
EK 28	Grow fairly well on a variety of soils	Earle 1928

inhibition. Protein and secondary metabolites of leaves from a variety of plants have been examined and were found to be promising as markers, such as dehydrin (Lopez et al. 2003), superoxide dismutase (Zang and Komatsu 2007), ASR (ABA-water stress-ripening-induced) protein (Riccardi et al. 1998), and actin depolymerizing factor (Salekdeh et al. 2002). Following PEG treatment, a 18.kDa soluble protein was shown to be overexpressed in leaves of drought-tolerant variety "K86–161" compared to the sensitive cultivar "Khon Kaen 1," which can serve as a potential biochemical marker for drought tolerance in sugarcane (Jangpromma et al. 2007). Among eight clones of sugarcane subjected to withholding irrigation for 21 days at 90 days after planting, CP92–675, HoCP01–523, TCP89–3505, and TCP87–3388 were screened as tolerant based on chlorophyll content and leaf-relative water content, which emerged as the best parameters for assessment of drought tolerance (Silva et al. 2010). Usually, small leaf area, low desiccation, low leaf-rolling scores, and high drought tolerance index (DTI) are associated with the drought-tolerant varieties (F141, Co396, Co285, B63340, and KD-01) (Ishaq et al. 2012). Abbas et al. (2013) sorted out two drought-tolerant sugarcane genotypes, e.g., CPHS-35 and S-2003-US-694, which had the lowest lipid peroxidation (malondialdehyde content) and showed highest membrane stability at PEG 12.5%.

Sugarcane hybrids, ISH-007 and ISH-135 had shown tolerance to water stress, waterlogging, and salinity. At the same time, cv. POJ 2727 adapted to dry rocky lands and cv. Cavengiri and cv. Co-210 were suited to dry lands (Srivastava and Srivastava 2016). The better performance for cane and commercial cane sugar yield of CoVC 99134 (135.8 t/ha and 17.63 t/ha) and CoVC 99263 (132.3 t/ha and 16.71 t/ha) under drought condition have been reported by Meena et al. (2013).

7.5 GENETIC AND BIOCHEMICAL BASIS OF DROUGHT TOLERANCE

Drought tolerance is a highly complex trait with appreciably high nonadditive gene action and reflects high genotype (G) × environment (E) interaction (Mahadevaiah et al. 2021). In fact, the physiological and biochemical responses of plants to drought stress result from the signal perception and transduction and activation of stress-inducible gene expression. ABA regulates an array of genes that coordinates cellular responses conferring plant resistance toward water-deficit stress (Sah et al. 2016). About two-thirds of gene expression during drought stress is ABA-regulated (Xu et al. 2018). Barring the ABA-independent TFs, such as DREB2, DREB3, and ZHDH; most of TFs, e.g., MYB2, MYC2, CBF, B-box zinc finger (BBX), and AREB are regulated by level of ABA in the cytosol. In turn, these crucial TFs switch ON the drought-inducible genes by binding with dehydration-responsive elements (DRE) and ABA-responsive elements (ABRE) to help the plants survive under moisture stress (Chinnusamy et al. 2004).

H_2O_2 acts as a secondary messenger (at its low concentration) to activate stress-inducible genes that regulate key physiological processes such as photosynthesis, stomatal conductance, growth and development, and programmed cell death (Gill and Tuteja 2010). H_2O_2 activates Ca^{2+} in the plasmamembrane and cytosol to carry forward the signal-transduction process, and it also activates the expression of post-translation protein modifiers, such as the SUMO proteins during drought and heat stress (Kurepa et al. 2003). In sugarcane, superoxide dismutase (SOD) has been reported as the antioxidant enzyme with a well-defined role during drought stress (Azevedo et al. 2011). Drought stress increases the SOD gene activity, particularly, Cu/Zn-SOD than the Mn-SOD and Fe-SOD (Ashraf 2009) to counteract the impact of the ROS. Other stress-responsive genes for antioxidant enzymes that are found in plants include ascorbate peroxidase (APX), catalase (CAT), glutathione-S-transferase (GST), polyamines, ascorbate (ASC), and phenolic compounds (Tsaniklidis et al. 2021).

Gene families comprising heat-shock proteins (HSPs), late-embryogenic abundant proteins (LEAs), are the major players in combating drought stress. Among HSPs, the members of HSP70 family act as molecular chaperones, besides having role an enhanced membrane thermostability, RWC, chlorophyll content, and photosynthesis rate (Augustine et al. 2015), repair of damaged proteins, and reducing the protein denaturation (Bossis and Melchior 2006) under moisture stress. Ubiquitin group of stress-inducible genes encode LEA proteins (Survila et al. 2010), which characteristically hold water molecules (phydrophilic) in leaf tissues of sugarcane (Hanumappa and Nguyen 2010). Dehydrin-a group II LEA protein, maintain high RWC in tolerant cultivars due to its role in plasmamembrane stability (Chen et al. 2019). Similar is the role of aquaporines (water channel proteins) to maintain water potential (de Andrade et al. 2016), enhancing sucrose storage in sugarcane. MYB and E3 SUMO-Protein ligase SIZ2 are key TFs that enhance biosynthesis of aquaporin proteins (AQPs) and flavonoids antioxidants (Catala et al. 2007). Besides, over expression (as high as 10 fold) of Δ1-Pyrroline-5-Carboxylate synthase (P5CS) gene meant for proline biosynthesis protects sugarcane-tolerant cultivars against osmotic stress (Li et al. 2018), in addition to maintaining cytosolic pH and cell redox levels, as well as reducing protein denaturation. ATPase, a candidate gene for ion transport across plasmamembrane and those which are involved in biosynthesis of osmolytes (betaine, proline and sugars), detoxifying enzymes (catalase, peroxidise, hydrolase, glutathione S-transferase, glutamate dehydrogenase), and proteins for the protection of macromolecules (osmotin, LEA protein, chaperons, antifreeze proteins and mRNA binding protein) or regulatory proteins (DREB, bZIP, CDP kinase, receptor protein kinase, MAP kinase, and ribosomal protein kinase) (Agarwal et al. 2006) have a significant role for drought tolerance in sugarcane. The sugarcane drought-responsive gene, scdr 1, is reported to be involved in moisture, salt, and oxidative stress tolerance in sugarcane (Prabowo et al. 2019) and also validated in transgenic tobacco (Begcy et al. 2019).

7.6 GENOMICS AND BIOINFORMATICS RESOURCES

Sugarcane nuclear genome is much more complex, while its chloroplast genome is much smaller and simpler. The chloroplast genome sequence (141,182 bp long) of sugarcane was made available in 2004 independently by Asano et al. (2004) and Calsa Júnior et al. (2004) (data base: http://chloroplast.cbio.psu.edu/). Recently, Xu et al. (2019) sequenced Cp genome of *S. officinarum* (cv.

var. Badila) and *S. spontaneum* (cv. Yunnan 83–184). The length of *S. officinarum* Cp genome is 141,187 bp, which is 7 bp longer than *S. spontaneum*, with the same GC content (38.44%) and annotated gene number (134). However, genomic sequence analysis of 26 regions in the chloroplast clearly distinguished *S. spontaneum* from the other five species of *Saccharum* including *S. barberi* (Takahashi et al. 2005). In contrast, less variation was detected in the mitochondria genomes of *S. officinarum*, *S. barberi*, and *S. sinense* than that of *S. spontaneum* as revealed from identical RFLP profiling of the mitochondrial genome among 18 *S. officinarum* clones and 15 of the 17 *S. robustum* clones (Zhang et al. 2014). At the nuclear-genome level, *S. spontaneum* was also shown to be quite divergent from rest of the *Suchharum* species based on targeted region amplified polymorphism (TRAP) (Alwala et al. 2006), sequence-related amplified polymorphism (SRAP) markers (Andru et al. 2008), amplified fragment-length polymorphism (AFLP) (Selvi et al. 2006), randomly amplified polymorphic DNA (RAPD) (Nair et al. 1999), and SSR markers (Selvi et al. 2003). However, phylogenetic analysis based on SSR profiling of 30 accessions from five species, *S. barberi*, *S. officinarum*, *S, robustum*, *S. sinense*, and *S. spontaneum*, indicated that *S. robustum* might be the ancestor of *S. officinarum* (Brown et al. 2007).

A reference sequence is essential to analyzing the variations between the sugarcane varieties. Some 20 years ago, *Sorghum* genome was used as a template to assemble and select the sugarcane chromosome fragments to sequence because the genomic structures of sugarcane and sorghum are very similar. Recently, Garsmeur et al. (2018) generated the first bacterial artificial chromosome (BAC)-based sugarcane monoploid genome sequence of cv. R570 using multiple parallel genome-sequencing initiatives with the *Sorghum* genome as a reference. This constituted of 4535 bacterial artificial chromosomes (BACs) sequences that were colinear to the gene-rich portion of the *Sorghum* genome. The final assembly consisted of 382 Mb of high-quality sequence, which contains 25,316 predicted protein-coding gene models. The next milestone was the allele-defined genome reference of a tetraploid *S. spontaneum* genotype cv. AP85–441 (Zhang et al. 2018) and cv. US851008 (Nascimento et al. 2019), and recently, in a Brazilian cv. SP80–3280 (Souza et al. 2019) covering 4.24 Gb, which includes the complete sequence of 373,869 genes. Genomic information is being helpful to sugarcane breeders to develop varieties with improved sugar yield, quality traits, and variety of biotic and abiotic tolerance.

Recent advances in DNA-sequencing technologies and strategies for genome assembly are making the sugarcane genome more tractable. Now, the long read-sequencing (Hoang et al. 2017) techniques, e.g., PacBio single-molecule real-time (SMRT) sequencing, the Illumina TruSeq, and the Oxford Nanopore sequencing have allowed the generation of a more complete set of sugarcane gene transcripts for genetic research. However, on the applications side, next-generation sequencing (NGS) technologies played a major role in the discovery of single nucleotide polymorphic (SNP) markers and the development of low to high throughput genotyping platforms (Davey et al. 2011). The two mainstream high throughput genotyping platforms are the SNP microarray and genotyping by sequencing (GBS). Mining sequence variations (enabled by NGS) particularly SNPs at targeted site among elite sugarcane varieties can pave the way for identifying alleles related to complex traits including drought tolerance. In this context, the comparative genome analysis has the added benefit of curation by gene alignment with the closest reference genome particularly *Sorghum bicolour* (Paterson et al. 2009). But it is sure that a combination of targeted GBS and array-based genotyping methods can increase the accuracy of genomic selection and marker-assisted breeding (Manimekalai et al. 2020).

Sugarcane unigene sequence clusters are now publicly available by NCBI (http://www.ncbi.nlm.nih.gov/sites/entrez?db=unigene) and by Plant GDB (http://www.plantgdb.org/). The publically available mapping resource for sugarcane has been developed by CIRAD as part of TropGENE-DB (http://tropgenedb.cirad.fr/) data base (Ruiz et al. 2004). The high-density genetic maps have been published by different researchers (Aitken et al. 2005, Garcia et al. 2006, Aitken et al. 2014). The database has a search capability for molecular markers, QTL, and clones, as well as associated phenotypic data, particularly pest and disease ratings. At present, there is a wealth of molecular data available for sugarcane, especially transcript clusters and gene-expression profiles.

The transcript clusters are easy to access and analyze. Besides, the data underpinning gene-expression profiles can also be easily downloaded for reanalysis, as and when required.

7.7 GENOME COMPLEXITY

Sugarcane genome has very large (~10 Gb) and complex genome comprising 100–130 chromosomes. The genome complexity in terms of chromosome number variation primarily stems from natural hybridization between allied species. Besides, a typical cytological peculiarity of 2n(S.o) + n(S.s) chromosome transmission into F_1 plants following interspecific crosses of two parental species *S. officinarum* and *S. spontaneum* was confirmed by genomic *in situ* hybridization (GISH) (Yu et al. 2018). Further, the random sorting of chromosomes in each back cross during nobilization process evidently resulted a mixture of euploid and aneuploid chromosome sets among progenies. Thus, the modern sugarcane cultivars contain different ploidies of homo/homeolog groups, and hence, they are referred to as poly-aneuploids. Besides, these species differ in their monoploid genomic size (S.o: 930 Mb and S.s: 750 Mb, respectively, about twice that of rice: ~430 Mbp and comparable to sorghum: ~760 Mbp) and basic chromosome numbers due to their transposable elements (TEs) and some large chromosomal rearrangements (Garsmeur et al. 2018). The role of TEs (transposons and retrotransposons) for genome expansion in grasses strongly supports similar cytological peculiarity in sugarcane (de Setta et al. 2014). In fact, the transposase-mediated chromosome repatterning with high copy number (8–12) of homologous genes, different allelic dosages, and a significant repeat content (Souza et al. 2011) at other sites further complicates the understanding of genome functioning in sugarcane (Figueira et al. 2012). Besides, the complex chromosome organization also complicates the recombination events, making the selection process slower for identification of elite genotypes.

7.8 GENOMIC TOOLS FOR PRECISION BREEDING

Genetic variability is the main driving force for adaptive responses against drought and other stresses. Conventional breeding primarily harnesses genetic variability in sugarcane for a variety of development. Conventional breeding takes 10 to15 years ,and its progress is still limited by the sugarcane's complex genome (Butterfield et al. 2001). Modern sugarcane are hybrids from the cross between *S. spontaneum* and *S. officinarum*, and the high sugar content is considered to the attribution of *S. officinaurm*, while the *S. spontaneumin* is the donor for stress tolerance in the nobilization breeding process. Conventional breeding faces a major setback due to its characteristic nonsynchronous flowering and low sexual seed viability. In order to introgress abiotic stress tolerance, intergeneric crosses *S. officinarum* or the hybrid *S. officinarum* × *S. spontaneum* with *Erianthus, Miscanthus, Bambusa, Zea mays, Sorghum and Imperata* species were designed to eliminate the natural barrier to crossing, but resulted in only limited success. Sugarcane is an excellent material for inducing mutants using *in vitro* mutagenesis with EMS and screening for drought tolerance under PEG-induced moisture stress. Once a desirable drought-tolerant clone with improved yield and quality features is recovered, it can be multiplied using the conventional method of vegetative propagation and put in the cultivation programme. Besides, recent genomic tools, e.g., high throughtput genotyping platforms, transcriptome analysis, marker-assisted breeding, *in vitro* mutagenesis and genome engineering proved to be rewarding. Figure 7.1 illustrates the detailed genomic-assisted precision breeding approaches compared to traditional methods being used for improving drought tolerance in sugarcane.

7.8.1 High Throughput Genotyping Platforms

A number of SNP genotyping platforms, e.g., low-throughput gel-based approach, the cleaved amplified polymorphic sequence (CAPS) marker approach (Thiel et al. 2004), PCR-based

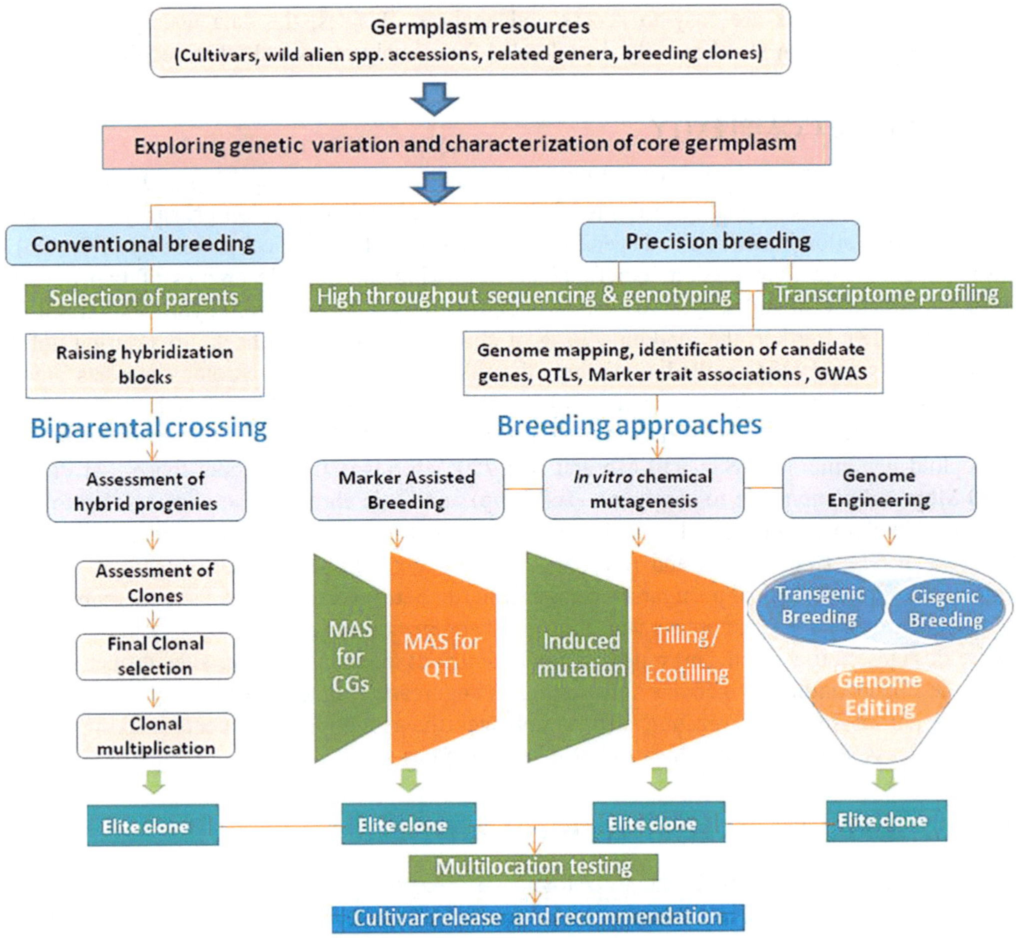

FIGURE 7.1 Schematic breeding strategies for development drought tolerant sugarcane varieties.

fluorescently-labeled high-throughput methods, high-resolution melting (HRM) curve analysis, TaqMan, and KASP assay (Martino et al. 2010), fixed array systems such as Illumina Infinuium (Mason et al. 2017), Affymetrix Axiom (Allen et al. 2017), NGS-enabled genotyping by sequencing (GBS), and hybridization-based SNP array (Thomson 2014, Scheben et al. 2017) are now available, but these differ in reaction chemistry, signal detection, SNP identification, and assay flexibility. In general, the above SNP genotyping options comprise three broad categories: whole-genome re-sequencing (WGR), reduced-representation sequencing (RRS), and SNP arrays. WGR and RRS are based on next-generation sequencing (NGS) technology, while SNP arrays is hybridization based and relies on allele-specific oligonucleotide probes (Nielsen et al. 2011). The available high throughput genotyping in most of the crop plants is based on either Illumina Infinuium Bead array or Affymetrix Axiom technology. A set of 76K SNPs from 14 sugarcane accessions (Yang et al. 2017), and 84K SNPs from 151 sugarcane clones (Balsalobre et al. 2017) are now available. Sugarcane requires many-fold higher NGS read depth (100x) than normal diploids (4~6x) for sequencing large number samples. Recently, McNeil et al. (2017) developed a low-cost SNP panel for sugarcane breeding.

7.8.2 Transcriptome Analysis

ESTs are basically short DNA coding sequences (100 to 800 nucleotides on cistrons) produced by sequencing; they are randomly selected from the cDNA libraries. Thus, an EST database can directly point to genes. The publicly available sugarcane ESTs were assembled into tentative consensus sequences (virtual transcripts), singletons, and mature transcripts, referred to as the Sugarcane Gene Index (SGI; http://compbio.dfci.harvard.edu/tgi/cgi-bin/tgi/gimain.pl?gudb=sofficinarum). Such a comprehensive, functional EST map of the sugarcane genome has been a tremendous help for developing "perfect markers" associated with key QTLs. Expressed sequence tags (ESTs) have been identified from meristems (Carson and Botha 2000), stem (Casu et al. 2004), and red-rot infected tissue of sugarcane (Gupta et al. 2010). Transcriptome analysis of RNA-Seq libraries from top and bottom internodes of 20 different genotypes, including commercial cultivars and introgression lines, are available (Kasirajan et al. 2018). But, the Brazilian SUCEST database (http://sucest.lad.ic.unicamp.br/en), comprises 237,954 ESTs (derived fron 26 cDNA libraries), which are organized into 43,141 putative, unique sugarcane transcripts (26,803 contigs and 16,338 singletons) known as the largest sugarcane assembled EST sequences.

A variety of transcriptomic analysis techniques are now available. However, the SuperSAGE (Matsumura et al. 2003) approach, a modification of the serial analysis of gene expression (SAGE) (Terauchi et al. 2008), proved to be ideal for screening of large number of ESTs using *Sorghum* genome as reference due to the higher homology (95%) in the genic regions (Wang et al. 2010). The long-read isoform sequencing (Iso-Seq) is widely used in transcriptome sequencing because this technique offers a better alternative for sequencing more complete transcriptomes and successfully predicting and validating gene models. Among them, hybridization-based approaches using DNA arrays have been the most successful for large-scale gene-expression profiling in sugarcane underwater stress (Rodrigues et al. (2011). Li et al. (2016) used cDNA arrays to study the gene-expression profile in leaves of sugarcane plants subjected to different water-stress conditions.

Transcriptome analysis offers detection of novel genes, including TFs involved in cell metabolism, hormone biosynthesis, stress perception, signal transduction, and activation of stress-inducible genes in a drought-tolerant variety. About 36% of the differentially expressed genes were repressed in sensitive varieties since up to 93.3% of DEGs are up-regulated in tolerant cultivars in sugarcane (Ferreira et al. 2017). However, qRT-PCR must be carried out for assessing the status of expression of the candidate genes. In sugarcane, around 10,000 unidentified sugarcane coding genes (Xu et al. 2018) are yet to be annotated. Kido et al. (2012) identified many potential target candidates, including the classical genes of the osmotic stress (as P5CS) in the drought-stress response. Belesini et al. (2017) analyzed the transcriptome profiles of the two sugarcane cultivars "SP81–3250" (drought-tolerant) and the "RB855453" (drought-sensitive), which led to explore involvement of diverse TFs/genes, namely ascorbate peroxidase, MYB, E3 SUMO-Protein ligase SIZ2, key enzyme for biosynthesis of flavonoids such as co-enzyme A ligase and aquaporin (Belesini et al. 2017) and several undescribed genes. Pereira-Santana et al. (2017) carried out a transcriptomic analysis of a Mexican variety "Mex 69–290" in response to osmotic stress, which led to the identification of DEGs in transcriptional regulation, carbohydrate catabolism, oxide reduction, flavonoid, and other secondary metabolite biosynthesis. Xu et al. (2018) carried out transcriptome sequencing for six contrasting sugarcane genotypes, revealing the up-regulation of unigenes related to peroxidase, antioxidant activity, and signal transduction in response to drought stress in the tolerant genotype. The changes in differential-expressed genes (DEGs) are survival strategies plants employ to escape unfavourable growth conditions (Li et al. 2020). More recently, Liu et al. (2018) conducted an experiment on wild type sugarcane "*S. narenga*" treated with drought stress and identified 3389 DEGs (1772 up-regulated and 1617 down-regulated) involved in various metabolic pathways. Additionally, Vantini et al. (2015) conducted a comparative study for screening the gene-expression profiles of two different drought-tolerant sugarcane varieties against drought stress through DNA-amplified, fragment-length polymorphism (AFLP). The AFLP results revealed that 173 fragments of tolerant cultivars had the altered expression pattern in response to water stress. A sugarcane homologue of early response to drought protein 4 (ERD4) (Seki et al. 2002), was identified using the published amino acid sequence of ERD4 in a TBLASTN search of the sugarcane EST subset of the NCBI. Changes in the expression level of the sugarcane ERD4 (scERD4) by real-time PCR analysis were increased to 4.7 fold after drought treatment (McQualter and Dookun-Saumtally 2007). Nawae et al. (2020) treated sugarcane cv. KPS01–12 and UT12 with mild and moderate water stress and found differentially expressed genes in various biological processes. KPS01–12 had higher expression of genes that were involved in water retention, antioxidant secondary metabolite biosynthesis, and oxidative and osmotic stress response than UT12.

Post-transcriptional gene regulation by expression of antisense transcripts (Lembke et al. 2012) and miRNAs (Ferreira et al. 2012) under moisture stress reflects a more complicated mode of gene expression. In fact, a battery of miRNAs (about 22 nucleotides long, non-coding regulatory RNA) are induced in response to moisture stress, and these mostly target different transcription factors, proteins involved in tolerance to oxidative stress, cell modification, as well as hormone signaling (Gentile et al. 2015).

Interestingly, subsets of DEGs that are induced in response to drought also overlap the genes meant for sucrose accumulation. Vantini et al. (2015) observed differentially expressed genes (chaperones, heat-shock proteins, antioxidant enzymes, and protease inhibitor proteins) between tolerant and sensitive sugarcane varieties along time (1, 3, 5, and 10 days after withholding water) in root tissues. The trehalose phosphatase synthase (TPS) (for biosynthesis of trehalose) and serine/threonine kinase receptors also showed higher expression in the tolerant variety. Besides, the members (TIPs, NIPs, SIPs) of the gene families meant for aquaporins (Maurel et al. 2008) are shown to be up-regulated in sugarcane roots under water stress (da Silva et al. 2013).

7.8.3 Marker-assisted Breeding

7.8.3.1 Candidate Gene Approach

Sugarcane varieties harbouring stress-responsive candidate genes proved to offer climate resilience. However, the use of limited clones of *Saccharum* species has narrowed down the genetic base due to intensive selection pressure, evidently slowing down the pace of varietal improvement. Candidate genes provide a productive approach for establishing the basic and immediate responses of plants to drought. Effective identification and understanding of genes contribute to improve plant-drought resistance. In addition, an allele-defined genome of *S. spontaneum* is available (Zhang et al. 2018), offering the opportunity to perform the identification and phylogeny of various gene families in sugarcane at a genome-wide scale. Besides, the BAC-based sugarcane monoploid genome sequence of cv. R570 (Garsmeur et al. 2018), having high synteny with the *Sorghum* genome, may serve as an excellent source for exploring annotated candidate genes.

TFs are known to play a major role in abiotic stress tolerance. The expression of seven TF genes, namely, WRKY, NAC, bZIP, DREB, G2 like, Homeobox, and TUB, showed significant difference between the drought-stress tolerant and susceptible varieties (Dapanage and Bhat 2019). Among these TFs, bZIP showed the highest expression, both under moderate (22.39 fold) and severe stress (13.45 fold) conditions. Besides, HSPs (heat-shock proteins), LEAs (late embryogenesis proteins), RAB (proteins responsive to abscisic acid), osmotin, choline oxidase and annexin (Nair 2011), sugarcane ethylene-responsive factor SodERF3 (Trujillo et al. 2009), dehydrins (Iskandar et al. 2011), P5CS, and TPS (genes for accumulation of proline and trehalose) (Guimaraes et al. 2008), early response to dehydration protein 4 (ERD4) (McQualter and Dookun-Saumtally 2007) are some of the examples of genes identified in response to drought-water deficit. Also, a number of candidate genes underlying cold, chilling salinity or high temperature and heat stress often cross-talk, displaying their role in drought tolerance. Utilization of these stress-induced genes may inculcate stress tolerance in elite sugarcane cultivars.

SOD gene is reported to help the plant during rehydration when the superoxide anions increase in the cells and tissues. During drought, auxin-related gene expression has established the activation of genes, including IGS, which is a IAA biosynthesis gene (Aloni et al. 2003). Simon and Hemaprabha (2010) identified donors carrying the drought-responsive genes superoxide dismutase (SOD) and indole 3-glycerol phosphate synthase (IGS) using gene-specific molecular markers in

RT PCR. The drought-tolerant parent (Co 740) showed 618bp and 340bp gene-specific bands for SOD and IGS, respectively, while there was absence of band in drought-sensitive variety Co775. These candidate genes hold promise to provide drought tolerance of sugarcane using marker-assisted selection. In an another study, Priji and Hemaprabha (2014) identified 11 sugarcane-specific drought-responsive candidate genes (DREB 1A, DRF 1, Hep2, HRD, NAC 2, NIT 1, PIN 3, SHN 1, S1Z 1, Snac 1, and Wrky 38) belonging to ABA-independent pathway from stress-tolerant *S. spontaneum*, *S. barberi*, *S. sinense*, *S. robustum*, and *Erianthus* species. The above genes belong to drought-responsive regulatory factors, dehydration responsive element transcription factors for drought and salt stress, auxin-mediated regulatory factors, and genes encoding protective factors. The drought-tolerant species clones exhibited 8–11 genes, while only a few of the *S. officinarum* clones harbour maximum, up to three such candidate genes, possibly introgressed from other tolerant species. Interestingly, all above candidate genes were uniformly present in drought-tolerant species clones viz., Iritty 2 (*S. spontaneum*, coimbatore), and SES 600 (*S. spontaneum*), NG 77-59 (*S. robustum*), Kewali and Khatuia (*S. barberi*), Ikhri (*S. sinense*), and IK 76-62 and IK 76-48 (*Erianthus* sp.). Hence, it is certain that these genetic stocks can be used for marker-assisted introgression-breeding programmes.

The candidate genes, or DNA sequences with predicted function, are used to design genic markers to associate with phenotypes expressed in segregating populations or genetic stocks (Huh et al. 2001). A drought-related EST databases, microarray analysis and mutagenesis approach yielded a large number of valuable candidate genes for verification of their association with drought-tolerant traits. TRAP (targeted region amplified polymorphism) markers are designed from ESTs of functional genes. Although TRAPs were considered less effective than SSRs in detecting polymorphism (Liu et al. 2005), these offer unique polymorphism when ESTs do not contain microsatellite sequences. Several TRAP markers, representing DREBs, showed association with leaf-relative water content (Sharma 2009). Fawaz et al. (2013) used dehydration binding factor and aquaporin gene-based TRAP markers that differentiated drought-tolerant sugarcane genotypes EI 8–129 and EI 24-3 from most sensitive ones NCO310 and G2000-79. Further, a few genic marker loci (SSR 9) showed significant association with multiple physiological parameters (Sharma 2009), suggesting the cross-talk in the stress-signaling pathways (Fujita et al. 2006).

AFLP and SSR markers are often used to explore polymorphic loci in the diploid genome, but these do not allow estimation of the number of copies of each allele or dosage (aaaaaaaa: nulliplex, Aaaaaaaa: simplex, AAaaaaaa: duplex, and so on up to AAAAAAAA: octoplex) at a given polymorphic locus in complex polyploids like sugarcane. Besides, the homoeolog dominance, increased gene and genome dosage, and large genome size typically complicate the identification of molecular markers for complex traits, including drought tolerance (Borrill et al. 2015). Therefore, high throughput genotyping platforms for exploring SNPs assume great importance, though identification of SNPs is difficult in sugarcane. Garcia et al. (2013) demonstrated a successful genetic analysis in a highly polyploid sugarcane genome by the quantitative analysis of single-nucleotide polymorphism (SNP) allelic dosage. The SNP databases are a potential source to explore ideal markers for identifying genes or haplotypes associated with complex traits (Cuervo-Alarcon et al. 2021). SNPs are huge in number and are amenable for high throughput screening and maximum genome coverage, making the genome-wide association study more effective and meaningful for development of climate resilient cultivars. Raw SNP datasets are often typically inaccurate and incomplete. Therefore, SNP-filtering criteria and checking for quality control (QC) are a must to avoid bias and false signals in genotype–phenotype association (Pavan et al. 2020).

7.8.3.2 Marker-Trait Association(MTA)

The erstwhile discussed genome complexity in sugarcane complicates the application of conventional genetics and breeding techniques. However, use of molecular markers helps keep the selection process more simple and accurate. Molecular markers associated with agronomic traits and moisture-stress tolerance could significantly reduce the time and cost involved in developing new varieties, as well as exploring potential donors for drought tolerance. In sugarcane, a high density of markers may not be needed to detect marker-trait associations (Rafalski 2002), provided the breeding history of the parents used for hybridization is known and small number of cycles of inter-crossing followed by vegetative propagation are carried out to develop clones. But, finding a marker associated with a trait of interest is more complicated and laborious than in diploid species. In sugarcane, the mapping progenies are derived from biparental crosses between highly heterozygous outbreed parents in which there are different numbers of alleles per locus, resulting in a mixture of marker-segregation patterns in the progeny (Pastina et al. 2010). To date, approximately 22 linkage maps are available in sugarcane. The majority of published sugarcane linkage maps have been constructed based on single-dose markers (SDM) and on the double pseudo-testcross strategy, which results in two individual maps, one for each parent (Pastina et al. 2010).

Despite the genomic complexity of sugarcane, genome-wide association studies (GWAS) may be used to identify marker-trait associations (MTAs) and then assist breeders in better managing crosses and selecting superior genotypes in breeding programs. For example, Yang et al. (2019) identified resistance loci to orange rust and yellow leaf virus diseases in sugarcane through GWAS (Yang et al. 2019). Barreto et al. (2019) reported 23 MTAs related to soluble solid content, stalk height, stalk number, stalk weight, and cane yield traits in sugarcane (*Saccharum* spp.) using GWAS with a multi-locus mixed model. In sugarcane, a number of QTLs have been identified for stalk number and suckering (Jordan et al. 2004), sugar content (Aitken et al. 2006b), and cane yield-related traits (such as cane weight, number and diameter) (Aitken et al. 2008).

Various molecular markers, such as random amplified polymorphic DNA (RAPD), inter-simple sequence repeat (ISSR), and amplified fragment-length polymorphism (AFLP) were used to investigate sugarcane genetic variability (Adawy et al. 2013). The AFLP is one of the best currently available molecular markers and shows distinct advantages, including reproducibility and the high number of markers. Khaled et al. (2017) used a combination of RAPD and ISSR (termed as R-ISSR) to assist selection for drought tolerance in sugarcane. In addition, the DNA sequence of the RAPD marker OPAK 12724 was used to generate SCAR marker, which was effective enough to sort 23 genotypes into 13 tolerant and 10 susceptible to drought stress (Srivastava et al. 2012). In an another study, eight AFLP primer combinations were used to estimate the level of polymorphism among drought-tolerant sugarcane genotype Co. 997 and drought-susceptible genotype Co. 775, which amplified a total number of 886 amplicons, where 55 were polymorphic representing 6.2% polymorphism (Khaled et al. 2018). These polymorphic AFLP markers can be used as genetic markers to assess the sugarcane genotypes for drought tolerance and to identify QTLs. Besides, phylogenetic trees developed by SSR (Prasitsom et al. 2019) and ISSR (da Costa et al. 2011) markers were shown to be potent enough for grouping the drought-tolerant test varieties into single clusters. Similarly, Khaled et al. (2017) identified three SSR markers that clearly distinguished drought-tolerant (Sp8032-80, Co997, and BOT41) genotypes from sensitive (Co775, F153, and F161) ones in sugarcane. Primer SSR15 produced one 123 bp amplicon in sensitive varieties, which could be used as marker for drought sensitivity (negative marker); however, SSR80 produced two bands of amplicon size 24 bp (negative marker) and 35 bp (positive marker) in drought-sensitive and tolerant genotypes, respectively. This revealed the ability of using SSR technique as marker-assisted selection for drought tolerance in sugarcane.

QTLs are stretches of DNA that are most likely to contain the gene(s) for the trait in question. Hence, genic markers can also serve to identify potential genomic regions explaining substantial contribution to the genetic variation in complex traits including drought tolerance. In fact, the polymorphism produced by EST-SSR is more likely to be associated with the trait of interest than SSR markers developed from genomic DNA. Similarly, TRAP markers are also useful as these are also designed from EST sequences. Sharma (2009) revealed significant association of 21 EST-SSR markers and 11 TRAP markers with QTLs either or both of chlorophyll content and leaf-relative water content in a set of 80 sugarcane genotypes under drought stress. However, Racedo et al. (2016) carried out GWAS using DArT (diversity arrays technology) and TRAP (target region amplified polymorphism) markers and revealed multi-QTLs for cane yield and sucrose content. As many as 20 and 11 markers were associated for cane yield and sucrose content, respectively, over two cropping seasons. Jackson et al. (2010) identified a significant negative effect (–39.4t/ha cane yield) QTL linked to AFLP marker Accctc58, explaining 13% phenotypic variation and two significant positive-effect (27.1t/ha and 24.4t/ha cane yield) QTLs linked to AFLP markers Accctc56 and Accctc51, explaining 6% and 5% phenotypic variations, respectively, for cane yield in a cross ROC25 × YN02–356.

SNPs are in fact the markers of choice. However, identification of SNPs is difficult in a highly polyploidy species like sugarcane. Recently, only a few significantly associated SNPs have been detected for agronomic traits and abiotic stress tolerance (Yang et al. 2018, Fickett et al. 2019) in sugarcane. However, to overcome it, the genomic region-specific sequencing following SeqSNP and target SNP-seq approaches appear to be interesting to accelerate high throughput SNP detection and to capture SNPs linked to the agronomic (Zhang et al. 2010) and drought stress-related traits. The recently developed low-cost 45K SNP Affymetrix Axiom chip (McNeil et al. 2017) for sugarcane has made it easy to construct high-density linkage maps and identification of target QTLs for agronomic traits and stress tolerance. Further, an integrated genetic linkage comprising a set of 688 SNP, AFLP and SSR markers is now available in sugarcane (Costa et al. 2013) making the MAS meaningful for improving complex traits.

Besides, genomic selection plays an important role in precision breeding in sugarcane, particularly when the target traits are highly complex with high G × E interaction. Genomic selection (GS) is a relatively new breeding method in which individuals are selected based on their predicted breeding values that are calculated from genome-wide DNA marker profiles (Meuwissen et al. 2001). Gouy et al. (2013) evaluated the potential of GS for sugarcane breeding in two different panels consisting of 167 clones each. GS is a novel approach of marker-assisted selection using high-density markers covering the whole genome, thus enabling all QTLs in linkage disequilibrium with at least one marker. Superior individuals are selected based on genomic-estimated breeding values, which take into account the effects of multiple genes controlling a target trait. Thus, the strategy offers rapid rates of genetic gain by unbiased prediction of clonal performance, even in a crop like sugarcane having a large complex genome. In this context, SNPs are the markers of choice since they provide maximum genome coverage. The SNP database, coupled with high throughput phenotyping methods, can further improve genomic-prediction accuracy, particularly when dealing with moisture stress.

7.8.4 *In vitro* Mutagenesis

The induced mutations play a key role in the improvement of plant varieties (Jain 2005). Low mutation frequency, a requirement of large size of mutagenic population, and difficulty detecting mutants are still a hindrance to conventional mutation breeding. However, transcriptome study revealed that mutagenesis of plant tissues can be a highly preferable approach to induce significant changes in gene-expression patterns (Batista et al. 2008) throughout the genome within a manageable population size, and a single plant may carry a number of mutations in different genes (Serrat-Gurrera 2016). Other advantages include high mutation frequency, uniform mutagen treatment, application of selective agents to homogeneous cell population, use of single cell system (somatic embryos) to avoid chimeras, requirement of less space and time to handle large population, and keeping the plant material disease free (Penna et al. 2012). Success has been achieved to induce drought tolerance using *in vitro* mutagenesis in a variety of crop plants (Oladosu et al. 2016).

Chemical mutagenesis provides an easy and cost-effective way to produce a high density of novel nucleotide diversity within plants having a complex genome. Various researchers have been exploring the potential of induced mutation for its application in sugarcane improvement using chemicals and physical mutagens. Mutagenesis is most effective when applied in actively dividing meristematic cells (Koch et al. 2012). Sugarcane being highly responsive to *in vitro* culture, mutations are effectively induced in single cells and callus tissues (Mahlanza et al. 2013). Of course, the rate of mutation in the *in vitro* mutagenesis depends on the genotype, proper choice of the mutagen, concentration of the mutagen, pH of the solution, incubation period and the type of tissue used (Suprasanna et al. 2012).

The structure and size of the genome do not affect the effectiveness of EMS mutagenesis. *Arabidopsis* and *Zea mays* have a similar gene-mutational density after EMS treatment even though the *Zea mays* genome size is 20-fold larger than that of *arabidopsis* (Goll and Bestor 2002). This suggests that EMS mutagenesis is ideal to induce random point mutations at high frequencies in crops with large genome size such as sugarcane. Several missense mutations, such as A to G change, which results in asparagine to aspartic acid and G to A change result in serine to asparagine, have been reported in sugarcane EMS mutants (Koetle et al. 2018). Changes in amino acid can be very important in the plant response to water stress. For instance, asparagine in young and mature sugarcane culm acts as an osmoprotectant during sucrose accumulation (Ferreira et al. 2017). Genetic variation generated with EMS-induced mutagenesis is a random process, which makes it difficult to produce an intended DNA sequence (Shirasawa et al. 2016). Abnormalities such as lack of chlorophyll, chimera, and stunted growth are common in mutants produced from EMS (Masoabi et al. 2017). Callus tissues are the most common target for EMS mutagenesis (Khalil et al. 2018). This is followed by a high-throughput screening based on the phenotyping of a large number of calli pieces (Manchanda et al. 2018) although there may be high rate of escapes (Rutherford et al. 2017). Consequently, there is a lack of correlation between tolerances of cells in culture versus at the whole-plant level (Snyman et al. 2016). For instance, only about 16% of EMS-derived osmotic-tolerant putative mutant were tolerant when screened *ex vivo* in pots (Masoabi et al. 2017). Khalil et al. (2018) also observed a similar trend where only about 11% of EMS-derived mutants were tolerant to drought when screened under field conditions.

However, *in vitro* mutagenesis can be integrated with molecular marker techniques for high throughput screening of mutants (Shu 2009). EMS mainly induces C–to-T substitutions resulting in point mutation with C/G to T/A transitions (IRGSP 2005). TILLING technology is a cost-effective reverse-genetics approach for ensuring high mutagenesis rates as well as rapid detection of chemically induced point mutation in a target region. It effectively combines with high-throughput detection assays and PCR screening technologies. The key enzyme of TILLING technology is the "CEL-I" that cleave DNA heteroduplexes with high specificity at DNA distortions and base mismatch sites (Szurman-Zubrzycka et al. 2016). EMS mutagenesis of in vitro induced calli is reported to be a new method for rapidly obtaining TILLING mutant populations (Serrat-Gurrera 2016) amenable for direct screening of *in vitro* regenerated plants. It also saves time, reduces the greenhouse space and work load during the generation of mutant plant populations, and ensures high mutagenesis rates. Among molecular markers, simple sequence repeats (SSRs) are important due to their high polymorphism, co-dominant inheritance, extensive genome coverage, high reproducibility, and ease of use (Haq-UI et al. 2016). Khalil et al. (2018) carried out EMS (0.1%) induced *in vitro* mutagenesis in sugarcane cv. ROC22 and successfully detected two drought tolerant mutants (i.e., MR22-15 and MR22-20) using SSR marker "sep-23." Further, they identified the mutation sites in the genome of both the mutants using Cel-1 digestion of SSR-amplified products followed by capillary electrophoresis.

7.8.5 Genome Engineering

7.8.5.1 Transgenic Breeding

Scope for transgenic approach: Genetic-engineering techniques make possible to clone and characterize useful genes and also to improve commercially important traits in elite sugarcane clones that subsequently lead to the development of an ideal cultivar. Hence, genetic engineering turns out to be a more directed alternative to conventional plant breeding (Pillay 2013). Success of the transformation depends on a transformation system that minimizes somaclonal variation, stable integration of the transgene into the genome of the target tissue, strong promoter driving high-level expression of the transgene and selection of transformed cells (Rashid & Lateef 2016). Plant-regeneration systems via direct embryogenesis (Snyman et al. 2006) may be suitably used for the purpose. Genetic transformation in sugarcane has been reported for a number of traits, e.g., herbicide resistance (Leibbrandt and Snyman 2003), resistance to pests and pathogens (Arvinth et al. 2010), reduction of browning of sugarcane juice (Vickers et al. 2005), and resistance to drought stress (Molinari et al. 2007). Besides, genetic modification underlying altered plant architecture, plant growth, enhanced nitrogen-use efficiency, enhanced water-use efficiency, drought, and salinity tolerance (Reis et al. 2014) have been attempted in sugarcane. In Brazil, there have been a number of field trials for abiotic traits, including drought tolerance (Matsuoka et al. 2009). Genetically engineered sugarcane, developed by the Indonesian public research for drought stress tolerance, was approved in Indonesia for food use and cultivation in 2013. Besides, Brazil came forward for field trials for GE sugarcane for a number of traits (www.ogtr.gov.au.). In 2011, a transgenic sugarcane variety containing the transgene DREB2A has been released in Brazil; it offers tolerance against heat, drought, and salinity. Besides, a drought-tolerant transgenic sugarcane variety harboring a bacterial gene for betaine production has been released for commercial cultivation in Indonesia.

Engineering of stress proteins or the enzymes of biosynthetic pathways associated with stress responses has evolved as important methods for improving drought-stress tolerance (Hong et al. 2000). Hundreds of drought stress-induced genes have been identified, and some of these have been cloned. Plant genetic engineering approaches allow development of drought-tolerant germplasm. Structural genes (key enzymes for osmolyte biosynthesis, such as proline, glycinebetaine, mannitol and trehalose, redox proteins and detoxifying enzymes, stress-induced LEA proteins) and regulatory genes, including dehydration–responsive element binding (DREB) factors, zinc-finger proteins, and NAC transcription factor genes, are being used. Identification of differentially expressed candidate genes with regulatory function to water deprivation can pave the way for production of drought-tolerant transformed plants (Kirch et al. 2005).

Transgenic techniques used in sugarcane: The method chosen for sugarcane transformation is dependent on the type of tissue being modified. Transformation efficiency of seeds (Mayavan et al. 2013), apical meristems (Khan et al. 2013), axillary bud explants (Manickavasagam et al. 2004), young leaf whorl (Eldessoky et al. 2011), somatic embryogenic callus (de Alcantara et al. 2014), protoplasts (Arencibia et al. 1995), and cell aggregates of suspension culture (Efendi and Matsuoka 2011) have been reported. Among these, embryogenic callus are the preferred explant for transformation owing to their high regeneration response (Taparia et al. 2012). The genetic transformation of sugarcane protoplasts and cell suspension has been carried out by using polyethylene glycol (Aftab and Iqbal 2001) and electroporation (Rakoczy-Trojanowska 2002). Microprojectile delivery systems can produce a large number of transiently expressing sugarcane suspension cultures (Rani 2012). Bower and Birch (1992) reported production of transgenic plants by microprojectile bombardment of embryogenic callus using either gene gun or particle inflow gun. A pressurized helium gas is used to accelerate the gold particle coated with DNA for bombardment into the target tissue (Rashid and Lateef 2016). Prior to bombardment, the target tissue is placed onto a conditioning medium containing osmoticum that will allow easy penetration of the DNA into the cells and their nuclei (Slater et al. 2008). However, *Agrobacterium*-mediated genetic transformation is a method of choice due to reduced transgene silencing (Kumar et al. 2014) and its simple methodology (Brumbley et al. 2008) and a high efficiency of transgene integration with few copies (Kumar et al. 2013).

Gene promoters: DNA sequence upstream of the transcribed region at promoter site stabilizes RNA polymerases to start transcription. C4 plant species, e.g., s*orghum* and *zea mays* having high genomic synteny (with sugarcane), can be used for map-based identification of regulatory framework involving cis regulatory elements and promoters (Waclawovsky et al. 2010). A number of researchers used promoters, e.g., emu, maize adh 1, CaMV 35S, maize ubiquitin promoter, TMV 35S, and Rab17 for construction of gene cascade in sugarcane genetic transformation (Reis et al.

2014). AVP1 (arabidopsis vacuolar pyrophosphatase-1) gene driven under 35S CaMV promoter was used for genetic transformation against drought and salinity tolerance in sugarcane (Kumar et al. 2014). Reis et al. (2014) used the stress-inducible ABA-responsive Rab17 gene promoter to drive AtDREB2A CA gene expression in sugarcane. Role of TFs in activating stess-inducible promoters for gene expression has been elucidated in the foregoing section. The stress-inducible rd29A promoter is *in vogue* useful compared to 35S CaMV (a constitutive promoter) for improving drought, cold, and salt stress without any penalty on growth (Kasuga et al. 2004). In sugarcane, the MYB transcriptional factor with its promoter PScMYBAS1 also imparted positive regulation for abiotic stress tolerance (dehydration, salt, cold and wounding) (Prabu and Prasad 2012). Many researchers preferred constitutive expression of the CBF genes. The arabidopsis CBF4 gene transferred to sugarcane under the control of the maize ubiquitin promoter, and the nos terminator was reported to improve drought tolerance (McQualter and Dookun-Saumtally 2007). But, transgenic plants overexpressing AtDREB1A using 35S CaMV constitutive promoter showed growth retardation in tobacco. The problem, however, was overcome by using abiotic stress-inducible promoter rd29A (Kasuga et al. 2004). In contrast, constitutive overexpression of CBF3 and ABF3 in rice led to the development of transgenic rice with elevated tolerance to drought and salinity without any growth inhibition or any phenotypic aberrations (Oh et al. 2005). Thus, the requirement of constitutive or inducible promoter is transgene as well as crop specific. Overexpression of SodERF3 is a sugarcane ethylene-responsive factor with CaMV 35S promoter-enhanced salt and drought tolerance in tobacco plants (Trujillo et al. 2008). Besides, under drought conditions, both AtBBX29 and SINAC2 (TFs) overexpression in sugarcane-enhanced drought tolerance (Mbambalala 2020) using ubiquitin promotor. Transgenic plants overexpressing AtBBX29 are able to maintain relative water content (RWC) comparable to control, and they also showed higher chlorophyll fluorescent rates, stomatal conductance, and scavenging activity of antioxidants while SINAC2 overexpression enhanced root elongation under drought stress.

Transgenics for modulating osmotic stress: Until now, only a few scientists have tried to develop drought-tolerant transgenic sugarcane plants. Kumar et al. (2013) used *Agrobacterium*-strain EHA105 harboring pGreen0029 vector with a*rabidopsis* vacuolar pyrophosphatase (AVP1) gene for genetic transformation against drought and salinity tolerance in sugarcane. Zhang et al. (2006) reported the overexpression of *Grifola frondosa* trehalose synthase (*Tsase*) gene under CaMV35S promoter for improving drought tolerance in sugarcane. Similarly, Wang et al. (2005) developed the transgenic sugarcane plants harboring grifola frondosa synthase gene, which was confirmed by PCR and southern analysis. They reported some transgenic plants with multiple phenotypic alterations, and some plants demonstrated improved tolerance to osmotic stress. In another study, Molinari et al. (2007) evaluated the stress-inducible production of proline in transgenic sugarcane. Their results suggested that stress-inducible proline accumulation in transgenic sugarcane plants under water deficit stress acts as a component of antioxidant defense system rather than as osmotic adjustment mediator. Transgenic plants overexpressed six stress-responsive genes, namely RDA29A, RDA29B, COR47, RD20, GSTF6, and P5CS1 (Jiang et al. 2014), which, among other functions, are involved in proline synthesis and cellular dehydration. Raza (2013) recovered drought-tolerant sugarcane transgenic lines with elevated RWC in the leaf, leaf-water potential, osmotic potential, turgor potential, photosynthetic rate, and transpiration rate. In Indonesia, three P5CS transgenic sugarcane varieties tolerant to drought stress received environmental safety and variety-release approvals from the ministry of agriculture in 2013 (James 2013).

Transgenics using TFs: A number of studies have been done to improve abiotic stress tolerance using transcriptional factors. Transcription factors (TFs) are proteins that regulate the gene expression through a unique recognition of cis-acting elements located on the promoter region of target genes (Nakashima et al. 2009). A single TF can regulate the expression of many target genes. Hence, the TFs are essential in relation to plant growth and development, e.g., cell differentiation, development of tissue and organs, responses to hormones and environmental factors, metabolic networks, and disease resistance (Gonzalez 2015). Hence, they are potential targets for gene manipulation to enhance drought tolerance in crop species. Mbambalala (2020) studied the role of two TFs namely, BBX (B-Box zinc finger) and NAC2 (NAM, ATAF, and CUC) in enhancing drought tolerance in sugarcane. DREBs are the important TFs in the (ABA)-independent pathway. Transcriptome study revealed SsDREB1F, SsDREB1L, SsDREB2D, and SsDREB2F were up-regulated under drought stress in S. spontaneum (Li et al. 2021). Besides, genetic transformation of sugarcane for drought tolerance has been attempted using arabidopsis bax inhibitor-1 gene (Ramiro et al. 2016). Sugarcane transgenics expressing MYB transcription factors show improved glucose release with a significant decrease in acid-insoluble lignin (Poovaiah et al. 2016). OsNAC5 TFs was associated as a positive regulator of drought stress-responsive genes (Takasaki et al. 2010).

Four major TFs gene families (WRKY, NAC, MYB, and AP2/ERF) that modulate stress-responsive gene expression under biotic and abiotic stresses (Javed et al. 2020) are reported to be the ideal choice for genetic engineering for abiotic stress tolerance, including moisture stress (Wang et al. 2016). In sugarcane, a WRKY-IIc TF gene ScWRKY3 overexpressed in response to salinity, artificial drought stress caused by PEG (Wang et al. 2018). Typical NAC TFs are characterized by the presence of NAC domain for DNA binding at N terminal and a diverse C-terminal (Olsen et al. 2005). Their expressions are mostly tissue and species-specific. In sugarcane, SsNAC23 expression was induced in response to temperature fluctuations (Ditt et al. 2011). ANAC019, ANAC055, and ANAC072 genes are reported to be involved in drought stress in arabidopsis (Pandey and Shukla 2015). Overexpression of Os NAC6 and OsNAC10 (root-specific) exhibited to improve root dynamics and to enhance drought tolerance in rice under field conditions (Jeong et al. 2010). Another observation revealed that a sugarcane ScMYB2 was participated in an ABA-mediated leaf senescence-signaling pathway and acted as a positive regulator in respond to PEG-mediated drought stress (Guo et al. 2017). AP2/ERF is a broad group of TFs comprising five subgroups, e.g., AP2, RAV (related to ABI3/VP1), DREB (dehydration-responsive element-binding protein), ERF (ethylene-responsive factors), and others. Until now, numerous AP2/ERF TFs were identified in plant species, including six AP2/ERF TFs in sugarcane and 30 AP2/ERF TFs in arabidopsis. Expression of SodERF3 was induced by ABA, salt stress, and wounding in sugarcane. Besides, drought resistance was reported to be positively regulated by AtDREB2A CA (de Souza 2019), DREB 1(Donde et al. 2014), and SodERF3 (Trujillo et al. 2009) in sugarcane.

7.8.5.2 Cisgenic Approach

Cisgenic plants have great potential to overcome major challenges faced by traditional breeding techniques (Schouten et al. 2006) and regulatory barriers regarding the evaluation of transgenic plants in the field (Holme et al. 2013). During cisgenesis and intragenesis, plants must only be transformed with genetic material derived from the species itself or closely related species capable of sexual hybridization (Holme et al. 2013). Although cisgenic plants may not require the same level of regulation as transgenic plants, their application for crop improvement has been restricted by lack of knowledge of the regulatory sequences required (Espinoza et al. 2013). Begcy et al. (2019) found two sugarcane genes (Scdr1 and Scdr2) encoding proteins with unknown function that confer tolerance to drought stress in tobacco plants. Scdrl overexpression had a positive impact in maintaining high RWC and photosynthetic rate under drought; however, its major impacts were observed under salt stress. Findings of Begcy et al. (2012) suggest that Scdrl has the potential to produce sugarcane varieties with enhanced tolerance to both drought and salt stress. Characterising the role of Scdr1 and Scdr2 in sugarcane exposed to drought stress is essential because they may represent new sources of cisgenic sugarcane plant. The erstwhile-mentioned SodERF3, a novel sugarcane ethylene responsive Factor (ERF) that enhances drought and salt tolerance may be tried in sugarcane for overexpression by gene stacking. HSP70 stabilizes the proteins by binding to its protein substrates and prevents unwanted denaturation. The *Agrobacterium*-mediated transformed plants of the sugarcane hybrid Co 86032 showed drought and salinity tolerance by overexpression of the EaHSP70 gene from the wild relative of sugarcane (*Erianthus arundinaceus*) (Augustine et al. 2015). Similarly, an expansin gene (EaEXPA1 for cell wall loosening) from *Erianthus arundinaceus* was overexpressed in the commercial sugarcane variety Co 86032 through *Agrobacterium*-mediated transformation. The transgenic plants had shown significantly higher relative water content, cell membrane thermostability, chlorophyll content, photosynthetic efficiency (Fv/Fm), and gas-exchange parameters than untransformed control plants. Besides, some transgenic lines transformed with P5CS gene from *S. officinarum* showed significantly higher in SoP5CS expression level, proline accumulation, abscisic acid content, superoxide dismutase activity, and relative water content but lower MDA content and chlorophyll (SPAD value) decrease as compared with the WT plants (Li et al. 2018).

7.8.5.3 Genome Editing

The conventional transgenic technique is highly time consuming and labour intensive, and it requires strict regulatory clearances for general cultivation owing to genetic modification with foreign gene(s). On the other hand, the traditional mutation breeding creates genetic changes at random genomic sites still at very low frequency, besides causing genomic instability. However, in the recent genomic-engineering era, genome-editing tools, e.g., zinc-finger nucleases (ZNFs), transcription activator-like effector nucleases (TALLENs) and clustered regularly interspaced short palindromic repeat-associated Cas9 nuclease (CRISPR-Cas9) system hold promise to replace several bottlenecks of traditional crop breeding (Kumar and Jain 2015). In recent years, CRISPR-Cas9 is widely used because of its simplicity, efficiency, and versatility. The sgRNA recognizes the target site of the genome, while the Cas9 causes double-stranded breaks (DSBs) at the targeted site of DNA. The HNH domain of Cas9 cuts the complementary strand of CRISPR RNA (crRNA), and the RucV-like domain cleaves the opposite strand of the double-stranded DNA. Consequently, the double-strand breaks (DSBs) are instantly repaired either by nonhomologous end-joining (NHEJ) or homology-directed repair (HDR) resulting indels/gene knockout or gene replacement/ foreign DNA knock-ins, respectively (Kumar and Jain 2015).

Designing the target sequence-specific gRNAs for CRISPR/Cas9 system is the major setbeck in sugarcane owing to its genome complexity and nonavailability of complete annotated genome sequences. However, the RNA-Seq data of *S. officinarum* or the allele-defined genome of *S. spontaneum* (as a reference genome) can be used to design precise gRNAs targeting allelic forms of the gene of interest. Besides, the knowledge of next-generation sequencing (NGS), gene-editing systems (Shan et al. 2013), gene silencing (Yin et al. 2014), and overexpression methods (Saad et al. 2013) hold an advantage to provide short sequence libraries that can be used during gRNA design. In addition, dimeric RNA-guided fokI nucleases (RFNs) can be used to induce high genome-editing frequencies and reduced off-target effects (Bortesi and Fischer 2015).

The genome-editing system have been reported for precision breeding for a variety of plant species, including rice (Ren et al. 2017), maize (Zong et al. 2017), and even in the hexaploid wheat (Gil-Humanes et al. 2017). TALLEN has been demonstrated in sugarcane (Jung and Altpeter 2016), and CRISPR/Cas9 has also been opted as method of choice for targeted genome editing in the crop (Ferreira et al. 2017). Multiplex genome editing has been demonstrated, targeting simultaneously a maximum up to 14 distinct genome loci with no detectable off-target events (Peterson et al. 2016). However, only few workers ventured CRISPR/Cas9 technique to target a gene related to drought stress. The CRISPR-Cas9 edited maize gene 'AGOS8' (a negative regulator of ethylene response) had shown drought tolerance without any yield penalty (Shi et al. 2017) in maize. On bio-safety issues, the CRISPR/Cas9 edited mutants do not face as many harsh regulatory hurdles as compared to transgenics for release as commercial variety in many countries (Liang et al. 2017). It is expected that future researchers would pay more focus on CRISPR/Cas9 genome editing to improve drought tolerance in sugarcane.

7.9 CONCLUSION

Huge production of sugarcane is the key for sustainability of sugar industries. The crop is highly water consumptive, but the prevalent harsh environment with particularly irratic rainfall and long drought spell drastically affects the cane yield and juice quality. Breeding a drought-tolerant sugarcane variety is really difficult owing to the complex genetic basis of the trait and, more emphatically, the complex polyploid nature of the crop. Traditional breeding so far attempted has not made any major progress in this direction. However, the availability of high throughput genome-sequencing platforms and modern bioinformatic tools made it easy to understand its complex genome, and explore and annotate novel genes involved in a variety of metabolic pathways in response to moisture stress. Sugarcane being highly responsive to *in vitro* culture, it is quite amenable for *in vitro* mutagenesis, *in vitro* screening, genetic transformation, and genome editing. Besides, being a clonally propagated crop, any drought-tolerant F_1 hybrid, genetic segregant, or mutant with high cane yield and better juice quality can serve as elite initial material for quick varietal development, regardless of its genotypic (alletic) constitution. It is expected that traditional breeding integrated with advanced biotechnological tools, e.g., transcriptome analysis, marker-assisted selection, genetic transformation supported by appropriate promoters and TFs, cisgenic gene stacking and genome editing using CRISPR-Cas9 system may be a game changer for accelerating the pace of precision breeding toward drought tolerance in sugarcane.

REFERENCES

Abbas S.R., Gardazi S.D.A., Sabir S.M. et al. (2013). Screening of drought tolerant genotypes of sugarcane through biochemical markers against polyethylene glycol. *International Journal of Scientific and Engineering Research*, 4(7): 980–988.

Adawy S.S., Diab A.A., Atia M.A. et al. (2013). Construction of genetic linkage map with chromosomal assigment and quantitative trait loci associated with some important agronomic traits in cotton. *GM Crops Food*, 4: 36–49.

Aftab F. and Iqbal J. (2001). PEG-Mediated somatic hybridization studies in sugarcane (*Saccharum* Spp. Hybrid CVS. CoL-54 and CP-43/33). *Pakistan Journal of Botany*, 33(3): 233–238.

Agarwal P.K., Agarwal P., Reddy M.K. et al. (2006). Role of DREB transcription factors in abiotic and biotic stress tolerance in plants. *Plant Cell Reports*, 25(12):1263–1274. doi: 10.1007/s00299-006-0204-8

Aitken K. et al. (2006a). AFLP analysis of genetic diversity within *Saccharum officinarum* and comparison with sugarcane cultivars. *Australian Journal of Agricultural Research*, 57(11):167–184, 10.1071/AR05391

Aitken K., Jackson P.A., and McIntyre C.L. (2006b). Quantitative trait loci identified for sugar related traits in a sugarcane (*Saccharum* spp.) cultivar x *Saccharum officinarum* population. *Theoretical and Applied Genetics*, 112(7):1306–1317.

Aitken K. et al. (2008). Genetic control of yield related stalk traits in sugarcane. *Theoretical and Applied Genetics*, 117(7): 1191–1203. 10.1007/s00122-008-0856-6

Aitken K.S., Mcneil M.D., Hermann S. et al. (2014). A comprehensive genetic map of sugarcane that provides enhanced map coverage and integrates high-throughput diversity array technology (DArT) markers. *BMC Genomics* 15:152. doi: 10.1186/1471-2164-15-152

Aitken K., Jackson P., and Mcintyre C. (2005). A combination of AFLP and SSR markers provides extensive map coverage and identification of homo(eo)logous linkage groups in a sugarcane cultivar. *Theoretical and Applied Genetics*, 110:789–801. doi: 10.1007/s00122-004-1813-7

Aitken K.S., and Jackson, P.A., C.L. (2006). Quantitative trait loci identified for sugar related traits in a sugarcane (Saccharum spp.) cultivar × Saccharum officinarum population. *Theoretical and Applied Genetics* 112:1306–1317.10.1007/s00122-006-0233-2

Allen A.M., Winfield M.O., Burridge A.J. et al. (2017). Characterization of a wheat breeders' array suitable for high-throughput SNP genotyping of global accessions of hexaploid bread wheat (*Triticum aestivum*). *Plant Biotechnology Journal*, 15:390–401. doi: 10.1111/pbi.12635

Aloni R., Schwalm K., Langhans M. et al. (2003). Gradual shifts in sites of free-auxin production during leaf-primordium development and their role in vascular differentiation and leaf morphogenesis in *Arabidopsis*. *Planta*, 216:841–853.

Alwala S., Andru S., Arro J.A., et al. (2006). Target region amplification polymorphism (TRAP) for assessing genetic diversity in sugarcane germplasm collections. *Crop Science*, 46:448–455.

Andru S., Kimbeng C.A., Edmé S.J. et al. (2008). Sequence related amplified polymorphism (SRAP) markers for assessing genetic relationship and diversity in sugarcane germplasm collection. *Plant Genetic Research*, 6:222–231.

Anjum S.A., Xie X.Y., Wang L.C. et al. (2011). Morphological, physiological and biochemical responses of plants to drought stress. *African Journal of Agricultural Research*, 6(9):2026–2032.

Anonymous (2011). *Vision-2030: IISR- Perspection Plan*, Indian Institute of Sugarcane Research, pp. 41.

Arencibia A., Molina P.R., de la Riva G. et al. (1995). Production of transgenic sugarcane (*Saccharum officinarum* L.) plants by intact cell electroporation. *Plant Cell Reports*, 14(5): 305–309. 10.1007/BF00232033

Arvinth S. et al. (2010). Genetic transformation and pyramiding of aprotinin-expressing sugarcane with cry1Ab for shoot borer (*Chilo infuscatellus*) resistance. *Plant Cell Reports*, 29(4):383–395. 10.1007/s00299-010-0829-5

Asano T., Tsudzuki T, Takahashi S. et al. (2004). Complete nucleotide sequence of the sugarcane (*Saccharum officinarum*) chloroplast genome: a comparative analysis of four monocot chloroplast genomes. *DNA Research*, 11: 93–99.

Ashraf M. (2009). Biotechnological approach of improving plant salt tolerance using antioxidants as markers. *Biotechnology Advances*, 27(1): 84–93.

Ashraf M. and Harris P.J. (2013). Photosynthesis under stressful environments: an overview. *Photosynthetica*, 51(2):163–190.

Augustine S.M., Narayan J.A., Syamaladevi D.P. et al. (2015). *Erianthus arundinaceus* HSP70 (EaHSP70) overexpression increases drought and salinity tolerance in sugarcane (*Saccharum* spp. hybrid). *Plant Science*, 232: 23–34.

Azevedo R.A., Carvalho R.F., and Cia M.C. (2011). Sugarcane under pressure: An overview of biochemical and physiological studies of abiotic stress. *Tropical Plant Biology*, 4: 42–51.

Balsalobre T.W.A., Da Silva P.G., Margarido G.R.A. et al. (2017). GBS-based single dosage markers for linkage and QTL mapping allow gene mining for yield-related traits in sugarcane. *BMC Genomics*, 18:72. doi: 10.1186/s12864-016-3383-x

Barreto F.Z., Rosa J.R.B.F., Balsalobre T.W.A. et al. (2019). A genome-wide association study identified loci for yield component traits in sugarcane (*Saccharum* spp.). *PLoS ONE*, 14: e0219843.

Batista R., Saibo N., Lourenco T. et al. (2008). Microarray analyses reveal that plant mutagenesis may induce more transcriptomic changes than transgene insertion. *Proceedings of the National Academy of Sciences*, 105(9): 3640–3645.

Begcy K., Mariano E.D., Lembke C.G. et al. (2019). Overexpression of an evolutionarily conserved drought-responsive sugarcane gene enhances salinity and drought resilience. *Annals of Botany*, 124: 691–700.

Begcy K., Mariano E.D., Gentile A. et al. (2012). A novel stress-induced sugarcane gene confers tolerance to drought, salt and oxidative stress in transgenic tobacco plants. *PloS one*, 7(9): e44697. 10.1371/journal.pone.0044697

Belesini A., Carvalho F., Telles B. et al. (2017). De novo transcriptome assembly of sugarcane leaves submitted to prolonged water-deficit stress. *Genetics and Molecular Research*, 16: gmr16028845.

Blum A. (2005). Drought resistance, water-use efficiency, and yield potential—are they compatible, dissonant, or mutually exclusive? *Crop and Pasture Science*, 56:1159–1168. doi: 10.1071/AR05069

Bortesi L. and Fischer R. (2015). The CRISPR/Cas9 system for plant genome editing and beyond. *Biotechnol. Advances*, 33: 41–52.

Borrill P, Adamski N, and Uauy C. (2015). Genomics as the key to unlocking the polyploid potential of wheat. *New Phytology*, 208:1008–1022.

Bossis G. and Melchior F. (2006). SUMO: regulating the regulator. *Cell Division*, 1: 1–8.

Boutraa T., Akhkha A., Al-Shoaibi A.A. et al. (2010). Effect of water stress on growth and water use efficiency (WUE) of some wheat cultivars (*Triticum durum*) grown in Saudi Arabia. *Journal of Taibah University for Science*, 3(1): 39–48.

Bower R. and Birch R.G. (1992). Transgenic sugarcane plants via micro-projectile bombardment. *Plant Journal*, 2: 409–416. 10.1111/j.1365-313X.1992.00409.x

Bremer G. (1923). A cytological investigation of some species and species-hybrids witbin the genus Saccharum. *Genetica*, 5: 97–148.

Brown J.S., Schnell R.J., Power E.J. et al. (2007). Analysis of clonal germplasm from five Saccharum species: *S. barberi, S. robustum, S. officinarum, S. sinense* and *S. spontaneum*. A study of inter- and intra species relationships using microsatellite markers. *Genetic Resources and Crop Evolution*, 54: 627–648.

Brumbley S.M., Snyman S.J., Gnanasambandam A. et al. (2008). Sugarcane. In: *Compendium of transgenic crop plants: Transgenic sugar, tuber and fiber crops*. (Eds.) C Kole and TC Hall: Blackwell, Oxford, pp. 1–58, 10.1002/9781405181099.k0701

Butterfield M., D'hont A., and Berding N. (2001). The sugarcane genome: a synthesis of current understanding, and lessons for breeding and biotechnology. In: Proceedings of South African sugar Technologist Association, A-75, pp. 1–5.

Calsa Júnior T., Carraro D.M., Benatti M.R. et al. (2004). Structural features and transcript-editing analysis of sugarcane (*Saccharum officinarum* L.) chloroplast genome. *Current Genetics*, 46: 366–373.

Carson D.L. and Botha F.C. (2000). Preliminary analysis of expressed sequence tags for sugarcane. *Crop Science*, 40: 1769–1779.

Casu R. et al. (2004). Identification of differentially expressed transcripts from maturing stem of sugarcane by in silico analysis of stem expressed sequence tags and gene expression profiling. *Plant Molecular Biology*, 54(4): 503–517.

Catala R., Ouyang J., Abreu I.A. et al. (2007). The Arabidopsis E3 SUMO ligase SIZ1 regulates plant growth and drought responses. *The Plant Cell*, 19: 2952–2966.

Costa, T. H.F., Masarin, F., Bonifácio, T. O., Milagres, A. M.F., & Ferraz, A. (2013). The enzymatic recalcitrance of internodes of sugar cane hybrids with contrasting lignin contents. *Industrial Crops and Products*, 51:202–21110.1016/j.indcrop.2013.08.078

Cha-um S., Wangmoon S., Mongkolsiriwatana C. et al. (2012). Evaluating sugarcane (*Saccharum* sp.) cultivars for water deficit tolerance using some key physiological markers. *Plant Biotechnology*, 29: 431–439. doi:10.5511/plantbiotechnology.12.0726a

Chen Y., Li C., Zhang B. et al. (2019). The role of the late embryogenesis-abundant (LEA) protein family in development and the abiotic stress response: A comprehensive expression analysis of potato (*Solanum tuberosum*). *Genes*, 10: 1–16.

Chinnusamy V., Schumaker K., and Zhu J.K. (2004). Molecular genetic perspectives on cross-talk and specificity in abiotic stress signalling in plants. *Journal of Experimental Botany*, 55(395): 225–236.

Cox M., Hogarth M., and Smith G. (2000). Cane breeding and improvement. In: *Manual of cane growing*. (Eds.)M. Hogarth and P Allsopp: Bureau of Sugar Experimental Stations, Indooroopily, Australia, pp. 91–108.

Cuadrado A. et al. (2004). Genome remodelling in three modern *S. officinarum* × *S. spontaneum* sugarcane cultivars. *Journal of Experimental Botany*, 55:(398):847–854. 10.1093/jxb/erh093

Cuervo-Alarcon L., Arend M., Müller M. et al. (2021). A candidate gene association analysis identifies SNPs potentially involved in drought tolerance in European beech (*Fagus sylvatica* L.). *Scientific Reports*, 11: 2386. 10.1038/s41598-021-81594-w

D'Hont A. et al. (1998). Determination of basic chromosome numbers in the genus *Saccharum* by physical mapping of ribosomal RNA genes. *Genome*, 41(2): 221–225. 10.1139/g98-023

da Costa M.L.M., Amorim L.L.B., Onofre A.V.C. et al. (2011). Assessment of genetic diversity in contrasting sugarcane varieties using inter-simple sequence repeat (ISSR) markers. *American Journal of Plant Sciences*, 2: 425–432.

Davey J.W., Hohenlohe P.A., Etter P.D. et al. (2011). Genome-wide genetic marker discovery and genotyping using next generation sequencing. *National Review of Genetics*, 12(7): 499–510.

da Silva M.D., Silva R.L.O., Costa F.N. (2013). Expression analysis of sugarcane aquaporin genes under water deficit. *Journal of Nucleic Acids*, 2013: ID 763945. doi: 10.1155/2013/763945

Daniels J. and Roach B.T. (1987). Taxonomy and evolution. In: *Sugarcane improvement through breeding*. (Ed.)DJ Heinz: Elsevier, Amsterdam, The Netherlands. pp. 7–84.

Dapanage and Bhat S. (2019). Quantitation of differential expression of transcription factors under moisture deficit stress in sugarcane. *African Journal of Biotechnology*, 18(8):163–170.

Das K. and Roychoudhury A. (2014). Reactive oxygen species (ROS) and response of antioxidants as ROS-scavengers during environmental stress in plants. *Frontiers in Environmental Science*, 2; 1–13.

de Alcantara G.B., Dibax R., de Oliveira R.A. et al. (2014). Plant regeneration and histological study of the somatic embryogenesis of sugarcane (*Saccharum* spp.) cultivars RB855156 and RB72454. *Acta Scientiarum Agronomy*, 36(1):1–6. 10.4025/actasciagron.v36i1.16342

de Andrade, Nobile L.M., Ribeiro P.M. et al. (2016). Characterization of PIP2 aquaporins in *Saccharum* hybrids. *Plant Gene*, 5: 31–37.

de Setta N., Monteiro-Vitorello C.B., and Metcalfe C.J. (2014). Building the sugarcane genome for biotechnology and identifying evolutionary trends. *BMC Genomics*, 15(1): 540.

de Souza W, De Oliveira N.G. et al. (2019). Field evaluation of AtDREB2A CA overexpressing sugarcane for drought tolerance. *Journal of Agronomy and Crop Science*, 205(6):1–9. doi: 10.1111/jac.12341

Ditt R.F., Gentile A., Tavares R.G. et al. (2011). Analysis of the stress-inducible transcription factor SsNAC23 in sugarcane plants. *Scientific Agriculture*, 68(4): 454–461.

Donde R., Samal K.C., and Rout G.R. (2014). Studies on *Agrobacterium* mediated genetic transformation of rice (*Oryza sativa* L.) for drought tolerance using Dreb 1 gene. *International journal of Agricultural Science*, 4(12):341–350.

Earle F.S. (1928). *Sugarcane and its culture*. John Wiley, New York, 1928, p. 355.

Efendi E. and Matsuoka M. (2011). An efficient Agrobacterium-mediated transformation method for sugarcane (*Saccharum officinarum* L.). In: Proceedings of the Annual International Conference Syiah Kuala University, Banda Aceh, Indonesia, November 29–30, 2011.

Eldessoky D.S., Ismail R.M., Abdel-Hadi A.H.A. et al. (2011). Establishment of regeneration and transformation system of sugarcane cultivar GT54-9 (C9). *GM Crops*, 2(2): 126–134. 10.4161/gmcr.2.2.17288

Espinoza C., Schlechter R, Herrera D. et al. (2013). Cisgenesis and Intragenesis: New tools for improving crops. *Biological Research*, 46(4):323–331. doi: 10.4067/S0716-97602013000400003

Food and Agriculture Organization, Corporate Statistical Database (FAOSTAT). (2019). Retrieved 10 October 2020. https://en.wikipedia.org/wiki/Sugarcane

Food and Agriculture Organization (2004). *Saccharum officinarum L.* Food and Agriculture Organization: www.fao.org/ag/AGP/AGPC/doc/GBASE/data/Pf000310.HTM

Farooq M., Basra S.M.A., Wahid A. et al. (2008). Physiological role of exogenously applied glycinebetaine to improve drought tolerance in fine grain aromatic rice (*Oryza sativa* L.). *Journal of Agronomy and Crop Science*, 194(5): 325–333.

Fawaz A.W., Nasr M.I., Hamdy M.E. et al. (2013). Genetic variability among sugarcane genotypes based on TRAP markers of candidate drought genes and some agronomic traits. *Assiut Journal of Agricultural Sciences*, 44(3): 46–62.

Ferreira T.H., Gentile A., Vilela R.D. et al. (2012). microRNAs associated with drought response in the bioenergy crop sugarcane (*Saccharum* spp.). *PLoS ONE*, 7:e46703. doi: 10.1371/journal.pone.0046703

Ferreira T.H., Tsunada M.S., Bassi D. et al. (2017). Sugarcane water stress tolerance mechanisms and its implications on developing biotechnology solutions. *Front. Plant Science*, 8: 1–18.

Fickett N., Gutierrez A., Verma M. et al. (2019). Genome wide association mapping identifies markers associated with cane yield components and sucrose traits in the Louisiana sugarcane core collection. *Genomics*, 111(6):1794–1801. doi: 10.1016/j.ygeno.2018.12.002

Figueira T., Okura V., Rodrigues Da Silva F. et al. (2012). A BAC library of the SP80-3280 sugarcane variety (*Saccharum* sp.) and its inferred microsynteny with the *Sorghum* genome. *BMC Research Notes*, 5: 185. 10.1186/1756-0500-5-185

Fujita M., Fujita Y., Noutoshi Y. et al. (2006). Crosstalk between abiotic and biotic stress responses: A current view from the point of convergence in the stress signaling networks. *Current Opinion in Plant Biology*, 9: 436–442.

Garcia A.A., Kido E.A., Meza A.N. et al. (2006). Development of an integrated genetic map of a sugarcane (*Saccharum* spp.) commercial cross, based on a maximum-likelihood approach for estimation of linkage and linkage phases. *Theoretical and Applied Genetics*, 112: 298–314. doi: 10.1007/s00122-005-0129-6

Garcia A., Mollinari M., Marconi T. et al. (2013). SNP genotyping allows an in-depth characterisation of the genome of sugarcane and other complex autopolyploids. *Scientific Reports*, 3: 3399. 10.1038/srep03399

Garsmeur O., Droc G., Antonise R. et al. (2018). A mosaic monoploid reference sequence for the highly complex genome of sugarcane. *Nature Communication*, 9: 2638. 10.1038/s41467-018-05051-5

Gentile A., Li D., Mattos R.S. et al. (2015). MicroRNAs and drought responses in sugarcane. *Frontiers in Plant Science*, 6:58. doi: 10.3389/fpls.2015.00058

Gil-Humanes J., Wang Y., Liang Z. et al. (2017). High-efficiency gene targeting in hexaploid wheat using DNA replicons and CRISPR/Cas9. *Plant Journal*, 89:1251–1262. doi: 10.1111/tpj.13446

Gill, S.S. and Tuteja, N. (2010). Reactive oxygen species and antioxidant machinery in abiotic stress tolerance in crop plants. *Plant Physiology and Biochemistry*, 48(12): 909–930.

Goll M.G. and Bestor T.H. (2002). Histone modification and replacement in chromatin activation. *Genes and Development*, 16: 1739–1742.

Gomathi R., Krishnapriya V., Arunkumar R. et al. (2020). Physiological traits imparting drought stress tolerance to promising sugarcane (*Saccharum* spp.) clones. *Plant Physiology Reports, Online fast*, 10.1007/s40502-020-00536-2

Gonzalez D.H. (ed.). (2015). *Plant transcription factors: evolutionary, structural and functional aspects.* Elsevier Inc. Academic Press. pp.85

Gouy M., Rousselle Y., Bastianelli D. et al. (2013). Experimental assessment of the accuracy of genomic selection in sugarcane. *Applied Genetics*, 126: 2575–2586.

Guimaraes E.R., Mutton M.A., Mutton M.J.R. et al. (2008). Free proline accumulation in sugarcane under water restriction and spittlebug infestation. *Scientia Agricola*, 65: 628–633.

Guo J., Ling H., Ma J. et al. (2017). A sugarcane R2R3-MYB transcriprion factor gene is alternatively spliced during drought stress. *Scientific Reports*, 7: 41922.

Gupta V., Raghuvanshi S., Gupta A. et al. (2010). The water-deficit stress- and red-rot-related genes in sugarcane. *Functional & Integrative Genomics*, 10(2): 207–214. 10.1007/s10142-009-0144-9

Hanumappa M. and Nguyen H.T. (2010). Genetic approaches toward improving heat tolerance in plants. In: *Genes for Plant Abiotic Stress*. (Eds.) Jenks, M.A. and Wood, A.J.: Wiley-Blackwell, Ames, pp. 221–247.

Haq-UI S., Kumar P., Singh R.K. et al. (2016). Assessment of functional EST-SSR markers (Sugarcane) in cross-species transferability, genetic diversity among Poaceae plants, and bulk segregation analysis. *Genetics Research International*, 2016:1–16. doi: 10.1155/2016/7052323

Hemaprabha G., Simon S., Lavanya L., Sajitha B., and Venkataramana S. (2012). Evaluation of drought tolerance potential of elite genotypes and progenies of sugarcane (*Saccharum* sp. hybrids). *Sugar Tech*, 15(1): 9–16. doi: 10.1007/s12355-012-0182-9

Hoang N.V., Furtado A., Mason P.J. et al. (2017). A survey of the complex transcriptome from the highly polyploid sugarcane genome using full-length isoform sequencing and de novo assembly from short read sequencing. *BMC Genomics*, 18:395. doi: 10.1186/s12864-017-3757-8

Hoarau J.Y., Offmann B., D'Hont A. et al. (2001). Genetic dissection of a modern sugarcane cultivar (*Saccharum* spp.). I. Genome mapping with AFLP markers. *Theoretical and Applied Genetics*, 103(1): 84–97.

Hogarth D.M. (1987). Genetics of Sugarcane. In: *Sugarcane improvement through breeding*. (Ed.)D. J. Heinz: Elsevier, Amsterdam, pp. 255–271.

Holme I.B., Wendt T., and Holm P.B. (2013). Intragenesis and cisgenesis as alternatives to transgenic crop development. *Plant Biotechnology Journal*, 11: 395–407.

Hong Z., Lakkineni K., Zhang K., and Verma D.P.S. (2000). Removal of feedback inhibition ofD1-pyrroline-5-carboxylatesynthetase results in increased proline accumulation andprotection of plants from osmotic stress. *Plant Physiology*, 122: 1129–1136.

Huh J., Kang B., and Nahm S. (2001). A candidate gene approach identified phytoene synthase as the locus for mature fruit color in red pepper (*Capsicum* spp). *Theoretical and Applied Genetics*, 102: 524–530.

Hussain S., Khaliq A., and Mehmood U. (2018). Sugarcane production under changing climate: effects of environmental vulnerabilities on sugarcane diseases, insects and weeds. In: *Climate Change and Agriculture*. (Eds.)S. Kumari, M. Hiloidhari, S.N. Naik, and R.J. Dahiya: InTechOpen, United Kindom, pp. 1–17.

Inman-Bamber N.G. (2004). Sugarcane water-stress criteria for irrigation and drying off. *Field Crops Research*, 89 (1): 107–122.

IRGSP: International Rice Genome Sequencing Project (2005). The map-based sequence of the rice genome. *Nature*, 436(7052): 793–800.

Ishaq M.N. Olaoye G., and Akinsanya T.O. (2012). Screening sugar-cane germplasm for drought tolerance in Nigeria. *Plant Genetic Resources News letter*, 154:48–54.

Iskandar H.M., Casu R.E., Fletcher A.T. et al. (2011). Identification of drought-response genes and a study of their expression during sucrose accumulation and water deficit in sugarcane culms. *BMC Plant Biology*, 11: Article No.12; doi: 10.1186/1471-2229-11

Jackson P.A., Fan Y.H., Li Q.W. et al. (2010). Utilisation of wild canes from China. In: Proceedings of the International Society of Sugarcane Technology. 27:1–10.

Jackson P.A. (2005). Breeding for improved sugar content in sugarcane. *Field Crop Research*, 92(2-3):277–290. 10.1016/j.fcr.2005.01.024

Jain S.M. (2005). Major mutation assisted plant breeding programme supported by FAO/IAEA. *Plant Cell, Tissue and Organ Culture*, 2: 113–123.

James C. (2013). Global Status of commercialized biotech/GM crops. ISAAA Brief No. 46, 2013, Ithaca, NY. pp. 7–23.

Jangpromma N., Kitthaisong S., Daduang S. et al. (2007). 18 kDa protein accumulation in sugarcane leaves under drought stress conditions. *KMITL Science and Technology Journal*, 7(S1): 44–54.

Javed T., Shabbir R., Ali A. et al. (2020). Transcription factors in plant stress responses: Challenges and potential for sugarcane improvement. *Plants*, 9: 491. doi: 10.3390/plants9040491

Jeong J.S., Kim Y.S., Baek K.H. et al. (2010). Root-specific expression of OsNAC10 improves drought tolerance and grain yield in rice under field drought conditions. *Plant Physiology*, 153: 185–197.

Jiang G., Jiang X., Lü P. et al. (2014). The rose (Rosa hybrida) NAC transcription factor 3 gene, RhNAC3, involved in ABA signalling pathway both in rose and *Arabidopsis*. *PloS one*, 9(10): e109415.

Jordan D.R. et al. (2004). Markers associated with stalk number and suckering in sugarcane colocate with tillering and rhizomatousness QTLs in *Sorghum*. *Genome*, 47(5): 988–993.

Jung J.H., and Altpeter F. (2016). TALEN mediated targeted mutagenesis of the caffeic acid O-methyltransferase in highly polyploid sugarcane improves cell wall composition for production of bioethanol. *Plant Molecular Biology*, 92: 131–142. doi: 10.1007/s11103-016-0499-y

Kasirajan L., Hoang N.V., Furtado A. et al. (2018). Transcriptome analysis highlights key differentially expressed genes involved in cellulose and lignin biosynthesis of sugarcane genotypes varying in fiber content. *Scientific Reports*, 2018:8. 10.1038/s41598-01830033-4

Kasuga M., Miera S., Shinozaki K. et al. (2004). A combination of *Arabidopsis* DREB1A gene and stress inducible rd29A promoter improved drought and low-temperature stress tolerance in tobacco by gene transfer. *Plant Cell Physiology*, 5(3): 346–350. 10.1093/pcp/pch037

Khaled K.A., El-Arabi N.I., Sabry N.M. et al. (2018). Sugarcane genotypes assessment under drought condition using amplified fragment length polymorphism. *Biotechnology*, 17: 120–127. doi: 10.3923/biotech, 17:120–127.

Khaled K.A.M., El sherbeny S.E.E., and Abdelhadi A.A. (2017). R-ISSR as marker assisted selection for drought tolerance in sugarcane. *Journal of Agricultural. Chemistry and Biotechnology*, 8(2): 91–97.

Khalil F., Naiyan X., Tayyab M. et al. (2018). Screening of EMS-induced drought-tolerant sugarcane mutants employing physiological, molecular and enzymatic approaches. *Agronomy*, 8; 11–13.

Khan S.A., Hanif Z., Irshad U. et al. (2013). Genetic transformation of sugarcane variety HSF-240 with marker gene GUS. *International Journal of Agriculture and Biology*, 15: 1258–1264.

Khonghintaisong J., Songsri P., Toomsan B. et al. (2018). Rooting and physiological trait responses to early drought stress of sugarcane cultivars. *Sugar Tech*, 20: 396–406.

Kido E.A., Neto J.R.C.F., de Oliveira Silva R.L. et al. (2012). New insights in the sugarcane transcriptome responding to drought stress as revealed by supersage. *The Scientific World Journal*, Article ID 821062. doi:10.1100/2012/821062

Kirch H.H., Schlingensiepen S., Kotchoni S. et al. (2005). Detailed expression analysis of selected genes of the Aldehyde Dehydrogenase (ALDH) Gene Superfamily in *Arabidopsis thaliana. Plant Mol Biology*, 57(3): 315–332. 10.1007/s11103-004-7796-6

Koch A.C., Ramgareeb S., Rutherford R.S. et al. (2012). An *in vitro* mutagenesis protocol for the production of sugarcane tolerant to the herbicide imazapyr. *In Vitro Cellular Developmental Biology-Plant*, 48: 417–427.

Koetle M.J., Evans D.L., Singh V. et al. (2018). Agronomic evaluation and molecular characterisation of the acetolactate synthase gene in imazapyr tolerant sugarcane (Saccharum hybrid) genotypes. *Plant Cell Reports*, 37: 1201–1213.

Kumar T., Uzma Khan M.R., Abbas Z. et al. (2013). Genetic improvement of sugarcane for drought and salinity stress tolerance using *Arabidopsis* Vacuolar Pyrophosphatase(AVP1) gene. *Molecular Biotechnology*, 56: 199. 10.1007/s12033-013-9695-z

Kumar T., Uzma Khan M.R., and Ali G. (2014). Genetic improvement of sugarcane for drought and salinity stress tolerance using *Arabidopsis* Vacuolar Pyrophosphatase (AVP1) gene. *Molecular Biotechnology*, 56(3): 199–209. 10.1007/s12033-013-9695-z

Kumar S and Singh J. (2012). Sugarcane (*Saccharum* spp.) In: *Breeding of field crops*. (Ed.)D.N. Bharadwaj: Agrobios, Jodhpur, pp. 709–733.

Kumar V. and Jain M. (2015). The CRISPR-Cas system for plant genome editing: advances and opportunities. *Journal of Experimental Botany*, 66: 47–57. doi: 10.1093/jxb/eru429

Kurepa J., Walker J.M., Smalle J. et al. (2003). The small ubiquitin-like modifier (SUMO) protein modification system in *Arabidopsis*: accumulation of SUMO1 and-2 conjugates is increased by stress. *Journal of Biology and Chemistry*, 278: 6862–6872

Lakshmanan P., Geijskes R.J., Aitken K.S. et al. (2005). Sugarcane biotechnology: the challenges and opportunities. *In Vitro Cellular and Developmental Biology-Plant*, 41(4): 345–363.

Lakshmanan P. and Robinson N. (2014). Stress physiology: Abiotic stresses. In: *Sugarcane: Physiology, Biochemistry and Functional Biology*. (Eds) P. H. Moore and F. C. Botha: Chichester: JohnWiley & Sons, Inc., pp. 411–434.

Leibbrandt N.B. and Snyman S.J. (2003). Stability of gene expression and agronomic performance of a transgenic herbicide-resistant sugarcane line in South Africa. *Crop Science*, 43(2): 671–677. 10.2135/cropsci2003.6710

Lembke C.G., Nishiyama M.Y., Sato P.M. et al. (2012). Identification of sense and antisense ranscripts regulated by drought in sugarcane. *Plant Molecular Biology*, 79: 461–477. doi: 10.1007/s11103-012-9922-1

Li Z., Wang G., Xihui L.X. et al. (2021). Genome-wide identification and expression profiling of DREB genes in *Saccharum spontaneum. BMC Genomics*, 22:456. 10.1186/s12864-021-07799-5

Li C., Nong Q., Solanki M.K. et al. (2016). Differential expression profiles and pathways of genes in sugarcane leaf at elongation stage in response to drought stress. *Scientific Reports*, 6:25698. doi: 10.1038/srep25698

Li J., Phan T.T., Li Y.R. et al. (2018). Isolation, transformation and overexpression of sugarcane SoP5CS gene for drought tolerance improvement. *Sugar Tech*, 20: 464–473.

Li P., Chai Z., Lin P. et al. (2020). Genome-wide analysis of AP2/ERF transcription factors in sugarcane *Saccharum spontaneum* L. reveals functional divergence during drought, salt stress and plant hormones treatment. *BMC Genomics (Online)*, 10.21203/rs.3.rs-19836/v1

Liang Z., Chen K., Li T. et al. (2017). Efficient DNA-free genome editing of bread wheat using CRISPR/Cas9 ribonucleoprotein complexes. *Nature Communications*. 8:14261. doi: 10.1038/ncomms14261

Liu Z.H., Anderson J.A., Hu J. et al. (2005). A wheat intervarietal genetic linkage map based on microsatellite and target region amplified polymorphism markers and its utility for detecting quantitative trait loci. *Theoretical and Applied Genetics*, 111:782–794.

Liu X., Zhang R., Ou H. et al. (2018). Comprehensive transcriptome analysis reveals genes in response to water deficit in the leaves of *Saccharum narenga* (Nees ex Steud.) hack. *BMC Plant Biology*, 18: 250.

Lopez C.G., Banowetz G.M., Peterson C.J. et al. (2003). Dehydrin expression and drought tolerance in seven wheat cultivars. *Crop Science*, 43: 577–582.

Machado R., Ribeiro R., Marchiori P. et al. (2009). Biometric and physiological responses to water deficit in sugarcane at different phenological stages. *Pesquisa Agropecuária Brasileira*, 44: 1575–1582.

Machado Jr G.R. and Burnquist W.L. (1986). *Variety notes (Fourth revision)*. Copersucar Technology Center, Piracicaba, Sao Paulo, Brazil, 1986, pp. 78.

Mahadevaiah C., Hapase P., Sreenivasa V. et al. (2021). Delineation of genotype × environment interaction for identification of stable genotypes for tillering phase drought stress tolerance in sugarcane. *Scientific Reports*, 11: 18649. 10.1038/s41598-021-98002-y

Mahlanza T., Rutherford R.S., and Snyman S.J. (2013). *In vitro* generation of somaclonal variant plants of sugarcane for tolerance to *Fusarium sacchari*. *Plant Cell Reports*, 32: 249–262.

Manchanda P., Kaur A., and Gosal S.S. (2018). Somaclonal variation for sugarcane improvement. In: *Biotechnologies of crop improvement*. (Eds.)S.S. Gosal, S.H. Wani Volume 1: Cellular Approaches, Springer International Publishing, Cham, pp. 299–326. Ed's.

Manickavasagam M., Ganapathi A., Anbazhagan V.R. et al. (2004). *Agrobacterium*-mediated genetic transformation and development of herbicide-resistant sugarcane (*Saccharum* species hybrids) using axillary buds. *Plant Cell Reports*, 23: 134–143. doi: 10.1007/s00299-004-0794-y

Manimekalai R., Suresh G., Kurup H.G. et al. (2020). Role of NGS and SNP genotyping methods in sugarcane improvement programs.Critical Reviews in Biotechnology. *Online First*. 10.1080/07388551.2020.1765730

Marcos F.C.C., Silveira N.M., Marchiori P.E.R. et al. (2018). Drought tolerance of sugarcane propagules is improved when origin material faces water deficit. *PLoS ONE*, 13(12): e0206716. 10.1371/journal.pone.0206716

Martino A., Mancuso T., and Rossi A.M. (2010). Application of high-resolution melting to large-scale, high-throughput SNP genotyping: a comparison with the TaqMan method. *Journal of Biomolecular Screening*, 15: 623–629. doi: 10.1177/1087057110365900

Masoabi M., Lloyd J., Kossmann J. et al. (2017). Ethyl methanesulfonate mutagenesis and *In vitro* polyethylene glycol selection for drought tolerance in sugarcane (*Saccharum* spp.). *Sugar Tech*, 20: 50–59.

Mason A.S., Higgins E.E., Snowdon R.J. et al. (2017). A user guide to theBrassica60K Illumina Infinium™ SNP genotyping array. *Theoretical and Applied Genetics*, 130: 621–633. doi: 10.1007/s00122-016-2849-1

Matsuoka S., Ferro J., and Arruda P. (2009). The Brazilian experience of sugarcane ethanol industry. *In vitro Cellular and Developmental Biology-Plant*, 45:372–381.

Matsumura H., Reich S.A., Ito A, et al. (2003). Gene expression analysis of plant host-pathogen interactions by SuperSAGE. *Proceedings of the National Academy of Sciences of the United States of America*, 100(26): 15718–15723.

Maurel C., Verdoucq L., Luu D.T. et al. (2008). Plant aquaporins: membrane channels with multiple integrated functions. *Annual Review.of Plant Biology*, 59:595–624. doi: 10.1146/annurev.arplant.59.032607.092734

Mayavan S., Subramanyam K., Arun M. et al. (2013). *Agrobacterium tumefaciens*-mediated in planta seed transformation strategy in sugarcane. *Plant Cell Reports*, 32: 1557–1574. doi: 10.1007/s00299-013-1467-5

Mbambalala N. (2020). *Analysis of the role of transcription factors in enhancing drought tolerance in sugarcane (Saccharum spp.)*. M.Sc. Thesis, Faculty of Sciences, Stellenbosch University, South Africa.

McNeil M.D., Piperidis G., Bhuiyan S. et al. (2017). Development of a high-throughput, low-cost SNP genotyping panel for sugarcane breeding. In: Proceedings of the 2017 Conference of the Australian Society of Sugar Cane Technologists. (Ed.)M.C. Cox: Cairns, Queensland, Australia, pp. 304–311.

McQualter R.B. and Dookun-Saumtally A. (2007). Expression profiling of abiotic-stress-inducible genes in sugarcane. *Proceedings of Australian Society of Sugarcane Technology*, 29: 878–888.

Meena M.R., Murthy N., Kumar R. et al. (2013). Genotypic response of sugarcane under induced moisture deficit conditions. *Vegetos*, 26 (1): 229–232. doi:10.5958/j.2229-4473.26.1.033

Meuwissen T., Hayes B., and Goddard M. (2001). Prediction of total genetic value using genome-wide dense marker maps. *Genetics*, 157: 1819–1829.

Molinari H.B.C., Marur C.J., Daros E. et al. (2007). Evaluation of the stress-inducible production of proline in transgenic sugarcane (*Saccharum* spp.): osmotic adjustment, chlorophyll fluorescence and oxidative stress. *Physiologia Plantarum*, 130:218–229. Doi: 10.1111/j.1399-3054.2007.00909.x

Munawarti A., Taryono T., Semiarti E. et al. (2013). Tolerance of accessions of glagah (*Saccharum Spontaneum*) to drought stress and their accumulation of proline. *American Journal of Agricultural and Biological Science*, 8(1):1–11.

Nair N.V. (2011). Sugarcane varietal development programmes in India: an overview. *Sugar Tech*, 13(4): 275–280.

Nair, N. V., Nair, S., Sreenivasan T., et al. (1999). Analysis of genetic diversity and phylogeny in Saccharum and related genera using RAPD markers.*Genetic Resources and Crop Evolution*, 46:73–79.

Nakashima K., Ito Y., and Yamaguchi-Shinozaki K. (2009). Transcriptional regulatory networks in response to abiotic stresses in *Arabidopsis* and grasses. *Plant Physiology*, 149(1): 88–95.

Nascimento L.C., Yanagui K., Jose J. et al. (2019). Unraveling the complex genome of *Saccharum spontaneum* using polyploid gene assembler. *DNA Research*, 26:205–216. 10.1093/dnares/dsz001

Nawae W., Shearman J.R., Tangphatsornruang S. et al. (2020). Differential expression between drought-tolerant and drought-sensitive sugarcane under mild and moderate water stress as revealed by a comparative analysis of leaf transcriptome. *Peer J*, 8:e9608. doi: 10.7717/peerj.9608

Nielsen R., Paul J.S., Albrechtsen A. et al. (2011). Genotype and SNP calling from next-generation sequencing data. *Nature Reviews Genetics*, 12: 443–451. doi: 10.1038/nrg2986

Oh S.J., Song S.I., Kim Y.S. et al. (2005). Arabidopsis CBF3/DREB1A and ABF3 in transgenic rice increased tolerance to abiotic stress without stunting growth. *Plant Physiology*, 138: 341–351.

Oladosu Y., Rafii M.Y., Abdullah N. et al. (2016). Principle and application of plant mutagenesis in crop improvement: a review. *Biotechnology and Biotechnology Equipments* 30: 1–16.

Olsen A.N., Ernst H.A., Leggio L.L. et al. (2005). NAC transcription factors: Structurally distinct, functionally diverse. *Trends Plant Science*, 10: 79–87.

Pandey V. and Shukla A. (2015). Acclimation and tolerance strategies of rice under drought stress. *Rice Science*, 22: 147–161.

Parent B., Hachez C., Redondo E. et al. (2009). Drought and abscisic acid effects on aquaporin content translate into changes in hydraulic conductivity and leaf growth rate: A trans-scale approach. *Plant Physiology*, 149(4): 2000–2012.

Pastina M.M., Pinto L.R., Oliveira K.M. et al. (2010). Molecular mapping of complex traits. In: *Genetics, genomics and breeding of sugarcane*. (Eds) R Henry and C Kole: Science Publishers, Enfield, pp. 117–148.

Paterson A.H., Bowers J.E., Bruggmann R. et al. (2009). The *Sorghum bicolor* genome and the diversification of grasses. *Nature*, 457: 551.

Pavan S., Delvento C., Ricciardi L. et al. (2020). Recommendations for choosing the genotyping method and best practices for quality control in crop genome-wide association studies. *Frontiers in Genetics*, 11:447. doi: 10.3389/fgene.2020.00447

Peleg Z., Reguera M., Tumimbang E. et al. (2011). Cytokinin-mediated source/sink modifications improve drought tolerance and increase grain yield in rice under water-stress. *Plant Biotechnology Journal*. 9: 747–758.

Penna S., Sonawane B. and Patade V. (2012). *In vitro* mutagenesis and selection for crop improvement: Strategies and prospects. In: *Bioremediation, Biodiversity and Bioavailability*, Global Science Book 6 (Special Issue): 6–14.

Pereira-Santana A., Alvarado-Robledo E.J., Zamora-Briseño J.A. et al. (2017). Transcriptional profiling of sugarcane leaves and roots under progressive osmotic stress reveals a regulated coordination of gene expression in a spatiotemporal manner. *PLoS ONE*, 12: e0189271.

Peterson B.A., Haak D.C., Nishimura M.T. et al. (2016). Genome-wide assessment of efficiency and specificity in CRISPR/Cas9 mediated multiple site targeting in *Arabidopsis*. *PLoS ONE*, 11:e162169. doi: 10.1371/journal.pone.0162169

Pillay E. (2013). *In vitro culture and genetic transformation of selected ancestral and commercial sugarcane germplasm*, A dissertation submitted for the degree of Master of Science in the School of Life Science, University of KwaZulu-Natal, Durban, South Africa.

Poovaiah C.R., Bewg W.P., Lan W. et al. (2016). Sugarcane transgenics expressing MYB transcription factors show improved glucose release. *Biotechnology for Biofuels*, 9: 143. 10.1186/s13068-016-0559-1

Prabowo H., Kurniawan F.Y., Aristya, G.R. et al. (2019). Identification of SCDR 1 and P5CS genes in cultivar of environmental stress tolerant superior sugarcane (*Saccharum officinarum* L.). *Thai Journal of Agricultural Science*, 52(2):93–104.

Prabu G. and Prasad D.T. (2012). Functional characterization of sugarcane MYB transcription factor gene promoter(PScMYBAS1) in response to abiotic stresses and hormones. *Plant. Cell Reports*, 31: 661–669.

Prasitsom C., Jubsab N., Klomsa-ard P. et al. (2019). Selection of SSR Markers for drought resistant sugarcane in Thailand. *International Journal of Agricultural Technology*, 15(6):997–1010.

Priji P.J. and Hemaprabha G. (2014). Sugarcane-specific drought responsive candidate genes belonging to ABA-independent pathway identified from tolerant and susceptible clones of Saccharum and Erianthus species. *Indian Journal of Genetics*, 74(1): 64–72.

Racedo J., Gutiérrez L., Perera M.F. et al. (2016). Genome-wide association mapping of quantitative traits in a breeding population of sugarcane. *BMC Plant Biology*, 16:142. doi: 10.1186/s12870-016-0829-x

Rafalski A. (2002). Applications of single nucleotide polymorphisms in crop genetics. *Current Opinion in Plant Biology*, 5:94–100.

Rakoczy-Trojanowska M. (2002). Alternative methods of plant transformation -A short review. *Cellular & Molecular Biology Letters*, 7: 849–858.

Ramiro D.A., Melotto-Passarin D.M., Barbosa Md. A. et al. (2016). Expression of Arabidopsis bax inhibitor-1 in transgenic sugarcane confers drought tolerance. *Plant Biotechnology Journal* 14: 1826–1837. doi: 10.1111/pbi.12540

Rani K.S., Surinder K. and Gosal S.S. (2012). Genetic augmentation of sugarcane through direct gene transformation with Osgly II gene construct. *Sugar Tech*, 14(3):229–236. doi: 10.1007/s12355-012-0149-x

Rashid A.H.A. and Lateef D.D. (2016). Novel techniques for gene delivery into plants and its applications for disease resistance in crops. *American Journal of Plant Sciences*, 7: 181–193. 10.4236/ajps.2016.71019

Raza G. (2013). *Engineering Drought Tolerance in Sugarcane (Saccharum officinarum L.)*. Ph. D Thesis, Quaid-i-Azam University, Islamabad, Pakistan.

Reis R.R., Cunha B.A.D.B., Martins P.K. et al. (2014). Induced over-expression of AtDREB2A CA improves drought tolerance in sugarcane. *Plant Science*, 221–222: 59–68.

Ren B., Yan F., Kuang Y. et al. (2017). A CRISPR/Cas9 toolkit for efficient targeted base editing to induce genetic variations in rice. *Science China Life Sciences*, 60: 516–519. doi: 10.1007/s11427-016-0406-x

Riccardi F., Gazeau P., de Vienne D. et al. (1998). Protein changes in response to progressive water deficit in maize. quantitative variation and polypeptide identification. *Plant Physiology*, 117 (4): 1253–1263.

Rodrigues F.A., Da Graça J.P., De Laia M.L. et al. (2011). Sugarcane genes differentially expressed during water deficit. *Biologia Plantarum*, 55: 43–53. doi: 10.1007/s10535-011-0006-x

Ruiz M., Rouard M., Raboin L.M. et al. (2004). TropGENE-DB, a multi-tropical crop information system. *Nucleic Acids Research*, 32: D364–D367.

Rutherford R.S., Maphalala K.Z., Koch A.C. et al. (2017). Field and laboratory assessments of sugarcane mutants selected in vitro for resistance to imazapyr herbicide. *Crop Breeding and. Applied Biotechnology*, 17: 107–114.

Saad A.S., Li X., Li H.P. et al. (2013). A rice stress-responsive NAC gene enhances tolerance of transgenic wheat to drought and salt stresses. *Plant Science*, 203-204: 33–40. 10.1016/j.plantsci.2012.12.016

Sah S.K., Reddy K.R. and Li J. (2016). Abscisic acid and abiotic stress tolerance in crop plants. *Frontiers in Plant Science*, 7: 571.

Sahoo K.K., Tripathi A.K., Pareek A. et al. (2013). Taming drought stress in rice through genetic engineering of transcription factors and protein kinases. *Plant Stress*, 7(1): 60–72.

Salekdeh G.H., Siopongco J., Wade L.J. et al. (2002). A molecular mapping of QTLs for drought related traits at seedling stage under PEG induced stress conditions in rice. *Field Crops Research*, 76:199–219.

Sales C.R.G., Ribeiro R.V., Machado D.F.S.P. et al. (2012). Gas exchange and carbohydrate balance in sugarcane plants under root stressful conditions. *Bragantia*, 71: 319–327. doi: 10.1590/S0006-87052012 000300001

Scheben A., Batley J., and Edwards D. (2017). Genotyping-by-sequencing approaches to characterize crop genomes: choosing the right tool for the right application. *Plant Biotechnology Journal*, 15: 149–161. doi: 10.1111/pbi.12645

Schouten H.J., Krens F.A., and Jacobsen E. (2006). Do cisgenic plants warrant less stringent oversight? *Nature Biotechnology*, 24: 753.

Seki M., Narusaka M., Abe H. et al. (2001). Monitoring the expression pattern of 1300 Arabidopsis genes under drought and cold stresses by using a full-length cDNA microarray. *Plant Cell*, 13: 61–72.

Seki, M., Narusaka, M., Ishida, J., Nanjo, T., Fujita, M., Oono, Y., Kamiya, A., Nakajima, M., Enju, A., Sakurai, T., Satou, M., Akiyama, K., Taji, T., Yamaguchi-Shinozaki, K., Carninci, Piero, Kawai, J., Hayashizaki, Y., & Shinozaki, K. (2002). Monitoring the expression profiles of 7000 Arabidopsis genes under drought, cold and high-salinity stresses using a full-length cDNA microarray. *The Plant Journal*, 31:279–29210.1046/j.1365-313x.2002.01359.x

Selvi A., Nair N.V., Balasundaram N. et al. (2003). Evaluation of maize microsatellite markers for genetic diversity analysis and fingerprinting in sugar cane. *Genome*, 46: 394–403.

Selvi, A., Nair, N. V., Noyer, J. L., Singh, N. K., Balasundaram, N., Bansal, K. C., Koundal, K. R., and Mohapatra, T. (2006). AFLP analysis of the phenetic organization and genetic diversity in the sugarcane complex, Saccharum and Erianthus. *Genetic Resources and Crop Evolution*, 53: 831–84210.1007/s10722-004-6376-6

Serrat-Gurrera X. (2016). Applied biotechnology to improve Mediterranean rice varieties. A doctoral thesis submitted to University of Barcelona, Bercelona.

Shan Q., Wang Y., Li J. et al. (2013). Targeted genome modification of crop plants using a CRISPR-Cas system. *Nature Biotechnology*, 31: 686–688.

Sharma V. (2009). *Identification of drought-related quantitative trait loci (QTLs) in sugarcane (Saccharum spp.) using genic markers*. Ph.D. Thesis, Graduate Studies of Texas A&M University, USA.

Sharma P., Jha A.B., Dubey R.S. et al. (2012). Reactive oxygen species, oxidative damage, and anti-oxidative defence mechanism in plants under stressful conditions. *Journal of Botany*, 2012: Article ID217037.

Shi J., Gao H., Wang H. et al. (2017). ARGOS8 variants generated by CRISPR-Cas9 improve maize grain yield under field drought stress conditions. *Plant Biotechnology Journal*, 15: 207–216. doi: 10.1111/pbi.12603

Shirasawa K., Hirakawa H., Nunome T. et al. (2016). Genome-wide survey of artificial mutations induced by ethyl methanesulfonate and gamma-rays in tomato. *Plant Biotechnology Journal*, 14: 51–60.

Shu Q.Y. (2009). A summary of the international symposium on induced mutations in plants. In: *Induced plant mutations in the genomics era*. (Ed.)QY Shu: Food and Agriculture Organization of the United Nations, Rome, pp. 15–18.

Silva M.S., Sharma V., Jifon J.L. et al. (2010). Assessment of chlorophyll and leaf relative water content as indicators of drought tolerance on sugarcane initial growth. *Proceedings of the International Society of Sugarcane Technology*, 27:1–9.

Simon S. and Hemaprabha G. (2010). Identification of two new drought specific candidate genes in sugarcane (Saccharum spp.). *Electronic Journal of Plant Breeding*, 1(4):1164–1170.

Slater A., Scott N., and Fowler M.R. (2008). Techniques for Plant Transformation. In: *Plant Biotechnology: The genetic manipulation of plant*. (Eds.) Slater A., Scott N. and Fowler M.R.: Second edition, Oxford University Press, Oxford, pp. 54–76.

Snyman S.J., Meyer G.M., Richards J.M. et al. (2006). Refining the application of direct embryogenesis in sugarcane: effect of the developmental phase of leaf disc explants and the timing of DNA transfer on transformation efficiency. *Plant Cell Reports*, 25: 1016–1023.

Snyman S.J., Mhlanga P., Watt M.P. et al. (2016). Rapid screening of sugarcane Plantlets for In vitro mannitol-induced stress. *Sugar Tech*, 18: 437–440.

Snyman, S.J., 2004. Sugarcane transformation. In: *Transgenic Crops of the World: Essential Protocols*. (Ed.) Curtis, I.S.: Springer, Netherlands, Dordrecht, pp. 103–114.

Solomon S. (2014). Sugarcane agriculture and sugar industry in India: At a glance. *Sugar Tech*, 16: 113–124. doi: 10.1007/s12355-014-0303-8

Solomon S. (2016). Sugarcane Production and Development of Sugar Industry in India. *Sugar Tech*, 18(6):1–15. doi: 10.1007/s12355-016-0494-2

Souza G.M., Berges H., Bocs S., Casu R., D'Hont A., Ferreira J.E., et al., 2011, The sugarcane genome challenge: Strategies for sequencing a highly complex genome. *Tropical Plant Biology*, 4: 145–156. 10.1007/s12042-011-9079-0

Souza G.M., Van Sluys M.A., Lembke C.G. et al. (2019). Assembly of the 373K gene space of the polyploid sugarcane genome reveals reservoirs of functional diversity in the world's leading biomass crop. *Gigascience*, 8(12):giz129. doi: 10.1093/gigascience/giz129

Sreenivasan T.V. (2004). Improving indigenous sugarcane of India. *Sugar Tech*, 6(3): 107.

Sreenivasan T.V. et al. (1987). "Cytogenetics". In: *Sugarcane Improvement through Breeding*. (Ed.)D.J. Heinz: Elsevier, Amsterdam, Netherlands, pp. 211–253.

Shrivastava A.K., Shahi H.N., Kulshreshtha N. et al. (2001). Nutrient uptake characteristics of *Saccharum* species. In: *XIV International Plant Nutrition Colloquium*, Springer, Hannover, Germany, pp. 218.

Sreenivasan, T.V., Amalraj, V.A., and Jebadhas, A.W. (2001). Catalogue on sugarcane genetic resources IV. *Erianthus species*, SBI, Coimbatore, India, pp. 98.

Srivastava A.K. and Srivastava S. (2016). Diversity of the germplasm of Saccharum species and related genera available for use in directed breeding programmes for sugarcane improvement. *Current Science*, 111(3):475–482.

Srivastava M.K., Li C.N., and Li Y.R. (2012). Development of sequence characterized amplified region (SCAR) marker for identifying drought tolerant sugarcane genotypes. *Australian Journal of Crop Science*, 6(4): 763–767.

Suprasanna P., Jain S.M., Ochatt S.J. et al. (2012). Applications of in vitro techniques in mutation breeding of vegetatively propagated crops. In: *Plant Mutation Breeding and Biotechnology*. (Eds.) QY Shu, BP Forster, and H Nakagawa: Joint FAO/IAEA Division of Nuclear Techniques in Food and Agriculture, International Atomic Energy Agency, Vienna, Austria, pp. 371–385.

Survila M., Heino P., and Palva E.T. (2010). Genes and gene regulation for low-temperature tolerance. In: *Genes for plant abiotic stress*. (Eds.)M.A. Jenks and A.J. Wood: Wiley-Blackwell, New York, pp. 185–207.

Szurman-Zubrzycka M., Chmielewska B., Gajewska P. et al. (2016). Mutation detection by analysis of DNA heteroduplexes in TILLING populations of diploid species. In: *Biotechnologies for Plant Mutation Breeding: Protocols*. (Eds.)J. Jankowicz-Cieslak, T. Tai, J. Kumlehn, and B. Till: Springer: Cham, Switzerland, pp. 281–303.

Takahashi S., Furukawa T., Asano T. et al. (2005). Very close relationship of the chloroplast genomes among Saccharum species. *Theoretical and Applied Genetics*, 110: 1523–1529.

Takasaki H., Maruyama K., Kidokoro S. et al. (2010). The Abiotic stress-responsive NAC-type transcription factor OsNAC5 regulates stress-inducible genes and stress tolerance in rice. *Molecular Genetics and Genomics*, 284(3):173–183. doi:10.1007/s00438-010-0557-0

Taparia Y., Fouad W.M., Gallo M. et al. (2012). Rapid production of transgenic sugarcane with the introduction of simple loci following biolistic transfer of a minimal expression cassette and direct embryogenesis. *In VitroCellular and Developmental Biology- Plant*, 48: 15–22. 10.1007/s11627-011-9389-9

Taratima W., Ritmaha T., Jongrungklang N. et al. (2020). Effect of stress on the leaf anatomy of sugarcane cultivars with different drought tolerance(*Saccharum officinarum*, Poaceae). *Revista de Biología Tropical*, 68(4): 1159–1170.

Tardieu F. (2012). Any trait or trait-related allele can confer drought tolerance: just design the right drought scenario. *Journal of Experimental Botany*, 63: 25–31. doi: 10.1093/jxb/err269

Terauchi R., Matsumura H., Kruger D.H. et al. (2008). SuperSAGE: the most advanced through comparative genomics. In: *The Handbook of Plant Functional Genomics: Concepts and Protocols*. (Eds.)G. Kahl and K. Meksem: Wiley-VCH, Weinheim, Germany, pp. 37–54.

Thiel T., Kota R., Grosse I. et al. (2004). SNP2CAPS: a SNP and INDEL analysis tool for CAPS marker development. *Nucleic Acids Research*, 32: e5. doi: 10.1093/nar/gnh006

Thomson M.J. (2014). High-throughput SNP genotyping to accelerate crop improvement. *Plant Breeding and Biotechnology*, 2: 195–212. doi: 10.9787/PBB.2014.2.3.195

Trujillo L.E., Sotolongo M., Menendez C. et al. (2008). SodERF3, a novel sugarcane ethylene responsive factor (ERF), enhances salt and drought tolerance when overexpressed in tobacco plants. *Plant Cell Physiology*, 49(4): 512–525.

Trujillo L.E. et al. (2009). Engineering drought and salt tolerance in plants using SodERF3, a novel sugarcane ethylene responsive factor. *Biotecnologia Aplicada*, 26(2): 168–171.

Tsaniklidis G., Chatzistathis T., Fanourakis D. et al. (2021). Leaf antioxidant machinery stimulation by *Meloidogyne javanica* infestation: a case study on *Cucumis melo* seedlings. *Plant Stress*, 1:100002. 10.1016/j.stress.2021.100002 in press.

Vantini J., Dedemo G., Gimenez D.J. et al. (2015). Differential gene expression in drought-tolerant sugarcane roots. *Genetics and. Molecular Research*, 14: 7196–7207.

Vasantha S., Gomathi R., and Brindha C. (2017). Growth and nutrient composition of sugarcane genotypes subjected to salinity and drought stresses. *Communications in Soil Science and Plant Analysis*, 48(9): 989–998.

Vickers J.E., Grof C.P.L., Bonnett G.D. et al. (2005). Effects of tissue culture, biolistic transformation and introduction of PPO and SPS gene constructs on performance of sugarcane clones in the field. *Australian Journal of Agricultural Research*, 56(1): 57–68. 10.1071/AR04159

Vickers, J. E., Grof, C. P. L., Bonnett, G. D., Jackson, P. A., Knight, D. P., Roberts, S. E., & Robinson, S. P. (2005). Overexpression of polyphenol oxidase in transgenic sugarcane results in darker juice and raw sugar. *Crop Science*, 45: 354–362.10.2135/cropsci2005.0354

Waclawovsky A.J., Sato P.M., Lembke C.G. et al. (2010). Sugarcane for bioenergy production: An assessment of yield and regulation of sucrose content. *Plant Biotechnology Journal*, 8: 263–276. 10.1111/j.1467-7652.2009.00491.x

Wang Z.Z., Zhang S.Z., Yang B.P. et al. (2005). Trehalose synthase gene transfer mediated by Agrobacterium tumefaciens enhances resistance to osmotic stress in sugarcane. *Sugar Tech*, 7(1): 49–54. 10.1007/BF02942417

Wang H., Wang H., Shao H. et al. (2016). Recent advances in utilizing transcription factors to improve plant abiotic stress tolerance by transgenic technology. *Frontiers in Plant Science*, 7: 67.

Wang J., Roe B., Macmil S. et al. (2010). Microcollinearity between autopolyploid sugarcane and diploid sorghum genomes. *BMC Genomics*, 11: 261.

Wang L., Liu F., Zhang X. et al. (2018). Expression characteristics and functional analysis of the ScWRKY3 gene from sugarcane. *International Journal of Molecular Science*, 19: 4059.

Xu F., Lilian H.L., Gao S. et al. (2019). Comparative Analysis of two Sugarcane Ancestors Saccharum o_cinarum and S. spontaneum based on Complete Chloroplast Genome Sequences and Photosynthetic Ability in Cold Stress. *International Journal of Molecular Sciences*, 20: (3828):1–26. doi:10.3390/ijms20153828

Xu S., Wang J., Shang H. et al. (2018). Transcriptomic characterization and potential marker development of contrasting sugarcane cultivars. *Scientific Reports*, 8: 1–11. doi: 10.1038/s41598-018-19832-x

Yang X., Sood S., Luo Z. et al. (2019). Genome-wide association studies identified resistance Loci to orange rust and yellow leaf virus diseases in sugarcane (*Saccharum* spp.). *Phytopathology*, 109: 623–631.

Yang X., Luo Z., Todd J. et al. (2018). Genome-wide association study of multiple yield components in a diversity panel of polyploid sugarcane (*Saccharum* spp.). *Plant Genome*, 13(1): e20006. doi: 10.1101/387001

Yang X., Song J., You Q. et al. (2017). Mining sequence variations in representative polyploid sugarcane germplasm accessions. *BMC Genomics*, 18:594. doi: 10.1186/s12864-017-3980-3

Yin H., Chen C.J., Yang J. et al. (2014). Functional genomics of drought tolerance in bioenergy crops. *Crit. Rev. Plant Science*, 33: 205–224

Yu F., Wang P., Li X. et al. (2018). Characterization of chromosome composition of sugarcane in nobilization by using genomic in situ hybridization. *Molecular Cytogenetics*, 11: 35. 10.1186/s13039-018-0387-z

Zang X. and Komatsu S. (2007). A proteomics approach for identifying osmotic-stress-related proteins in rice. *Phytochemistry*, 68 (4): 426–437.

Zhang Z., Liu J., Ding X. et al. (2010). Best linear unbiased prediction of genomic breeding values using a trait-specific marker-derived relationship matrix. *PLoS ONE*, 5:12648. doi: 10.1371/journal.pone.0012648

Zhang H.X., Zhang B.P., Fang C.L. et al. (2006). Expression of the GrifolaTrehalose synthase gene and improvement of drought tolerance in sugarcane (*Saccharum officinarum* L.). *Journal of Integrative Plant Biology*, 48:453–459. doi: 10.1111/j.1744-7909.2006.00246.x

Zhang J., Zhang X., Tang H. et al. (2018). Allele-defined genome of the autopolyploid sugarcane Saccharum spontaneum L. *Nature Genetics*, 50:1565–1573. 10.1038/s41588-018-0237-2

Zhang J., Zhou M., Walsh J. et al. (2014). Sugarcane genetics and genomics. In: *Sugarcane- Physiology, Biochemistry, and Functional Biology*. (Eds.) Paul H. Moore and Frederik C. Botha. C: John Wiley & Sons, Inc. pp. 623–643.

Zong Y., Wang Y., Li C. et al. (2017). Precise base editing in rice, wheat and maize with a Cas9- cytidine deaminase fusion. *Nature Biotechnology*, 35(5):438–440. doi: 10.1038/nbt.3811

8 Enhancing Abiotic Stress Tolerance of Sugarcane: Genome-Editing Approach

N. Prabhakaran
T. Stanes & Company Ltd, Coimbatore, Tamil Nadu, India

G Boobalakrsihan
Division of Vegetable Science, Indian Agricultural Research Institute, Pusa Campus, New Delhi, India

Pandiyan Muthuramalingam
Department of Horticultural Science, Gyeongsang National University, Jinju, Korea

S. Radhesh Krishnan
T. Stanes & Company Ltd, Coimbatore, Tamil Nadu, India

CONTENTS

8.1 INTRODUCTION

Sugarcane is a valuable cash crop of subtropical and tropical countries like Brazil, India, Thailand, China, Australia, Pakistan, the Philippines, Cuba, Colombia, and the United States used to make both sugar, bio-fibers, and ethanol. It has recently earned importance as an imperative bioenergy crop (Khan et al. 2017). In India, sugarcane is grown under varied agro-climatic conditions. Among the most important industrial cash crops, sugarcane (Saccharum spp.) contributes significantly to the global sugar industry and biofuel production. In a wide range, the crop is subjected to numerous environmental stress conditions (Table 8.1). Genetic engineering and biotechnology have helped cultivate and improve it since prehistoric times with natural selection and conventional breeding, as well as using methods of genetic engineering and biotechnology more recently. However, sugarcane's heterogeneity, complex polyaneuploid genome, and susceptibility to different biotic and abiotic stresses are obstacles that require us to pay more attention to sugarcane crop improvement.

The crop faces various biotic and abiotic stresses that impact productivity in a significant way. Sugarcane is one of the most efficient plants for converting sunlight energy into chemical energy,

DOI: 10.1201/9781003292425-8

TABLE 8.1
Abiotic Stress Experienced by the Sugarcane Crop in Various Sugarcane Zones in India

Zone	Area Covered	Abiotic Stress Experienced by the Crop
North-west	Punjab, Haryana, western, Central Uttar Pradesh, Rajasthan and Uttarakhand	High and low temperatures, frost, salinity and alkalinity
North-central	Eastern Uttar Pradesh, Bihar and West Bengal	Relatively less pre-monsoon rainfall and high rainfall in later part of the monsoon, floods
North-east	Assam and Nagaland	Floods, water logging, less development of water resources and irrigation, acid soils
East coastal	Orissa, coastal Andhra and Tamil Nadu	Cyclonic winds before summer and winter monsoons uproot and damage cane, ingression of sea water, temperatures favorable for growth but not the ripening
Peninsular	Maharashtra, Karnataka, Gujarat, Madhya Pradesh, interior Andhra Pradesh, peninsular area of Tamil Nadu	Variability in soil, seepage and poor drainage, now severe drought, salinity and acid soils in some parts (Marathwada and Vidarbha in Maharashtra and parts of Karnataka)

Source: Shrivastava et al. (2016).

which can further be used to produce biofuels (Lu and Mosier 2008). Sugarcane's complicated polyploid DNA has hampered its improvement through traditional breeding methods. Sugarcane has seen significant advancements thanks to transgenic technology. Allopolyploid hybrids of saccharum give an ideal material for genetic-modification research because of their duplicated genomes. Sugarcane has been employed in the research of duplicated genes (process of producing new genetic material during molecular evolution) using transgenic procedures via *Agrobacterium*-mediated transformation and the biolistic approach (Arvinth et al. 2010). Sugarcane is a model organism for research into hybridization, genome remodeling, chromosomal deletion, and the influence of gene dosage, higher polyploidy chromosomal bivalent pairing, allelic variation, and more.

Environmental stresses limit plant growth and crop productivity. Numerous factors, like growth-hormone regulation, deficiency in nutrients and enzymes, and an unstable ROS activity, could be hindering the developmental attributes of sugarcane (Figure 8.1). Drought is the main lethal abiotic stress affecting crop productivity worldwide. Sugarcane is a relatively high-water-demanding crop, and its growth is highly sensitive to water deficit. It's a source for production of sugar and ethanol. Sugarcane produces 8–12 tons of cane per ML of irrigated water, and water deficits will directly affect approximately 60% of productivity losses (Ferreira et al. 2017).

CRISPR was to begin with a recognition in E. coli in 1987 and was detailed as a resistant instrument to fight against attacking viral and plasmid DNA. Be that as it may, both of them require a complex development, which has obliged their large-scale application in plants (Ishino et al. 1987).

In the field of genome editing (GE), DNA is inserted or removed from an organism to modify its genetic makeup. The four gene-editing nuclease families are meganucleases, transcription activator-like effector nuclease (TALEN), zinc-finger nuclease (ZFN), and clustered regularly interspaced short palindromic repeat (CRISPR/Cas9 technologies). In prokaryotes, CRISPR/Cas9 is the adaptive immunity against invading genomes or viral predators, established by memorizing the previous infections and integrating the spacers (short snippets of the invading genomes) into the CRISPR locus. When reinfection occurs, crRNAs (CRISPR RNAs) are used by Cas nucleases to target invader sequences. By targeting DNA sequences through programmed RNA, CRISPR/CAS9 has opened up new possibilities for genome editing. A TALEN-mediated GE strategy for increasing bioethanol production in sugarcane has proven successful despite the widely prevalent CRISPR genome-editing technology currently in use. As a complex polyploid crop with a genome size of

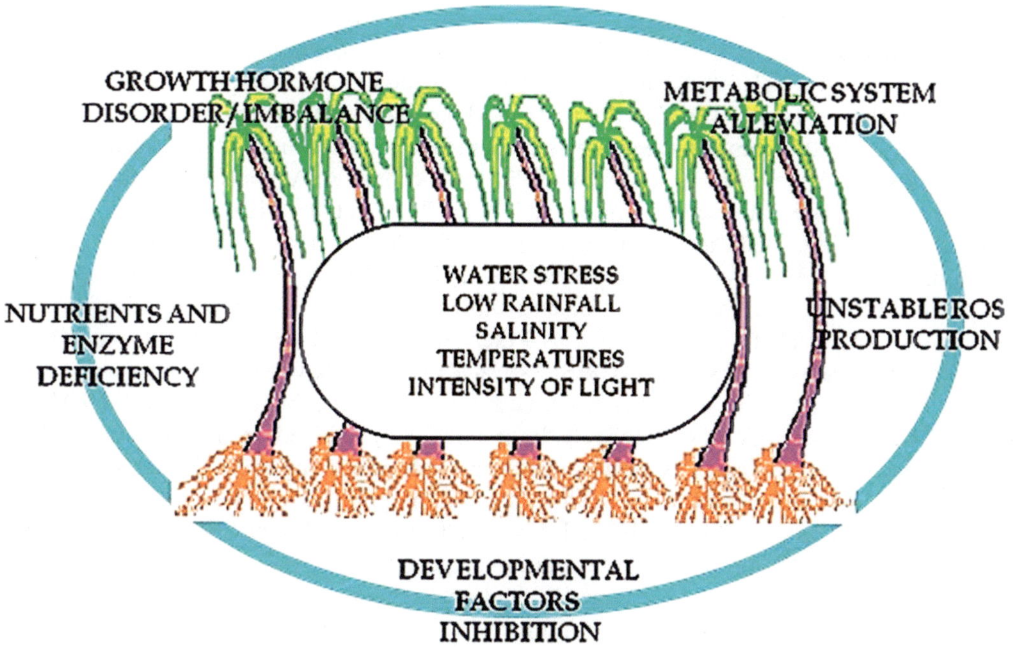

FIGURE 8.1 Different forms of stress and its impact on developmental stage in sugarcane.

about 10 GB and 8 to 12 identical gene copies, sugarcane presents several challenges when it comes to genome editing. Molecular improvement of sugarcane is hindered by transgene-silencing at both transcriptional and post-transcriptional levels. Like other GE nucleases, Cas9 may have off-target effects, leading to mutations that are not intended. gRNA-Cas9 complexes can also cut off target DNA sequences in addition to cutting target DNA. If gRNA–Cas9 complexes with short ends are selected among specific promoters, high mutation rates can be achieved with minimal off-target effects. To overcome off-target cleaving issues, modified variants of Cas9 can be used. The use of CRISPR-associated nucleases and Cas9 variants can soon be a powerful tool with which to enable successful GE in sugarcane and other polyploid crops, even though several challenges remain.

Abiotic stresses like dry season, saltiness, and temperature limits antagonistically influence development and usefulness of rural yields. Cold is among the major abiotic stresses, which essentially decrease yield and influence pretty much every viewpoint of the physiology and natural chemistry of plants (Josine et al. 2011; Sanghera et al. 2011). Low temperature (LT), including chilling (0–10 °C) and freezing (<4 °C), is known to affect plants' endurance and geological conveyance. While calm plants don't show freezing resilience, they are known to be chilling lenient (Josine et al. 2011). Openings to chilling temperatures increment their freezing resistance by an interaction known as 'chilly acclimation' (Levitt 1980; Thomashow 1999). As opposed to this idea, plants from tropical/subtropical districts, like rice, maize, corn, cotton, and tomato, are chilling delicate and don't have the limit to cold adjust (Thomashow 1999). Besides, cold acclimation is related with adjustments in plant cell films, expanded degrees of responsive oxygen species (ROS), and initiation of ROS searching frameworks, proline amassing, checked changes in quality articulation, and biochemical pathways influencing photosynthesis (Sanghera et al. 2011). Low temperature forces weigh on a plant in two ways: the impacts of LT alone and lack of hydration of the cells and tissues at the point when cell water freezes (Beck et al. 2007). In particular, LT influences cell endurance, cell division, photosynthetic effectiveness, and water transport with resulting adverse consequence on plant development and usefulness (Beck et al. 2007). As ordinary cell capacities are upset during abiotic stress, a fast and wide reconstructing at the subatomic level is needed to react to these disturbances. This reconstructing is the consequence of transcriptional, post-transcriptional, and translational guidelines of the declaration of stress-responsive qualities (Jaglo et al. 1998; Skinner et al. 2005; Van Buskirk and Thomashow 2006; Chinnusamy et al. 2003 and 2007; Jeknić et al. 2014). Among the vital participants in the guideline of quality articulation in plants are miRNAs, which are bountiful, endogenous, little non-coding RNA atoms known to adjust post-transcriptional administrative cycles (Wang et al. 2011; Sunkar et al. 2012).

Genome altering opens new and extraordinary freedoms for specialists to improve crop creation. Until 2013, the zinc-finger nucleases (ZFNs) and record activator-like effector nucleases (TALENs) were the key instruments utilized for genome-altering applications. The coming of RNA-directed designed nucleases – the sort II bunched consistently interspaced short palindromic rehash (CRISPR)/Cas9 (CRISPR-related) framework from *Streptococcus* pyogenes – holds incredible potential since it is basic, viable, and more flexible than ZFNs and TALENs. CRISPR/Cas9 framework has effectively been effectively utilized in a few harvest plants. Utilization of these strategies is in its infancy in sugarcane. Jung and Altpeter (2016) have announced a TALEN-interceded approach for the initial opportunity to lessen lignin content in sugarcane to make it suitable for biofuel creation. Until now, this has been the main report depicting genome altering in sugarcane. Enormous genome size, polyploidy, low change proficiency, transgene hushing, and absence of high-throughput screening methods represent incredible difficulties for genome altering in sugarcane, which would be talked about exhaustively in this survey.

8.2 DROUGHT STRESS

Changes in the life cycle, modulation of growth and development to match water supply, regulation of whole-plant functions to balance resource allocation for growth and stress adaptation, environmental stresses that limit plant growth and crop productivity, and evolution of stress-signal perception for rapid and long-term expression of stress tolerance are all examples of drought-tolerance strategies that plants have evolved. Drought is a main lethal abiotic stress, affecting crop productivity worldwide. Biotechnology and molecular breeding approaches can help increase crop output in drought-stricken areas.

Despite advances in our understanding of stress responses and the availability of molecular tools and tactics, developing drought-tolerant crops remains a serious issue (Wang et al., 2016). Sugarcane is a relatively high water-demanding crop, and its growth is highly sensitive to water deficits (Lakshmanan and Robinson, 2014). It's as source for the production of sugar and ethanol. Sugarcane produces 8–12 tons of cane per ML of irrigated water, and water deficit will directly affect approximately 60% of productivity losses. The drought-responsive mechanism is very typically regulated in sugarcane, as studied by Ferreira et al. (2017) (Figure 8.2). The drought responsiveness of miRNAs varies depending on the species. Even members of the same family have been shown to react differently to drought stress. miRNA expression under drought stress in sugarcane, its expression may modify their micro-transcriptome in different ways to cope with different phases and intensities of drought stress, and these responses may differ depending on genetic background (Gentile et al. 2013). The difficulty of identifying and exploding large effect genes and alleles for developing drought-tolerant verities. Shi et al. (2017) have developed a corn variety through CRISPR/Cas genome-editing approach, which has improved yield under drought stress. Improved expression of ARGOS8 (negatively regulates ethylene responses) in genome-edited plants showed enhanced drought tolerance.

Drought stress alters the plant growth in various metabolic pathways. An imbalance in hormones, enzymes, macromolecules, ROS production, decreased level in nutrient uptake, and photosynthesis significantly contribute in plant growth and development.

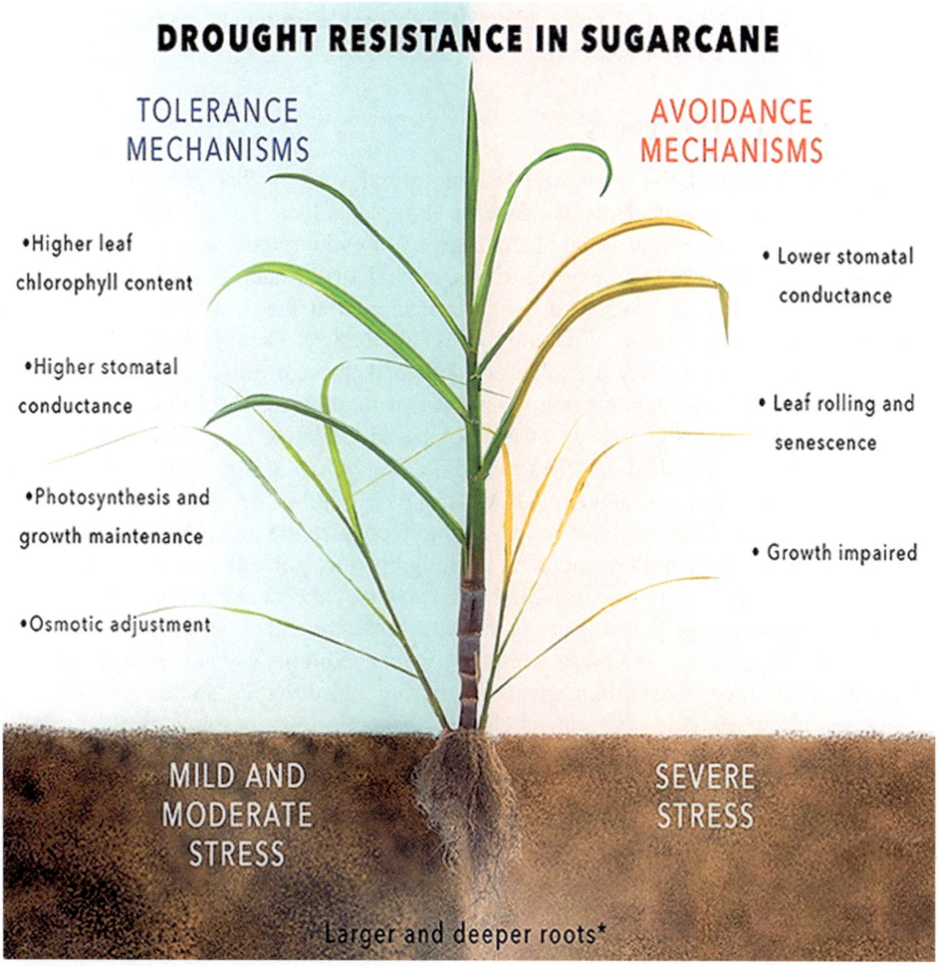

FIGURE 8.2 Scheme of sugarcane drought-response mechanisms.

Source: Ferreira et al. 2017.

Drought (water stress) is one of the most important environmental stresses and occurs for several reasons, including low rainfall, salinity, high and low temperatures, and high light intensity, among others. Drought stress is a multidimensional stress and causes changes in the physiological, morphological, biochemical, and molecular traits in plants. Many plants have improved their resistance mechanisms to tolerate drought stress, but these mechanisms are varied and depend on the plant species. Typically, mechanisms involved in plant tolerance to drought follow a general plan: maintaining cell homeostasis in water-deficit situations, which is possible by increasing the water inlet to the cells. Drought avoidance is other common drought-resistance mechanism in annual plants. With this mechanism, escape from stress conditions is the main strategy for plant growth under drought conditions. Dry spell pressure in plants is described by diminished leaf-water potential and turgor pressure, stomatal conclusion, and diminished cell development and expansion (Farooq et al., 2009b). Dry spell pressure lessens the plant development by affecting different physiological and biochemical capacities, such as photosynthesis, chlorophyll combination, supplement digestion, particle take-up and movement, breath, and sugars digestion (Jaleel et al., 2008; Farooq et al., 2009b; Li et al., 2011). Plants may experience water shortage during dry spells, yet low temperature may likewise cause turgor stress at the cell level

(Thomashow, 1994; Janska et al., 2009; Yadav, 2010) inferable from helpless root pressure-driven conductance and decreased root action (Aroca et al., 2003).

8.3 HEAT AND COLD STRESS

Low temperature is one of the most well-known natural stresses that genuinely influence the development and advancement of plants. Nevertheless, plants have pliancy with all components, empowering them to endure and, some of the time, even endure antagonistic natural conditions. MicroRNAs (miRNAs) are little noncoding RNAs, around 18–24 nucleotides long, and have been progressively perceived as controllers of quality articulation at the post-transcriptional even out and can impact an expansive scope of natural cycles (Figure 8.3). There is developing proof in the writing that reconstructing quality articulation intervened through miRNAs is a significant protection component in plants, empowering them to react to stresses. Until this point, various examinations have set up the significance of miRNA-based guideline of quality articulation under low temperature stress. Individual miRNAs can adjust the outflow of different mRNA targets, and, subsequently, the control of a solitary miRNA can possibly influence various organic processes. Numerous practical investigations have endeavored to recognize the miRNA–target collaborations and have expounded the job of a few miRNAs in chilly pressure guidelines. This audit sums up the current comprehension of miRNA-intervened adjustment of the declaration of key qualities as hereditary and administrative pathways engaged with low-temperature stress reactions in plants.

Plants face a mix of various abiotic stresses under field conditions that are deadly to establishing development and creation. Concurrent events of chilling and dry-spell stresses in plants, due to exceptional and fast worldwide environment changes, can adjust the morphological, physiological, and atomic reactions. Both of these anxieties antagonistically influence the plant development and yields because of actual harms, physiological and biochemical disturbances, and atomic changes. As a general rule, the co-event of chilling and dry-season mix is surprisingly more devastating for crop creation rather than a singular pressure condition. Plants accomplish different normal and diverse physiological and atomic-defensive methodologies for resilience under chilling and dry season stresses. Plant reactions to a mix of chilling and dry season stresses are interesting from those to individual pressure. In the current survey, we summed up the new proof on plant reactions to chilling and dry spell weights on shared just as extraordinary premise and attempted to find a

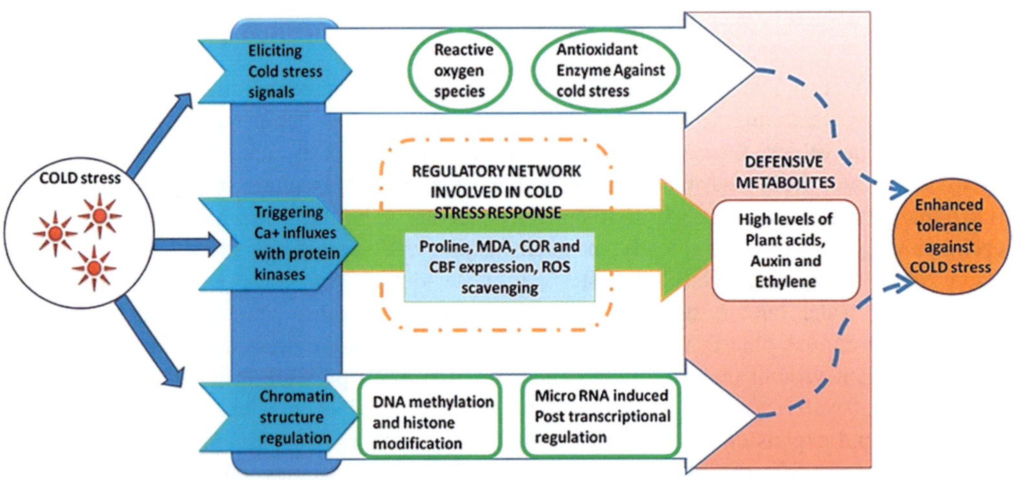

FIGURE 8.3 Mitigation of cold stress in sugarcane crop.

Adapted from Megha et al., 2018.

repeating theme possibly hidden these responses. We addressed the conceivable cross talk between plant reactions to these burdens and examined the potential administration systems for directing the components of plant resilience to dry season and additionally chilling anxieties. Until now, different novel methodologies have been tried in limiting the adverse consequences of consolidate stresses. Regardless of the principle upgrades, there is as yet a major opportunity to improve in the blend of dry spell and chilling resilience. Consequently, future investigations utilizing biotechnological and atomic methodologies ought to be completed to foster hereditarily designed plants with upgraded resilience against these pressure factors.

Heat stress is defined as the temperature that slows or stops plant growth, irreversibly damaging the plant. Excessive heating causes excessive transpiration, leading to dehydration.

The amount of soluble sugars, free proline, and glycine was observed to increase linearly. However, none of these metabolites have osmotic significance and, therefore, cannot be considered as defensive substances against desiccation caused by high temperature. An increase in the total phenol content induced by thermal stress may be related to the prevention of oxidative damage. Heat stress increases the production of transcripts for peroxidase and superoxide dismutase, as well as their enzymatic activities required to protect the plant cells against oxidative damage. Thermo tolerance is usually associated with the accumulation of trehalose. A mutation that suppresses the gene for trehalose-6-phosphate (T6P) synthase (TPS) transforms the resistant plant into heat sensitivity. Trehalose is an osmotic of high solubility in water that stabilizes cytoplasmic proteins, avoiding their thermal denaturation.

Cold is a major abiotic stress that limits the production of tropical and subtropical crops in new production areas. Sugarcane (*Saccharum* spp.) comes from the tropics but is grown mainly in the subtropics, where it often encounters cold stress. Sugarcane is highly sensitive to freezing, either under natural or experimentally induced conditions. Eggleston et al. (2004) demonstrated remarkable changes in the main indicators of freeze deterioration, such as increased juice viscosity or decreased pol-filterability. As demonstrated by Thiebaut et al. (2012), the production of miRNAs (miR319) was provoked by subjecting sugarcane plants to a prolonged cold period.

8.4 NUTRITION-RELATED STRESSES

Despite its importance as a food crop, sugarcane has a large ecological impact, meaning there is much to be improved. The targeted genes provide an easy-to-identify phenotype, as well as being useful in future research. The reduction in nitrogen fertilizer requirements might be caused by altering the chlorophyll content of sugarcane. This tall, dense plant is able to get a lot of sunlight on its top leaves, which shade lower leaves. In combination with less fertilizer and the same amount of light, light can penetrate deeper into the leaves with less chlorophyll, resulting in more biomass and less light (Oz et al. 2021).

Supplement status is a natural component that can impact development rate, number of green leaves per mother shoot, leaf region, and turner thickness of sugarcane (Chatterjee et al. 1998; Shukla et al. 2009). Subsequently, supplement irregularity is probably the most established subject in proper sugarcane science (Kobus, 1900). The particle stress brought about by abundance aluminum (Al) and iron (Fe) on sugarcane could be reduced with augmentations of phosphorus (P) and potassium (K). Metabolic (ethyleneand Ca^{+}) pathways were researched, looking for the actuation of qualities expanding the resilience of sugarcane to potassium inadequacy. Zeng et al. (2015) have played out a transcriptome profiling of sugarcane plants exposed to pressure deficiency in potassium and have observed a recognizable expansion in the movement of peroxidase enzyme, clearly identified with stress processes.

In spite of the fact that dry-spell lenient sugarcane genotypes display higher free-proline content than drought-delicate plants (Guimarães et al. 2008), more studies are required to affirm that such an aspect is balanced by P in sugarcane. In one more example of opposition against abiotic stress, K and Si-upgraded salt tolerance in salt-touchy sugarcane genotypes (CPF 243 and SPF 213) coming about in diminished Na^+ fixation and increased K^+ with an improvement in K^+/Na^+ proportion (Ashraf et al. 2010). It is additionally fascinating to state the work by Sudama et al. (1998), who showed that the utilization of Kat the hour of sugarcane planting submerged stress conditions altogether expanded the stomatal diffusive resistance, subsequently diminishing happening rate and increasing the leaf-water potential, sugarcane length, sucrose content in the juice, and sugarcane yield.

8.5 CONCLUSIONS AND FUTURE PROSPECTS

Successful developments in the sphere of the genetic manipulation of plants using molecular tools, as compared to classical breeding, have proven beneficial for obtaining sustainable and higher yields with lower susceptibility toward biotic and abiotic stresses. Besides advances in breeding technologies, such as molecular marker-assisted breeding, genetic transformations, genome editing, and multiple omics technologies, we have seen a great potential for sugarcane genetic improvement. AFLP, RAPD, and SSR markers have been widely used to improve the adaptability of sugarcane to abiotic stressors, such as drought, extreme temperature, and salinity, among other stressful condition. Modern biotechnologies can lead to the release of genetically improved cultivars with better physiological and morphological traits and improved field performance under varying environmental conditions and in a shorter time than conventional breeding.

Future trends and expectations should be aimed at cracking the current main hurdles (biotic and abiotic stressors) to plants. Modern genome-editing tools (CRISPR/Cas9, base editing) can provide a permanent solution by developing stress tolerant/resistant varieties, although further studies are needed to reach these goals. The integration of all these approaches will lead to the sustainable production of the sugarcane crop, through the effective management of environmental stresses under the present scenario of changing climate. An important step toward crop improvement would be to promote a transparent dialogue between molecular biologists and plant physiologists, on the one hand, and between farmers, breeding companies, and the public, on the other, to jointly resolve the economic, sociological, legal, and ethical hurdles.

REFERENCES

Aroca R., Vernieri P., Irigoyen J.J., Sancher-Diaz M., Tognoni F., and Pardossi A. (2003). Involvement of abscisic acid in leaf and root of maize (Zea mays L.) in avoiding chilling-induced water stress. *Plant Sci.* 165, 671–679. doi: 10.1016/S0168-9452(03)00257-7

Arvinth S., Arun S., Selvakesavan R.K., Srikanth J., Mukunthan N., Kumar P.A., Premachandran M.N., and Subramonian N. (2010). Genetic transformation and pyramiding of aprotinin-expressing sugarcane with cry1Ab for shoot borer (Chiloinfuscatellus) resistance. *Plant Cell Rep.* 29, 383–395.

Ashraf M., Rahmatullah A.R., Bhatti A.S., Afzal M., Sarwar A., Maqsood M.A., and Kanwal S. (2010). Amelioration of salt stress in sugarcane (Saccharumofficinarum L.) by supplying potassium and silicon in hydroponics. *Pedosphere* 20, 153–162.

Beck E.H., Fettig S., Knake C., Hatrtig K., and Bhattrai T. (2007). Specific and unspecific responses of plants to cold and drought stress. *Journal of Biosciences* 32, 501–510.

Chatterjee C., Nautiyal N., and Dube B.K. (1998). Effects of potassium concentration on biochemistry, yield and sucrose concentration in sugarcane. *Sugar Cane* 5, 12–15.

Chinnusamy V., Ohta M., Kanrar S., Lee B.H., Hong X., Agarwal M., and Zhu J.K. (2003). ICE1: a regulator of cold-induced transcriptome and freezing tolerance in Arabidopsis. *Genes and Development* 17, 1043–1054.

Chinnusamy V., Zhu J., and Zhu J.K. (2007). Cold stress regulation of gene expression in plants. *Trends in Plant Science* 12, 444–451.

Devarumath R.M., Kalwade S.B., Kulkarni P.A., Sheelavanthmath S.S., and Suprasanna P. (2013). *Integrating OMICS Approaches in Sugarcane Improvement*. OMICS Applications in Crop Science; CRC Press: New York, NY, USA, pp. 191–250.

Eggleston G., Legendre B., and Tew T., (2004). Indicators of freeze-damaged sugarcane varieties which can predict processing problems. *Food Chem.* 87, 119–133.

Farooq M., Wahid A., Kobayashi N., Fujita D., and Basra S.M.A. (2009b). Plant drought stress: effects, mechanisms and management. *Agron. Sustain. Dev.* 29, 185–212. doi: 10.1051/agro:2008021

Ferreira T.H.S., Tsunada M.S., Bassi D., Araújo P., Mattiello L., Guidelli G.V., Righetto G.L., Gonçalves V.R., Lakshmanan P., and Menossi M. (2017). Sugarcane Water Stress Tolerance Mechanisms and Its Implications on Developing Biotechnology Solutions. *Front. Plant Sci.* 8:1077. doi: 10.3389/fpls.2017.01077

Gentile A., Ferreira T.H., Mattos R.S., Dias L.I., Hoshino A.A., Carneiro M.S., et al. (2013). Effects of drought on the microtranscriptome of field-grown sugarcane plants. *Planta* 237, 783–798.

Guimarães E.R., Mutton M.A., Mutton M.J.R., Ferro M.I.T., Ravaneli G.C., and Silva J.A. (2008). Freeproline accumulation in sugarcane under water restriction and spittlebug infestation. *SciAgric* 65, 628–633.

Ishino Y., Shinagawa H., Makino K., Amemura M., and Nakata A. (1987). Nucleotide sequence of the iap gene, responsible for alkaline phosphatase isozyme conversion in Escherichia coli, and identification of the gene product. *J. Bacteriol.* 169, 5429–5433.

Jaglo K.R., Gilmour S.J., Zarka D.G., Schabenberger O., and Thomashow M.F. (1998). Arabidopsis CBF1 overexpression induces COR genes and enhances freezing tolerance. *Science* 280, 104–106.

Jaleel C.A., Gopi R., Sankar B., Gomathinayagam M., and Panneerselvam R. (2008). Differential responses in water use efficiency in two varieties of Catharanthus roseus under drought stress. *Comp. Rend. Biol.* 331, 42–47. doi: 10.1016/j.crvi.2007.11.003

Janska A., Mars P., Zelenkova S., and Ovesna J. (2009). Cold stress and acclimation–what is important for metabolic adjustment? *Plant Biol.* 12, 395–405. doi: 10.1111/j.1438-8677.2009.00299.x

Jeknić Z., Pillman K.A., Dhillon T., Skinner J.S., Veisz O., Cuesta-Marcos A., and Stockinger E.J. (2014). Hv-CBF2A overexpression in barley accelerates COR gene transcript accumulation and acquisition of freezing tolerance during cold acclimation. *Plant Molecular Biology* 84, 67–82.

Josine T.L., Ji J., Wang G., and Guan C.F. (2011). Advances in genetic engineering for plants abiotic stress control. *African Journal of Biotechnology* 10, 5402–5413.

Jung J.H., and Altpeter F. (2016). TALEN mediated targeted mutagenesis of the caffeic acid O-methyltransferase in highly polyploidy sugarcane improves cell wall composition for production of bioethanol. *Plant Mol.Biol.* 92, 131–142.

Khan M.T., Seema N., Khan I.A., and Yasmine S. (2017). Applications and potential of sugarcane as an energy crop. In: *Agricultural research updates*. Nova Science Publishers, New York, pp. 1–24.

Khan M.T., Khan I.A., and Yasmeen S. (2019). Genetically modified sugarcane for biofuels production: Status and perspectives of conventional transgenic approaches, RNA interference, and genome editing for improving sugarcane for biofuels. In: *SugarcaneBiofuels*. Springe, Cham, Switzerland, pp. 67–96.

Kobus J.D. (1900). *After van Dillewijn botany of sugar cane*. Chronica Bot Co, p. 227.

Lakshmanan P., and Robinson N. (2014). Stress physiology: Abiotic stresses. In: P.H. Moore and F.C. Botha (ed) *Sugarcane: Physiology, Biochemistry, and Functional Biology*, John Wiley & Sons, Inc., Chichester, pp. 411–434.

Levitt J. (1980). *Responses of Plants to Environmental Stresses*, p. 497. Academic Press, New York.

Li C., Jiang D., Wollenweber B., Li Y., Dai T., and Cao W. (2011). Waterlogging pretreatment during vegetative growth improves tolerance to waterlogging after anthesis in wheat. *Plant Sci.* 180, 672–678. doi: 10.1016/j.plantsci.2011.01.009

Lu Y.L., and Mosier N. (2008). Current technologies for fuel ethanol production from lignocellulosicplant biomass. In: W. Vermerris (ed) *Genetic improvement of bioenergy crops*. Springer, New York, pp. 161–182.

Megha S., Basu U., and Kav N.N.V. (2018). Regulation of low temperature stress in plants by microRNAs. *Plant, Cell and Environment* 41, 1–15

Oz M.T., Altpeter A., Karan R., Merotto A., and Altpeter F. (2021). CRISPR/Cas9-mediated multi-allelic gene targeting in sugarcane confers herbicide tolerance. Front Genome Ed.; 3. doi: 10.3389/fgeed.2021.673566

Sanghera G.S., Wani S.H., Hussain W., and Singh N.B. (2011). Engineering cold stress tolerance in crop plants. *Current Genomics* 12, 30–43.

Shi J., Gao H., Wang H., Lafitte H.R., Archibald R.L., Yang M., Hakimi S.M., Mo H., and Habben J.E. (2017). ARGOS 8 variants generated by CRISPR-Cas9 improve maize grain yield under field drought stress conditions. *Plant Biotechnol. J.* 15, 207–216.

Shrivastava A.K., Srivastava T.K., Srivastava A.K., Misra V., Srivastava S., Singh V.K., and Shukla S.P. (2016). *Climate change induced abiotic stresses affecting sugarcane and their mitigation.* ICAR-Indian Institute of Sugarcane Research Lucknow, pp. 108.

Shukla S.K., Yadav R.L., Singh P.N., and Singh I. (2009). Potassium nutrition for improving stubble bud sprouting, dry matter partitioning, nutrient uptake and winter initiated sugarcane (Saccharum spp. hybrid complex) ratoon yield. *Eur J Agron* 30, 27–33.

Skinner J.S., Zitzewitz J., Szucs P., Marquez-Cedillo L., Filichkin T., Amundsen K., Stockinger E.J., and Thomashow M.F. (2005). Structural, functional, and phylogenetic characterization of a large CBF gene family in barley. *Plant Molecular Biology* 59, 533–551.

Sudama S., Tiwari T.N., Srivastava R.P., Singh G.P., and Singh S. (1998). Effect of potassium on stomatalbehaviour, yield and juice quality of sugarcane under moisture stress conditions. *Indian J Plant Physiol* 3, 303–305.

Sunkar R., Li Y.F., and Jagadeeswaran G. (2012). Functions of microRNAs in plant stress responses. *Trends in Plant Science* 17, 196–203.

Thiebaut F., Rojas C.A., Almeida K.L., Grativol C., Domiciano G.C., Lamb C.R.C., et al., 2012. Regulation of miR319 during cold stress in sugarcane. *Plant Cell Environ.* 35, 502–512.

Thomashow M.F. (1999). Plant cold acclimation: freezing tolerance genes and regulatory mechanisms. *Annual Review of Plant Physiology and Plant Molecular Biology*. 50, 571–599.

Thomashow M.F. (1994). Arabidopsis thaliana as a model for studying mechanisms of plant cold tolerance. In: E.M. Meyerowitz and C.R. Somerville (eds) *Arabidopsis*. Cold Spring Harbor Laboratory Press, Cold Spring Harbor, NY, pp. 807–834.

Van Buskirk H.A., and Thomashow M.F. (2006). Arabidopsis transcription factors regulating cold acclimation. *Physiologia Plantarum* 126, 72–80.

Wang T.Z., Chen L., Zhao M.G., Tian Q.Y., and Zhang W.H. (2011). Identification of drought-responsive microRNAs in Medicago truncatula by genome-wide high-throughput sequencing. *BMC Genomics* 12, 367.

Wang H., Wang H., Shao H., and Tang X. (2016). Recent advances in utilizing transcription factors to improve plant abiotic stress tolerance by transgenic technology. *Frontiers in plant science*, 7, 67.

Yadav S.K. (2010). Cold stress tolerance mechanisms in plants. *A review. Agron. Sustain. Dev.* 30, 515–527. doi: 10.1051/agro/2009050

Zeng Q., Ling Q., Fan L., Li Y., Hu F., Chen J., Huang Z., Deng H., Li Q., and Qi Y. (2015). Transcriptomeprofiling of sugarcane roots in response to low potassium stress. *PLoS ONE* 10, e0126306.

9 Techniques Used for the Identification of Defense-related Protein in Sugarcane in Response to Biotic Stress

Smriti Mall
Molecular Plant Pathology Lab., Department of Botany, DDU Gorakhpur University Gorakhpur, Uttar Pradesh, India

CONTENTS

9.1 INTRODUCTION

Sugarcane (*Saccharum* spp. hybrids) belongs to the family *Poaceae*, and, on average, accounts for nearly 85% of global sugar production. The International Sugar Organization (2020) data showed that in the last year, leading countries like Brazil, China, India, Thailand, etc., produced 75% of sugar worldwide. Processing of sugar production, growers, energy, and food companies, are facing so many social and environmental problems related to sugar production and sustainability (Huang et al., 2018). However, pathogen infections cause constant losses in yield. The expression of several proteins, viz., glycoproteins, chitinases, peptidase, etc., induced by the host plant in response to pathogens is a way by which the host is able to develop its own defense mechanisms to fight pathogens and destroy them. Due to biotic stresses, sugarcane production is constantly challenged

DOI: 10.1201/9781003292425-9

(Azevedo et al., 2011). All pathogenic infections come in the category of biotic stress; in the sugarcane crop, the highlighting pathogens are sugarcane grassy shoot disease (*Phytoplasma*); sugarcane mosaic virus (ScMV) and yellow leaf virus (ScYLV); red rot disease caused by (*Colletotrichumfalcatum*); sugarcane smut (*Sporisoriumscitamineum*); stripe (*Acidovoraxavenae*), leaf scald (*Xanthomonas albileneans*), etc., and also attack by potential insect vectors (Rao et al., 2017; Meng et al., 2020; Singh et al., 2021).

Plant defense mechanisms are either constitutive or inducible to biotic stress. Defense-related proteins (DR) are mostly detected in different hosts after their response to infection (Van Loon et al., 2006). "Pathogenesis-related"(PR) proteins are formed only when the sugarcane crop is infected by pathogens; scientists, while working on plant-microbe interactions, found that if no infection occurred, then no such proteins are expressed (Van Loon et al., 2006; Broekaert et al., 1995). Li et al. (2015) reported that some other well-known proteins also participated in plant-defense mechanisms. For example, nucleotide-binding site leucine-rich repeat proteins, able to recognize a broad variety of insects and pathogens (Legaz et al., 1998), and another protein called WRKY protein consist of a large family of transcription factors that identify the Wbox-type DNA sequence (TTGACC/T), which plays an important role as the promoter of different plant-defense genes (Rushton et al., 1996; Santos et al., 2015; Sauza et al., 2017). In this chapter review, defense-related proteins in sugarcane crop against pathogenic infections are detailed, with their mechanisms of action, highlighted pathogen, and biotechnological perspectives.

9.2 DIFFERENT TECHNIQUES USED IN IDENTIFICATION OF EXPRESSED PROTEINS IN SUGARCANE AGAINST INFECTION

In sugarcane crops, some of the important techniques used today to identify proteins involved in the defense-related mechanisms are described in the details below:

iTRAQ, or isobaric tagby mass spectrometry, is able to determine the amount of proteins from different sources. The reagents that are used for isobaric labeling consist of a reporter group that is charged, a reactive peptide group, and a neutral balance group. An iTRAQ reagent isobaric tag is covalently linked to peptides, reactive with each lysine-side chain and N-terminus group of a peptide, labeling all peptides in a given sample digest (source:https://www.creative-proteomics.com/blog/index.php).

In an iTRAQ experiment, every sample is reduced, blocked by cysteine and digested with trypsin. In a single tube, all samples are labeled with a different iTRAQ tag. Finally, combine all iTRAQ reagent-labelled samples and make the single mixture to perform liquid chromatography.

9.3 IN SUGARCANE CROP IDENTIFICATION OF INDUCED PROTEINS IN RESPONSE TO INFECTION BY *SPORISORIUM SCITAMINEUM* BY QUANTITATIVE PROTEOMICS

At Sugarcane Research Institute, Nanning, China, Singh et al. (2021) conducted a study by using 02 sugarcane varieties one is resistant (GT29) and two are susceptible (Yacheng71–374 and ROC22) to *Sporisorium scitamineum.*

Quantification of protein from sugarcane varieties is performed by using a Qubit Protein Assay Kit. Some 100 µg of protein was taken from each sample and mixed in 1 µL of 2% SDS solution and 2 µL of reducing reagent from the quantitative kit (iTRAQ) for 1 h at 60°C. Samples were alkylated after incubation, with cysteine-blocking reagent 1 µL at 25°C for 30 min and digested by 1 µg µL−1 trypsin at 20:1 (protein: trypsin) and incubated for 16 hrs at 37°C. All samples were tagged with unique iTRAQ tags by following instructions of iTRAQ reagent-8-plex kit.

9.3.1 Liquid Chromatography Orbitrap Mass Spectrometry Analysis

From the first-dimensional peptide fractionation, fractions produced were dehydrated by vacuum centrifugation and desalted, and then re-suspended in mobile phase A for liquid chromatography mass spectophotometer analysis. C18 reversed-phase capillary column were loaded with 5 μL peptide samples with nano HPLC. Peptide elution was performed through the HPLC. The High-Resolution LTQ Orbitrap Elite Mass Spectrometer was used to analyzed eluted peptides.

9.3.2 Protein Identification and Data Processing

Su et al. (2015) used mass spectrophotometry to collect the data using Proteome Discoverer 2.2 software and the search engine Mascot 2.3.02 and compared with data of *Saccharum officinaruma* available in NCBI accession ID PRJNA383616.

Proteins were recognized as differently expressed proteins (DEPs) when a ratio with a p-value <<0.83 as down-regulated and and >1.20 up-regulated, (quantitative changes) occurred (Romero et al., 1998; Su et al., 2015). The recognized DEPs were categorized functionally like biological processes, cellular component, and molecular function. And through gene ontology, in addition to important pathways involved in sugarcane, smut-resistant proteins database and Kyoto Encyclopedia of Genes and Genomes database were studied.

9.3.3 Real-Time PCR (qRT-PCR)

By following *S. scitamineum* inoculation at intervals of different days, the appearence patterns of the SuSOD, SuCAT,SuGluD1, SuPAL, and SuChi, genes were analyzed by using Trizol reagent (Tiangen, China) and RNA was extracted from the leaf tissues. Some 1 μg of DNase I was used to prepared cDNA. qRT-PCR using SYBR Premix Ex TapTMII was performed with a real-time PCR-detection system. 20 μL volume of reactions mixture prepared by adding of 2 μL template (10 × diluted cDNA), 0.8 μL of each primer (10 μM), 6.4 μL ddH2O, along with distilled water and 10 μL SYBR Premix Ex Tap™ II, was used as a template for blank control. The standard qRT-PCR program followed, as suggested by (Niu et al. 2015). Using melting-curve analysis, the specificity of the reaction was checked and the relative gene-expression quantification were completed by the Ct target gene reference gene method (Livak and Schmittgen, 2001). So, in this way, the spectra were quantified from both sugarcane varieties, respectively.

9.3.4 Bioinformatics Analysis of DEPs in Sugarcane

After *S. scitamineum* infection in both sugarcane varieties, categories that are functional of all the DEPs were detected and characterized by using the gene ontology (GO) tool and further grouped into three categories: cellular, molecular, and biological functions.

Between GT versus GC, the DEPs in GT29 were categorized into 37 functional groups. Within this, 15 GO categories only accounted for biological process, 04 GO for cellular component, and 18 GO for molecular function. Whereas between YT versus YC, the DEPs in Yacheng71–374 were charcterized into 33 functional groups. 07 GO accounted for biological process, 06 GO for cellular component, and 20 GO for molecular functions. Further, KEGG pathway analysis was conducted to study the role of the DEPs in the important metabolic pathways in both smut-resistant and smut-susceptible varieties of sugarcane after *S. scitamineum* infection.

On that basis, a total of seven metabolic pathways were recognized; three out of seven shows enriched pathways related to ribosome, photosynthesis, and plant-pathogen interaction in GT29. And only three metabolic pathways, i.e., secondary metabolites biosynthesis, photosynthesis, and phenylpropanoid biosynthesis, were identified in Yacheng71–374. In the smut-resistant variety, more proteins are needed to run the metabolic pathways as compared to smut-susceptible sugarcane variety. The differences in the expressed protein in the photosynthesis pathway of resistant and susceptible sugarcane varieties after inoculation by *S. scitamineum* was understandable. KEGG enrichment analysis in the resistant variety revealed that another two enriched pathways were ribosome and plant-pathogen interactions. In the resistant variety, a total of 25 differentially expressed proteins were observed to be associated with the ribosome pathway. In the pathogen-plant interaction pathway, up-regulation of two proteins, i.e., calcium-dependent protein kinase (CDPK) and heat-shock protein (HSP90), was noticed; it plays a major role in plants under biotic stress condition.

9.4 IDENTIFICATION OF DEFENSE PROTEINS IN SUGARCANE IN RESPONSE TO *XANTHOMONAS ALBINEANS*

Meng et al. (2020) studied two sugarcane cultivars differing in resistance to leaf scald, LCP 85–384 (resistant) from U.S.A. and ROC20 (susceptible)collectedfrom China.

9.4.1 Proteomic Profiling

On the basis of iTRAQ analysis, 6891 proteins were recognized among proteins from all samples during screening against the *Saccharum* spp. unigene database. GO category proteins were annotated for biological process, oxidation-reduction process, the carbohydrate metabolic, and proteolysis process. Proteins present in the cell component were distributed into 10 different subcellular locations: ribosome, membrane, cytoplasm, intracellular, nucleus, proteasome core complex, integral component of the membrane, photosystem II, photosystem II oxygen-evolving complex and nucleosome. GO category proteins are annotated for molecular work, and most proteins were also attributed to the other functions, like ATP binding, protein binding, and nucleic acid binding. The KEGG pathway analysis shows proteins were attributed in 107 pathways, such as amino acid metabolism, carbohydrate metabolism, etc.

9.5 DETECTION OF EXPRESSED PROTEINS AGAINST *X. ALBILINEANS*

In two sugarcane varieties after infection with *Xanthomonas albilineans* and based on a (up-regulation) and (down-regulation) proteins, a total of 285 DEPs were recognized. Out of these, some DEPs were downregulated, and some DEPs were upregulated. For resistant cultivar, 172 DEPs found were upregulated and 59 were downregulated. For susceptible variety, 192 DEPs were recognized after inoculation and the same number of DEPs (96) was up and downregulated. Seventy nine DEPs were shared by both cultivars. Among those, 32 were downregulated and 45 were upregulated, whereas two proteins were upregulated in LCP 85–384 but downregulated in ROC20. The proteins detected by KEGG mostly participated in metabolic pathways of the selected sugarcane varieties. Proteins participating in the biosynthesis of secondary metabolites were upregulated in resistant cultivar LCP 85–384. Some purine and pyrimidine metabolic pathways, amino sugar, nucleotide sugar, and amino acid metabolism pathways were significantly enriched in both varieties. Proteins involved with defense-response pathways were essentially expressed in the leaf-scald resistant cultivar (LCP 85–384) and included DEPs associated with ubiquitin proteolysis-mediated pathway, biosynthesis phenylpropanoid pathway, and glutathione metabolism.

9.5.1 (PPIs) Protein–Protein Interactions Network

Interaction between proteins and their functions form the base of the cellular machinery. Their interlinked connectivity is now essential to understanding the biological functioning of an organism, but very limited information is available on protein–protein interactions. The **STRING database** moto is to collect, score, and combine all available sources of protein-interaction information and used for computational predictions. The latest version of STRING (11.0) doubles the number of organisms in their databases. In the latest version, the important change is an option to entire upload of genome-wide datasets as input, and users are able to visualize subsets as interaction networks and to perform gene-set enrichment analysis on the entire input. Gene ontology and KEGG are the well-known implementations of STRING used for classification of proteins. The STRING resource is available online at https://string-db.org/ (Szklarczyk et al., 2019; Ntambo et al., 2019).

This database is a large repository including functional interactions, stable complexes of proteins, and regulatory interactions for protein–protein interaction networks. PPIs of the 285 sugarcane DEPs were annotated. DEPs interacted with each other in the two sugarcane varieties inoculated with *X. albilineans*. A huge number of these proteins were enriched in the biological secondary metabolic, glutathione and glycolysis/gluconeogenesis pathways.

9.5.2 Detection of Defense Genes Triggered by *X. albilineans* in Sugarcane

A total of 35 representative DEPs from two sugarcane var., LCP 85–384 or ROC20, were characterized, including ubiquitin-mediated proteolysis (4 DEPs), phenylpropanoid biosynthesis (9 DEPs), MAPK signaling pathway (1 DEP), glutathione metabolism (3 DEPs), zeatin biosynthesis (1 DEP), pantothenate, and CoA biosynthesis (1DEP). In LCP, 85–384 of the DEPs involved one recognized protein: ubiquitin-activating enzyme was upregulated whereas one unrecognized protein was downregulated. Two proteins were upregulated in ROC20 in ubiquitin proteolysis process. And in LCP 85–384, in the glutathione metabolism, two proteins were upregulated and one protein was downregulated. In LCP 85–384 and ROC20, the sucrose non-fermenting-related protein kinases, serine kinaseare isoform participated in the mitogen protein kinase-signaling pathway, which was upregulated and the CoA biosynthesis pathway and cis-zeatin O-glucosyltransferase of the zeatin biosynthesis pathway were downregulated.

9.5.3 Transcript Profiling by qRT-PCR

For transcript profiling, seven candidate genes were selected as the differently expressed proteins encoded by these genes were participated in two main Kyoto Encyclopedia of Genes and Genomes pathways. By using qRT-PCR analysis, the transcription level of these seven genes was examined. In PCR amplification assay, these seven genes are ranged between 98% and 101%. With the help of protein profiling, it will be easier for us to understand functioning of different expressed proteins.

In comparison to control plants in ROC20, genes coding for the photosystem I apoprotein, the ubiquitin-activating enzyme, and a non-specific lipid-transfer protein and were all downregulated. Whereas UDP-glycosyltransferase were highly upregulated in resistant cultivar. In ROC20 genes coding for the cytochrome P450, argonaute protein cytosolic glyceroldehyde-3-phosphate dehydrogenase were significantly upregulated.

In LCP 85–384 sugarcane-resistant variety with reponse to *X. albilineans,* the UBA1 gene was upregulated and the same gene was almost not recognized in ROC20. Seven candidate genes that are coding for different metabolic pathways are used as a source for further study of the pathways in sugarcane crop against *X. albilineans* infection. This study generates the first global proteomic dataset of sugarcane against *X. albilineans* infection.

9.6 TRANSCRIPTOMIC ANALYSIS OF SUGARCANE AGAINST *LEIFSONIA XYLI*

Leifsonia xyli, an obligate pathogen, causes ratoon-stunting disease in sugarcane crops; this destructive disease occurs around the world, especially in sugarcane-growing countries. Infected sugarcane crops show some changes in physiological responses. Transcriptomes of *Leifsonia*-affected and healthy sugarcane leaves and stalks were analysed to gain more information into the biochemical and physiological pathways in sugarcane. The unigenes, which are identified in both healthy and infected sugarcane, showed a significant number of expressed genes, especially in diseased plants. Due to infection in sugarcane, crop modifications are mainly reported at metabolic pathways viz., plant-pathogen interaction, biosynthesis of phytohormone, and photosynthesis

(Zhu et al., 2021). Also found were that protein phosphorylation and cell-wall defense pathways are responsible for infection and play a role in the development of disease. Differentially expressed genes that participate in photosynthetic processes were reported to cause significant reduction in energy metabolism, energy capture, and chloroplast structure. Also due to pathogenic activity, modifications in gibberellic acid response along with the inhibition of photosynthetic processes resulted in growth retardation in ratoon-stunting diseased plants. In response to infection, expressed proteins are also involved in biosynthesis of secondary metabolite, cell wall, and protein dephosphorylation. Ratoon-stunting disease occurs globally and causes economic losses due to low sugarcane yield.

9.7 PROTEOMIC ANALYSIS: *ACIDOVORAX AVENAE* CAUSING RED STRIPE IN SUGARCANE

Acidovorax avenae causal agent of red stripe in sugarcane causes devastating effect on sugarcane production. Two sugarcane varieties, ROC22 resistant and MT11–610 susceptible, were studied for examination of proteins that were expressed in defense response against the pathogen by using isobaric tag-based technique (iTRAQ). A total of 1962 (DEPs) were identified. DEPs were involved in many metabolic activities as single metabolism, catalytic activity, and the reduction-oxidation process revealed by GO analysis. These proteins, on the basis of KEGG analysis, are involved in pathways like biosynthesis of secondary metabolites and phenylpropanoids. In ROC22, a resistant variety of fold increases as compared to susceptible MT11–610 variety. On the basis of defense proteins information, it could be possible to develop resistant varieties of sugarcane against red stripe disease (Zhou et al., 2021). MicroRNA-gene regulation has been reported to enhance tolerance against different biotic stresses (Gupta et al., 2014). In sugarcane varieties against infection by *Acidovor axavenae,* several micro RNAs were recognized that have good efficiency for use in the improvement of protein defence-related pathways (Thiebaut et al., 2012).

9.8 CONCLUSION

From the literature, isobaric labeled-based study enhances our knowledge of the physiological roles of disease-related proteins in sugarcane cultivars. The differently expressed proteins that were detected and identified are predicted to be part of many biological synthesis of metabolites, as well as metabolic pathways. Such study provides updated information regarding usefulness of protein profiling by using iTRAQ-based molecular analysis and characterization of sugarcane red stripe, leaf scald, ratoon-stunting, and smut-induced changes, including physiological as well as biochemical variations. From this review chapter, we are able to understand the applications of different techniques like Nano Liquid Chromatography-Orbitrap Mass spectrophotometry analysis, proteomic profiling, DEPs, qRT-PCR prepared set of databases etc., against sugarcane crop under different biotic stress conditions. Understanding of defense-related proteins in sugarcane crops promote us to develop resistant varieties and formulate control measures to protect from different pathogens for obtaining high yield of sugar in sugarcane-growing countries.

REFERENCES

Azevedo R.A., Carvalho R.F., Cia M.C., and Gratão P.L. (2011). Sugarcane under pressure: An overview of biochemical and physiological studies of abiotic stress. *Trop Plant Biol*, 4:42–51.

Broekaert W.F., Terras F., Cammue B., and Osborn R.W. (1995). Plant defensins: Novel antimicrobial peptides as components of the host defense system. *Plant Physiol*, 108:1353.

Gupta O.P., Sharma P., Gupta R.K., and Sharma I. (2014). Current status on role of miRNAs during plant-fungus interaction. *Physiol Mol Plant Pathol*, 85:1–7.

Huang N., Ling H., Su Y.C., Liu F., Xu L.P., and Su W.H. (2018). Transcriptional analysis identifies major pathways as response components to Sporisorium scitamineum stress in sugarcane. *Gene*, 678:207–218.

Legaz M.E., de Armas R., Pinon D., and Vicente C. (1998). Relationships between phenolics-conjugated polyamines and sensitivity of sugarcane to smut (Ustilago scitaminea). *J Exp Bot*, 49:1723–1728.

Li X., Kapos P., and Zhang Y. (2015). NLRs in plants. *CurrOpin Immunol*, 32:114–121.

Livak K.J. and Schmittgen T.D. (2001). Analysis of relative gene expression data using real-time quantitative PCR and the 2– ΔΔCT method. *methods*, *25*(4): 402–408.

Meng J.Y., Ntambo M.S., Rott P.C., Fu H.Y., Huang, M.T., Zhang, H.L., and Gao, S.J. (2020). Identification of differentially expressed proteins in sugarcane in response to infection by Xanthomonas albilineans using iTRAQ quantitative proteomics. *Microorganisms*, 8(1): 76.

Ntambo M.S., Meng J.Y., Rott P.C., Henry R.J., Zhang H.L., and Gao S.J. (2019). Comparative transcriptome profiling of resistant and susceptible sugarcane cultivars in response to infection by Xanthomonas albilineans. *Int. J. Mol. Sci*, 20: 6138.

Niu X., Chen, M., Huang, X., Chen, H., Tao, A., Xu, J., and Qi, J. (2017). Reference gene selection for qRT-PCR normalization analysis in kenaf (Hibiscus cannabinus L.) under abiotic stress and hormonal stimuli. *Frontiers in plant science*, 8:771.

Rao G.P., Thorat V., Manimekalai R., Tiwari A.K., and Yadav, A. (2017). A century progress of research on phytoplasma diseases in India. *Phytopathogenic Mollicutes*, 7(1):1–38.

Romero G.O., Simmons C., Yaneshita M., Doan M., Thomas B.R., and Rodriguez R.L. (1998). Characterization of rice endo-betaglucanase genes (Gns2-Gns14) defines a new subgroup within the gene family. *Gene*, 223:311–320.

Rushton P.J., Torres J.T., Parniske M., Wernert P., Hahlbrock K., and Somssich I.E. (1996). Interaction of elicitor-induced DNAbinding proteins with elicitor response elements in the promoters of parsley PR1 genes. *EMBO J*, 15:5690–5700.

Santos F.R.C., Pinto L.R., Carlini-Garcia L.A., Gazaffi R., Mancini M.C., Goncalves B.S., Medeiros C.N.F., Perecin D., Garcia A.A.F., and Souza A.P. (2015). Marker-trait association and epistasis for brown rust resistance in sugarcane. *Euphytica*, 203:533–547.

Su Y., Xu L., Wang S., Wang Z., Yang Y., Chen Y., and Que Y. (2015) Identification, phylogeny, and transcript of chitinase family genes in sugarcane. *Sci Rep*, 5:10708.

Singh P., Singh, R.K., Song Q.Q., Li H.B., Guo D.J., Malviya M.K., and Li Y.R. (2021). Comparative analysis of protein and differential responses of defense-related gene and enzyme activity reveals the long-term molecular responses of sugarcane inoculated with Sporisorium scitamineum. *Journal of Plant Interactions*, 16(1):12–29.

Szklarczyk D., Gable A.L., Lyon D., Junge A., Wyder S., Huerta-Cepas J., Simonovic M., Doncheva N.T., Morris J.H., Bork P., Jensen L.J., and Mering C.V. (2019). STRING v11: protein–protein association networks with increased coverage, supporting functional discovery in genome-wide experimental datasets. *Nucleic Acids Research*, 47 (D1):D607–D613.

Thiebaut F., Grativol C., Carnavale-Bottino M., Rojas C.A., Tanurdzic M., Farinelli L., Martienssen R.A., Hemerly A.S., and Ferreira P.C.G. (2012). Computational identification and analysis of novel sugarcane microRNAs. *BMC Genomics*, 13:290.

Van Loon L.C., Rep M., and Pieterse C.M. (2006). Significance of inducible defense-related proteins in infected plants. *Annu Rev Phytopathol*, 44:135–162.

Zhu K., Yang L.T., Li C.X., Lakshmanan P., Xing Y.X., and Li Y.R. (2021). A transcriptomic analysis of sugarcane response to Leifsoniaxyli subsp. xyli infection. *Plos one*, 16(2):e0245613.

Zhou J.R., Sun H.D., Ali A., Rott P., and Javed P. (2021). Quantitative proteomic analysis of the sugarcane defense responses incited by Acidovorax avenae subsp. Avenae causing red stripe. *Industrial Crops and Products, Elsevier*, 162:113275

10 Physiological and Molecular Stress Responses in Sugarcane

Abdelaleim Ismail ElSayed
Biochemistry Department, Faculty of Agriculture, Zagazig University, Zagazig, Egypt

Ahmad A. Omar
Biochemistry Department, Faculty of Agriculture, Zagazig University, Zagazig, Egypt

Citrus Research and Education Center, University of Florida, IFAS, Lake Alfred, FL, USA

CONTENTS

10.1 INTRODUCTION

In recent decades, climate change has been regarded as a severe problem that impacts agricultural productivity and human health. Climate changes are projected to result in extreme temperatures, floods, and drought. Many reports have proposed that drought stress can increase soil salinity and reduce plant growth and crop productivity. Sugarcane (*Saccharum* spp.) is a remarkable crop in the world's tropical regions (D'Hont et al. 2008). It is an important source of food and bioenergy, grown in many tropical and subtropical countries (Waclawovsky et al. 2010). However, sugarcane is a salt sensitive and has symptoms of toxicity, low sprouting, nutritional imbalance, and overall reduced growth, which results in a low biomass rate (Patade, Suprasanna, and Bapat 2008). The physiological effects of salinity on sugar cane were investigated (Patade, Bhargava, and Suprasanna 2011; Patade, Suprasanna, and Bapat 2008). Nevertheless, research into the growth and physiological effect of saline on sugarcane is much less than on other plants.

Abiotic stressors are one of the leading causes of crop productivity losses worldwide (Bray, Bailey-Serres, and Weretilnyk 2000), negatively impacting crop adaptability and production. Drought is the most prominent stress in this setting, and it is regarded as an incredibly critical component in sugarcane production losses (Wang, Vinocur, and Altman 2003). Several plant biotechnology projects have been launched to improve drought stress resistance in crop plants using genetic engineering and traditional breeding (Wang et al. 2003). Despite tremendous advances in understanding the physiological and molecular responses of plants to water deficits as a result of breeding efforts, there is still a significant difference in yields between normal and stress conditions (Cattivelli et al. 2008).

DOI: 10.1201/9781003292425-10

Drought-tolerance strategies have evolved in plants, including life cycle changes, growth, and development modulation to match water supply, regulation of whole plant functions to balance resource allocation for growth and stress adaptation, and evolution of stress-signal perception for rapid and long-term expression of stress tolerance (Hu and Xiong 2014; You and Chan 2015). The growing body of information has aided in the identification of essential genes linked to drought tolerance and the ability to maintain growth in water-stressed crops, such as sugarcane (Augustine et al. 2015; Hu and Xiong 2014; Ramiro et al. 2016). In addition, biotechnology and molecular breeding approaches can help increase crop production in water shortage regions. However, despite advances in our understanding of stress responses and the availability of molecular strategies and features, developing drought-tolerant crops remains a critical problem (Hu and Xiong 2014; Wang et al. 2016). This situation is due to the difficulties of identifying and exploiting significant effect genes and alleles and the related selection features for generating drought-tolerant cultivars suited for commercial agricultural production (Hu and Xiong 2014; Wang et al. 2016). Sugarcane is a crucial crop for producing ethanol and bioenergy, a clean alternative to fossil fuels, and its spread into rainfed areas necessitates more drought-tolerant plants. Exploring plant acclimation and signaling systems, on the other hand, could be a valuable technique for enhancing crop establishment and early growth in such new places.

10.2 MORPHOLOGICAL AND PHYSIOLOGICAL RESPONSES OF SUGARCANE TO WATER STRESS

Sugarcane plants' morphological and physiological responses differ depending on genotype, stress duration (rapid or gradual), stress intensity (severe or mild), and tissue type (Inman-Bamber, Lakshmanan, and Park 2012; Graça et al. 2010). As previously stated, water stress has a significant impact on both cane and sugar yield. However, it has been observed that there is a lot of potentially exploitable genetic diversity for cane and sugar output under water stress (Basnayake et al. 2012). Leaf rolling, stomatal closure, stalk and leaf growth inhibition, leaf senescence, and reduced leaf area are the most typical water-stress reactions in sugarcane (Inman-Bamber, Lakshmanan, and Park 2012).

Furthermore, cell division and elongation are disrupted under water stress (Machado et al. 2009), with stem and leaf elongation being the most severely affected growth processes. Sugarcane has a C4 photosynthetic metabolism, making it a tropical crop. Stomatal conductance (*gs*), transpiration rate (*E*), internal CO_2 concentration (Ci), and photosynthetic rate all decrease when exposed to mild water stress, owing to stomatal constraints (Basnayake et al. 2015; da Silva et al. 2012; Endres et al. 2010; Graça et al. 2010; Medeiros et al. 2013). When sugarcane plants are exposed to mild to moderate dryness, this, along with stalk and leaf development suppression, is the most common initial adaptation (Inman-Bamber and Smith 2005). Nonstomatal constraints caused by water stress have also been described as a source of photosynthesis suppression in sugarcane (Ribeiro et al. 2013). It's worth noting that sugar accumulation in the leaves affects the photosynthetic rate (McCormick, Cramer, and Watt 2008). Low leaf-sugar levels promote photosynthesis in nonstressed conditions, whereas excessive sugar content moderates carbon fixation (Goldschmidt and Huber 1992). Increased amounts of certain sugars, such as trehalose, can help plants cope with water deficits by minimizing cell membrane damage ((Delorge et al. 2014). Sugarcane roots were shown to have the ability to collect trehalose when they were drought stricken (Delorge et al. 2014). Sales et al. (2012) found that increased starch hydrolysis resulted in more significant amounts of soluble sugars, which helped maintain carbon supply even when CO_2 fixation was low, allowing for faster recovery after stress.

10.3 SUGARCANE DEFENSE MECHANISM TO DROUGHT

Understanding the nature of the agricultural problem from both a genetic and a crop production standpoint is essential for generating effective and financially viable remedies (Blum 2005). By definition, stress is any restriction of normal functioning and development that plants face during their life cycles. Plants have evolved several adaptive methods to live and flourish under stress, which include the notions of escape, avoidance, and tolerance in general terms. Plant genetics are responsible for physiological and morphological changes (Ciarmiello et al. 2011). Drought affects *Arabidopsis* (Catala et al. 2007), rice (Geisler 2010), and sugarcane gene expression in more than 1000 ways (Kido et al. 2012; Li et al. 2016; Xu et al. 2018). Plants use differentially expressed genes (DEGs) to escape unfavorable growing environments as a survival strategy (Iskandar et al. 2011; Kido et al. 2012; Li et al. 2016; Li et al. 2020).

Rocha et al. (2007) discovered a large number of DEGs with unknown functions on sugarcane exposed to drought stress, phosphate deficiency, herbivory, methyl jasmonate, abscisic acid, and two nitrogen-fixing endophytic bacteria. Furthermore, Begcy et al. (2012); and Begcy et al. (2019) performed additional research in tobacco plants to evaluate the involvement of sugarcane drought-responsive 1 (*Scdr1*) and sugarcane drought-responsive 2 (*Scdr2*), two DEGs with unknown functions under drought stress. *Scdr1* and *Scdr2* overexpression in transgenic tobacco plants improved drought, salinity, and oxidative stress tolerance (Begcy et al. 2012; Begcy et al. 2019). In sugarcane exposed to drought stress, an increase in DEGs-activated antioxidant enzymes, such as peroxidase (Li et al. 2016; Xu et al. 2018), transcriptional factors (TFs) such as MYB (Figure 10.1), NAC (Carrillo-Bermejo et al. 2020), and E3 SUMO-Protein ligase SIZ2 (Belesini et al. 2017; Guo et al. 2017). MYB and E3 SUMO-protein ligase SIZ2 are important transcription factors involved in manufacturing co-enzymes such A ligase, which boost the production of flavonoids and aquaporin proteins (AQPs) (Catala et al. 2007). The E3-SIZ1 ligase enzyme is required to accumulate small ubiquitin-related modifier (SUMO) proteins (Belesini et al. 2017; Catala et al. 2007). Through their unique ability to bind and alter the activity, localization, and reverse the actions of preexisting

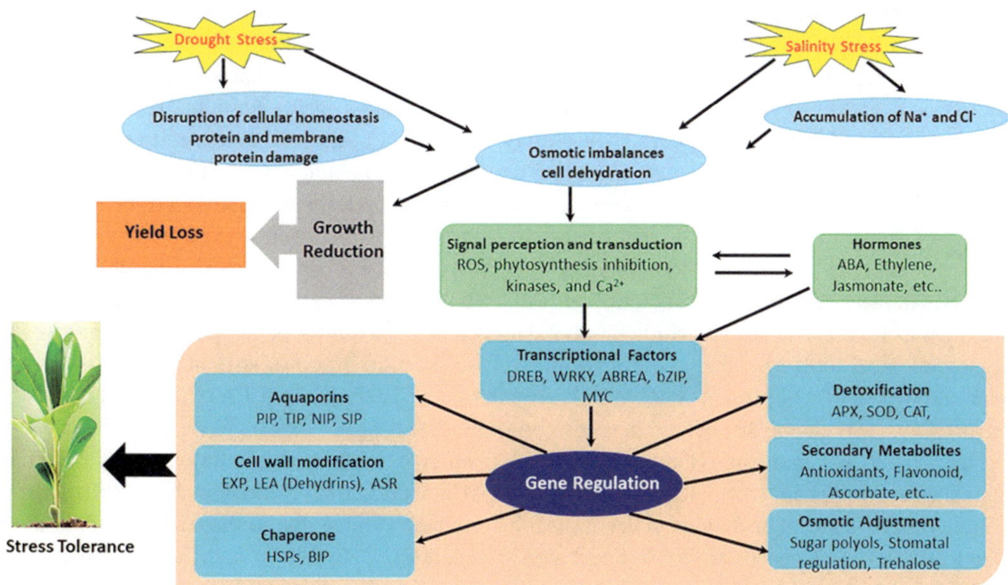

FIGURE 10.1 Regulation of gene expression in response to drought and salinity. Drought and salt stress share similar signaling pathways. Abbreviations: ABA – Abscisic Acid; Ca_2^+ – Calcium ion; DREB – dehydration-responsive element-binding protein; bZIP – Basic Leucine Zipper; MYC – Master Regulator of Cell; PIP – Plasma membrane Intrinsic Proteins; TIP- Tonoplast Intrinsic Protein; NIP – Nodulin-26 like Intrinsic Protein; SIP – Small basic Intrinsic Protein; HSP – heat shock proteins; BiP – Binding immunoglobulin protein; Exp – Expansin Proteins; *LEA* – late embryogenesis abundant proteins; ASR –abscisic acid stress and ripening-induced; SOD – superoxide dismutase; APX – Ascorbate peroxidase; CAT – catalase. The graph was adapted according to Dlamini (2021).

proteins or DNA, SUMO post-translational modifications of proteins play a vital role in most physiological processes (Bossis and Melchior 2006; Kurepa et al. 2003).

10.4 DROUGHT STRESS-REGULATED GENES AND PROTEIN

Several genes are either overexpressed (up-regulated) or repressed in response to drought stress (Geisler 2010). Heat-shock proteins (HSP) are a family of proteins that play an essential role in how plants respond to environmental stresses (Murphy 2011). During drought, salinity, and heat stresses, heat-shock protein 70 (HSP70) is up-regulated (Geisler 2010). High membrane thermostability, relative water content, gas-exchange parameters, chlorophyll content, and photosynthetic efficiency are all improved when HSP70 is overexpressed (Augustine Narayan et al. 2015). HSPs work as molecular chaperones, mending damaged proteins, preventing denaturation, and regulating heat-shock transcription-factor activity (Bossis and Melchior 2006). Heat and drought stress are typical in the field, which restricts HSP's activity (Li and Assmann 2010). During drought stress, high temperatures frequently cause malfunction and denaturation of functional proteins, including HSP (Survila, Heino, and Palva 2010; Wang et al. 2014). As a result, plants have evolved various interconnected stress (drought and heat) response mechanisms that serve complementary and sometimes overlapping roles in stress protection for proteins and other plant parts (Murphy 2011; Wang et al. 2016). For example, overexpression of *SUMO1* in *Arabidopsis* causes a decrease in small heat-shock protein (sHSP) levels (Cohen-Peer et al. 2010). Other research suggests that SUMO and HSPs work together to help refold broken proteins (Mishra et al. 2017; Vaseva et al. 2012). Several studies have found that certain environmental stresses, such as heat shock, cold, oxidative stress, and dehydration, significantly increase the conjugation of *SUMO1* and *SUMO2* to protein substrates (Bossis and Melchior 2006; Kurepa et al. 2003).

Plant responses to environmental factors and signaling cascades are influenced by SUMOylation, according to functional genomics research (Rabara, Tripathi, and Rushton 2014). (Garcia Tavares et al. 2018) regulated *ScGAI* expression in sugarcane and suggested that sugarcane *OTS* overexpression is involved in SUMOylation and *ScGAI* suppression during drought stress. Sugarcane plants with up-regulated *ScGAI* genes grow slowly, have a lot of tillers, have less carbon allocated to the stem, and have a huge number of tillers and a reduction in the amount of carbon allocated to the stem. As a result, regulating *ScGAI* expression in sugarcane under water stress may be critical for achieving plants with increased height, internode length, and carbon allocation to the stem (Garcia Tavares et al. 2018). Given the importance of the SUMOylation process in regulating physiological processes in plants that are crucial determinants of drought-stress tolerance, elucidating the molecular mechanisms involved in SUMO conjugation in sugarcane is critical for developing new markers and biotechnological tools to aid in its improvement. The ubiquitin group of stress-targeting genes encodes another family of proteins known as late embryogenesis abundant (LEA) (Survila, Heino, and Palva 2010). LEA can be discovered in seeds and vegetative tissues (Survila, Heino, and Palva 2010) and leaf tissues in sugarcane (Hanumappa and Nguyen 2010). The hydrophilic nature of LEA distinguishes them (Chen et al. 2019; Chen et al. 2015; Hundertmark and Hincha 2008). During droughts, LEA collects and binds water molecules on their surface due to their hydrophilic nature (Chen et al. 2019; Chen et al. 2015; Sugiharto et al. 2002). Dehydrin proteins are group II LEA proteins that improve the plant's ability to keep water and ions within its cell surface. As a result, drought-tolerant plants have a higher relative water content than susceptible plants (Augustine et al. 2015; Chen et al. 2019).

The ability of the cell to withhold water during water stress is dependent on the cell's ability to regulate potential water equilibrium by altering cell membrane permeability. Aquaporins (AQPs) are important participants in the mechanism that allows water to pass through the cell membrane (Afzal et al. 2016; Feltrim et al. 2021). AQPs maintain the water potentials of cells in sugarcane (De Andrade et al. 2016) and play a vital role in the sugar storage process in sugarcane vacuoles (Papini-Terzi et al. 2009). AQPs are core membrane proteins with six transmembrane helices that allow water or small uncharged molecules to pass through the osmotic gradient selectively (Kruse, Uehlein, and Kaldenhoff 2006). Plasma membrane intrinsic proteins (PIPs), tonoplast intrinsic proteins (TIPs), NOD26-like intrinsic proteins (NIPs), and small basic intrinsic proteins (SIPs) are the four subfamilies of AQPs. PIPs are further classified into two categories, PIP1 and PIP2, based on the lengths of amino and carboxyl termini (Afzal et al. 2016; Kruse, Uehlein, and Kaldenhoff 2006). da Silva et al. (2013) confirmed the up-regulation of PIP1, NIP, and SIP isoforms (PIP1-1, NIP3-1, and SIP1-2, respectively) in sugarcane roots during drought stress. De Andrade et al. (2016) also looked at the expression profiles of PIP2 isoforms (ShPIP2;1, ShPIP2;5, and ShPIP2;6) in sugarcane leaves when they were under drought stress. Once 21 days of water stress, ShPIP2;1, ShPIP2;5, and ShPIP2;6 were solely up-regulated, and they were exclusively down-regulated after the soil was rehydrated (De Andrade et al. 2016). Even though the purpose of PIP2 in sugarcane is unknown, its reaction to water deficiency suggests that it is involved in the regulation of the plant's water status (Feltrim et al. 2021; Ferreira et al. 2017). Low temperature causes overexpression of PIP2;2, PIP2;5, and PIP2;6 in eucalyptus globulus (Feltrim et al. 2021). PIP2 appears to be important in sucrose storage during drought stress in *Arabidopsis* (Wu et al. 2013). However, more research is needed to fully understand the role of AQPs in sugarcane under osmotic stress. This will provide us new breeding possibilities for water stress resistance and sucrose storage.

10.5 PROTEOMICS STUDIES ON SUGARCANE RESPONSE TO SALINITY STRESS

Sugarcane crop development is also hampered by soil salinity. Pacheco et al. (2013) used two-dimensional difference gel electrophoresis and mass spectrometry (2-DE and MS) analysis (UPLC–ESI–Q-ToF) to investigate differentially delayed root proteome responses to salt stress in sugarcane varieties to better understand the molecular mechanisms of tolerance to salinity stress. They discovered that a tolerant variety had the highest accumulation of proteins involved in growth, development, carbohydrate and energy metabolism, reactive oxygen species metabolism, protein protection, and membrane stabilization after two days. Murad et al. (2014) performed a proteomic analysis of two commonly used *Saccharum* spp. cultivars (RB867515 and RB855536) grown under salt stress using 2DE and MS analysis, which revealed that four proteins, including fructose 1,6-bisphosphate aldolase, germin-like protein, glyceraldehyde 3-phosphate dehydrogenase, and heat-shock protein 70 were differential. Passamani et al. (2017) used a label-free proteomics approach (ESI–LC–MS/MS analysis) to investigate the proteomic effects of salt stress in micropropagated shoots of two sugarcane cultivars (CB38-22 and RB855536). The results showed that the RB855536 cultivar exposed to salt stress had a higher abundance of proteins involved in nonenzymatic antioxidant mechanisms, ion transport, photosynthesis, and some proteins (calcium-dependent protein kinase photosystem I, phospholipase D, and glyceraldehyde-3-phosphate dehydrogenase). These findings indicated that various proteins play essential roles in acquiring ionic and osmotic homeostasis during salt stress in sugarcane.

10.6 TRANSCRIPTOMICS STUDIES ON SUGARCANE RESPONSE TO ABIOTIC STRESSES

The study of sugarcane transcriptome responses to abiotic stimuli can reveal defense mechanisms and potential molecular techniques for generating resistant cultivars. The involvement of transcription factors (TFs) (WRKY, MYB, bZIP, AP2/DREBP, and zinc-finger-like proteins) in plants as regulatory agents of metabolic pathways involved in plant defences against environmental stresses has been widely investigated (Song et al. 2011). Drought, cold, and excessive salinity, for example, significantly affect plant growth and thus productivity, and drought stress has been identified as one of the main limiting factors in sugarcane growth and production over the world (McQualter and Dookun-Saumtally 2007). Belesini et al. (2017) used the Illumina HiScanSQ System and HiSeq 2500 platforms to investigate the transcriptome profiles of the two sugarcane cultivars "SP81-3250" (drought-tolerant) and "RB855453" (drought-sensitive) under a variety of drought-stress conditions. They discovered ascorbate peroxidase, MYB, E3SUMO-protein ligase SIZ2, an important enzyme for biosynthesis of flavonoids such as co-enzyme A ligase, and aquaporin were all activated in drought-tolerant cultivars (Belesini et al. 2017). Abiotic stress tolerance is mediated by these genes/TFs (Ambawat et al. 2013; Guo et al. 2017; Roy 2016; Sakuraba et al. 2015). Many stress-induced kinases have been discovered in drought-sensitive cultivars (Belesini et al. 2017), including receptor-like protein kinases (RLK), which may play a role in stress perception, bHLH transcription factors, 1-aminocyclopropane-1-carboxylate (ACC) oxidase from the ethylene biosynthetic pathway, and several undescribed genes. Pereira-Santana et al. (2017) also employed a high-throughput sequencing method to investigate the transcriptional profile of Mexico's second most significant sugarcane variety, "Mex 69–290," in response to osmotic stress. They discovered that osmotic stress caused the cultivar to increase gene expression involved in transcriptional control, carbohydrate catabolism, oxide-reduction, flavonoid biosynthesis, and another secondary metabolite biosynthesis. ABA, water deprivation, and heat stress-related genes were also increased. The HT-SuperSAGE approach was used by da Silva et al. (2017) to evaluate four drought-resistant and susceptible cultivars. They discovered that when the sensitive cultivars were stressed by irrigation reduction, 9831 triggered unitags from the roots of the tolerant cultivars were regulated differently. Ethylene stress attenuation (ACCD), root growth (-EXP8), protein degradation, oxidative detoxification (TRX), fatty acid synthesis (ACC), amino acid transport (AAT), and carbohydrate metabolism [glycolysis (PFK, TPI, FBA), pentose phosphate pathway (TKT)] are just a few of the genes involved in sugarcane metabolism (da Silva et al. 2017). Liu et al. (2018) conducted an experiment on drought-stressed wild type sugarcane "*Saccharum narenga*" and found 3389 DEGs (1772 up-regulated and 1617 downregulated). These DEGs were found to be implicated in biological processes such as the metabolic route, blue light response, and single transduction of plant hormones (Liu et al. 2018). Furthermore, Vantini et al. (2015) used DNA-amplified fragment length polymorphism (AFLP) to evaluate the gene expression profiles of two drought-tolerant sugarcane cultivars against drought stress. According to the AFLP data, 173 pieces of tolerant cultivars displayed a changed expression pattern in response to water stress.

REFERENCES

Afzal, Z., T.C. Howton, Y. Sun, and M.S. Mukhtar. 2016. "The roles of aquaporins in plant stress responses." *Journal of Developmental Biology* 4 (1):9. 10.3390/jdb4010009

Ambawat, S., P. Sharma, N.R. Yadav, and R.C. Yadav. 2013. "MYB transcription factor genes as regulators for plant responses: an overview." *Physiology and Molecular Biology of Plants* 19 (3):307–321. 10.1007/s12298-013-0179-1

Augustine, S.M., J.A. Narayan, D.P. Syamaladevi, C. Appunu, M. Chakravarthi, V. Ravichandran, and N. Subramonian. 2015. "Erianthus arundinaceus HSP70 (EaHSP70) overexpression increases drought and salinity tolerance in sugarcane (*Saccharum* spp. hybrid)." *Plant Science* 232:23–34. 10.1016/j.plantsci.2014.12.012

Augustine, S.M., J.A. Narayan, D.P. Syamaladevi, C. Appunu, M. Chakravarthi, V. Ravichandran, N. Tuteja, and N. Subramonian. 2015. "Overexpression of EaDREB2 and pyramiding of EaDREB2 with the pea DNA helicase gene (PDH45) enhance drought and salinity tolerance in sugarcane (*Saccharum* spp. hybrid)." *Plant Cell Reports* 34 (2):247–263. 10.1007/s00299-014-1704-6

Basnayake, J., P.A. Jackson, N.G. Inman-Bamber, and P. Lakshmanan. 2012. "Sugarcane for water-limited environments. Genetic variation in cane yield and sugar content in response to water stress." *Journal of Experimental Botany* 63 (16):6023–6033. 10.1093/jxb/ers251

Basnayake, J., P.A. Jackson, N.G. Inman-Bamber, and P. Lakshmanan. 2015. "Sugarcane for water-limited environments. Variation in stomatal conductance and its genetic correlation with crop productivity." *Journal of Experimental Botany* 66 (13):3945–3958. 10.1093/jxb/erv194

Begcy, K., E.D. Mariano, A. Gentile, C.G. Lembke, S.M. Zingaretti, G.M. Souza, and M. Menossi. 2012. "A novel stress-induced sugarcane gene confers tolerance to drought, salt and oxidative stress in transgenic tobacco plants." *PLoS ONE* 7 (9):e44697. 10.1371/journal.pone.0044697

Begcy, K., E.D. Mariano, C.G. Lembke, S.M. Zingaretti, G.M. Souza, P. Araújo, and M. Menossi. 2019. "Overexpression of an evolutionarily conserved drought-responsive sugarcane gene enhances salinity and drought resilience." *Annals of Botany* 124 (4):691–700. 10.1093/aob/mcz044

Belesini, A.A., F.M.S. Carvalho, B.R. Telles, G.M. De Castro, P.F. Giachetto, J.S. Vantini, S.D. Carlin, J.O. Cazetta, D.G. Pinheiro, and M.I.T. Ferro. 2017. "De novo transcriptome assembly of sugarcane leaves submitted to prolonged water-deficit stress." Embrapa Informática Agropecuária-Artigo em periódico indexado (ALICE) 16 (2):1–20. doi: https://www.alice.cnptia.embrapa.br/alice/handle/doc/1082783 (Last Accessed November 29, 2021).

Blum, A. 2005. "Drought resistance, water-use efficiency, and yield potential—are they compatible, dissonant, or mutually exclusive?" *Australian Journal of Agricultural Research* 56 (11):1159–1168. doi: https://www.publish.csiro.au/cp/pdf/AR05069 (Last Accessed November 29, 2021).

Bossis, G. and F. Melchior. 2006. "SUMO: regulating the regulator." *Cell Division* 1 (1):1–8. 10.1186/1747-1028-1-13

Bray, E.A., J. Bailey-Serres, and E. Weretilnyk. 2000. "Responses to abiotic stresses." In *Biochemistry and Molecular Biology of Plants*, edited byW. Gruissem, B.B. Buchannan and R.L. Jones, 1158–1203. Rockville: American Society of Plant Physiologists.

Carrillo-Bermejo, E.A., S.D. Gamboa-Tuz, A. Pereira-Santana, M.A. Keb-Llanes, E. Castaño, L.J. Figueroa-Yañez, and L.C. Rodriguez-Zapata. 2020. "The SoNAP gene from sugarcane (*Saccharum officinarum*) encodes a senescence-associated NAC transcription factor involved in response to osmotic and salt stress." *Journal of Plant Research* 133 (6):897–909. 10.1007/s10265-020-01230-y

Catala, R., J. Ouyang, I.A. Abreu, Y. Hu, H. Seo, X. Zhang, and N-H. Chua. 2007. "The Arabidopsis E3 SUMO ligase SIZ1 regulates plant growth and drought responses." *The Plant Cell* 19 (9):2952–2966. 10.1105/tpc.106.049981

Cattivelli, L., F. Rizza, F-W. Badeck, E. Mazzucotelli, A.M. Mastrangelo, E. Francia, C. Marè, A. Tondelli, and A.M. Stanca. 2008. "Drought tolerance improvement in crop plants: an integrated view from breeding to genomics." *Field Crops Research* 105 (1-2):1–14. 10.1016/j.fcr.2007.07.004

Chen, Y-S., S-F. Lo, P-K. Sun, C-A. Lu, T-H. D. Ho, and S-M. Yu. 2015. "A late embryogenesis abundant protein HVA 1 regulated by an inducible promoter enhances root growth and abiotic stress tolerance in rice without yield penalty." *Plant Biotechnology Journal* 13 (1):105–116. 10.1111/pbi.12241

Chen, Y., C. Li, B. Zhang, J. Yi, Y. Yang, C. Kong, C. Lei, and M. Gong. 2019. "The role of the late embryogenesis-abundant (LEA) protein family in development and the abiotic stress response: a comprehensive expression analysis of potato (*Solanum tuberosum*)." *Genes* 10 (2):148. 10.3390/genes10020148

Ciarmiello, L.F., P. Woodrow, A. Fuggi, G. Pontecorvo, and P. Carillo. 2011. "Plant genes for abiotic stress." In *Abiotic Stress in Plants–Mechanisms and Adaptations*, edited byA. Shanker, 283–308. Rijeka, Croatia: InTech.

Cohen-Peer, R., S. Schuster, D. Meiri, A. Breiman, and A. Avni. 2010. "Sumoylation of *Arabidopsis* heat shock factor A2 (HsfA2) modifies its activity during acquired thermotholerance." *Plant Molecular Biology* 74 (1-2):33–45. doi: https://link.springer.com/content/pdf/10.1007/s11103-010-9652-1.pdf (Last Accessed November 29, 2021).

D'Hont, A., G.M. Souza, M. Menossi, M. Vincentz, M-A. Van-Sluys, J. Christophe Glaszmann, and E. Ulian. 2008. "Sugarcane: a major source of sweetness, alcohol, and bio-energy." In *Genomics of Tropical Crop Plants*, edited by P.H. Moore and R. Ming, 483-513. New York, USA: Springer.

da Silva, M.D., R.L. de Oliveira Silva, J.R.C.F. Neto, A.M. Benko-Iseppon, and E.A. Kido. 2017. "Genotype-dependent regulation of drought-responsive genes in tolerant and sensitive sugarcane cultivars." *Gene* 633:17–27. 10.1016/j.gene.2017.08.022

da Silva, M.D., R.L. Silva, C.F. de Oliveira, J.R. Neto, A.C.R. Guimarães, D.T. Veiga, S.M. Chabregas, W.L. Burnquist, G. Kahl, A.M. Benko-Iseppon, and E.A. Kido. 2013. "Expression analysis of sugarcane aquaporin genes under water deficit." *Journal of Nucleic Acids* 2013:1–14. 10.1155/2013/763945

da Silva, P.P., L.B. Soares, J.G. da Costa, L. da Silva Viana, J.C.F. de Andrade, E.R. Gonçalves, J.M. dos Santos, G.V. de Souza Barbosa, V.X. Nascimento, and A.R. Todaro. 2012. "Path analysis for selection of drought tolerant sugarcane genotypes through physiological components." *Industrial Crops and Products* 37 (1):11–19. 10.1016/j.indcrop.2011.11.015

De Andrade, L.M., P.M. Nobile, R.V. Ribeiro, J.F. De Oliveira, C. Nebó, A.V. de Oliveira Figueira, L.T.M. Frigel, D. Nunes, D. Perecin, M. dos Santos Brito, and R.C. de Matos Pires. 2016. "Characterization of PIP2 aquaporins in Saccharum hybrids." *Plant Gene* 5:31–37. 10.1016/j.plgene.2015.11.004

Dlamini, Ph. J. 2021. " Drought stress tolerance mechanisms and breeding effort in sugarcane: A review of progress and constraints in South Africa." *Plant Stress* 2 (2021) 100027. 10.1016/j.stress.2021.100027

Delorge, I., M. Janiak, S. Carpentier, and P. Van Dijck. 2014. "Fine tuning of trehalose biosynthesis and hydrolysis as novel tools for the generation of abiotic stress tolerant plants." *Frontiers in Plant Science* 5:147. 10.3389/fpls.2014.00147

Endres, L., J.V. Silva, V.M. Ferreira, and G.V. Souza Barbosa. 2010. "Photosynthesis and water relations in Brazilian sugarcane." *The Open Agriculture Journal* 4 (1):31–37. 10.2174/1874331501004010031

Feltrim, D., L. Pereira, M.G. de Santana Costa, T.S. Balbuena, and P. Mazzafera. 2021. "Stem aquaporins and surfactant-related genes are differentially expressed in two Eucalyptus species in response to water stress." *Plant Stress* 1:100003. 10.1016/j.stress.2021.100003

Ferreira, T.H.S., M.S. Tsunada, D. Bassi, P. Araújo, L. Mattiello, G.V. Guidelli, G.L. Righetto, V.R. Gonçalves, P. Lakshmanan, and M. Menossi. 2017. "Sugarcane water stress tolerance mechanisms and its implications on developing biotechnology solutions." *Frontiers in Plant Science* 8:1077. 10.3389/fpls.2017.01077

Garcia Tavares, R., P. Lakshmanan, E. Peiter, A. O'Connell, C. Caldana, R. Vicentini, J.S. Soares, and M. Menossi. 2018. "ScGAI is a key regulator of culm development in sugarcane." *Journal of Experimental Botany* 69 (16):3823–3837. 10.1093/jxb/ery180

Geisler, M. 2010. "Transcription and signaling factors in the drought response regulatory network." In *Genes for Plant Abiotic Stress*, edited byM.A. Jenks and A.J. Wood, 55–79. Ames, Iowa, USA: Wiley-Blackwell.

Goldschmidt, E.E., and S.C. Huber. 1992. "Regulation of photosynthesis by end-product accumulation in leaves of plants storing starch, sucrose, and hexose sugars." *Plant Physiology* 99 (4):1443–1448. 10.1104/pp.99.4.1443

Graça, J.P., F.A. Rodrigues, J.R.B. Farias, M.C.N. Oliveira, C.B. Hoffmann-Campo, and S.M. Zingaretti. 2010. "Physiological parameters in sugarcane cultivars submitted to water deficit." *Brazilian Journal of Plant Physiology* 22 (3):189–197. 10.1590/S1677-04202010000300006

Guo, J., H. Ling, J. Ma, Y. Chen, Y. Su, Q. Lin, S. Gao, H. Wang, Y. Que, and L.P. Xu. 2017. "A sugarcane R2R3-MYB transcription factor gene is alternatively spliced during drought stress." *Scientific Reports* 7 (1):1–11. 10.1038/srep41922

Hanumappa, M. and H.T. Nguyen. 2010. "Genetic approaches toward improving heat tolerance in plants." In *Genes for plant abiotic stress*, edited byM.A. Jenks and A.J. Wood, 221–260. Ames, Iowa, USA: Wiley-Blackwell.

Hu, H. and L. Xiong. 2014. "Genetic engineering and breeding of drought-resistant crops." *Annual review of plant biology* 65:715–741. 10.1146/annurev-arplant-050213-040000

Hundertmark, M. and D.K. Hincha. 2008. "LEA (late embryogenesis abundant) proteins and their encoding genes in Arabidopsis thaliana." *BMC Genomics* 9 (1):1–22. 10.1186/1471-2164-9-118

Inman-Bamber, N.G., P. Lakshmanan, and S. Park. 2012. "Sugarcane for water-limited environments: Theoretical assessment of suitable traits." *Field Crops Research* 134:95–104. 10.1016/j.fcr.2012.05.004

Inman-Bamber, N.G. and D.M. Smith. 2005. "Water relations in sugarcane and response to water deficits." *Field Crops Research* 92 (2-3):185–202. 10.1016/j.fcr.2005.01.023

Iskandar, H.M., R.E. Casu, A.T. Fletcher, S. Schmidt, J. Xu, D.J. Maclean, J.M. Manners, and G.D. Bonnett. 2011. "Identification of drought-response genes and a study of their expression during sucrose accumulation and water deficit in sugarcane culms." *BMC Plant Biology* 11 (1):1–14. 10.1186/1471-2229-11-12

Kido, E.A., J.R.C. Ferreira Neto, R.L.O. Silva, V. Pandolfi, A.C.R. Guimaraes, D.T. Veiga, S.M. Chabregas, S. Crovella, and A.M. Benko-Iseppon. 2012. "New insights in the sugarcane transcriptome responding to drought stress as revealed by supersage." *The Scientific World Journal* 2012. 10.1100/2012/821062

Kruse, E., N. Uehlein, and R. Kaldenhoff. 2006. "The aquaporins." *Genome Biology* 7 (2):1–6. 10.1186/gb-2006-7-2-206

Kurepa, J., J.M. Walker, J. Smalle, M.M. Gosink, S.J. Davis, T.L. Durham, D-Y. Sung, and R.D. Vierstra. 2003. "The small ubiquitin-like modifier (SUMO) protein modification system in Arabidopsis: accumulation of SUMO1 and-2 conjugates is increased by stress." *Journal of Biological Chemistry* 278 (9):6862–6872. 10.1074/jbc.M209694200

Li, C., Q. Nong, M.K. Solanki, Q. Liang, J. Xie, X. Liu, Y. Li, W. Wang, L. Yang, and Y. Li. 2016. "Differential expression profiles and pathways of genes in sugarcane leaf at elongation stage in response to drought stress." *Scientific Reports* 6 (1):1–11. 10.1038/srep25698

Li, P., Z. Chai, P. Lin, C. Huang, G. Huang, L. Xu, Z. Deng, M. Zhang, Y. Zhang, and X. Zhao. 2020. "Genome-wide analysis of AP2/ERF Transcription Factors in sugarcane Saccharum spontaneum L. Reveals Functional Divergence During Drought, Salt Stress and Plant Hormones Treatment." *Research Squre*:1–33. 10.21203/rs.3.rs-19836/v1

Li, S., and S. Assmann. 2010. "Genetic determinants of stomatal function." In *Genes for Plant Abiotic Stress*, edited by M.A. Jenks and A.J. Wood, 5–33. Ames, Iowa, USA: Wiley-Blackwell.

Liu, X., R. Zhang, H. Ou, Y. Gui, J. Wei, H. Zhou, H. Tan, and Y. Li. 2018. "Comprehensive transcriptome analysis reveals genes in response to water deficit in the leaves of Saccharum narenga (Nees ex Steud.) hack." *BMC Plant Biology* 18 (1):1–16. 10.1186/s12870-018-1428-9

Machado, R.S., R.V. Ribeiro, P.E.R. Marchiori, D.F.S.P. Machado, E.C. Machado, and M.G.A. Landell. 2009. "Biometric and physiological responses to water deficit in sugarcane at different phenological stages." *Pesquisa Agropecuária Brasileira* 44 (12):1575–1582. 10.1590/S0100-204X2009001200003

McCormick, A.J., M.D. Cramer, and D.A. Watt. 2008. "Regulation of photosynthesis by sugars in sugarcane leaves." *Journal of Plant Physiology* 165 (17):1817–1829. 10.1016/j.jplph.2008.01.008

McQualter, R.B. and A. Dookun-Saumtally. 2007. "Expression profiling of abiotic-stress-inducible genes in sugarcane." XXVI Congress, International Society of Sugar Cane Technologists, Durban, South Africa, 29 July-2 August, 2007.

Medeiros, D.B., E.C. Silva, R.J. Nogueira, C. Mansur, M.M. Teixeira, and M.S. Buckeridge. 2013. "Physiological limitations in two sugarcane varieties under water suppression and after recovering." *Theoretical and Experimental Plant Physiology* 25:213–222. https://www.scielo.br/j/txpp/a/4qt4C3bBNn7bB5NkpxsWHLF/?format=pdf&lang=en (Last Accessed November 29, 2021).

Mishra, N., L. Sun, X. Zhu, J. Smith, A. Prakash Srivastava, X. Yang, N. Pehlivan, N. Esmaeili, H. Luo, and G. Shen. 2017. "Overexpression of the rice SUMO E3 ligase gene OsSIZ1 in cotton enhances drought and heat tolerance, and substantially improves fiber yields in the field under reduced irrigation and rainfed conditions." *Plant and Cell Physiology* 58 (4):735–746. 10.1093/pcp/pcx032

Murad, A.M., H.B.C. Molinari, B.S. Magalhaes, A.C. Franco, F.S.C. Takahashi, N.G. de Oliveira-, O.L. Franco, and B.F. Quirino. 2014. "Physiological and proteomic analyses of *Saccharum* spp. grown under salt stress." *PLoS One* 9 (6):e98463. 10.1371/journal.pone.0098463

Murphy, D. 2011. "Plants, biotechnology and agriculture." In *Crop Improvement in the 20th Century*, edited byD. Murphy, 194–219. Wallingford, UK: CABI.

Pacheco, C.M., M.C. Pestana-Calsa, F.C. Gozzo, C. Mansur, R.J. Nogueira, M. Menossi, and T. Calsa Junior. 2013. "Differentially delayed root proteome responses to salt stress in sugar cane varieties." *Journal of Proteome Research* 12 (12):5681–5695. 10.1021/pr400654a

Papini-Terzi, F.S., F.R. Rocha, R.Z.N. Vêncio, J.M. Felix, D.S. Branco, A.J. Waclawovsky, L.E.V. Del Bem, C.G. Lembke, M.D.L. Costa, and M.Y. Nishiyama. 2009. "Sugarcane genes associated with sucrose content." *BMC Genomics* 10 (1):1–21. 10.1186/1471-2164-10-120

Passamani, L.Z., R.R. Barbosa, R.S. Reis, A.S. Heringer, P.L. Rangel, C. Santa-Catarina, C. Grativol, C.F.M. Veiga, G.A. Souza-Filho, and V. Silveira. 2017. "Salt stress induces changes in the proteomic profile of micropropagated sugarcane shoots." *PLoS One* 12 (4):e0176076. 10.1371/journal.pone.0176076

Patade, V.Y., S. Bhargava, and P. Suprasanna. 2011. "Salt and drought tolerance of sugarcane under iso-osmotic salt and water stress: growth, osmolytes accumulation, and antioxidant defense." *Journal of Plant Interactions* 6 (4):275–282. 10.1080/17429145.2011.557513

Patade, V.Y., P. Suprasanna, and V.A. Bapat. 2008. "Effects of salt stress in relation to osmotic adjustment on sugarcane (Saccharum officinarum L.) callus cultures." *Plant Growth Regulation* 55 (3):169–173. 10.1007/s10725-008-9270-y

Pereira-Santana, A., E.J. Alvarado-Robledo, J.A. Zamora-Briseño, J.T. Ayala-Sumuano, V.M. Gonzalez-Mendoza, F. Espadas-Gil, L.D. Alcaraz, E. Castaño, M.A. Keb-Llanes, and F. Sanchez-Teyer. 2017. "Transcriptional profiling of sugarcane leaves and roots under progressive osmotic stress reveals a regulated coordination of gene expression in a spatiotemporal manner." *PloS one* 12 (12):e0189271. 10.1371/journal.pone.0189271

Rabara, R.C., P. Tripathi, and P.J. Rushton. 2014. "The potential of transcription factor-based genetic engineering in improving crop tolerance to drought." *Omics: a Journal of Integrative Biology* 18 (10):601–614. 10.1089/omi.2013.0177

Ramiro, D.A., D.M. Melotto-Passarin, M.A. Barbosa, F. Santos, S.G.P. Gomez, N.S. Massola Junior, E. Lam, and H. Carrer. 2016. "Expression of Arabidopsis Bax Inhibitor-1 in transgenic sugarcane confers drought tolerance." *Plant Biotechnology Journal* 14 (9):1826–1837. 10.1111/pbi.12540

Ribeiro, R.V., R.S. Machado, E.C. Machado, D.F.S.P. Machado, J.R. Magalhães Filho, and M.G.A. Landell. 2013. "Revealing drought-resistance and productive patterns in sugarcane genotypes by evaluating both physiological responses and stalk yield." *Experimental Agriculture* 49 (2):212–224. 10.1017/S0014479712001263

Rocha, F.R., F.S. Papini-Terzi, M.Y. Nishiyama, R.Z.N. Vêncio, R. Vicentini, R.D.C. Duarte, V.E. de Rosa, F. Vinagre, C. Barsalobres, and A.H. Medeiros. 2007. "Signal transduction-related responses to phytohormones and environmental challenges in sugarcane." *BMC Genomics* 8 (1):1–22. 10.1186/1471-2164-8-71

Roy, S. 2016. "Function of MYB domain transcription factors in abiotic stress and epigenetic control of stress response in plant genome." *Plant Signaling & Behavior* 11 (1):e1117723. 10.1080/15592324.2015.1117723

Sakuraba, Y., Y-S. Kim, S-H. Han, B-D. Lee, and N-C. Paek. 2015. "The Arabidopsis transcription factor NAC016 promotes drought stress responses by repressing AREB1 transcription through a trifurcate feed-forward regulatory loop involving NAP." *The Plant Cell* 27 (6):1771–1787. 10.1105/tpc.15.00222

Sales, C.R.G., R.V. Ribeiro, D.F.S.P. Machado, R.S. Machado, V.L. Dovis, and A.M.M.A. Lagôa. 2012. "Gas exchange and carbohydrate balance in sugarcane plants under root stressful conditions." *Bragantia* 71 (3):319–327. 10.1590/S0006-87052012000300001

Song, S., T. Qi, H. Huang, Q. Ren, D. Wu, C. Chang, W. Peng, Y. Liu, J. Peng, and D. Xie. 2011. "The Jasmonate-ZIM domain proteins interact with the R2R3-MYB transcription factors MYB21 and MYB24 to affect Jasmonate-regulated stamen development in Arabidopsis." *The Plant Cell* 23 (3):1000–1013. 10.1105/tpc.111.083089

Sugiharto, B., N. Ermawati, H. Mori, K. Aoki, K. Yonekura-Sakakibara, T. Yamaya, T. Sugiyama, and H. Sakakibara. 2002. "Identification and characterization of a gene encoding drought-inducible protein localizing in the bundle sheath cell of sugarcane." *Plant and Cell Physiology* 43 (3):350–354. 10.1093/pcp/pcf039

Survila, M., P. Heino, and E.T. Palva. 2010. "Genes for plant tolerance to temperature extremes." In *Genes for Plant Abiotic Stress*, edited by M.A. Jenks and A.J. Wood, 185–209. Ames, Iowa, USA: Wiley-Blackwell.

Vantini, J.S., G.C. Dedemo, D.F. Jovino Gimenez, L.F. Fonseca, R.I. Tezza, M.A. Mutton, J.A. Ferro, and M.I.T. Ferro. 2015. "Differential gene expression in drought-tolerant sugarcane roots." *Genet. Mol. Res* 14 (2):7196–7207. 10.4238/2015.June.29.13

Vaseva, I., J. Sabotič, J. Šuštar-Vozlič, V. Meglič, M. Kidrič, K. Demirevska, and L. Simova-Stoilova. 2012. "The response of plants to drought stress: the role of dehydrins, chaperones, proteases and protease inhibitors in maintaining cellular protein function." In *Droughts: New Research*, edited byD.F. Neves and J.D. Sanz, 1–45. Hauppauge, NY, USA: Nova Science Publishers, Inc.

Waclawovsky, A.J., P.M. Sato, C.G. Lembke, P.H. Moore, and G.M. Souza. 2010. "Sugarcane for bioenergy production: an assessment of yield and regulation of sucrose content." *Plant Biotechnology Journal* 8 (3):263–276. 10.1111/j.1467-7652.2009.00491.x

Wang, W., B. Vinocur, and A. Altman. 2003. "Plant responses to drought, salinity and extreme temperatures: towards genetic engineering for stress tolerance." *Planta* 218 (1):1–14. 10.1007/s00425-003-1105-5

Wang, X., J. Cai, F. Liu, T. Dai, W. Cao, B. Wollenweber, and D. Jiang. 2014. "Multiple heat priming enhances thermo-tolerance to a later high temperature stress via improving subcellular antioxidant activities in wheat seedlings." *Plant Physiology and Biochemistry* 74:185–192. 10.1016/j.plaphy.2013.11.014

Wang, X., X. Cai, C. Xu, Q. Wang, and S. Dai. 2016. "Drought-responsive mechanisms in plant leaves revealed by proteomics." *International Journal of Molecular Sciences* 17 (10):1706. 10.3390/ijms17101706

Wu, X.N., C.S. Rodriguez, H. Pertl-Obermeyer, G. Obermeyer, and W.X. Schulze. 2013. "Sucrose-induced receptor kinase SIRK1 regulates a plasma membrane aquaporin in Arabidopsis." *Molecular & Cellular Proteomics* 12 (10):2856–2873. 10.1074/mcp.M113.029579

Xu, S., J. Wang, H. Shang, Y. Huang, W. Yao, B. Chen, and M. Zhang. 2018. "Transcriptomic characterization and potential marker development of contrasting sugarcane cultivars." *Scientific Reports* 8 (1):1–11. 10.1038/s41598-018-19832-x

You, J. and Z. Chan. 2015. "ROS regulation during abiotic stress responses in crop plants." *Frontiers in plant science* 6:1092. 10.3389/fpls.2015.01092

11 Avoidance and Tolerance Strategy of Sugarcane in Drought Stress: Response Mechanism

Varucha Misra, Ashutosh Kumar Mall, and A.D. Pathak
ICAR-Indian Institute of Sugarcane Research, Lucknow, India

CONTENTS

11.1 INTRODUCTION

Drought is just one of many abiotic stresses that influence the growth and development of sugarcane. The occurrence of drought during sugarcane production causes heavy losses. This loss poses a vital threat to food security due to low cane production and productivity. Reduction in available water level in soil and incessant water loss from plants through transpiration are the major reasons for drought conditions (Misra et al. 2022). The sugarcane production and productivity losses are augmented by the occurrence of drought stress (Misra et al. 2020, Misra et al. 2016a, Misra et al. 2016b). Silva et al. (2013) found the reason behind this problem as low-deficit rainfall or high temperatures or a combination of these two which results in hindrance to proper cane growth and development. Studies found that drought occurrence during the different stages of the cane growth caused loss of 70% in yield and productivity of the crop (Gosal et al. 2009, Morison et al. 2008, Swami et al. 2018). Basnayake et al. (2012) found that the productivity losses

DOI: 10.1201/9781003292425-11

extend to 60% in sugarcanes. The variation in losses of productivity depends on the impact and stage of the growth on which drought strikes the sugarcane crop (Mall et al. 2019).

Sugarcane is a long extent crop consisting of four differentiated stages, *viz.*, germination, tillering, grand growth and maturity (Gascho and Shih 1983). Jaleel et al. (2009) revealed that drought stress is a crucial factor when it occurs during the initial stage of plant growth. Ramesh (2000) found that the initial stage of sugarcane growth, known as the formative phase (that includes tillering and grand growth stage), was the most crucial for the plants' water requirements. During this phase, the plant's major water demand is for utilization in later cane yield production (Singh and Rao 1987). This is the reason behind the association of water relation and photosynthetic response of sugarcane under drought stress to be used as a distinguishing feature for screening tolerant genotypes for such conditions (Mall et al. 2022). Drought stress alters the growth, morphology, and physiology of sugarcane (Mall et al. 2019, Mall et al. 2017, Mall et al. 2020). Silva et al. (2008) showed that physiological processes occurring during drought stress in canes hinder plant photosynthesis and growth. This further contributes to low biomass production and yield of sugarcane (Zhao and Li 2015). The frequent occurrence of drought has caused breeders to develop sugarcane genotypes/varieties for coping up with drought-stress conditions. Many of the varieties, in this perspective, have been commercially released by government institutes but still require much research. The improvement of sugarcane for drought stress needs a proper understanding of the strategies, which the cane has been bestowed naturally.

The chapter throws an insight on drought tolerance and avoidance mechanism in sugarcane crop.

11.2 STRATEGIES OF DROUGHT TOLERANCE IN SUGARCANE

Avoidance and tolerance are the two prominent strategies for sugarcane crops to develop drought tolerance (Levitt 1980). A number of protective mechanisms are involved in the stress-avoidance strategy that helps in preventing the harmful effects of drought on sugarcane. On the other hand, stress tolerance strategy helps the plant become accustomed to these conditions. Drought tolerance can be defined as a plant mechanism that causes the plant to sustain its metabolism by lowering the tissue-water potential (Mitra 2001). This strategy, once adapted by the plant, is assessed by the cell membrane enhancement under these conditions. Leaf area maintenance and proper growth during the vegetative period are traits important for drought tolerance in plants. In general, the roots of the plant are the key organs that help in adapting any strategy by the plant for drought condition (Abobatta 2019).

11.2.1 Protective Mechanisms of Drought-Avoidance Strategy

When drought strikes the formative phase of sugarcane, mainly due to high temperatures and elevated evaporation rate, leaf rolling, lower leaf area, reduction in stalk and leaf growth, leaf senescence, and leaf rolling are some of the noticeable responses in this crop (Srinivasan 1993, Rao et al. 2007, Dhansu et al. 2021), but these responses were dependent on the genotype, duration of stress, and intensity at which it influences the cane (Misra et al. 2020a, Kumar et al. 2021).

11.2.2 Response of Stomatal Conductance and Its Mechanism under Drought Stress

Generally, when plants are exposed to drought stress, the stomata in the leaves are closed to avoid the loss of water through transpiration (Pirasteh-Anosheh et al. 2016). There are two reasons behind the closing of the stomata, one being the water loss through evaporation directly from guard cells with no metabolic action and other being lower availability of water in soil and high temperatures. Taiz and Zeiger (2002) showed that the plant does so as to adopt avoidance strategy for drought stress. Stomatal closure as a response to drought stress helps in reducing the photosynthesis rate with a reduction in the carbon dioxide level. This causes the photorespiration rate to increase. Ribeiro et al. (2013) showed that in sugarcane, stomatal alterations in response to water availability under drought conditions occur at a speedy pace. This helps in avoiding the unnecessary turgor loss in leaves and also prevents further reduction in the water contents of the leaves. Pirasteh-Anosheh et al. (2016) found that due to this partial substrate, replenishment was assured and carboxylating function of RuBisCO gets maintained. The closing of the stomata is dependent on the turgidity loss of the two guard cells contiguous to the opening. Induction of several signals has been reported by the closure of the stomata, the most important being the ABA signaling. Studies found that secondary messengers (Ca^{2+}, O^2, etc) and nitrous oxide were also involved during the signaling pathway and help in closing the stomata under drought-stress conditions (Manzoni et al. 2011, Arve et al. 2013). Arve et al. (2011) revealed that even the passive loss in turgidity was also responsible for such an event. Hydropassive and hydroactive are the two mechanisms involved in stomatal closure during drought stress.

11.2.2.1 Hydropassive Mechanism

During drought conditions, the production of abscisic acid (ABA) in plant roots and leaves has been observed. ABA is transported to the guard cells by ATP binding cassette (ABC) transporters, where signals were activated for stomata closure. The ABC transporters are situated in the plasma membrane of the leaves. Mutava et al. (2015) showed that during drought conditions, when these transporters are removed, then reduction in ABA was seen, resulting in the opening of stomata and reduction in stress-tolerance capability. The signals induced by ABA are known to be accepted by receptors (Kang et al. 2010, Kim et al. 2010, Mutava et al. 2015). Arve et al. (2013) revealed the association of the opening of the stomata under drought conditions with ion and water transportation via channel proteins. These proteins are located on the plasma and vacuole membrane of the leaves. Induction of ABA promotes the production of reactive oxygen species (ROS). The production of ROS acts as a trigger point for the nitrous oxide production, calcium ions influx, and inhibition of membrane proton pumps. For the induction of ABA in stomata closing, H^+ ATPases plays role in hyperpolarizing of the plasma membranes. The secondary messengers, particularly calcium ions, were increased under drought conditions, and these ions activate anion channels creating efflux of anions from cells. This efflux causes depolarization of plasma membrane and further generates potassium-ion efflux, which moves out from potassium ion channels. These channels were located in vacuoles and plasma membranes. Concurrently, the production of calcium ions also inhibits the potassium ions from entering the channels. Malate produces in plants and also converts to starch under drought stress for lowering the turgor and osmotic pressure. Depolarisation of plasma membrane with lowering of turgor pressure and cell volume results in stomatal closure (Kim et al. 2010, Arve et al. 2011, Arve et al. 2013). Mittler and Blumwald (2015) reported the generation of a loop for mediating stomatal closure by direct relation of ROS with ABA production. Higher ABA causes higher ROS production under drought stress condition. Studies showed that ROS production is observed to be responsible for biosynthesis and inhibition of ABA (Saed-Moucheshi et al. 2014, Rauf et al. 2015). Stomatal closure, induction of photorespiration (due to low carbon dioxide) causes higher accural of hydrogen peroxide (Mittler and Blumwald 2015).

11.2.2.2 Hydroactive Mechanism

High ABA content in leaves during drought stress causes the regulation of the stomata closing. On the other hand, in roots of plants exposed to drought stress, the hydraulic conductivity gets enhanced due to ABA production. This enhancement aids in the uptake of water and transportation of solutes (Arve et al. 2011, Arve et al. 2013).

11.2.3 Response of Leaf Rolling and Its Mechanism under Drought Stress

Alteration in movement of leaves during drought stress is a common adaptive approach (Begg 1980, Ehleringer and Forseth 1980). Leaves are the main organ that indicates the water shortage level through assimilation and transpiration (Yang et al. 2021). Leaf rolling, paraheliotropism, and leaf folding are some of the leaf movements seen in adaptive response for plants under stress conditions. In sugarcane, leaf rolling is the prominent response and effective protective mechanism (O'Toole and Chang 1978; Misra et al. 2020a) in which leaves are protected from photodamage due to high light radiations (Corlett et al. 1994). Begg (1980) found that the leaf orientation changes due to lower radiation interception. The leaf rolling in turn provides flexibility in changing water availability to sugarcane, thus causing reduction in radiation load on the leaf canopy, whenever required. Leaf rolling in sugarcane under drought conditions provides two major adaptive benefits. First, the injury caused by an increase in leaf temperature due to high light radiation is minimized. This occurs due to a reduction in leaf area by which less radiation is intercepted by leaf tissue (Begg 1980). Second, the rate of transpiration is also lowered due to higher humidity and boundary-layer resistance, resulting in water conservation in sugarcane under drought conditions (Oppenheimer 1960). The leaf rolling in sugarcane under drought condition was mainly due to a loss of turgor pressure due to deficit water conditions in the upper epidermis of the leaves (Misra et al. 2020). Studies showed that leaf rolling is considered a selection criterion for scoring drought tolerance in plants (Loresto et al. 1976, Loresto and Chang 1981).

For adapting to drought stress, reduced leaf area, larger leaf thickness, and higher leaf tissue density are some other traits (Werner et al. 1999). Reduction in leaf area has also been known as avoidance trait because it helps in further curtailing the water loss. Association of leaf area with photosynthesis and yield has been well known under drought conditions. Under drought stress, leaf turgor pressure and photosynthesis, responsible for reduction in leaf area, were reduced (Rucker et al. 1995). In sugarcane under drought stress, the leaves were known to be generally elongated a bit (Misra et al. 2020).

11.2.4 Response of Leaf Senescence and Its Mechanism under Drought Stress

Leaf senescence is another response trait by plants under drought stress conditions. Nutrient remobilization comes into action when leaf senescence due to drought stress occurs in sugarcanes. In this situation, the plant consumes the stored nutrients, thus helping the rest of the plant parts to cope up with stress. Leaf senescence and abscission go hand in hand as an avoidance strategy for drought stress. This helps in sustaining water balance in the plant by preventing heavy losses occurring during transpiration. Leaf senescence induced by drought stress in plants is differentiated by the alteration at cellular, molecular, and biochemical levels. Munne-Bosch and Algere (2004) found that loss in chlorophyll content, protein breakdown, lipid peroxidation, chromatin condensation, etc., were some of the changes seen in plants due to drought-induced leaf senescence. De Souza et al. (1997) revealed that prolonged drought stress speeds the senescence of leaves, resulting in the demise of tissues and ultimately the effected leaf fall. This is mostly seen in old and mature leaves. Brevedan and Egli (2003) showed that if the plants were exposed to drought stress for a relatively shorter period and later re-watered, even under such a condition, the impact of drought on senescence was not altered. Under drought stress, genotypes having decreased leaf senescence have frequently been considered as a tolerant mechanism.

11.2.5 Response of Photosynthesis and Their Role under Drought Stress

At a certain stage, drought stress in sugarcane causes significant changes in physiological activities of plants (Wei-Hua et al. 2003, Ali et al. 2019). The photosynthetic activity has been known to be very sensitive toward drought stress; under this stress, a lower rate of carbon dioxide and severe injury to chloroplast structure may occur (Zhang et al. 2015, Boyer et al. 1997). Because leaves are the site for transpiration and photosynthesis processes, when under drought stress, they may become much more sensitive. Stomatal conductance and transpiration are the two other physiological activities that are known to be dependent on the photosynthesis process (da Silva et al. 2012). Sugarcane leaves had deteriorated chloroplast under drought stress (Zhang et al. 2015), but the severity was dependent on the nature of sugarcane genotype (sensitive/tolerant). The sensitive genotype had severe distortion in the chloroplast structure under drought stress, while it was not the case with tolerant genotypes (Zhang et al. 2015). Rodrigues et al. (2009) showed that drought stress causes disparity in photosynthetic activities of different sugarcane genotypes. Studies reported that more effect was seen in the photosynthetic systems of sensitive cultivars than tolerant ones (Ribeiro et al. 2013, Sales et al. 2013). Marcos et al. (2018) reported that plants facing drought for three to four cycles had higher photosynthesis and stomatal conductance than the ones facing one or two water cycles.

11.2.5.1 Sugarcane Drought-Tolerance Accustomed Strategy

Rise in chlorophyll content, higher stomatal conductance, increased photosynthesis, and maintenance of growth and osmotic adjustment are some of the drought-tolerance traits in sugarcane. Zhang et al. (2015) showed that under drought stress, the sugarcane chloroplast structure gets loosened.

11.2.5.2 Accumulation of Solutes, Secondary Metabolites, and Antioxidant Activity as Response to Drought Tolerance in Sugarcane

Studies revealed that due to the chemical signals activated by drought stress conditions, the accrual of compatible solutes showed an increase in antioxidant capacity (Sairam et al. 2002, Moussa and Abdel-Aziz 2008). In general, for maintaining the turgor pressure under drought stress, plants produce and accumulate solutes (organic or inorganic in nature) in the cytosol of the cell that results in increasing the osmotic pressure. The rise in osmotic pressure creates a gradient for the uptake of water in plants under such a condition (Rhodes and Samaras 1994). Association of compatible solutes and antioxidant ability of sugarcane under drought stress act as a tolerant capability for coping with lowering of reduced water level in the soil (Moussa and Abdel-Aziz 2008). The compatible solutes so produced have been known to have no interference with structural and functional properties of macromolecules (Hoekstra et al. 2001, Rogiers et al. 2011b). Shao et al. (2006) found that the production of proline under drought stress conditions is also a physio-biochemical indicator. Higher proline content is correlated with more drought tolerance in sugarcane (Molinari et al. 2007, Vendruscolo et al. 2007), thus confirming the preservation of protein structure and its functions (Wingler 2002, Verlues et al. 2006). Marcos (2018) found that reduced proline content in sugarcane under drought conditions also indicates that the plant may be exposed to less stressful conditions. The proline serves various functions in sugarcane under drought stress conditions. These are adjustment of osmotic pressure, utilization of C and N stored in the recovery period from drought, scavenging of ROS, and stabilization of proteins along with membranes (Kishor et al. 2005). Mittler (2002) showed that the protective system of plants under drought stress includes many enzymatic and nonenzymatic compounds that avoid oxidative damage occurring through ROS scavenging. The signaling pathway by which drought stress stimulates the production of proline synthesis involves molecules like abscisic acid, phospholipase, and calcium (Parre et al. 2007, Savoure et al. 1997, Strizhov et al. 1997, Yoo et al. 2005). Sharma et al. (2011) showed that the protection strategy induced by ABA needs proline metabolism in plants exposed to drought stress. The molecular mechanism of this solute is still not well known, particularly during stress conditions, but due to its osmolytic property and water-balance capability, proline plays an important role in such conditions (Delauney and Verma 1993). In general, to overcome the stress by plant through proline, two strategies are involved: either the plant starts accumulating the proline, or the proline metabolic flux protects the plant. The proline metabolic flux so generated causes protection to cells either by activating other signaling pathways, maintaining the energy in the cells. or balancing the $NADP^+$/NADPH ratio (Liang et al. 2013). The increase in synthesis of free amino acids in sugarcane under drought stress had no correlation with proteolysis due to no significant reduction in soluble protein content; however, it could be due to the impact of drought stress (Medeiros et al. 2012). Studies reported that canes grown under drought conditions have high proline content in leaves and root portions (Queiroz et al. 2008, Cha-um and Kirdmanee 2009). Ashraf and Foolad (2007) revealed that production of proline in sugarcanes under drought conditions can be considered as a preliminary reaction of stress tolerance. This free amino acid also assists in altering the osmotic potential acting as a tolerant trait (Kavi Kishor et al. 2005). Stabilization of membrane was also seen by proline accumulation under drought stress with the release of oxygen-free radicals (Ashraf and Foolad 2007, Silva et al. 2013). Guimarães et al. (2008) showed that association of proline with drought-stress tolerance in sugarcane. In transgenic canes, P5CS genes under drought conditions, have shown a significant and direct association with proline, yield, and photosystem II (Molinari et al. 2007). In such canes, the proline works as an antioxidant defense system constituent, besides functioning as an osmoregulator in general conditions. Iskandar et al. (2011) revealed that, in mature canes among the various amino acids accumulation under drought conditions, proline stands the least in content.

Production of ROS was also seen to increase under drought stress in sugarcane. ROS is considered as a crucial signaling molecule produced as a toxic product of aerobic metabolism. The accrual of this product was highly seen when abiotic stress conditions prevail (Huang et al. 2012). The production of this metabolic by-product was due to the low carbon dioxide levels under drought conditions in relation to stomatal closure and inequity between photochemical and biochemical reactions in drought-stress plant leaves (Du et al. 1996, Chaves et al. 2009, Machado et al. 2013). Drought causes increases in ROS, resulting in an imbalance of redox potential of cells, interference in the electron transport chain, and initiation of oxidative stress due to ROS toxic action (Gratão et al. 2008). During drought, the balance between ROS scavenging (involving enzymatic and non enzymatic compounds) and production gets disturbed, causing unexpected enhancement in redox potential. This unexpected increase further causes damage to cell constituents like lipids, DNA, and proteins (Mittler 2002, Anjum et al. 2011b).

Increase hydrogen peroxide levels have been reported with the increase in lipid peroxidation under severe drought conditions when the sugarcane plants were in the initial growth phase (Boaretto et al. 2014). At times, the plant's scavenging capacity exceeds ROS levels, and lipid peroxidation tends to increase that effect on the cell's physiological activities. Also, studies have shown the association of drought tolerance with reduced lipid peroxidation content in canes exposed to drought conditions (Sales et al. 2015). The amount of lipid peroxidation has also been considered as a tolerant parameter for screening drought-tolerant sugarcanes (Abbas et al. 2014). Malondialdehyde, a product of lipid oxidative, is accountable for damage by cell membranes involving alteration in intrinsic properties of membrane, such as ion transport, loss in enzymatic activity, and protein cross linking. All of these alterations lead to cell death (Sharma et al. 2012). Initiation, propagation, and termination are the three steps of lipid peroxidation. At initiation, production of fatty acid radical is seen. This generated hydroxyl radical combines with methylene groups of polyunsaturated fatty acids for production of hydroperoxides, lipid peroxy radicals, and other conjugated dienes (Smirnoff 1995). Aldehydes were readily produced when these hydroperoxides were decomposed (Stadman 1986).

The antioxidant enzymatic activity in sugarcane under drought condition helps in providing tolerance to plants. Superoxide dismutase (SOD), acsorbate dismutase (APX), and catalase (CAT) were found to be activated during drought stress as tolerance traits. These enzymes have been reported as an indicator for the assessment of stress conditions (Noctor and Foyer 1998). The response of these antioxidant enzymes in sugarcane under drought stress conditions has been often used to measure the tolerance capability against oxidative stress (Cia et al. 2012, Hemaprabha et al. 2013, Patade et al. 2011). Sales et al. (2013) showed that the rise in superoxide dismutase and ascorbate peroxidase was related to the revival of leaf gas exchange in sugarcane plants when the drought stress was released. Madhav et al. (2017) found that Co 95020 and Co 87025 had better performance under drought stress with a tolerance trait for oxidative stress.

11.3 DROUGHT TOLERANCE METABOLISM

For a plant to lose its water content, transpiration through stomata is the most efficient and major pathway. Association of carbon dioxide and water loss is important in relation to temperature and hygrometry. The process of transpiration initiates when the opening of the stomata for exchange of gases takes place. Because sugarcane belongs to C_4 family assimilation of carbon dioxide, during day, that assimilation takes place through oxaloacetic acid in mesophyll cells, which later on reaches to bundle sheath cells. Decarboxylation of C_4 molecules with the release of carbon dioxide takes place in the tissue. The free carbon dioxide then enters into the calvin cycle by Rubisco enzyme (Moneveux and Belhassen 1996).

11.4 ABA-MEDIATED SIGNALING AND PRODUCTION IN CANES UNDER DROUGHT STRESS

Abscisic acid (ABA) expression in sugarcane under drought conditions has also been associated with the alteration at transcriptional level (Ding et al. 2012). This hormone is a signaling regulator having essential roles in different plant processes. Verlues et al. (2006) reported that the hormone can further enhance the making of protective osmolytes functioning for the maintenance of cell membrane structure with defense systems for photosynthetic apparatus (Fleta-Soriano et al. 2015). In ABA-mediated signaling, three components are involved: pyrabactin resistance receptors (PYR), protein phosphatase 2C (PP2C), and sucrose non-fermenting-related protein kinase 2 (SNF1-SnRK2). When the ABA is present in the plant under drought conditions, a complex formation of PYR/pyrabactin resistance-like (PYL)/regulatory component of ABA receptors (RCAR)-PP2C occurs that causes the inhibition of PP2C activity. This inhibition stimulates the SnRK2 activation. The activated SnRK2 causes phosphorylation of substrate proteins by downstream processing, like transcription factors (TFs), leading to transcription of ABA responsive genes (Sah et al. 2016).

11.5 GENE EXPRESSION IN SUGARCANE UNDER DROUGHT-STRESS TOLERANCE

Screening and identification of drought-tolerant genotypes is important for agriculture and economical reasons. The need to identify genes related to drought tolerance in sugarcane will help augment the knowledge of adaption strategies under drought stress. Such tolerant enhancing genes further help in new varieties through transformation (Nepomuceno et al. 2001, Lenka et al. 2011, Kido et al. 2012). Deshmukh et al. (2014) reported the comprehensive acquaintance of plant response under drought stress at the molecular level. This response acts as a management tool for drought stress. For sugarcane tolerance under drought stress conditions, genes encoding functional proteins and genes encoding regulatory proteins were the two categories to which genes expressing drought tolerance belong. The genes expressing functional proteins for drought stress involve chaperones or proteases, detoxification enzymes, transporters, water channels, etc. The genes expressing regulatory proteins involve proteins having either signaling response or stimulating transcription factors (Shinozaki and Yamaguchi-Shinozaki 2007). Zingaretti et al. (2012) found that due to signal transduction, plants under drought stress were able to process that stimulated signals were for adaptation or avoidance or none for such conditions. Complex coping pathways have also been developed by plants exposed to drought stress. Plant growth-promoting bacteria (PGPB) alters the gene expression for providing drought tolerance. This is a general mechanism of PGPB for plants under drought stress (Naylor and Coleman-Derr 2018).

The genes expressing drought tolerance in sugarcane have been described in Table 11.1.

TABLE 11.1
Gene Expression in Drought Stress Condition for Providing Tolerance to Sugarcane

Gene	Function	Reference
ScMYB2S1 ScMYB2S2	ABA mediated leaf senescence signaling pathway	Guo et al. 2017
SoDip22	Adapt drought stress in bundle sheath cells Induction of ABA mediated signaling pathway	Sugiharto et al. 2002
Bet A	Act as an osmoprotectant Accumulate drought stress	Sugiharto 2017
Ea-DREB2B	Regulation of other stress inducible genes Provide tolerance to drought	Augustine et al. 2015
Scdr2	Resilience to drought stress	Begcy et al. 2019
Scdr1 (Sugarcane drought-related-1)	Unknown function	Begcy et al. 2012
AtDREB2A CA		deSouza et al 2020
SCL target gene	Activate other genes	Gentile et al 2015
12 DREB1 genes 6 DREB2 genes	Regulation of drought stress	Li et al. 2021
Ea-DREB2B gene	Regulates the expression of a few stress-inducible genes	Zhao et al. 2020
PDH45	Confers drought tolerance	Augustine et al. 2015
IGS and PIN1	ABA independent- Auxin dependent pathway	Simon and Hemprabha, 2012
DREB	ABA independent pathway of regulatory mechanism	
SOD, DHAR, cAPX, GST1 and GSHS1	Genes involved in plant defense mechanism	

11.6 CONCLUSION

Sugarcane is a well-known crop grown for sugar production because it contains high amounts of sucrose. It is also used as a raw material for biofuel production. The occurrence of drought causes higher losses in growth, production, and productivity of this crop due to its water-loving capability. The climate change occurring all through the globe has caused frequent droughts. Droughts have a significant impact on the sugarcane crop, resulting in morpho-physiological modifications to prevent or adapt to the situation. In sugarcane, drought tolerance is developed through two key techniques, *viz.*, avoidance and tolerance. In general, when plants are stressed by drought, the stomata in their leaves close to prevent water loss through transpiration. Hydropassive and hydroactive are the two mechanisms involved in stomatal closure during drought stress. Drought stress causes changes in leaf mobility, which is a frequent adaptive strategy. Reduced leaf area, bigger leaf thickness, and increased leaf tissue density are some other features that help plants adapt to drought stress. Another response feature of plants under drought stress is leaf senescence. When sugarcane leaves senescence, due to initiation of drought stress, nutrient remobilization kicks in. Reduced leaf senescence genotypes have been proposed as a tolerant mechanism in drought-stressed plants.

Under these conditions, abscisic acid levels in the leaves rise, affecting stomata control and even stimulating higher ROS generation. Drought-induced changes in abscisic acid (ABA) expression in sugarcane have also been linked to transcriptional changes. Higher accural of hydrogen peroxide is caused by stomatal closure and induction of photorespiration (because of low carbon dioxide). According to studies, the accumulation of compatible solutes was linked to an increase in antioxidant capacity due to molecular signals generated by drought stress. In general, there are two mechanisms for plants to overcome stress *via* proline: either the plant accumulates proline, or the proline metabolic flux protects the plant. By activating other signalling pathways, preserving energy in the cells, or balancing the NADP+/NADPH ratio, the proline metabolic flux provides protection to cells. Under drought conditions, the antioxidant enzymatic activity in sugarcane aids plant tolerance. Drought-tolerant sugarcane genotypes must be identified and screened for agricultural and commercial reasons. The need to identify genes associated with drought tolerance in sugarcane will aid in expanding our understanding of drought adaptation strategies. During sugarcane growth stage, water relations and photosynthetic responses to water-deficiency stress could be valuable in identifying drought tolerant genotypes.

REFERENCES

Abbas S., Ahmad S., Sabir S., and Shah A. (2014). Detection of drought tolerant sugarcane genotypes (*Saccharum officinarum*) using lipid peroxidation, antioxidant activity, glycine-betaine and proline contents. *Journal of Soil Sciences and Plant Nutrition*, 14:233–243. 10.4067/s0718-95162014005000019

Abobatta W.F. (2019). Drought adaptive mechanisms of plants- A review. *Advances in agriculture and environmental Sciences: Open Access*, 2(1):62–65.

Ali M.A., Pan G., Cheng K., Williams S.A., Ogle S.M., Yeluripati J.B., Begum K., Kuhnert M., Parton, W.J., and Smith P. (2019). Modelling greenhouse gas emissions and mitigation potentials in fertilized paddy rice fields in Bangladesh. *Geoderma*, 341:206–215. 10.1016/j.geoderma.2019.01.047

Anjum S.A., Wang L., Farooq M., Khan I., and Xue L. (2011b). Methyl jasmonate induced alteration in lipid peroxidation, antioxidative defence system and yield in soyabean under drought. *Journal of Agronomy Crop Science*, 197:296–301.

Arve L.E., Terfa M.T., Gislerod H.R., Olsen J.E., and Torre S. (2013). High relative air humidity and continuous light reduce stomata functionality by affecting the ABA regulation in rose leaves. *Plant Cell Environment*, 36:382–392.

Arve L.E., Torre S., Olsen J.E., and Tanino K.K. (2011). Stomatal responses to drought stress and air humidity. *In*: *Abiotic stress in plants – mechanisms and adaptations*. A. Shanker(Ed.) InTech Publication, pp. 267–280.

Ashraf M., and Foolad M.R. (2007). Roles of glycine betaine and proline in improving plant abiotic stress resistance. *Environmental and Experimental Botany*, 59:206–216.

Augustine S.M., Narayan J.A., Syamaldevi D.P., Appunu C., Chakravarthi M., Ravichandran V., Tuteja N., and Subramonian N. (2015). Overexpression of EaDREB2 and pyramiding of EaDREB 2 with pea DNA helicase gene (PDH45) enhance drought and salinity tolerance in sugarcane. *Plant Cell reports*, 34:247–263 10.1007/s00299-014-1704-6

Basnayake J., Inman-Bamber N.G., Lakshmanan P., and Jackson (2012). Sugarcane for water-limited environments. Genetic variation in cane yield and sugar content in response to water stress. *Journal of Experimental Botany*, 63(16):6023–6033.

Begcy K., Mariano E.D., Lembke C.G., Zingaretti S.M., Souza G.M., Araujo P., and Menossi M. (2019). Overexpression of an evolutionarily conserved drought-responsive sugarcane genes enhances salinity and drought resilience. *Annals of Botany*, 124(4):691–700 10.1093/aob/mcz044

Begcy K., Mariano E.D., Gentile A., Lembke C.G., Zingaretti S.M., and Souza G.M. (2012). A novel stress-induced sugarcane gene confers tolerance to drought, salt and oxidative stress in transgenic tobacco plants. *PLoS One*, 7: e44697. 10.1371/journal.pone.0044697

Begg J.E. (1980). Morphological adaptation of leaves to water stress. *In*: *Adaptation of plants to water and high temperature stress*. N.C. Turner, P.J. Kramer(Eds.) John Wiley & Sons, New York, pp. 33–42.

Boaretto L.F., Carvalho G., Borgo L., Creste S., Landell M.G., Mazzafera P., and Azevedo M.G.A. (2014). Water stress reveals differential antioxidant responses of tolerant and non-tolerant sugarcane genotypes. *Plant Physiology and Biochemistry*, 74:165–175. 10.1016/j.plaphy.2013.11.016

Boyer J.S., Chin Wong S., and Farquhar G.D. (1997). CO_2 and water vapor exchange across leaf cuticle (epidermis) at various water potentials. *Plant Physiology*, 114:185–191. 10.1104/pp.114.1.185 PMID: 12223698.

Brevedan R.E., and Egli D.B. (2003). Short periods of water stress during seed filling, leaf senescence and yield of soyabean. *Crop Physiology and Metabolism*, 43(6):2083–2088.

Cha-um S., and Kirdmanee C. (2009). Effect of osmotic stress on proline accumulation, photosynthetic abilities and growth of sugarcane plantlets (*Saccharum officinarum* L.). *Pakistan Journal of Botany*, 40:2541–2552.

Chaves M.M., Flexas J., and Pinheiro C. (2009). Photosynthesis under drought and salt stress: regulation mechanisms from whole plant to cell. *Annals of Botany*, 103:551–560.

Cia M.C., Guimarães A.C.R., Medici L.O., Chabregas S.M., and Azevedo R.A. (2012). Antioxidant responses to water deficit by drought-tolerant and-sensitive sugarcane varieties. *Annals of Applied Biology*, 161:313–324. 10.1111/j.1744-7348.2012.00575.x

Corlett J.E., Jones H.G., Masssacci A., and Masojidek J. (1994). Water deficit, leaf rolling and susceptibility to photoinhibition in field grown sorghum. *Physiology Plantarum*, 92:423–430.

da Silva P.P., Soares L., da Costa J.G., da Silva Viana L., de Andrade J.C.F., Gonc¸alves E.R., et al. (2012). Path analysis for selection of drought tolerant sugarcane genotypes through physiological components. *Industrial Crops and Products*, 37:11–19. 10.1016/j.indcrop.2011.11.015

De Silava N.D., Cholewa E., and Ryser P. (2012). Effects of combined drought and heavy metal stresses on xylem structure and hydraulic conductivity in red maple (*Acer rubrum* L.). *Journal of Experimental Botany*, 63(16):5957–5966.

de Souza, W., Oliveira, N.G., Vinecky, F., Ribeiro, A.P., Basso, M.F., Nakashima, K., Yamaguchi-Shinozaki, K., and Molinari, H.B.C. (2020). Development of drought tolerant sugarcane overexpressing the AtDREB2A CA gene. JIRCAS Working Report No. 91. 89–98.

De Souza, P.I., Egli, D.B., and Bruening, W.P. (1997). Water stress during seed fi lling and leaf senescence in soybean. *Agronomy Journal*, 89:807–812.

Delauney A.J., and Verma D.P.S. (1993). Proline biosynthesis and osmoregulation in plants. *Plant Journal*, 4:215–223.

Deshmukh R., Sonah H., Patil G., Chen W., Prince S., Mutava R., Vuong T., Valliyodan B., and Nguyen H.T. (2014). Integrating omic approaches for abiotic stress tolerance in soybean. *Frontiers Plant Sciences*, 5:244.

Dhansu P., Kulshreshtha N., Kumar R., Raja A.K., Pandey S.K., Goel V., and Ram B. (2021). Identification of drought tolerant co canes based on physiological traits, yield attributes and drought tolerance indices. 10.1007/s12355-021-00967-7

Ding Y., Fromm M., and Avramova Z. (2012). Multiple exposures to drought 'train' transcriptional responses in *Arabidopsis*. *Nature Communications*, 3:740.

Du Y.C., Kawamitsy Y., Nose A., Hiyane S., Murayama S., Wasano K., and Uchida Y. (1996). Effects of water stress on carbon exchange rate and activities of photosynthetic enzymes in leaves of sugarcane (*Saccharum* sp.). *Australian Journal of Physiology*, 23:719–726.

Ehleringer J., and Forseth I. (1980). Solar tracking by plants. *Science*, 210:1094–1098.

Fleta-Soriano E., Pintó-Marijuan M., and Munné-Bosch S. (2015). Evidence of drought stress memory in the facultative CAM, *Aptenia cordifolia*: possible role of phytohormones. *PLoS One*, 10:e0135391.

Gascho G.J., and Shih S.F. (1983). Sugarcane. *In*: *Crop water relations*. I.D. Teare and M.M. Peet (Eds.) John Wiley & Sons, New York, pp. 445–479.

Gentile A., Dias L. I., Mattos R. S., Ferreira T. H., and Menossi M. (2015). MicroRNAs and drought responses in sugarcane. *Frontiers in Plant Science*, 6 : 58. 10.3389/fpls.2015.00058.

Gosal S.S., Wani S.H., and Kang M.S. (2009). Biotechnology and drought tolerance. *Journal of Crop Improvement*, 23:19–54.

Gratão P.L., Monteiro C.C., Antunes A.M., Peres L.E.P., and Azevedo R.A. (2008). Acquired tolerance of tomato (*Lycopersicon esculentum* cv.Micro-Tom) plants to cadmium induced stress. *Annals of Applied Biology*, 153:321–333.

Guimarães E.R., Mutton M.A., Mutton M.J.R., Ferro M.I.T., Ravaneli G.C., and Silva J.A.D. (2008). Free proline accumulation in sugarcane under water restriction and spittlebug infestation. *Sci. Agric.* 65:628–633. 10.1590/S0103-90162008000600009

Guo J., Ling H., Ma J., Chen Y., Su Y., Lin Q., Gao S., Wang H., Que Y., and Xu L. (2017). A sugarcane R2R3-MYB transcription factor gene is alternatively spliced during drought stress. *Scientific Reports* 7:41922. 10.1038/srep41922

Hemaprabha G., Swapna S., Lavanya D.L., Sajitha B., and Venkataramana S. (2013). Evaluation of drought tolerance potential of elite genotypes and progenies of sugarcane (*Saccharum* sp. hybrids). *Sugar Tech* 15:9–16.

Hoekstra F.A., Golovina E.A., and Buitink J. (2001). Mechanisms of plant desiccation tolerance. *Trends in Plant Science*, 6:431–438.

Huang G.T., Ma S.L., Bai L.P., Zhang L., Ma H., Jai P., Liu J., Zhong M., and Guo Z.F. (2012). Signal transduction during cold, salt and drought stresses in plants. *Molecular Biology Reports*, 39:969–987. 10.1007/s11033-011-0823-1

Iskandar H.M., Casu R.E., Fletcher A.T., Schmidt S., Xu J., Maclean D.J., Manners J.M., and Bonnerr G.D. (2011). Identification of drought-response genes and a study of their expression during sucrose accumulation and water deficit in sugarcane culms. *BMC Plant Biology*, 11:12. doi: 10.1186/1471-2229-11-12

Jaleel C.A., Manivannan P., Wahid A., Farooq M., Al-Juburi H.J., Somasundaram R., and Panneerselvam R. (2009). Drought stress in plants: A review on morphological characteristics and pigments composition. *International Journal of Agriculture and Biology*, Institute of Agriculture. The University of Western Australia, 11(1): 100–105.

Kang J., Hwang J.U., Lee M., Kim Y.Y., Assmann S.M., Martinoia E., and Lee Y. (2010). PDR-type ABC transporter mediates cellular uptake of the phytohormone abscisic acid. *Proceedings of the National Academy of Sciences of the United States of America*, 107:2355–2360.

Kavi Kishor P.B., Sangam S., Amrutha R.N., Sri Laxmi P., Naidu K.R., Rao K.R.S.S., Sreenath Rao Reddy K.J., Theriappan P., and Sreenivasulu N. (2005). Regulation of proline biosynthesis, degradation, uptake and transport in higher plants: its implications in plant growth and abiotic stress tolerance. *Current Science*, 88:424–438.

Khaled K.A., El-Arabi N.I., Sabrey N.M., and El-Shrbing S. (2018). Sugarcane genotypes assessment under drought conditions using amplified fragment length polymorphism. *Biotechnology*, 17(3):120–127.

Kido E.A., Neto J.R.C.F., Silva R.L.D.O., Pandolfi V., Guimaraes A.C.R., Veiga D.T., Chabregas S.M., Crovella S., and Iseppon A. (2012). New insights in the sugarcane transcriptome responding to drought stress as revealed by supersage. *The scientific World Journal*, 5235:821062.

Kim T.H., Bohmer M., Hu H.H., Nishimura N., and Schroeder J.I. (2010). Guard cell signal transduction network: advances in understanding abscisic acid, CO_2, and Ca^{2+} signaling. *Annals of Review of Plant Biology*, 61:561–591.

Kishor P.B.K., Sangam S., Amrutha R.N., Laxmi P.S., Naidu K.R., Rao K.R.S.S., Rao S., Reddy K.J., Theriappan P., and Sreenivasulu N. (2005). Regulation of proline biosynthesis, degradation, uptake and transport in higher plants: Its implications in plant growth and abiotic stress tolerance. *Current Science*, 88:424–438.

Kumar D., Malik N., and Sengar R.S. (2021). Physiobiochemical insights into sugarcane genotypes under water stress. *Biological Rhythm Research*, 52:92–115.

Lenka S.K., Katiyar A., Chinnusamy V., and Bansal K.C. (2011). Comparative analysis of drought-responsive transcriptome in Indica rice genotypes with contrasting drought tolerance. *Plant Biotechnology Journal*, 9:315–327.

Levitt, J. (1980). Responses of plants to environmental stresses. In *Water, radiation, salt, and other stresses*. Vol. 2.Academic Press: New York, NY.

Li Z., Wang G., Liu X., Wang Z., Zhang M., and Zhang J. (2021). Genome –wide identification and expression profiling of DREB genes in *Saccharum spontaneum*. *BMC Genomics*, 22:456. 10.1186/s12864-021-07799-5

Liang X., Zhang L., Natrajan S.K., and Becker D.F. (2013). Proline mechanism of stress survival. *Antioxidants & Redox signaling*, 19(9):998–1011.

Loresto G.C., and Chang T.T. (1981). Decimal scoring systems for drought reaction and recovery ability in rice screening nurseries. *International Rice Research Newsletter*, 6(2):9–10.

Loresto G.C., Chang T.T., and Tagumpay O. (1976). Field evaluation and breeding for drought resistance. *Philippine Journal of Crop Sciences*, 1:36–39.

Machado D.F.S.P., Lagôa A.M.M.A., Ribeiro R.V., Marchiori P.E.R., Machado R.S., and Machado E.C. (2013). Baixa temperatura noturna e deficiência hídrica na fotossíntese de cana-de-açúcar. (Low high temperature and water deficit in sugarcane photosynthesis). *Brazilian Agricultural Research*, 48:487–495.

Madhav T.V., Bindu G.S.M., Lavanaya B., and Kumar M.V. (2017). Analysis of antioxidant enzyme activity in sugarcane varieties under moisture stress. *Current Journal of Applied Science and Technology*, 10.9734/CJAST/2017/34567

Mall A.K. and Misra V. (2017). Biotechnological approaches: Sustaining Sugarcane productivity and yield. In *Biotechnology for sustainability Achievements, challenges and perspectives*. S. Bhore, K. Marimutchu, M. Ravichandran (Eds.) AIMST University, Malaysia, pp. 387–398 (ISBN- 978-967-14475-3-6; eISBN 978-967-14475-2-9)

Mall A. K., Misra V., Pathak A. D., and Srivastava S. (2022). Breeding for Drought Tolerance in Sugarcane: Indian Perspective. *Sugar Tech.*10.1007/s12355-021-01094-z

Mall A.K., Misra V., Kumar A., and Pathak A.D. (2019). Drought: Status, losses and management in sugarcane crop. In: *Farm and development: technological perspectives.* P.K. Bharti and J. Ray (Eds.). Discovery Publishing House Pvt. Ltd., New Delhi, pp. 165–187.

Mall A. K., Misra V. Singh, B. D., Kumar M., and Pathak A. D. (2020). Drought Tolerance: Breeding Efforts in Sugarcane, *Agronomic Crops* pp. 157–17210.1007/978-981-15-0025-1_10.

Manzoni S., Vico G., Katul G., Fay P.A., Polley W., Palmroth S., and Porporato A. (2011). Optimizing stomatal conductance for maximum carbon gain under water stress: a meta-analysis across plant functional types and climates. *Functional Ecology*, 25:456–467.

Marcos F.C.C., Silveira N.M., Mokochinski J.B., Sawaya A.C.H.F., Marchiori P.E.R., Machado E.C., Souza G.M., Landell M.G.A., and Ribeiro R.V. (2018). Drought tolerance of sugarcane is improved by previous exposure to water deficit. *Journal of Plant Physiology*, 223:9–18.

Medeiros D.B., Silva E.C., Santos H.R.B., Pacheco C.M., Musser R.S., and Nogueira R.J.M.C. (2012). Physiological and biochemical response to drought stress in the Barbados cherry. *Brazilian Journal of Plant Physiology*, 24:181–192.

Misra V., Solomon S., Mall A.K., Prajapati C.P., Hashem A., Abd Allah E.F., and Ansari M.I. (2020a). Morphological assessment of water stressed sugarcane: A comparison of waterlogged and drought affected crop. *Saudi Journal of Biological Sciences*, 27:1228–1236.

Misra V., Solomon S., and Ansari M.I. (2016b). Impact of drought on post-harvest quality of sugarcane crop. *Advances in Life Sciences*, 20(5):9496–9505.

Misra V., Solomon S., Singh P., Prajapati C.P., and Ansari M.I. (2016a). Effect of water logging on post-harvest sugarcane deterioration. *Agrica*, 5:119–132.

Misra V., Solomon S., Ahmed A., Elsayed F.A., Mall A.K., Prajapati C.P., and Ansari M.I. (2020). Minimization of post-harvest sucrose losses in drought affected sugarcane using chemical formulation. *Saudi Journal of Biological Sciences*, 27(1):309–317.

Misra V., Solomon S., Mall A. K., Abid M., Abid Ali Khan M. M., and Ansari M.I. (2022). Drought Stress and Sustainable Sugarcane Production, *Microbial Biotechnology for Sustainable Agriculture, Volume 1, Microorganisms for Sustainability,* pp. 353–36810.1007/978-981-16-4843-4_11.

Mitra J. (2001). Genetics and genetic improvement of drought resistance in crop plants. *Current Science*, 80:758–763.

Mittler R. (2002). Oxidative stress, antioxidants and stress tolerance. *Trends in plant science* 7:405–410.

Mittler R., and Blumwald E. (2015). The roles of ROS and ABA in systemic acquired acclimation. *Plant Cell*, 10.1105/tpc.114.133090.

Molinari H.B.C., Marur C.J., Daros E., Campos M.K.F., Carvalho J.F.R.P., Bespalhok-Filho J.C., Pereira L.F.P., and Vieira L.G.E. (2007). Evaluation of the stress-inducible production of proline in transgenic sugarcane (*Saccharum* spp.) osmotic adjustment, chlorophyll fluorescence and oxidative stress. *Physiologia Plantarum*, 130:218–229.

Moneveux P., and Belhassen E. (1996). The diversity of drought adaptation in the wide. *Plant growth regulation*, 20:85–92.

Morison J.I.L., Baker N.R., Mullineaux P.M., and Davies W.J. (2008). Improving water use in crop production. *Philosophical Transactions of the Royal Society B: Biological Sciences*, 363(1491):639–658.

Moussa H.R., and Abdel-Aziz S.M. (2008). Comparative response of drought tolerant and drought sensitive maize genotypes to water stress. *Australian Journal of Crop Science*, 1:31–36.

Munne-Bosch S., and Algere L. (2004). Die and let live: leaf senescence contributes to plant survival under drought stress. *Functional Plant Biology*, 31(3):203–216.

Mutava R.N., Prince S.J.K., Syed N.H., Song L., Valliyodan B., Chen W., and Nguyen H.T. (2015). Understanding abiotic stress tolerance mechanisms in soybean: A comparative evaluation of soybean response to drought and flooding stress. *Plant Physiology and Biochemistry*, 86:109–120.

Naylor D., and Coleman-Derr D. (2018). Drought stress and root associated bacterial communities. *Front Plant Sciences*, 10.3389/fpls.2017.02223

Nepomuceno A.L., Neumaier N., Farias J.R.B., and Oyax T. (2001). Tolerância à seca em plantas. *Biotecnologia, Ciência e Desenvolvimento*, 4:12–18 (in Portuguese).

Noctor G., and Foyer C.H. (1998). Ascorbate and Glutathione: Keeping active oxygen under control. *Annual Review of Plant Physiology and Molecular Biology*, 49:249–270.

Oppenheimer H.R. (1960). *Plant water relationships in arid and semi-arid conditions.* INESCO, U.K., pp. 105–138.

O'Toole J.C., and Chang T.T. (1978). Drought and rice improvement in perspective. IRRI Research Paper Series, No. 14. International Rice Research Institute.

Parre E., Ghars M.A., Leprince A.S., Thiery L., Lefebvre D., Bordenave M., Richard L., Mazars C., Abdelly C., and Savoure A. (2007). Calcium signaling via phospholipase C is essential for proline accumulation upon ionic but not nonionic hyperosmotic stresses in *Arabidopsis. Plant Physiology*, 144:503–512.

Patade V.P., Bhargava S., and Suprasanna P. (2011). Salt and drought tolerance of sugarcane under iso-osmotic salt and water stress: Growth, osmolytes accumulation, and antioxidant defense. *Journal of Plant Interactions*, 6:275–282.

Pirasteh-Anosheh H., Saed-Moucheshi A., Pakniyat H., and Pessarakli M. (2016). Stomatal responses to drought stress. *In*: *Water stress and crop plants: a sustainable approach.* P. Ahmad (Ed.) John Wiley & Sons, Ltd. pp. 24–40.

Pirasteh-Anosheh H., Saed-Moucheshi A., Pakniyat H., and Pessarakli M. (2016). Stomatal responses to drought stress. In: *Water stress and Crop Plants: A sustainable approach.* P. Ahmad (Ed.) John wiley & Sons. pp. 24–40.

Queiroz C., Lopes M.L.M., Fialho E., and Valente-Mesquita V.L. (2008). Polyphenol oxidase: Characteristics and mechanisms of browning control. *Food Review of International*, 24(4):361–375.

Ramesh P. (2000). Effect of different levels of drought during the formative phase on growth parameters and its relationship with dry matter accumulation in sugarcane. *Journal of Agronomy and Crop Sciences*, 185:83–89.

Rao C.M., Kumar U.V., Naik R.B., Sekha D., and Katti G.S. (2007). Relative performance of sugarcane clones/varieties under soil moisture stress/drought during formative stage. *Indian Sugar*, 56(11):23–28.

Rauf S., Al-Khayri J.M., Zaharieva M., Monneveux P., and Khalil F. (2015). Breeding strategies to enhance drought tolerance in crops. *In*: *Advances in plant breeding strategies, agronomic, abiotic and biotic stress traits*. J.M. Al-Khayri, M. Jameel, M. Jain, D.V. Johnson (Eds.) Springer, Berlin, pp. 1–70.

Rhodes D., and Samaras Y. (1994). *Genetic control of osmoregulation in plants.In cellular and molecular physiology of cell volume regulation. Strange K.* Boca Raton, CRC Press. pp: 347–361.

Ribeiro R.V., MacHado R.S., MacHado E.C., MacHado D.F.S.P., Magalhães Filho J.R., and Landell M.G.A. (2013). Revealing drought-resistance and productive patterns in sugarcane genotypes by evaluating both physiological responses and stalk yield. *Experimental Agriculture*, 49:212–224. 10.1017/ S00144 79712001263

Rodrigues F.A., de Laia M.L., and Zingaretti S.M. (2009). Analysis of gene expression profiles under water stress in tolerant and sensitive sugarcane plants. *Plant Sciences*, 176:286–302. 10.1016/j.plantsci. 2008.11.007

Rogiers S.Y., Holzapfel B.P., and Smith J.P. (2011b). Sugar accumulation in roots of two grape varieties with contrasting response to water stress. *Annals of Applied Biology*, 159: 399–413.

Rucker K.S., Kvien C.K., Holbrook C.C., and Hook J.E. (1995). Identification of peanut genotypes with improved drought avoidance traits 1. *Peanut Sciences*, 22:14–18.

Saed-Moucheshi A., Pakniyat H., Pirasteh-Anosheh H., and Azooz M.M. (2014). Role of ROS as signaling molecules in plants. *In*: *Oxidative damage to plants antioxidant networks and signaling*. P. Ahmad (Ed.) Elsevier, Oxford.

Sah A.K., Reddy K.R., and Li J. (2016). Abscisic acid and abiotic stress tolerance in crop plants. *Frontier in Plant Science*, 7: Article 571 10.3389/fpls.2016.00571

Sairam R.K., Rao K.V., and Srivastava G.C. (2002). Differential response of wheat genotypes to long term salinity stress in relation to oxidative stress, antioxidant activity and osmolyte concentration. *Plant Science*, 163:1037–1046.

Sales C.R.G., Marchiori P.E.R., Machado R.S., Fontenele A.V., Machado E.C., Silveira J.A.G., and Ribeiro R.V. (2015). Photosynthetic and antioxidant responses to drought during sugarcane ripening. *Photosynthetica*, 53:547–554. 10.1007/s11099-015-0146-x

Sales C.R.G., Ribeiro R.V., Silveira J.A.G., Machado E.C., Martins M.O., and Lago^a A.M.M.A. (2013). Superoxide dismutase and ascorbate peroxidase improve the recovery of photosynthesis in sugarcane plants subjected to water deficit and low substrate temperature. *Plant Physiology and Biochemistry*, 73: 326–336. 10.1016/j.plaphy.2013.10.012 PMID: 24184453

Savoure A., Hua X.J., Bertauche N., Van Montagu M., and Verbruggen N. (1997). Abscisic acid-independent and abscisic acid-dependent regulation of proline biosynthesis following cold and osmotic stresses in *Arabidopsis thaliana. Mol Gen Genetics*, 254: 104–109.

Shao H.-B., Chen X.-Y., Chu L.-Y., Zhao X.-N., Wu G., Yuan Y.-B., Zhao C.-X., and Hu Z.-M. (2006) Investigation on the relationship of proline with wheat antidrought under soil water deficits. *Colloids and Surfaces B: Biointerfaces*, 53:113–119.

Sharma P., Jha A.B., Dubey R.S., and Pessarakli M. (2012). Reactive oxygen species, oxidative damage, and antioxidative defense mechanism in plants under stressful conditions. *Journal of Botany*, 26. 10.1155/2012/217037

Sharma S., Villamor J.G., and Verslues P.E. (2011). Essential role of tissue-specific proline synthesis and catabolism in growth and redox balance at low water potential. *Plant Physiology*, 157: 292–304.

Shinozaki K., and Yamaguchi-Shinozaki K. (2007). Genes networks involved in drought stress response and tolerance. *Journal of Experimental Botany*, 58(2):221–227.

Silva E.C., Albuquerque M.B., Azevedo Neto A.D., and Silva Junior C.D. (2013). Drought and its consequences to plants from individual to ecosystem. *In*: *Responses of organisms to water stress*. Şener Akıncı (Ed.). Croatia In Tech, pp. 17–47.

Silva M.D.A., Silva J.A.G.D., Enciso J., Sharma V., and Jifon J. (2008). Yield components as indicators of drought tolerance of sugarcane. *Sci. Agricola*, 65:620–627.

Simon S., and Hemprabha G. (2012). Sugarcane specific drought responsive candidate genes identified through differential expression in resistant and susceptible genotypes of sugarcane. *International Sugar Journal*, 114(1366):731–737.

Singh S., and Rao P.N.G. (1987). Varietal differences in growth characteristics in sugarcane. *The Journal of Agricultural Science*, 108:245–247.

Smirnoff N. (1995). Antioxidant systems and plant response to the environment. *In*: *Environment and plant metabolism: flexibility and acclimation*. V. Smirnoff (Ed.) Bios Scientific Publishers, Oxford U.K., pp. 217–243.

Srinivasan T.R. (1993). Management Practices for Drought Situation Lead Paper Presented in the 8th Sugarcane Research and Development Workers Meeting of Northern Karnataka held at MKN, Hubli.

Stadman E.R. (1986). Oxidation of proteins by mixed function oxidation systems: implication in protein turnover, ageing and neutrophil function. *Trends in biochemical sciences*, 11:11–12.

Strizhov N., Abraham E., Okresz L., Blickling S., Zilberstein A., Schell J., Koncz C., and Szabados L. (1997). Differential expression of two P5CS genes controlling proline accumulation during salt-stress requires ABA and is regulated by ABA1, ABI1 and AXR2 in *Arabidopsis*. *Plant Journal*, 12:557–569.

Sugiharto B. (2017). Biotechnology of drought tolerant sugarcane Intech Open, 10.5772/intechopen.72436

Sugiharto B., Ermawati N., Mori H., Aoki K., Yonekur-Sakakibara K., Yamaya T., Sugiyama T., and Sakakibara H. (2002). Identification and characterization of a gene encoding drought inducible protein localizing in the bundle sheath cell of sugarcane. *Plant Cell Physiology*, 43(3):350–354.

Swami D., Dave P., and Parathasarthy D. (2018). Agricultural susceptibility to monsoon variability: A district level analysis of Maharashtra India. *Sci. Total Environment*, 619:559–577.

Taiz, L., and Zeiger, E. (2002) *Plant physiology (Third Edition)*. Sinauer Associates, Inc., Publishers, Sunderland, pp. 67–86.

Vendruscolo E.C.G., Schuster I., Pileggi M., Scapim C.A., Molinari H.B.C., Marur C.J., and Esteves Vieira L.G. (2007). Stress-induced synthesis of proline confers tolerance to water deficit in transgenic wheat. *Journal of Plant Physiology*, 164:1367–1376.

Verlues P.E., Agarwal M., Katiyar-Agarwal S., Zhu J., and Zhu J.K. (2006). Methods and concepts in quantifying resistance to drought salt and freezing, abiotic stress that affect plant water status. *Plant Journal*, 45:523–539.

Wei-Hua G., Li B., Huang Y.-M., Hai-Xia Z., and Xin-Shi Z. (2003). Effects of different water stresses on eco-physiological characteristics of *Hippophae rhamnoides* seedlings. *Acta Botanica Sinia*, 45: 1238–1244. Available: http://www.chineseplantscience.com

Werner C., Correia O., and Beyschlag W. (1999). Two different strategies of Mediterranean macchia plants to avoid photoinhibitory damage by excessive radiation levels during summer drought. *Acta Oecologica*, 20:15–23.

Wingler A. (2002). The function of trehalose biosynthesis in plants. *Phytochemistry*, 60:437–440.

Yang X., Lu M., Wang Y., Wang Y., Liu Z., and Chen S. (2021). Response mechanism of plants to drought stress. *Horticulturae*, 7:50–86. 10.3390/ horticulturae7030050

Yoo J.H., Park C.Y., Kim J.C., Heo W.D., Cheong M.S., Park H.C., Kim M.C., Moon B.C., Choi M.S., Kang Y.H., Lee J.H., Kim H.S., Lee S.M., Yoon H.W., Lim C.O., Yun D.J., Lee S.Y., Chung W.S., and Cho M.J. (2005). Direct interaction of a divergent CaM isoform and the transcription factor, MYB2, enhances salt tolerance in *Arabidopsis*. *Journal of Biological Chemistry*, 280:3697–3706.

Zhang F.J., Zhang K.K., Du C.Z., Li J., Xing Y.X., Yang L.T., and Li Y.R. (2015). Effect of drought stress on anatomical structure and chloroplast ultrastructure in leaves of sugarcane. *Sugar Tech*, 17:41–48. 10.1007/s12355-014-0337-y

Zhao D., and Li Y.R. (2015). Climate change and sugarcane production: Potential Impact and mitigation strategies. *International Journal of Agronomy* Article Id 547386 10.1155/2015/547386

Zhao X., Jiang Y., Liu Q., Yang H., Wang Z., and Zhang M. (2020). Effects of drought tolerant EaDREB2B transgenic sugarcane on bacterial communities in soil. Frontier in Microbiology 10.3389/fmicb.2020.00704

Zingaretti S.M., Rodrigues F.A., Graca J.P.D., Pereira L.V.D.M., and Lourenco M.V. (2012). Sugarcane responses at water deficit conditions. *In Water Stress* InTech Open Rahman I.M.M., Hasegawa H., pp. 255–276.

ABBREVIATION USED

ABA	Abscisic Acid
ABC	ATP Binding Cassette
PGPB	Plant Growth Promoting Bacteria
PP2C	Protein Phosphatase 2C
PYL	Pyrabactin Resistance Like
PYR	Pyrabactin Resistance Receptors
RCAR	Regulatory Component of ABA Receptors
ROS	Reactive Oxygen Species
SNF1-SnRK2	Sucrose Nonfermenting Related Protein Kinase 2
TFs	Transcription Factors

12 High-Throughput Sequencing and SNP Markers-based Identification of Sugarcane Cultivars for Parentage Determination and Intellectual Property Protection

Bilal Saleem, Zeshan Haider, Muhammad Noman, and Muhammad Ramzan Khan
Functional Genomics and Bioinformatics Program, National Institute for Genomics and Advanced Biotechnology, National Agricultural Research Centre, Islamabad, Pakistan

CONTENTS

In Pakistan, about 56 sugarcane cultivars/hybrids are being grown. Pakistan has a traditional and limited capacity for sugarcane breeding. Besides other constraints, it is hampered by a lack of information about genetic makeup, genome sequence, and ancestor relationship of varieties/hybrid grown in Pakistan. Plant identification and varieties/hybrid certification are of utmost importance for breeding, seed market, and IPRs. The present criterion of DUS for plant variety protection may not be judged due to obscure genetic makeup, false positives, and huge reference collection of

DOI: 10.1201/9781003292425-12

major cereal crops, which makes comparison of morphological data almost impossible. In these scenarios, the implication of NGS–based SNP markers is the ultimate solution to this problem. Also, 45K, 100K, and 345K SNP array chips have been developed to be used for specific SNP markers-based genotyping/identification of different varieties. In SNP genotyping, the major issues are identifying a small set of variety-specific SNP that can be easily used for genotyping by PCR-based KASP markers and eventually developing a digital barcode for that variety. Around 56 varieties have been registered in Pakistan, and none of them is certified through a DNA profile. Furthermore, no repository or database of DNA-certified crop varieties with specific SNP haplotypes can identify new varieties and hybrids.

12.1 INTRODUCTION

Sugarcane belongs to the family *Poaceae* and the genus *Saccharum*. It is an important commercial crop due to its high level of sucrose and potential biofuel feedstock (Fickett et al. 2019). The agronomics hybrid sugarcane was obtained by crossing the *S. spontaneum* and *S. officinarum*. Recently, the manufacture of biodiesel and ethanol as an alternative for fossil fuels has attracted recent attention. It might be a valuable renewable source that helps alleviate the present energy crisis and reduce greenhouse gas emissions (Savage 2011). As a result, sugarcane has evolved into a biofuel crop used for food and fuel. Sugarcane is considered 80%, 45%, and 13% of sucrose production globally, the United States, and Pakistan, respectively (Wu & Altpeter 2015; Shah & Ali 2005). Sugarcane is considered a fourth main crop in Pakistan, contributing to the agricultural economy and accounting for 18% of large-scale industries and 1.9% of GDP (Malik 2013; Shaikh et al. 2021). The genome sequencing and analysis of sugarcane are very challenging due to a large number of chromosomes, excessive heterozygosity, a higher level of polyploidy, and frequent aneuploidy (Soltis et al. 2016).

Current sugarcane (2n=100–144) is a complicated aneupolyploid emerging via interspecies hybridization of *Saccharum spontaneum* (2n= 80) and *Saccharum officinarum* (2n= 40–128) (Meng et al. 2020; Moore et al. 2013; K. S. Aitken et al. 2014). *S. spontaneum* is utilized as a donor (male) to incorporate its plant vigor and efficient defense mechanism into the agronomic *S. officinarum* (Bremer 1961). Therefore, stabilizing the genome and reestablishing the sucrose trait required three to six generations of backcrossing because of the restitution of female chromosomes (Kearney 2001). Furthermore, the *in situ* genomic hybridization stated that *Saccharum spontaneum* and *Saccharum officinarum* are considered to share ~ 80% and ~10%, respectively, against the current sugarcane genome. In comparison, only ~10% of recombinant genetic material is present in hybrid sugarcane (Metcalfe et al. 2019). The sugarcane genome is monoploid and varies from 760 Mb to 926 Mb, comparable to the 760 Mb sorghum genome and about twice rice's 389 Mb genome (De Setta et al. 2014). Flow cytometry investigation of DNA sequences and genome length in several sugarcane cultivars found that sugarcane has a substantial genome of roughly 3.8–10.9 pg (Ghosal & De 2017).

In the past decade, the identification and utilization of DNA polymorphism with molecular markers have sped up and increased the accuracy of genetic analysis in the molecular breeding of plants (You et al. 2016). About 56 varieties of sugarcane are cultivated in different areas of Pakistan (Majid 2007). Due to limited capacity in the traditional breeding of sugarcane, parentage identification and hybrid authentication are the most significant for plant breeding, seed market, and intellectual property rights (IPRs) (Eaton 2007). The policy's aims of awarding plant variety IPRs are mainly based on an instrumentalist view of IPRs. New plant varieties are given legal protection under this approach, which encourages plant breeders to devote the resources, labor, and time required to enhance existing plant varieties by certifying that breeders are adequately compensated when they sell the proliferated material of those improved varieties (Helfer 2004).

Furthermore, parentage analysis assisted by molecular markers is the upmost available novel technique for plant breeders to extend genetic variability in sugarcane breeding and IPRs. It can help determine the identity of half-sib progenies and recreate the lineage in the outcrossing varieties (Bertan et al. 2007; Hill et al. 2008). Moreover, the emergence of molecular markers connected to diseases like yellow spot, rust-resistance gene, and yellow leaf virus (in sugarcane) promised sugarcane marker-assisted breeding (Manimekalai et al. 2020). Therefore, the plant-breeding practices have evolved dramatically with the incorporation of high-throughput marker techniques via next-generation sequencing (NGS). These novel technologies and computational bioinformatics have increased the viability of examining the complex polyploid genomes and transcriptomics of agronomics crops, i.e., *Triticum aestivum* (wheat), *Brassica napus*, *Gossypium* (cotton), and sugarcane. The low-density markers might create the wrong phenotypic association due to the complex sugarcane genome; therefore, high-density markers, i.e., single nucleotide polymorphisms (SNPs), are essential to characterize the genomic selection (Consortium 2014; De Setta et al. 2014; Chalhoub et al. 2014; Liu et al. 2015).

Nowadays, sugarcane breeding centers on genetic-enhancement programs aided by SNP markers. The SNPs are commonly used molecular markers worldwide because numerous SNPs per genotype could be screened and controllable to high-throughput screening. The creation of SNPs in sugarcane can overcome the present restrictions as they can be easily screened across the many genotypes in a high number throughput genome (Gayathri et al. 2019; Mirajkar et al. 2019). Recently, ffymetrix® Axiom® 45K, 100K, and 345K SNP array chips have been created for sugarcane, which has greatly enhanced the structure of high-density linkage maps and recognition of target QTLs for the agronomic trait in sugarcane (Meaburn et al. 2006). Moreover, the NGS-based long sequence read technologies, i.e., Illumina TruSeq, SMRT (single-molecule real-time), and Oxford Nanopore sequencing, can tackle the challenges in genome assembly (Eid et al. 2009; Kyriakidou et al. 2018; Branton et al. 2010). Additionally, NGS has been played a major role in identifying the SNPs and creating low to high throughput genotyping systems. Therefore, in this chapter, we have highlighted the significant influence of NGS-based different platforms in sugarcane genomics and genotyping. We have presented a brief sketch about the parentage and cultivars identification through various high-throughput sequencing applications and the vital role of IPRs in novel varieties development.

12.2 PARENTAGE DETERMINATION AND IPRS

In modern breeding programs, pedigree information is important for accurate selection from the available sugarcane germplasm. Earlier, morphological parameters were used to determine parentage (Kyaligonza et al. 2014). However, it had certain limitations specifically for similar progenies (Kyaligonza et al. 2014; Contreras et al. 2007). Before molecular technology, chromosomal polymorphism (Levine et al. 1980) and allozyme electrophoresis (Ellstrand 1984) were the main tools for parentage analysis. Parentage determination via DNA fingerprinting has a four-decade history. In the 1980s, it was first discovered in birds (Gibbs et al. 1990). Afterward, the shortcomings of the earlier techniques of parentage analysis led to the development of modern-day advanced molecular markers (Jones et al. 2010).

Molecular markers with the resolving capacity among sugarcane genotypes based on heritable variations are useful in parentage determination. Identifying such appropriate markers would promise higher accuracy in parentage analysis (Pemberton 2008). A handful of molecular markers with a high level of polymorphism or a large number of less polymorphic markers are required for a successful parentage study in crops, including sugarcane (Jones et al. 2010). Molecular markers-based parentage analysis is one of the techniques used in breeding for maximizing genetic gain. In the outcrossing crops, it can reconstruct pedigrees by resolving half-sib progenies. These reliable parental profiles and pedigrees are required for plant breeding these days. This information is helpful to decide about the effect of inbreeding depression, hybrid vigor, and existing divergence in progeny (Spanoghe et al. 2015; Tarn et al. 1992). In addition, it is also essential in resolving parental and recombinant genotypes for linkage analysis (Luo et al. 2000). Similarly, this information also validates relationships, breeding values, and genetic estimates (Gjedrem 2010; Braun & Wenzel 2004; Sun et al. 2003), resistance inheritance (Song et al. 2005), and association (Björn et al. 2010; Baldwin et al. 2011) and QTL mapping (Bink et al. 2002).

Pedigree reconstruction is essential for genetic studies and efficient breeding because it will aid in the designed breeding of open-pollinated crops by estimating genetic correlation and reliable heritability parameters (Norman et al. 2018; Blakeney 2020). Moreover, it also aids the breeding programs by ousting the mislabeled genotypes. Without a pedigree reconstruction and a thorough understanding of the life-history traits, designing effective strategies for the management and utilization of genotypes, as well as their conservation, would not be easy (Braun & Wenzel 2004). Parentage assessment in sugarcane is based on a reliable characterization of varieties. DNA can be a valuable tool in validating pedigree information and assigning potential parents to orphan varieties (Jones et al. 2010).

Intellectual property rights (IPRs) are rights conferred by law to some defined creativity in culture, industry, or science. IPRs provide an inducement for investing in innovations via certain commercial exclusivity for a specific duration during which the costs of the manufacturing or others could be recovered, in addition to creating a legal infrastructure for technology transfer. IPRs hold great importance in modern agriculture by making the assets secure for investment, generating tradeable assets, and making agricultural research attractive. The parentage-analysis technique is a powerful tool for plant breeders' rights protection (PBR). The PBR is a worldwide-recognized system that grants breeders intellectual property rights for new varieties or variations that have been produced or adapted for commercial or other end users. Breeders get endpoint royalties (EPR) under the PBR, which are used to fund and sustain breeding activities. The use of DNA profiling in wild populations eliminates challenges such as genetic mixes, mislabeling, mutations, and outcrossing, which frequently obstruct efficient selection in breeding programs. The use of highly informative, high-throughput, and consistently reproducible markers to verify genetic identity based on morphological features clearly and robustly aids in the safeguarding of PBR. Variety protection comes into effect following a registration process. The plant breeder under the plant variety rights legislation is free to produce/reproduce, sell, and import/export the material. These rights are valid for a period of 20 years for crops, including sugarcane.

12.3 HIGH THROUGHPUT SEQUENCING APPLICATIONS AND THEIR ROLE IN GENOTYPING

12.3.1 Whole-Genome Sequencing (WGS)

Whole-genome sequencing (WGS) is a powerful tool for understanding the genome of novel species. The short reads obtained after WGS are assembled via de novo assembly approaches for novel genome species. WGS provides the whole view of the genome, including coding and noncoding regions, to determine the genetic variations (SNPs). The variations in the noncoding region could be implicated in the genetic susceptibility to diseases. WGS can identify SNPs in exonic and intronic regions of the genome (Pryer et al. 2002). The WGS approach is 10 times more costly than exome sequencing. It is a suitable approach for species having their reference genomes already available. A mosaic-monoploid reference genome of the R570 sugarcane variety was published in 2018 (Garsmeur et al. 2018). Moreover, the genome sequences of CP96-1252 and SP80-3280 sugarcane hybrid cultivars have been reported in previous studies (Miller et al. 2017; Riaño-Pachón & Mattiello 2017). Therefore, the prevailing variations in different sugarcane species could be identified and restrained by using these existing references genomes of different sugarcane cultivars. So, the WGS is considered a fruitful method for determining the SNPs in the sugarcane genome.

12.3.2 Exome Sequencing

Exome sequencing, also known as whole-exome sequencing, is a cost-efficient method that determines genetic variants (SNPs). The exome sequencing includes sequencing all protein coding genes in the genome to determine the SNPs. Therefore, only desired genome regions are sequences in exome sequencing with fewer costs than WGS (Hehir-Kwa et al. 2015). Similarly, a method of exome sequencing is also recognized as exome capture that is used to sequence the target part of the genome. Exome capture involves sequencing the genomic region, which only includes the protein-coding part. This region is termed genomic space. The sequences that fall in the genomic space are called genomic sequences. These genomic sequences are then assigned to an array. DNA fragments that are affixed to an array are collected and sequenced (Sulonen et al. 2011). It is more cost-effective than whole-genome sequencing to identify the genetic variations in the genome (Rohland & Reich 2012).

In contrast, when the exome capture array for the barley was revealed in 2013, it paved the way for a slew of new research to employ this technique to identify genetic variants (Bayer et al. 2019). Similarly, this technology can be used to develop the exome capture array against sugarcane cultivars, and it can open novel doors in the research of this complex hybrid cultivar. At the same time, this approach requires an unbiased array containing the targeted genomic area and highly skilled employees to analyze the data from the sequencing centre (Chilamakuri et al. 2014).

12.3.3 RNA Sequencing

An effective approach for quantifying and sequencing the RNA sample using NGS is known as RNA-seq (RNA sequencing). Additionally, the RNA-seq may disclose the high number of SNPs that may be used for trait mapping and marker-assisted selection. For instance, a total of 708,125 SNPs were found in six sugarcane genotypes engaged in biparental crosses by transcriptome sequencing (Cardoso-Silva et al. 2014). These SNPs were subsequently recommended for use as priority resources for identifying markers and genome assembly (Balsalobre et al. 2017). Moreover, the approximately 442152 putative gene-based SNPs have been identified from the transcriptome of various sugarcane cultivars against drought stress, pookah boeng disease, and leaf abscission (Xu et al. 2018). Another earlier study showed that the estimated 261,000 efficient SNPs had been known from the massive yellow croaker using RNA-seq data (Xiao et al. 2015).

Furthermore, the previous studies have been shown that the RNA-seq data of two lines of maize cultivar: LM11 (sensitive to heat stress), CML25 (tolerant to heat stress), were used to identify the competitive allele-specific PCR (KASP) SNPs. The prediction showed that approximately 129,804 and 117,550 KASP SNPs were recognized for LM11 and CML25, respectively (Jagtap et al. 2020). However, high-density and polymorphic SNP markers are lacking in sugarcane cultivars, particularly KASP SNP markers that may be employed for automatic genotyping of sugarcane. To date, a revolution in computational biology has happened, and a significant number of sugarcane sequence data can be produced by RNA-seq that may provide a suitable pool of SNP loci to produce SNPs markers.

12.3.4 Genotyping by Sequencing (GBS)

This NGS technique was developed using the Illumina sequencing platform in the Buckler lab. The simultaneous SNP marker discovery and genotyping approach play an outstanding role in understanding the genomics and molecular biology of a wide range of plant species. It utilizes the restriction enzyme to digest each individual's genome in a population. Then, the digested fragments are ligated with one common and one unique adapter. After that, a PCR is carried out to amplify the shorter DNA fragments. Following the pooling of the PCR products, the sequencing is carried out using an NGS platform (Mahadevaiah et al. 2021; Zhang et al. 2017). GBS is a method of choice for genotyping plant species having the reference genome available or not available, as in the case of sugarcane, for which the complete genome is not present to date. The discovery of potential markers depends on the restriction enzymes. In the case of hexaploid bread wheat, two restriction enzymes were used for GBS (Poland et al. 2012). GBS was initially developed for

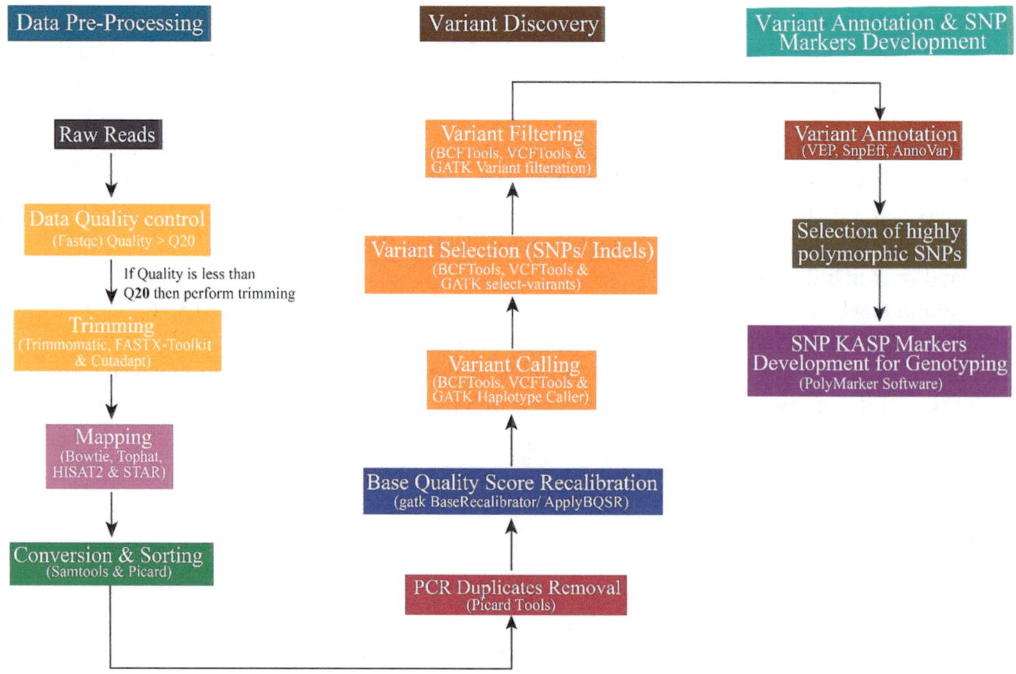

FIGURE 12.1 The workflow of the High Throughput Sequencing data analysis for SNP based genotyping. This pipeline starts from raw reads QC, Trimming, Mapping, Mapped Reads Sorting and Variant Calling, Variant Annotation, to the selection of highly polymorphic SNPS for development of KASP genotyping markers for varietal identification. Various tools involved in the data analysis are enlisted in parenthesis.

association studies in maize. It is now used for many other plant species with complex genome structures. Identifying SNPs through GBS compared to other techniques is less costly, making it a novel technique for genotyping different crops (Balsalobre et al. 2017). The SNPs obtained by this strategy are very high in numbers. Handling samples is more straightforward and less laborious for PCR and purification sets.

Various advanced NGS platforms have played their role in sequencing diploid and polyploid crops. The efficacy, robustness, and cost-effectiveness of short-read sequencing Illumina platform in polyploid genome analysis are better, even though in the case of sugarcane, a longer read length would be preferred due to the high level of repetitive regions in its genome (Deomano et al. 2020). The pacBio platform generates a read length of 15 kb, comparatively large in relation to the former platform. Due to the high sequencing cost of PacBio and other latest platforms, researchers from developing nations cannot benefit from them (Mendes de Paula et al. 2020). The brief workflow of HTS data analysis for SNPs extraction to develop the genotyping markers is shown in Figure 12.1.

12.4 SNP ARRAYS AND THEIR ROLE IN GENOTYPING

SNP arrays, along with other NGS strategies like GBS, are in frequent use for plant-breeding applications (Kim et al. 2021). The SNP array is a simple and straightforward approach for genotyping as compared to NGS tools because it requires less computational support and easy data analysis. However, the required SNPs to develop an array are to be discovered with the help of NGS techniques. In the case of sugarcane, which genome size comprises ~10 Gb, it requires a hefty amount of NGS data for the discovery of SNPs to represent on an SNP array. Once an SNP array is developed, genotyping can be done with assurity (Manimekalai et al. 2020). These SNP arrays can be any trait-specific array. Three SNP arrays have been developed for sugarcane, including the Axiom 100K (You et al. 2019), 45K(McNeil et al. 2017), and 345K (K. Aitken et al. 2016). Affymetrix Axiom 345K SNP array has been used to screen the large association-mapping population. Hence, with the advent of these SNP arrays, the genotyping of sugarcane cultivars is quite possible and less costly in comparison to NGS strategies. Besides genotyping, these high-density SNP arrays can also play their role in genomic selection (GS), diversity analysis and other plant breeding applications.

12.5 SNP MARKERS

Determination of nucleotide bases and its variation in the genome of plant species is crucial for understanding of the genetics and breeding for the development and fingerprinting of genotypes. Various molecular markers have been developed to date and can be used regarding their specific features to analyze the genetic variations. These molecular markers include RFLP, AFLP, SSRs, STRs, and SNPs. Most common and effective markers that are now being used frequently for genotyping are single nucleotide polymorphisms (SNPs) because of their high abundance and polymorphic nature in polyploid plant species like sugarcane. A summary of some of these markers follows:

RFLP stands for "restriction fragment length polymorphism" and was used for genetic fingerprinting before the discovery of SNPs markers. It utilizes restriction enzymes to cut DNA sequences on specific sites and produces nucleotide fragments of different lengths. Then those DNA fragments are viewed on the agarose gel.

STRs stands for "short tandem repeats." These occur in the form of repetitions in the intronic (noncoding) part of the DNA. It usually consists of two or more nucleotides. Due to the high polymorphic nature, the number of repetitions in a specific STR vary from individual to individual.

With the advent of next-generation sequencing (NGS) technologies, SNP markers' development occurs. This breakthrough is helping vastly in a variety of identifications to protect the breeder's IPRs. Besides the high-throughput genotyping strategy, it is also cost-effective compared to the medium-throughput simple sequence repeats (SSRs). The sequence alteration in the exonic (coding) region of the DNA may affect the phenotypes. The SNPs that involve in the phenotypic changes are of great functionality and may play a vital role in the development mechanism of a specific trait. In plant genomes, the ratio of SNP occurrence is one per 100–500 bp (Morgil et al. 2020). It can be varied among different species. For the identification of variation, the sequence of a genotype is compared with the reference genome and sequence variation on a specific position to be determined. If the reference genome of a species is not available, the de novo assembly approach can be applied, as in case of sugarcane, whose complete reference genome is not present; this strategy can be useful in such a scenario. Although the monoploid reference genome of sugarcane of almost 1 GB is published, because of high ploidy level and aneuploidy, there is a dire need of complete reference genome of this species for complete understanding at the molecular level. The affymetrix axiom SNP array data analysis pipeline is outlined in Figure 12.2. The raw probe intensities from affymetrix can be analyzed using the Axiom analysis suite for developing highly polymorphic SNP KASP genotyping markers for sugarcane cultivar identification and population genetic studies.

12.5.1 SNP Markers Validation by KASP Assay

Besides serving as an efficient method for high and medium-throughput SNP genotyping, the NGS and array-based genotyping also offer low cost per sample (Chen et al. 2021). Many different high-throughput (HTP) assays/platforms are available for SNP validation, including Infinium assay, TaqMan assay coupled with Open Array platform, and Kompetitive Allele-Specific PCR (KASP) assay (Mammadov et al. 2012). These advanced SNP validation assays and genotyping platforms vary from each other with respect to the throughput of samples to genotype, cost, chemistry, and number of SNPs to validate. High-density SNP markers are being used in genomic studies such as

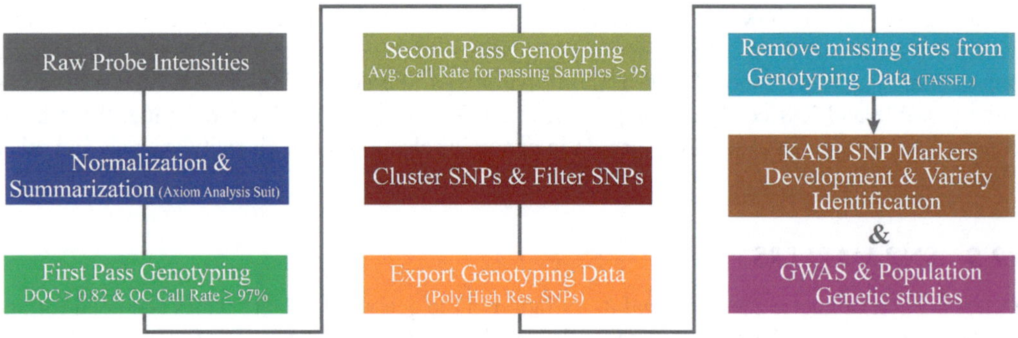

FIGURE 12.2 Overview of KASP SNP markers development from SNP array for variety identification. The raw probe intensities are normalized and summarized by Axiom analysis suite. The SNPs are clustered and filtered after first and second pass genotyping and exported as genotyping data. Using TASSEL, the missing sites from genotyping data are removed and highly polymorphic SNPs are used to develop KASP markers for fingerprinting of varieties.

Genome Wide Association Studies (GWAS), germplasm resource evaluation, and QTL mapping. For these studies, genotyping of a high number of samples is required at a few loci. So, the overall cost per sample increases and high-throughput genotyping platforms are not appropriate for these types of studies. In such a scenario, PCR-based KASP assay has emerged as an accurate, cost-effective, and feasible SNP genotyping method. Moreover, basic molecular lab instruments are required to carry this assay. Additionally, being the fully automated assay, the genotype calls improve, and the cost per sample reduces manyfold because of reduced reaction volume (Chen et al. 2021).

12.6 SUMMARY

Sugarcane is an important cash crop for sugar and biofuel production. The current sugarcane research aims at producing varieties with higher sugar contents and biomass. To strengthen the breeding programs for improved sugarcane varieties, parentage determination and DNA fingerprinting of available sugarcane cultivars is inevitable. In addition, varieties/hybrid certification is of utmost importance for breeding, seed market, and IPRs. To address this problem, we tried to explain the advanced techniques that can be used in precision breeding programs for high-value crop development. These high-throughput genotyping techniques include NGS-based applications such as WGS, RNA-Seq, Exome sequencing, GBS, and DNA SNP-array. The pipelines to develop KASP SNP genotyping markers from NGS and DNA array data have been detailed in this chapter.

REFERENCES

Aitken, K. et al. 2016. «Generation of a 345K sugarcane SNP chip» in. *Proc. Aust. Soc. Sugar Cane Technol* 29:1165–1172.

Aitken, K.S. et al. 2014. «Comparative mapping in the Poaceae family reveals translocations in the complex polyploid genome of sugarcane». *BMC Plant Biology* 14 (1): 1–15.

Baldwin, S.J. et al. 2011. «Association mapping of cold-induced sweetening in potato using historical phenotypic data». *Annals of Applied Biology* 158 (3): 248–256.

Balsalobre, T.W.A. et al. 2017. «GBS-based single dosage markers for linkage and QTL mapping allow gene mining for yield-related traits in sugarcane». *BMC Genomics* 18 (1): 1–19.

Bayer, M. et al. 2019. «Exome capture for variant discovery and analysis in barley» in. *Barley*, 283–310. Springer.

Bertan, I., de Carvalho, F.I.F., & Oliveira, Antonio Costa de. 2007. «Parental selection strategies in plant breeding programs». *Journal of Crop Science and Biotechnology* 10 (4): 211–222.

Bink, M.C.A.M. et al. 2002. «Multiple QTL mapping in related plant populations via a pedigree-analysis approach». *Theoretical and Applied Genetics* 104 (5): 751–762.

Björn, B. et al. 2010. «Population structure and linkage disequilibrium unravelled in tetraploid potato». *Theoretical and Applied Genetics* 121 (6): 1151–1170.

Blakeney, M. 2020. «Intellectual Property and Agricultural Innovation BT - Local Knowledge, Intellectual Property and Agricultural Innovation» in Blakeney, M. & Siddique, K.H.M. (arg.), 21–43. Singapore: Springer Singapore. 10.1007/978-981-15-4611-2_2. https://doi.org/10.1007/978-981-15-4611-2_2.

Branton, D. et al. 2008. «The potential and challenges of nanopore sequencing». Nat Biotechnology 26: 1146–1153.

Braun, A. & Wenzel, G. 2004. «Molecular analysis of genetic variation in potato (Solarium tuberosum L.). I. German cultivars and advanced clones». *Potato Research* 47 (1): 81–92. 10.1007/BF02731971.

Bremer, G. 1961. «Problems in breeding and cytology of sugar cane». *Euphytica* 10 (1): 59–78.

Cardoso-Silva, C.B. et al. 2014. «De novo assembly and transcriptome analysis of contrasting sugarcane varieties». *PloS One* 9 (2): e88462.

Chalhoub, B. et al. 2014. «Early allopolyploid evolution in the post-Neolithic Brassica napus oilseed genome». *Science* 345 (6199): 950–953.

Chen, Z. et al. 2021. «Development of genic KASP SNP markers from RNA-Seq data for map-based cloning and marker-assisted selection in maize». *BMC Plant Biology* 21 (1): 1–11.

Chilamakuri, C.S.R. et al. 2014. «Performance comparison of four exome capture systems for deep sequencing». *BMC Genomics* 15 (1): 1–14.

Consortium, International Wheat Genome Sequencing. 2014. «A chromosome-based draft sequence of the hexaploid bread wheat (Triticum aestivum) genome». *Science* 345 (6194): 1251788.

Contreras, R.N. et al. 2007. «Investigating parentage and hybridity of three azaleodendrons using amplified fragment length polymorphism analysis». *HortScience* 42 (3): 740–743.

Deomano, E. et al. 2020. «Genomic prediction of sugar content and cane yield in sugar cane clones in different stages of selection in a breeding program, with and without pedigree information». *Molecular Breeding* 40 (4): 1–12.

Eaton, D. 2007. «Intellectual property rights in plant breeding and biotechnology: a comparative institutional analysis» in. *11th Ann Conf Int Soc New Instit Econ (ISNIE), Reykjavik*. Libk. 25. Citeseer.

Eid, J. et al. 2009. «Real-time DNA sequencing from single polymerase molecules». *Science* 323 (5910): 133–138.

Ellstrand, N.C. 1984. «Multiple paternity within the fruits of the wild radish, Raphanus sativus». *The American Naturalist* 123 (6): 819–828.

Fickett, N. et al. 2019. «Genome-wide association mapping identifies markers associated with cane yield components and sucrose traits in the Louisiana sugarcane core collection». *Genomics* 111 (6): 1794–1801.

Garsmeur, O. et al. 2018. «A mosaic monoploid reference sequence for the highly complex genome of sugarcane». *Nature Communications* 9 (1): 1–10.

Gayathri, S. et al. 2019. «Identification of single nucleotide polymorphisms (SNPs) in the transcriptome of sugarcane variety Co 86032 exposed to oxidative stress». *Journal of Sugarcane Research* 9 (1): 71–85.

Ghosal, A.M.S.K & De, T.P.K.K. 2017. «Determination of DNA content and relative 2C genome sizes of some promising commercial varieties of sugarcane using flow cytometer». *Current Botany* 8: 66–73.

Gibbs, H Lisle et al. 1990. «Realized reproductive success of polygynous red-winged blackbirds revealed by DNA markers». *Science* 250 (4986): 1394–1397.

Gjedrem, Trygve. 2010. «The first family-based breeding program in aquaculture». *Reviews in Aquaculture* 2 (1): 2–15.

Hehir-Kwa, J.Y., Pfundt, R., & Veltman, J.A. 2015. «Exome sequencing and whole genome sequencing for the detection of copy number variation». *Expert Review of Molecular Diagnostics* 15 (8): 1023–1032.

Helfer, L.R. 2004. *Intellectual Property Rights in Plant Varieties*. FAO.

Hill, W.G., Salisbury, B.A., & Webb, A.J. 2008. «Parentage identification using single nucleotide polymorphism genotypes: application to product tracing». *Journal of Animal Science* 86 (10): 2508–2517.

Jagtap, A.B., Vikal, Y., & Johal, G.S. 2020. «Genome-wide development and validation of cost-effective KASP marker assays for genetic dissection of heat stress tolerance in maize». *International Journal of Molecular Sciences* 21 (19): 7386.

Jones, A.G. et al. 2010. «A practical guide to methods of parentage analysis». *Molecular Ecology Resources* 10 (1): 6–30.

Kearney, L. 2001. «Molecular cytogenetics». *Best Practice & Research Clinical Haematology* 14 (3): 645–668.

Kim, M. et al. 2021. «Genome-wide SNP discovery and core marker sets for DNA barcoding and variety identification in commercial tomato cultivars». *Scientia Horticulturae* 276: 109734.

Kyaligonza, V. et al. 2014. «Identification of F1 cassava (Manihot esculenta Crantz) progeny using microsatellite markers and capillary electrophoresis». *American Journal of Plant Sciences* 5:119–125.

Kyriakidou, M. et al. 2018. «Current strategies of polyploid plant genome sequence assembly». *Frontiers in Plant Science* 9: 1660.

Levine, L. et al. 1980. «Population genetics of Mexican Drosophila. V. A high rate of multiple insemination in a natural population of Drosophila pseudoobscura». *The American Naturalist* 116 (4): 493–503.

Liu, X. et al. 2015. «Gossypium barbadense genome sequence provides insight into the evolution of extra-long staple fiber and specialized metabolites». *Scientific Reports* 5 (1): 1–14.

Luo, Z.W. et al. 2000. «Predicting parental genotypes and gene segregation for tetrasomic inheritance». *Theoretical and Applied Genetics* 100 (7): 1067–1073.

Mahadevaiah, C. et al. 2021. «Genomic selection in sugarcane: Current status and future prospects». *Frontiers in Plant Science* 12:708233.

Majid, M.A. 2007. «Sugarcane variety composition in Pakistan». *Pakistan Sugar Journal (Pakistan)* 22(2):2–21.

Malik, K.B. 2013. «Cane and Sugar Production» 53 (9): 1689–1699.

Mammadov, J. et al. 2012. «SNP markers and their impact on plant breeding». *International Journal of Plant Genomics* 2012: 728398.

Manimekalai, R. et al. 2020. «Role of NGS and SNP genotyping methods in sugarcane improvement programs». *Critical Reviews in Biotechnology* 40 (6): 865–880.

McNeil, M.D. et al. 2017. «Development of a high-throughput, low-cost SNP genotyping panel for sugarcane breeding.» in. *Proceedings of the 2017 Conference of the Australian Society of Sugar Cane Technologists, Cairns, Queensland, Australia, 3-5 May 2017*, 304–311. Australian Society of Sugar Cane Technologists.

Meaburn, E. et al. 2006. «Genotyping pooled DNA using 100K SNP microarrays: a step towards genomewide association scans». *Nucleic Acids Research* 34 (4): e28–e28.

Mendes de Paula, T.O. et al. 2020. «Establishment of gene pools for systematic heterosis exploitation in sugarcane breeding». *Agronomy Journal* 112 (5): 3847–3858.

Meng, Z. et al. 2020. «Characterization of a Saccharum spontaneum with a basic chromosome number of x = 10 provides new insights on genome evolution in genus Saccharum». *Theoretical and Applied Genetics* 133 (1): 187–199.

Metcalfe, C.J. et al. 2019. «Flow cytometric characterisation of the complex polyploid genome of Saccharum officinarum and modern sugarcane cultivars». *Scientific Reports* 9 (1): 1–12.

Miller, J.R. et al. 2017. «Initial genome sequencing of the sugarcane CP 96-1252 complex hybrid». *F1000Research*6: 688.

Mirajkar, S.J. et al. 2019. «Sugarcane (Saccharum spp.): breeding and genomics» in. *Advances in Plant Breeding Strategies: Industrial and Food Crops*, 363–406. Springer.

Moore, P.H., Paterson, A.H., & Tew, T. 2013. «Sugarcane: the crop, the plant, and domestication». *Sugarcane: Physiology, Biochemistry, and Functional Biology*: 1–17.

Morgil, H., Gercek, Y.C., & Tulum, I. 2020. «Single nucleotide polymorphisms (SNPs) in plant genetics and breeding» in. *The Recent Topics in Genetic Polymorphisms*. IntechOpen.

Norman, P.E. et al. 2018. «Can parentage analysis facilitate breeding activities in root and tuber crops?» *Agriculture*. 10.3390/agriculture8070095.

Pemberton, J.M. 2008. «Wild pedigrees: the way forward». *Proceedings of the Royal Society B: Biological Sciences* 275 (1635): 613–621.

Poland, J.A. et al. 2012. «Development of high-density genetic maps for barley and wheat using a novel two-enzyme genotyping-by-sequencing approach». *PloS One* 7 (2): e32253.

Pryer, K.M. et al. 2002. «Deciding among green plants for whole genome studies». *Trends in Plant Science* 7 (12): 550–554.

Riaño-Pachón, D.M. & Mattiello, L. 2017. «Draft genome sequencing of the sugarcane hybrid SP80-3280». *F1000Research* 6: 861.

Rohland, N. & Reich, D. 2012. «Cost-effective, high-throughput DNA sequencing libraries for multiplexed target capture». *Genome Research* 22 (5): 939–946.

Savage, N. 2011. «The ideal biofuel: A biomass-based fuel needs to be cheap and energy dense. Gasoline sets a high standard». *Nature* 474 (7352): S9–S9.

De Setta, N. et al. 2014. «Building the sugarcane genome for biotechnology and identifying evolutionary trends». *BMC Genomics* 15 (1): 1–18.

Shah, A. & Ali, L. 2005. «Sugar industry dilemma in NWFP». *Science Technology and Development* 24 (3): 40–45.

Shaikh, S. et al. 2021. «Financial constraints and listed sugar mills on Pakistan Stock Exchange». *International Journal of Management (IJM)* 12 (3).

Soltis, D.E. et al. 2016. «Polyploidy: pitfalls and paths to a paradigm». *American Journal of Botany* 103 (7): 1146–1166.

Song, Y.-S. et al. 2005. «Mapping of extreme resistance to PVY (Ry sto) on chromosome XII using anther-culture-derived primary dihaploid potato lines». *Theoretical and Applied Genetics* 111 (5): 879–887.

Spanoghe, M. et al. 2015. «Investigation and development of potato parentage analysis methods using multiplexed SSR fingerprinting». *Potato Research* 58 (1): 43–65.

Sulonen, A.-M. et al. 2011. «Comparison of solution-based exome capture methods for next generation sequencing». *Genome Biology* 12 (9): 1–18.

Sun, G. et al. 2003. «RAPD and pedigree-based genetic diversity estimates in cultivated diploid potato hybrids». *Theoretical and Applied Genetics* 107 (1): 110–115.

Tarn, T.R. et al. 1992. «Breeding potatoes for long-day, temperate climates». *Plant Breeding Reviews* 9: 217–332.

Wu, H. & Altpeter, F. 2015. «Sugarcane (saccharum spp. hybrids)». *Methods in Molecular Biology (Clifton, N.J.)* 1224 (urtarrilak 1): 307–316. 10.1007/978-1-4939-1658-0_24.

Xiao, S. et al. 2015. «Functional marker detection and analysis on a comprehensive transcriptome of large yellow croaker by next generation sequencing». *PloS One* 10 (4): e0124432.

Xu, S. et al. 2018. «Transcriptomic characterization and potential marker development of contrasting sugarcane cultivars». *Scientific Reports* 8 (1): 1–11.

You, Q. et al. 2016. «Genetic diversity analysis of sugarcane germplasm based on fluorescence-labeled simple sequence repeat markers and a capillary electrophoresis-based genotyping platform». *Sugar Tech* 18 (4): 380–390.

You, Q. et al. 2019. «Development of an Axiom Sugarcane100K SNP array for genetic map construction and QTL identification». *Theoretical and Applied Genetics* 132 (10): 2829–2845.

Zhang, A. et al. 2017. «Effect of trait heritability, training population size and marker density on genomic prediction accuracy estimation in 22 bi-parental tropical maize populations». *Frontiers in Plant Science* 8: 1916.

13 Recent Advancements in Diagnostics and Management of Phytoplasma-infecting Sugarcane

Atul Kumar
Amity Institute of Biotechnology, Amity University, Uttar Pradesh Campus, Lucknow, India

Somnath Kadappa Holkar
ICAR-National Research Centre for Grapes, Pune, Maharashtra, India

CONTENTS

13.1 INTRODUCTION

Phytoplasma are the prokaryotic plant pathogens belonging to class *Mollicutes.* First discovered in Japan, they are obligate parasites that reside within the phloem of plants and lack cell walls (Doi et al. 1967; Rao 2021). Since the first record of the occurrence of phytoplasma disease in 1874 in India, 10 different groups of phytoplasma have been documented in more than 300 plant species serving as host; these include different ornamental plants and weeds, as well as many commercial crops, including sugarcane, across the globe (McCoy et al. 1989; Rao 2021). The aster yellows phytoplasma of 16SrI group is documented as one of the most common and widespread phytoplasmas, followed by 16SrII, 16SrVI, 16SrXI, and 16SrXIV groups (Rao 2021).

A large count of newly reported phytoplasma strains on several commercial crops has emerged during the last few decades, revealing that phytoplasmas are diverse groups (Rao 2021). Sugarcane (*Saccharum officinarum; Poaceae*) is one of the major cash crops worldwide, cultivated in more than 90 countries. Over the past 100 years, India has faced several constraints in sugarcane production, including low cane productivity, low/stagnant sugar recovery (~10%), and losses due to pests and diseases. Beside these constraints, diseases play a very pivotal role in reducing the worldwide production of sugarcane (Singh et al. 2009). However, several fungal, bacterial,

DOI: 10.1201/9781003292425-13

phytoplasma, and viral diseases are known to affect sugarcane production worldwide, whereas India has witnessed the damaging epidemics of several sugarcane diseases like smut, red rot, rust, leaf scald, wilt, and yellow leaf disease (YLD) (Singh et al. 2009; Viswanathan and Rao 2011).

Besides all these constraints, phytoplasma diseases are a major constraint in sugarcane production worldwide; the diseases include *Sugarcane grassy shoot phytoplasma* (SCGSP; Rao et al. 2008), *Sugarcane white leaf phytoplasma* (SCWLP) (Sdoodee 2001, Rao et al. 2008, Soufi et al. 2013), and *Sugarcane yellow phytoplasma* (SCYP; Cronje et al. 1998; Gaur *et al.* 2008; Kumar et al. 2015).

13.2 PHYTOPLASMA DISEASES OF SUGARCANE

Phytoplasmas infecting sugarcane are more diverse and are found to be associated with several diseases, including sugarcane white leaf disease (WLD), grassy shoot disease (GSD), green grassy shoot disease (GGSD), Ramu stunt disease, and leaf yellow disease. Among all of these, YLD, GSD, and WLD are reported to cause severe losses in sugarcane yield and yield-contributing parameters worldwide.

13.2.1 Grassy Shoot Disease (GSD)

GSD is an important disease of sugarcane and is caused by *Sugarcane grassy shoot phytoplasma* (SCGSP) (Sdoodee et al. 1999; Sdoodee 2001; Singh et al. 2002; Rao et al. 2008, 2012, 2014; Tiwari et al. 2012; Kumar et al. 2017; Yadav et al. 2017; Marcone et al. 2019; Nithya et al., 2020; Sharma et al. 2020; Rao 2021). In India, disease incidence in GSD-affected sugarcane cultivars (cvs.) was recorded up to 70% (Yadav et al. 2017). In India, GSD was reported for the first time in district Ahmadnagar Maharashtra (Chona 1958). Sugarcane affected with GSD exhibits various symptoms, including the transformation of the normal leaf into a thin leaf, along with the slender-adventitious tillers from the leaf base of the affected stools with shortened internodes and the development of side shoots from the base to the top. These shoots ultimately lead to a dense bunch of tillers, along with the leaf chlorosis, which causes a thin, narrow, and pale-yellow coloration of sugarcane leaves with reduced size and soft texture; the bunch of growing leaves with slender tillers give the grassy appearance, and therefore, it is known as "grassy shoots" (Rao et al. 2008, 2012, 2014; Tiwari et al. 2012; Yadav et al. 2017; Nithya et al., 2020; Sharma et al. 2020; Figure 13.1A). The ratoon crop is more prone to infection of GSD, and affected planta will lose the capability to produce millable canes (Rao et al. 2012, 2014; Tiwari et al. 2012; Kumar et al. 2017; Yadav et al. 2017).

13.2.2 White Leaf Disease (WLD)

White leaf disease (WLD) of sugarcane caused by *Sugarcane white leaf phytoplasma* (SCWLP) is the major disease of sugarcane in India (Rao et al. 2008, 2012, 2017, 2018; Kumar et al. 2018; Kumar et al. 2021; Rao 2021) and abroad (Kumarasinghe and Jones 2001; Soufi et al. 2013; Keerthipala 2016; Quan et al. 2020; Dayasena et al. 2021). WLD has a serious impact on sugarcane yield-contributing parameters, with recorded disease up to 35%–100% worldwide (Rao et al. 2005; Hanboonsong et al. 2006; Wang et al. 2014). In 1954, SCWLP was reported for the first time in Thailand (Mangelsdorf 1962). The WLD symptoms can be characterized by the formation of proliferated tillers with slender chlorotic shoots with smaller and narrower leaves and pronounced stunting; a soft texture also results, along with the partial or complete loss of green chlorophyl pigment of the leaves that ultimately gives a white colour appearance to leaves (Kumar et al. 2018; Rao et al. 2018; Quan et al. 2020; Dayasena et al. 2021; Figure 13.1B). The sugarcane plant affected severely by WLD can fail to produce millable canes that cause economically high yield losses (Rao et al. 2012; Quan et al. 2020; Dayasena et al. 2021).

(a) (b) (c)

FIGURE 13.1 (a): Field symptoms of YLD in sugarcane with yellowing of leaf and midrib. (b): Characteristic symptoms of WLD with complete chlorosis of sugarcane leaves. (c): Symptoms of GSD on sugarcane exhibiting narrowed leaves and short tillers.

13.2.3 Yellow Leaf Disease (YLD)

Worldwide, the first evidence of YLD was illustrated in Hawaii in *Saccharum officinarum* cv. H 65-7052, characterized by severe yellowing of leaves throughout the plantation (Schenck 1990, 97). One decade after its first record, the affiliated virus was diagnosed in YLD-infected sugarcane cvs. and was identified as a virus belonging to *Luteoviridae* and named as *Sugarcane yellow leaf virus* (SCYLV) (Schenck et al. 1997; Vega et al. 1997; Moonan et al. 2000; Scagliusi and Lockhart 2000; Smith et al. 2000). However, there was not any strict evidence found between the correlation of SCYLV and YLS; sometimes YLS symptoms were visible clearly, but SCYLV could not be detected. Later, phytoplasma was also isolated from YLD-infected plants of sugarcane and was established as a pathogen responsible for YLD (Cronje and Bailey 1999). Subsequently, identical visual symptoms of YLD were also documented in sugarcane in Florida (Comstock et al. 1994). Since then, symptoms of YLD have been recorded in exceeding than 20 countries, including Australia, Brazil, Cuba, Mauritius, Reunion, South Africa, and U.S.A. (Bailey et al. 1996; Schenck et al. 1997; Vega et al. 1997; Cronje et al. 1998).

YLD is characterized by yellowing of the leaf midribs followed by necrosis in leaf tissue from leaf tip to the base, wilting and stunt growth, subsequently leading to complete dieback of the plant (Borth et al. 1994; Schenck et al. 1997; Vega et al. 1997; Comstock et al. 1998; Kumar et al. 2015; Holkar al. 2016; Kumar et al. 2020a, b, c; Figure 13.1C). The most common symptoms were recorded as yellow discoloration of leaf and midrib in the cane of crops 6 to 8 months or older (Kumar et al. 2015; Holkar al. 2016; Kumar et al. 2020a b, c). Similar symptoms were also reported from Florida (Comstock et al. 1994), Africa (Bailey et al. 1996, where older reports had already noticed as 'yellow wilt'), and Brazil (Vega et al. 1997). In India, Viswanathan et al. (1999) documented the evidence of YLD for the first time in sugarcane as a serious constraint to the commercial sugarcane varieties *viz.*, CoPant 84211, Co 6304, CoC 86062, CoC 92061, CoA 05323, CoV 92102, CoA 92081, CoV 94102, CoV 06356, Co 86032, and Co 94012 (Rao et al. 2000; Viswanathan et al. 2002, 2014; Viswanathan and Rao 2011).

The widespread occurrence of YLD was recorded in various regions of India (Rao et al. 2000, 2001; Viswanathan 2002). Subsequently, the impact of YLD on sugarcane cvs. was recorded up to 34.15% based on different yield-contributing parameters, *viz.,* stalk thickness, weight, height, and number of internodes and juice quality (Viswanathan et al. 2002, 2014). However, association of phytoplasma with the YLD has also been established (Cronje and Bailey 1999). In India, Gaur et al. (2008) documented for the first time the association of SCYP with YLD of sugarcane. Since then, other groups of phytoplasma were also identified with YLD of sugarcane in India (Kumar et al. 2015, 2018, 2020a, b, c, 2021).

On the other hand, another phytoplasma-born disease called Ramu stunt disease of sugarcane was also documented during 1980s in the Ramu valley, Papua New Guinea; it caused severe losses in crop yield contributing parameters of commercial sugarcane cvs. *viz.,* Ragnar, cvs. Q90, and Yasawa (Eastwood 1990; Braithwaite 2010). In the late 1980s, it was believed that the causal agent of Ramu stunt was a virus, as suggested by Jones et al. (1989). Later, Cronje et al. (1999) reported the phytoplasma infection in Ramu stunt– affected sugarcane samples taken from the sugarcane plant, as well as from the putative vector *E. flavipes*. Ramu stunt is characterized by its most common symptom of pronounced stunting with a yellow, mottled striping on leaves, a short, erect, and stiff texture, and, sometimes, excessive tillering and grassy shoot appearance in leaves. This infection ultimately leads to the death of affected sugarcane plants after the appearance of the first symptoms.

13.3 DISEASE TRANSMISSION AND INSECT VECTORS

Some phytoplasma-carrying insects, *viz.,* leafhopper, planthopper, and psyllids, live freely in nature and feed the phloem tissues of phytoplasma-infected plants, picking up that phytoplasma and transmitting them to other crops. A dozen planthopper and leafhopper species have been identified as natural vectors, and more than 40 weed species have been found to be associated as natural hosts for many phytoplasma strains (Rao 2021). Recent reports stated that the phytoplasma transmission also depends on some antigenic protein found on their cell surface; it communicates with the particular insect-microfilament followed by its vector and is the major determining factor for phytoplasma diseases in different commercial crops, including sugarcane (Wilson and Turner 2010).

However, the SCGSP in sugarcane is mainly transmitted by a few vectors called leafhoppers, planthoppers (*Fulgoridae*), jumping plant lice (*Psyllidae*), and westwood fulgorid (*Proutista moesta*) (Maixner and Reinert 1999; Gatineau et al. 2001; Srivastava et al. 2006; Viswanathan & Rao 2011; Rao et al. 2014; Yadav et al. 2017; Rao 2021). Alternatively, the transmission of SCWLP occurs mainly by some species of plant vector leafhopper, viz., *Matsumuratettix hiroglyphicus* and *Deltocephalus Menoni* as well (Sdoodee 2001; Dayasena et al. 2021; Rao 2021). Moreover, the spread of Ramu stunt disease was suggested to be accomplished through an insect vector called leafhopper *Eumetopina flavipes* (Kunita et al. 1994). In the same manner, SCYP of sugarcane is also vectored by some phloem-feeding delphacid planthoppers – *Saccharosydne sachharivora* (family *Cicadellidae*) (Maixner and Reinert 1999; Gatineau et al. 2001). The primary transmission of SCYP occurs due to vegetative propagation, while secondary transmission occurs through the planthoppers in a circular manner from infected to healthy cane (Maixner and Reinert 1999).

13.4 DIAGNOSIS OF PHYTOPLASMAS

Through the introduction of molecular-advanced diagnostics tools into phytoplasmology, some sensitive and robust diagnosis methods have evolved, especially based on polymerase chain reaction (PCR) assays (Ahrens and Seemuller 1992). It became very easy to differentiate, characterize, and classify the phytoplasma pathogen based on their evolutionary molecular phylogenetic positions (Felsenstein 1985; Saitou and Nei 1987; Tamura et al. 2004; Kumar et al. 2016; Figure 13.2), sequence-identity matrix, similarity coefficient, and by restriction fragment-length polymorphism (RFLP) sequence analysis of 16S ribosomal DNA (16SrDNA) (Lee et al. 1993, 1998, 2000; Schneider et al. 1993; Seemuller et al. 1998). However, several phylogenetic groups of phytoplasma have been diagnosed using all of the above tools and established the aster yellows (AY) group, SCWL group, the X-disease group, and faba bean phyllody (FBP) group in sugarcane.

The phytoplasma infections in sugarcane can be seen in microscopic examination of phloem-tissues stained with the DNA-dye named fluorochrome 4’-6-diamidino-2-phenylindole (DAPI),

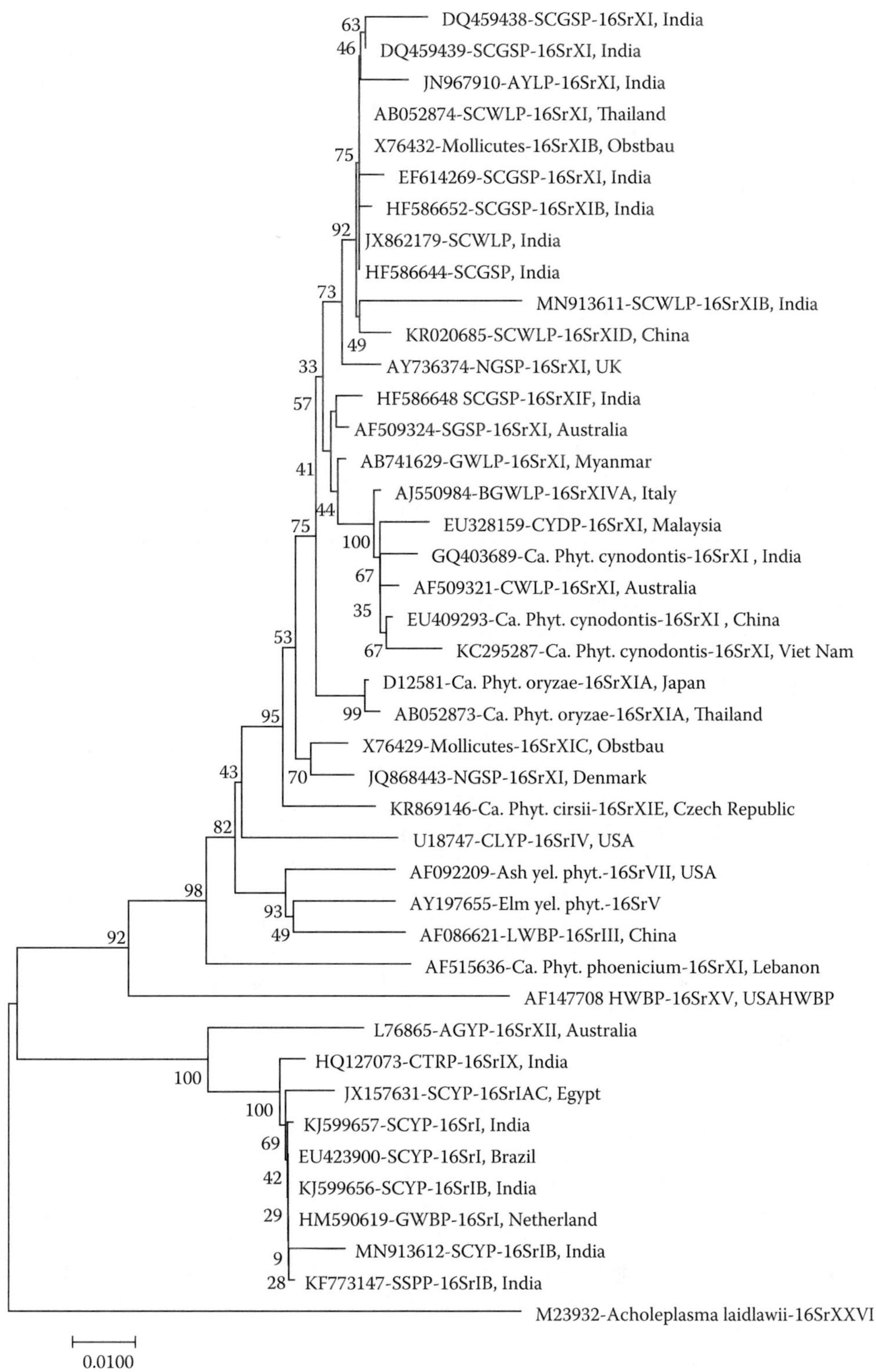

FIGURE 13.2 Molecular evolutionary positions of *Sugarcane white leaf phytoplasma* (SCWLP), *Sugarcane* grassy shoot *phytoplasma* (SCGSP), and *Sugarcane yellows phytoplasma* (SCYP) in a phylogenetic tree of 16S rDNA sequences retrieved from NCBI GenBank was inferred using the Neighbor-Joining method

(Felsenstein 1985; Saitou and Nei 1987; Tamura et al. 2004; Kumar et al. 2016). Other used abbreviations are as follows: Ca. = *'Candidatus*; IAYLP = *Indian arecanut yellow leaf phytoplasma*; NGSP = *Napier grass stunt phytoplasma*; SrGSP = *Sorghum grassy shoot phytoplasma*; GGWLP = *Goose grass white leaf phytoplasma*; BGWLP = *Bermuda grass white leaf phytoplasma*; CYDP = *Coconut yellow decline phytoplasma*; CWLP = *Cynodon white leaf phytoplasma*; CLYP = *Coconut lethal yellowing phytoplasma*; ASYP = *Ash yellows phytoplasma*; ELYP = *Elm yellows phytoplasma*; LWBP = *Loofah witches broom phytoplasma*; AGYP = *Australian grapevine yellows phytoplasma; CTRP = Catharanthus roseus phytoplasma; GLWBP = Gladiolus witches' broom phytoplasma and SSPP = Sesame phyllody phytoplasma.*

which is simple, rapid, and very economical in the detection of phytoplasma (Seemuller 1976; Sarindu and Clark 1993). Moreover, most of the time, the concentration of the phytoplasma population is very low in the plant tissue of host the organs of sugarcane. However, these microscopic tools have limitation in epidemiological studies and identification of phytoplasma as well. Hence, polyclonal-antisera have been successfully developed against partially purified antigenic preparations from SCWLP and SCGSP-affected plants and can be effectively used in enzyme-linked immunosorbent assay (ELISA) in the diagnosis of their respective phytoplasma antigens in crude tissue extracts of sugarcane diseased plant (Sarindu and Clark 1993; Viswanathan 1997, 2001). But sometimes, cross-reactions can be also seen in ELISA assay; these occur between these phytoplasma antigens with their reciprocal antisera. Moreover, these serological-based methods have also lacked some necessary specificity and sensitivity. Hence, these techniques are still not most widely implicated in phytoplasmology for diagnostic purposes (Seemuller et al. 1998; Adams et al. 2001).

On the other hand, randomly designed cloned-fragments of chromosomal as well as extrachromosomal-DNA of the SCWLP have been successfully used in dot and southern-blot hybridization assays as probes in the phytoplasma diagnosis, which displayed a broad range of detection (Klinkong and Seemiiller 1993; Nakashima et al. 1994, 2001; Nakashima and Hayashi 1995). However, these probes hybridized with total DNA of SCWLP-infected plants, along with other associated phytoplasmas including RYD, *Bermuda grass white leaf phytoplasma* (BGWLP), *Brachiaria white leaf phytoplasma* (BraWLP), and *Dactyloctenium white leaf phytoplasma* (DWLP) (Nakashima and Hayashi 1995; Nakashima et al. 2001). But sometimes, probes-based DNA-hybridization assays find no hybridization signals in DAPI-based fluorescence tests (Seemuller and Kirkpatrick 1996). Therefore, both dot and southern blot hybridization assays are now completely replaced by the powerful PCR-based assays in the diagnosis of sugarcane phytoplasmas and in the present decade. This nucleo-based PCR tool has been most widely employed worldwide with several advantages over other techniques, including relative semplicity, versatility, specificity, and high sensitivity (Nakashima et al. 1996, 2001; Wongkaew et al. 1997; Cronje et al. 1998; Arocha et al. 1999; Cronje and Bailey 1999; Sdoodee et al. 1999; Sdoodee 2001; Tran-Nguyen et al. 2000; Aljanabi et al. 2001). However, several universal as well as group-specific phytoplasma primers, amplifying to ribosomal and/or non-ribosomal DNA sequences, have been designed for PCR assay (Deng and Hiruki 1991; Schneider 1995; Gunderson and Lee 1996; Manimekalai et al. 2010). Another advanced tool based on oligo-capture secA gene of phytoplasma can be used in the robust diagnosis of all *Candidatus Phytoplasma* species (Hodgetts et al. 2008). These primers can be directly used in single-round PCR or by re-amplifying the DNA fragments from first-round amplification using some highly sensitive internal primers called nested PCR assay to overcome from low titre of phytoplasma infections (Tran-Nguyen et al. 2000; Aljanabi et al. 2001).

On the other hand, an accurate detection tool called the loop-mediated isothermal amplification (LAMP) was also developed to target the groEL gene and 16S rDNA of SCWL phytoplasma (Vu et al. 2016). Additionally, a simple DNA isolation tool that undoes the homogenizing and the phenol extraction by combining with that of LAMP assay could be originated as a simple, rapid, and convenient diagnostic method in the detection of phytoplasma (Tomlinson et al. 2010; Hodgetts et al. 2011; Bekele et al. 2011; Dickinson et al. 2015; Kogovsek et al. 2015).

13.5 HOST RANGE OF SUGARCANE PHYTOPLASMAS

Phytoplasmas are known to cause variable diseases in different grasses and weeds, which act as natural reservoirs for all important phytoplasma strains that play a key role in the spread of phytoplasmas and serve as natural alternative hosts for many insect vectors, causing severe yield losses in novel commercial crops (Seemuller et al. 1998; Blanche et al. 2003; Rao et al. 2007; Mall et al. 2010, 2011; Marcone et al. 2004). However, insect vectors are become one of the great concerns that could transmit phytoplasmas from many weed and grass species, growing near sugarcane fields (Blanche et al. 2003; Joomun et al. 2007, Mall et al. 2007).

Moreover, some sugarcane phytoplasma strains have been identified; they were found to closely relate to phytoplasma strains associated with many weed species, showing clustering in the same phylogenetic groups, including BraWLP from *Brachiaria* and BGWLP from Bermuda grass. This clearly proved that how dangerous they are for sugarcane. The BGWLP belonging to *Candidatus Phytoplasma cynodontis* of the 16SrXIV group, share very closely about 97.1% and 97.3% of sequence similarity, respectively, with SCGSP (Joomun et al. 2007; Rao et al. 2007). On the other hand, both these BraWL and BGWL phytoplasma also share a close relationship with that of SCWL phytoplasma (Marcone and Rao 2008). Similarly, *Areca palm phytoplasma* (AYLP), causing YLD in areca nut palm, has displayed 99% and 98% identity with SCWLP (16srXI group) and BGWLP (16srXIV subgroup), respectively (Manimekalai et al. 2010). These close relationships between AYLP with SCWLP and BGWLP can create a high risk of the further spread of respective phytoplasma on their alternative and collateral hosts.

13.6 INTEGRATED DISEASE MANAGEMENT APPROACHES

The identification of different phytoplasma diseases is totally based on the symptomatology, but sometimes, it is very difficult to identify these diseases using symptomatology alone due to lack of characteristic disease symptoms in affected canes. However, the actual identification and management of phytoplasma disease using symptomatology alone is not possible. Several phytoplasma diseases of sugarcane originated through the exchange of germplasm from one country to another, with the potential risk of introducing new diseases (Rao et al. 2012). Hence, to prevent such potential risk of spread of phytoplasma diseases in sugarcane, it become crucial for all countries to strictly follow quality inspection and quarantine facilities using different biotechnological and molecular diagnostic tools to reduce all risks (Rao et al. 2012). But using thermotherapy and chemotherapy had already failed in the control of phytoplasmas (Rao et al. 2012). Similarly, the application of resistant clones also had limited impact in the control of phytoplasma diseases due to the gradual evolution of pathogens over time (Rao et al. 2012).

In the present decade, several advancements have been occurred in biotechnology and molecular diagnostics. Hence, it became easy to identify and diagnose the phytoplasma in sugarcane at early crop stages for successful prevention and management of phytoplasma diseases, avoiding further spread in sugarcane (Rao et al. 2012). Alongside, it is also crucial to restrict the movement of phytoplasma outside from South-East Asia since most of the phytoplasma diseases like GSD, WLD, and GGSD are not reported throughout all sugarcane growing countries. Moreover, no single method is available to effectively work against different phytoplasma diseases of sugarcane forever. However, through the regular practice of phytoplasma-free seed-cane, appropriate cultural practices, and resistant sugarcane genotypes, the ideal management of phytoplasma-borne diseases in sugarcane should result (Rao 2006).

The YLD, GSD, and WLD are mainly transmitted through phytoplasma-infected cane. Treatment by moist hot-air treatment (MHAT) at 54°C for 2.5 hour at 99% humidity, along with hot water treatment at 54°C for 2 h, can be effective to some extent for the elimination of phytoplasma from infected seed canes (Frison and Putter 1993; Satyagopal et al. 2014). There are some successful reports of the elimination of SCYP by using molecular diagnostic tools and tissue-culture techniques (apical meristem tissue culture, auxiliary bud culture, and leaf rolls callus culture technique); these can be successfully used to manage phytoplasma from noble sugarcane varieties with a 100% success rate in the control of YLD (Parmessur et al. 2002; Viswanathan et al. 2018). The national certification system for tissue-culture-raised plants (NCS-TCP) is a national program for the commercial production of tissue-culture-raised plants and their distribution to the farmers. Accredited test laboratories (ATLs), as an important component of NCS-TCP for virus indexing and genetic fidelity testing, are working effectively for providing phytoplasma-free quality sugarcane planting materials (Kumar et al. 2017).

Several weeds have recently been identified to act as the natural reservoir for many phytoplasma strains. Thus, the removal of these weeds from sugarcane fields will restrict further phytoplasma-borne disease spread (Tran-Nguyen et al. 2000; Rao and Ford 2001; Joomun et al. 2007; Mall et al. 2011). Another novel approach, called cryotherapy, can also be used in the successful removal of phytoplasmas for both phytoplasma and virus-free plant production from different commercial crops, including banana, grapevine, Prunus, potato, and sweet potato (Brison et al. 1997; Helliot et al. 2002; Wang et al. 2003; Wang et al. 2006; Wang and Valkonen 2008; Feng et al. 2011). Thus, all these advanced diagnostic tools can be simultaneously applied to the production of pathogen-free plants with key advantages, including elimination of obligate parasites. On the other hand, long-term storage of plant germplasm in cryopreserved stocks could also be a way to overcome the inadvertent dispersal of many obligate virus and phytoplasma pathogens (Wang and Valkonen 2008; Wang et al. 2009).

13.7 CONCLUSION

Phytoplasma diseases are known to cause several constraints in global sugarcane production and are now widespread, with considerable economic yield losses worldwide. WLD and GSD, predominantly reported from the southeast Asian region, have the strict insect vector specificity. Whereas YLD of sugarcane is documented from almost all the continents where sugar cultivation occurs and is found to be associated with distinct phytoplasma strains, including AY, X-disease, and FBP groups known to cause several diseases in a wide range of wild species and commercial crops; it is listed with a low insect vector specificity. Sugarcane is known to be severely affected by three major phytoplasma diseases, *viz.,* yellow leaf disease (YLD) caused by *Sugarcane yellows phytoplasma* (SCYP), white leaf disease (WLD) caused by *Sugarcane white leaf phytoplasma* (SCWLP), and grassy shoot disease (GSD) caused by *Sugarcane grassy shoot phytoplasma* (SCGSP). In recent decades, YLD is emerged as a new disease of sugarcane that occurred due to single or mixed infection *Sugarcane yellow leaf virus* and/or SCY phytoplasma. Similarly, GSD and Ramu stunt disease of sugarcane have been also found to be associated with phytoplasma from Papua New Guinea and Thailand, respectively. Among all of these, YLD, GSD, and WLD have been evidenced as the most economically serious constraint, causing significant reduction in sugarcane yield-contributing parameters and sugar recovery worldwide.

The transmission and spread of all these phytoplasmas were occurring rapidly to the newer locations through infected seed cane and disease-vector delphacid planthoppers, *Saccharosydne sachharivora*. However, phytoplasma infecting sugarcane are more diverse, and hence, association of other diverse groups of phytoplasma with sugarcane cannot be overruled. However, it is an immediate need to systematically characterize and diagnose a large number of phytoplasma isolates at earlier stages in the sugarcane crop to more effectively manage the drastic losses occurred.

The identification and characterization of the associated phytoplasma through the characteristic disease symptoms at an early crop stage is not terribly reliable due to unspecific symptoms in diseased plants. However, through the introduction of recent advanced and highly improved molecular-diagnostic tools into the phytoplasmology, we can easily diagnose the phytoplasma infections in diseased plants and insect hosts at an early crop stage. The nested polymerase chain reaction (nested-PCR) is becoming one of the most advanced, reliable, and robust molecular-diagnostic tools due to its simplicity, specificity, and versatility. Through the application of these advancements, it has also been very easy to differentiate, characterize, and classify the different phytoplasma strains using morphological and molecular phylogenetic analysis of 16S ribosomal DNA (16SrDNA) gene sequence. However, through the utility of advancements in modern biotechnological and molecular diagnostic tools, we can easily diagnose and characterize the phytoplasma pathogens at very large scale in the screening of planting seed canes for the possible elimination and integrated disease management of different phytoplasma pathogens at early crop stage. We can also successfully use breeding programs in the future. This topic is totally focused on the different novel and advanced research efforts that have been made in the diagnosis and management of phytoplasma infecting sugarcane.

REFERENCES

Adams, A.N., Davies, D.L., and Kirby, M.J. 2001. Virus and phytoplasma detection in fruit trees. *Outlook Agriculture* 30: 45–54.

Ahrens, U. and Seemuller, E. 1992. Detection of DNA of plant pathogenic mycoplasma-like organisms by a polymerase chain reaction that amplifies a sequence of the 16S rRNA gene. *Phytopathology* 82: 828–832.

Aljanabi, S.M., Parmessur, Y., Moutia, Y., Saumtally, S., and Dookun, A. 2001. Further evidence of the association of a phytoplasma and a virus with yellow leaf syndrome in sugarcane. *Plant Pathology* 50: 628–636.

Arocha, Y., Gonzalez, L., Peralta, E.L., and Jones, P. 1999. First report of virus and phytoplasma pathogens associated with yellow leaf syndrome of sugarcane in Cuba. *Plant Disese* 83: 1177.

Bailey, R., Bechet, G., and Cronje, P. 1996. Notes on the occurrence of yellow leaf syndrome of sugarcane in southern Africa. *Proceedings of the South African Sugar Technologists Association* 70: 3–6.

Bekele, B., Hodgetts, J., Tomlinson, J., Boonham, N., Nikolić, P., Swarbrick, P., and Dickinson, M. 2011. Use of a real-time LAMP isothermal assay for detecting 16SrII and XII phytoplasmas in fruit and weeds of the Ethiopian Rift Valley. *Plant Pathology* 60(2): 345–355.

Blanche, K.R., Tran-Nguyen, T.T., and Gibb, K.S. 2003. Detection, identification and significance of phytoplasmas in grasses in northern Australia. *Australasian Plant Pathology* 52: 505–512.

Borth, W.B. 1994. Double-stranded RNA associated with sugarcane yellow leaf syndrome. *Sugar Cane*, 3: 5–8.

Braithwaite, K.S. 2010. Ramu stunt. In: *Pest and Diseases Image Library*.

Brison, M., De Boucaud, M.T., Pierronnet, A., and Dosba, F. 1997. Effect of cryopreservation on the sanitary state of a cv Prunus rootstock experimentally contaminated with Plum pox potyvirus. *Plant Science* 123: 189–196.

Chona, B.L. 1958. Presidential address. Some diseases of Sugarcane reported from India in recent years. *Indian Phytopathology* 11(1): 1–9.

Comstock, J.C., Irey, M.S., Lockhart, B.E., and Wang, Z.K. 1998. Incidence of yellow leaf syndrome in CP cultivars based on polymerase chain reaction and serological techniques. *Sugar Cane* 4: 21–24.

Comstock, J.C., Irvine, J.E., and Miller, J.D. 1994. Yellow leaf syndrome appears on the United States mainland. *Sugar Journal* 56: 33–35.

Cronje, C.P.R. and Bailey, R.A. 1999. Association of phytoplasmas with yellow leaf syndrome of sugarcane. In *Proceedings of the XXIII ISSCT Congress, New Delhi, India, 22-26 February, 1999. Volume 2* (pp. 373–381). Sugar Technologists' Association of India.

Cronje, C.P.R., Tymon, A.M., Jones, P., and Bailey, R.A. 1998. Association of a phytoplasma with a yellow leaf syndrome of sugarcane in Africa. *Annals of Applied Biology* 133(2): 177–186.

Cronje, C.P.R., Bailey, R.A., Jones, P., and Suma, S. 1999. The phytoplasma associated with Ramu stunt disease of sugarcane is closely related to the white leaf phytoplasma group. *Plant Disease* 83: 588.

Dayasena, Y.A., Panda, P., Thushari, A.N., and Rao, G.P. 2021. Geographical distribution and identification of phytoplasma strain associated with sugarcane white leaf disease in Sri Lanka. *Sugar Tech* 2: 1–8.

Deng, S. and Hiruki, C. 1991. Amplification of 16S rRNA genes from culturable and nonculturable mollicutes. *Journal of Microbiologicl Methods* 14: 53–61.

Dickinson, M. 2015. Loop-mediated isothermal amplification (LAMP) for detection of phytoplasmas in the field. *InPlant Pathology* (pp. 99–111). Humana Press, New York, NY.

Doi, Y., Teranaka, M., Yora, K., and Asuyama, H. 1967. Mycoplasma-or PLT group-like microorganisms found in the phloem elements of plants infected with mulberry dwarf, potato witches' broom, aster yellows, or paulownia witches' broom. Japanese Journal of Phytopathology 33(4): 259–266.

Eastwood, D. 1990. Ramu stunt disease. Development and consequences at Ramu Sugar Ltd. *Sugar Cane* 2: 15–19.

Felsenstein, J. 1985. Confidence limits on phylogenies: An approach using the bootstrap. *Evolution* 39: 783–791.

Feng, C., Yin, Z., Ma, Y., Zhang, Z., Chen, L., Wang, B., Li, B., Huang, Y., and Wang Q. 2011. Cryopreservation of sweet potato (Ipomoea batatas) and its pathogen eradication by cryotherapy. *Biotechnology Advances* 29(1): 84–93.

Frison E.A. and Putter C.A.J. 1993. FAO/IBPGR Technical guidelines for the safe movement of sugarcane germplasm. *Food and Agriculture Organization of the United Nations/ International Board for Plant Genetic Resources*, Rome, Italy

Gatineau, F., Larrue, J., Clair, D., Lorton, F., Richard-Molarc, M., and Boudon-Padieu, E. 2001. A new natural planthopper vector of stolbur phytoplasma in the genus *Pentastiridis* (Hemiptera: Cixiidae). *European Journal of Plant Pathology* 107: 263–271.

Gaur, R.K., Raizada, R., and Rao, G.P. 2008. *Sugarcane yellow leaf phytoplasma* associated for the first time with sugarcane yellow leaf syndrome in India. *Plant Pathology* 57: 772.

Gunderson, D.E. and Lee I.M. 1996. Ultrasensitive detection of phytoplasma by nested-PCR assays using two universal primer pairs. *Phytopathol Mediterr* 35 144–151.

Hanboonsong, Y., Ritthison, W., Choosai, C., and Sirithorn, P. 2006. Transmission of sugarcane white leaf phytoplasma by Yamatotettix flavovittatus, a new leafhopper vector. *Journal of Economic Entomology* 99(5): 1531–1537.

Helliot, B., Panis, B., Poumay, Y., Swenen, R., Lepoivre, P., and Frison, E. 2002. Cryopreservation for the elimination of cucumber mosaic and banana streak viruses from banana (Musa spp.). *Plant Cell Reports* 2: 1117–1122.

Hodgetts, J., Boonham, N., Mumford, R., Harrison, N., and Dickinson, M. 2008. Phytoplasma phylogenetics based on analysis of secA and 23S rRNA gene sequences for improved resolution of candidate species of '*Candidatus* Phytoplasma. *International Journal of Systematic and Evolutionary Microbiology* 58: 1826–1837.

Hodgetts, J., Tomlinson, J., Boonham, N., Gonzalez-Martin, I., Nikolić, P., Swarbrick, P., Yankey, E.N., and Dickinson, M. 2011. Development of rapid in-field loop-mediated isothermal amplification (LAMP) assays for phytoplasmas. *Bulletin of Insectology* 64: S41–S42.

Holkar, S.K., Gupta, H., Kumar, A., Baitha, A., Kumar, S., and Singh, J. 2016. Diagnosis of Yellow Leaf Disease of sugarcane from Sub-tropical India. Indian Phytopathological Society-MEZ and ICAR-IIPR, Kanpur, India, (29th-30th December, S5-PP-27), p177.

Jones, P., Antoniw, J.F., and Eastwood, D. 1989. Investigations into the aetiology of Ramu stunt, a new sugarcane disease in Papua New Guinea. *L'*A *gronomie Tropicale* 44: 179–184.

Joomun, N., Suamtally-Dookun, A., Saumtally, S., and Genshan, S. 2007. Sugarcane leaf yellows phytoplasmas in Mauritius: Molecular characterization, transmission and alternative hosts. *Proceedings of International Sugar Cane Technologists* 26: 1005–1013.

Keerthipala, A.P. 2016. Development of sugar industry in Sri Lanka. *Sugar Tech* 18(6): 612–626.

Klinkong, S. and Seemiiller, E. 1993. Detection and differentiation of the mycoplasma like organism associated with sugarcane white leaf disease using cloned extrachromosomal DNA probe. *Kasetsart Journal* 27: 98–103.

Kogovsek, P., Hodgetts, J., Hall, J., Prezelj, N., Nikolić, P., Mehle, N., Lenarčič, R., Rotter, A., Dickinson, M., Boonham, N., and Dermastia, M. 2015. LAMP assay and rapid sample preparation method for on-site detection of flavescence dorée phytoplasma in grapevine. *Plant Pathology* 64(2): 286–296.

Kumar, A., Holkar, S.K., Singh, R., and Singh, J. 2020c. Natural occurrence and diagnosis of yellow leaf disease affecting sugarcane genotypes under tropical and sub-tropical conditions in India. *Journal of Environmental Biology* 41(6): 1511–1520.

Kumar, A., Holkar, S.K., Singh, R., Singh, P.K., Mitra, S., Kumar, S., Basavaraj, Y.B., and Pathak, A.D. 2021. Molecular identification of 16SrXI-B subgroup of phytoplasma related strain with yellow leaf disease of sugarcane (Saccharum officinarum) in India. *Research on Crops* 1(2): 22.

Kumar, A., Holkar, S.K., and Singh, R. 2020a. Detection of phytoplasma in yellow leaf disease affected sugarcane genotypes cultivated in subtropical conditions in India. In: 2nd International Conference and

Exhibition on Sustainability-Innovation and Diversification in Sugar and Allied Industry, Pune, Maharashtra, India, (Jan 31-Feb 2), p. ASP20.

Kumar, A., Holkar, S.K., and Singh, R. 2020b. Screening of sugarcane genotypes against yellow leaf disease in tropical and sub-tropical conditions in India. In: 7th International Conference of Phytopathology in Achieving US Sustainable development goals, India, (January 16–20, PS10, 31), p 307.

Kumar, S., Stecher, G., and Tamura, K. 2016. MEGA7: Molecular evolutionary genetics analysis version 7.0 for bigger datasets. *Molecular Biology and Evolution* 33: 1870–1874.

Kumar, S., Rao, G.P., Jadon, V.S., and Baranwal, V.K. 2018. Genetic diversity of phytoplasmas associated with sugarcane grassy shoot and leaf yellows diseases in India. *Phytopathogenic Mollicutes* 8: 74–88.

Kumar, S., Jadon, V.S., and Rao, G.P. 2017. Use of secA gene for characterization of phytoplasmas associated with sugarcane grassy shoot disease in India. *Sugar Tech* 19(6): 632–637.

Kumar, S., Rao, G.P., Jadon, V.S., and Baranwal, V.K. 2018b. Genetic diversity of phytoplasmas associated with sugarcane grassyshoot and leaf yellows diseases in India. *Phytopathogenic Mollicutes* 8(2): 74–88.

Kumar, S., Tiwari, A.K., Holkar, S.K., Duttamajumder, S.K., and Rao, G.P. 2015. Characterization of a 16SrI-B subgroup phytoplasma associated with sugarcane leaf yellows disease in India. *Sugar Tech* 17(2): 156–161.

Kumarasinghe, N.C. and Jones, P. 2001. Identification of white leaf disease of sugarcane in Sri Lanka. *Sugar Tech* 3(1): 55–58.

Kunita, L.S., Young, G.R., Pais, E., Jones, P., and Nagaraja, H. 1994. Preliminary observation on Eumetopina sp. (Hemiptera: Delpahcidae) as a vector for Ramu stunt, a new sugarcane disease in Papu New Guinea. *Journal of Australian Entomological Society* 33: 185–186.

Lee, I.M., Davis, R.E., and Gundersen-Rindal, D.E. 2000. Phytoplasma: phytopathogenic mollicutes. *Annul Review in Microbiology* 54: 221–255.

Lee, I.M., Gundersen-Rindal, D.E., Davis, R.E., and Bartoszyk, I.M. 1998. Revised classification scheme of phytoplasmas based on RFLP analyses of 16S rRNA and ribosomal protein gene sequences. *International Journal of Systematic and Evolutionary Microbiology* 48(4): 1153–1169.

Lee, I.M., Hammond, R.W., Davis, R.E., and Gundersen, D.E. 1993. Universal amplification and analysis of pathogen 16S rDNA for classification and identification of mycoplasma like organisms. *Phytopathology* 83: 834–842.

Maixner, M. and Reinert, W.. 1999. Oncopsisalni (Schranck) (Auchenorrhyncha: Cicadellidae) as a vector of the *alder yellows phytoplasma* of *Alnus glutinosa* (L.) Gaertn. *European Journal of Plant Pathology* 105: 87–94.

Mall, S., Chaturvedi, Y., Rao, G.P., and Baranwal, V.K. 2011. Phytoplasma's diversity in India. *Bulletin of Insectology* 64: 77–78.

Mall, S., Chaturvedi, Y., Singh, M., Upadhyaya, P.P., and Rao, G.P. 2007. Molecular characterization of phytoplasmas infecting weeds in Eastern Uttar Pradesh, India. *Indian Journal of Plant Pathology* 25 (1&2): 77–80.

Mall, S., Rao, G.P., and Marcone, C. 2010. Phytoplasma on weeds: Detection, taxonomy and diversity. In: Gaur, R.K., Sharma, P., Pratap, R., Sharma, K.P., Sharma, M., Dwivedi, R. (Eds) *Recent Trends in Biotechnology and Microbiology*, Nova Science Publishers, Inc., NY, USA, pp. 87–108.

Mangelsdorf, A.J. 1962. A research program for the Thailand sugar industry. *Department of Agriculture*, Bangkok 1: 16.

Manimekalai, R., Soumya, V.P., Kumar, R.S., Selvarajan, R., Reddy, K. et al. 2010. Molecular detection of 16SrXI group phytoplasma associated with root (wilt) disease of coconut (*Cocus nucifera*) in India. *Plant Disease* 94: 636.

Marcone, C. 2019. Sugarcane grassy shoot phytoplasma (grassy shoot of sugarcane). *Crop Protection Compendium*, (57164).

Marcone, C. and Rao, G.P. 2008. 'Candidatus phytoplasm cynodontis': The causal agent of bermuda grass white leaf disease. In: Harrison, N.A., Rao, G.P., Marcone, C. (Eds) *Characterization, Diagnosis and Management of Phytoplasma*, Studium Press LLC, USA, pp. 353–364.

Marcone, C., Schneide, R.B., and Seemuller, E. 2004. 'Candidatus Phytoplasma cynodontis', the phytoplasma associated with Bermuda grass white leaf disease. *International Journal of Systematic and Evolutionary Microbiology* 54: 1077- 1082.

McCoy, R.E. 1989. Plant diseases associated with mycoplasma-like organisms. The mycoplasmas 5: 545–640.

Moonan, F., Molina, J., and Mirkov, T.E. 2000. *Sugarcane yellow leaf virus*: an emerging virus that has evolved by recombination between *Luteoviral* and *Poleroviral* ancestors. *Virology* 269: 156–171.

Nakashima, K., Chaleeprom, K., Wongkaew, P., and Sirithorn, P. 1994. Detection of mycoplasma-like organism associated with white leaf disease of sugarcane in Thailand using DNA probes. *JIRCAS Journal* 1: 57–67.

Nakashima, K. and Hayashi, T. 1995. Extrachromosomal DNAs of rice yellow dwarf and sugarcane white leaf phytoplasmas. *Annals of Phytopathological Society Japan* 61: 456–462.

Nakashima, K., Hayashi, T., Chaleeprom, W., Wongkaew, P., and Sirithorn, P. 1996. Complex phytoplasma flora in northeast Thailand as revealed by 16S rDNA analysis. *Annals of Phytopathological Society Japan* 62: 57–60.

Nakashima, K., Wungkaew, P., and Sirithorn, P. 2001. Molecular detection and characterization of sugarcane white leaf phytoplasmas. In: Rao, G.P., Ford, R.E., Tosir, M. Teakle, D.S. (eds.). *Sugarcane Pathology. Volume II: Virus and Phytoplasma Diseases. Enfield (NH)*, Science Publishers, Inc, USA, p. 157–175.

Nithya, K., Parameswari, B., and Viswanathan, R. 2020. Mixed infection of sugarcane yellow leaf virus and grassy shoot phytoplasma in yellow leaf affected Indian sugarcane cultivars. *The Plant Pathology Journal*, 36:364–377. 10.5423/ppj.oa.06.2020.0092.

Nithya, K., Parameswari, B., Bertaccini, A., and Rao, G.P. 2020b. Viswanathan R. Grassy shoot: The destructive disease of sugarcane. *Phytopathogenic Mollicutes* 10(1): 10–24.

Parmessur, Y., Aljanabi, S., Saumtally, S., and Dookun-Saumtally, A. 2002. *Sugarcane yellow leaf virus* and *Sugarcane yellows phytoplasma*: elimination by tissue culture. *Plant Pathology* 51: 561–566.

Quan, M.V., Liem, N.V., Man, L.Q., Huy, N.Q., Le, X.V., Cuong, H.V., Nguyen, D.T., Thao, H.T., Hung, N.V., Huy, N.D., and Weintraub, P.G. 2020. A new phytoplasma strain associated with the sugarcane white leafdisease in Vietnam. *Phytopathogenic Mollicutes* 10(1): 60–68.

Rao, G.P. 2006. Management of phytoplasma diseases of sugarcane. In: Gadewar, A.V., Singh, B.P. (Eds.) *Plant Protection in New Millennium (Vol II)*, Satish Serial Publishing House, New Delhi, pp 1–16.

Rao, G.P. 2021. Our understanding about phytoplasma research scenario in India. *Indian Phytopathology* 23: 1–31.

Rao, G.P., Alvarez, E., and Yadav, A. 2018. Phytoplasma diseases of industrial crops. 2018. *InPhytoplasmas: Plant Pathogenic Bacteria-I* (pp. 91–121). Springer, Singapore.

Rao, G.P. and Ford E. 2001. Vectors and phytoplasma diseases 6f sugarcane - an overview, . In: Rao, G.P., Ford, R.E., Tosir, M., Teakle, D.S. (eds). *Sugarcane Pathology. Volume II: Virus and Phytoplasma Diseases*. Science Publishers, Inc, Enfield (NH), USA, p. 267–318.

Rao, G.P., Mall, S., and Marcone, C. 2012. Recent biotechnological approaches in diagnosis and management of sugarcane phytoplasma diseases. *Recent Trends in Biotechnology and Microbiology* 6(2): 19–29.

Rao, G.P., Gaur, R.K., Singh, M., Srivastava, A.K., Virk, A.S., Singh, N., Patil, A.S., Viswanathan, R., and Jain, R.K. 2000. Existence of *sugarcane yellow leaf luteovirus* in India. *Sugar Tech* 2: 37–38.

Rao, G.P., Gaur, R.K., Singh, M., Viswanathan, R., Chandrasena, G., and Dharamwardhaanhe, M.N.N. 2001. Occurrence of *sugarcane yellow leaf virus* in India and Srilanka. *Proceedings of International Society of Sugar Cane Technologists* 24: 469–470.

Rao, G.P., Raj, S.K., Snehi, S.K., Mall, S., Singh, M., and Marcone, C. 2007. Molecular evidence for the presence of 'Candidatus Phytoplasma cynodontis', the Bermuda grass white leaf agent in India. *Bulletin of Insectology* 60: 145–146.

Rao, G.P., Singh, A., Singh, H.B., and Sharma, S.R. 2005. Phytoplasma diseases of sugarcane: characterization, diagnosis and management. *Indian Journal of Plant Pathology* 23: 1–21.

Rao, G.P., Srivastava, S., Gupta, P.S., Singh, A., Singh, M., and Marcone, C. 2008. Detection of *sugarcane grassy shoot phytoplasma* infecting sugarcane in India and its phylogenetic relationships to closely related phytoplasmas. *Sugar Tech* 10: 74–80.

Rao, G.P., Thorat, V., Manimekalai, R., Tiwari, A.K., and Yadav, A. 2017. A century progress of research on phytoplasma diseases in India. *Phytopathogenic Mollicutes* 7(1): 1–38.

Rao, G.P., Tiwari, A.K., Kumar, S., and Baranwal, V.K. 2014. Identification of sugarcane grassy shoot-associated phytoplasma and one of its putative vectors in India. *Phytoparasitica* 42(3): 349–354.

Saitou, N. and M. Nei. 1987. The neighbor-joining method: A new method for reconstructing phylogenetic trees. *Molecular Biology and Evolution* 4: 406–425.

Sarindu, N. and Clark, M.F. 1993. Antibody production and identity of MLOs associated with sugar-cane white leaf disease and Bermuda grass white leaf disease from Thailand. *Plant Pathology* 42: 396–402.

Satyagopal, K., Sushil, S.N., Jeyakumar, P., Shankar, G., Sharma, O.P., Boina, D., Sain, S.K., Chattopadhyay, D., Ram Asre, Kapoor, K.S., Sanjay Arya, Subhash Kumar, Patni, C.S., Chattopadhyay, C., Pandey, A., Pachori, R., Thakare, A.Y., Basavanagoud, K., Halepyati, A.S., Patil, M.B., and Sreenivas, A.G., 2014, 49.

Scagliusi, S.M.M. and Lockhart, B.E.L. (2000) Transmission, characterization, and serology of a *luteovirus* associated with yellow leaf syndrome of sugarcane. *Phytopathology* 90: 120–124.

Schenck, S., 1990. Yellow leaf syndrome – a new sugarcane disease. Hawaiian Sugar Planters Association. *Annual Report* 38–39.

Schenck, S., Hu, J.S., and Lockhart, B.E.L. 1997. Use of a tissue blot immunoassay to determine the distribution of *Sugarcane yellow leaf virus* in Hawaii. *Sugar Cane* 4: 5–8.

Schneider, B., Seemueller, E., Smart, C.D., and Kirkpatrick, B.C. 1995. Phylogenetic classification of plant pathogenic mycoplasma-like organisms or phytoplasmas, In: Razin, R. and Tully, G. (editors), *Molecular and Diagnostic Procedures in Mycoplasmology* 1: 369–380.Academic Press, San Diego.

Schneider, B., Ahrens, U., Kirkpatrick, B.C., and Seemiiller, E. 1993. Classification of plant-pathogenic mycoplasma-like organisms using restriction-site analysis of PCR-amplified 16S rDNA. *Journal of General Microbiology* 139: 519–527.

Sdoodee, R. 2001. Recent studies on white leaf and grassy shoot phytoplasmas of sugarcane. Sugarcane pathology. Virus and Phytoplasma Diseases 2: 235–244.

Sdoodee, R., Schneider, B., Padovan, A., and Gibbs, K. 1999. Detection and genetic relatedness of phytoplasmas associated with plant diseases in Thailand. *Journal of Biochemistry Molecular Biology and Biophysics* 3: 133–142.

Seemuller, E. 1976. Investigations to demonstrate mycoplasma like organisms in diseased plants by fluorescence microscopy. *Acta Horticulture* 67: 109–112.

Seemuller, E. and Kirkpatrick, B.C. 1996. Detection of phytoplasma infections in plants. In: Tully, J.G., Razin, S. (eds), *Molecular and Diagnostic Procedures in Mycoplasmology*. Vol. If. Academic Press, San Diego, USA, p. 299–311.

Seemuller, E., Marcone, C., Lauer, U., Ragozzino, A., and Giischl, M. 1998. Current status of molecular classification of the phytoplasmas. *Journal of Plant Pathology* 80: 3–26.

Sharma, P., Singh, J., Kumar, P., Kumar, M., Kumar, R., Tiwari, A.K., and Baranwal, V.K. 2020. Emerging incidence and diversity analysis of sugarcane grassy shoot phytoplasma in Western Uttar Pradesh, India. *Phytopathogenic Mollicutes* 10(2): 194–202.

Singh, V., Baitha, A., and Sinha, O. K. 2002. Transmission of grassy shoot diseaseof sugarcane by a leafhopper (Deltocephalus vulgaris Dash andViraktamath) Indian. *Journal of Sugarcane Technology* 17(1-2):60–63.

Singh, D., Tewari, A.K., Rao, G.P., Karuppaiah, R., Viswanathan, R., Arya, M., and Baranwal, V.K. 2009. RT-PCR/PCR analysis detected mixed infection of DNA and RNA viruses infecting sugarcane crops in different states of India its phylogenetic relationships to closely related phytoplasmas. *Sugar Tech* 11: 373–380.

Smith, G.R., Borg, Z., Lockhart, B.E., Braithwaite, K.S., and Gibbs, M.J. 2000. *Sugarcane yellow leaf virus*: a novel member of the Luteoviridae that probably arose by inter-species recombination. *Microbiology* 81(7): 1865–1869.

Soufi, Z., Sakuanrungsirikul, S., Wongwarat, T., Hamarn, T., Srisink, S., and Komor, E. 2013. Sugarcane yellow leaf symptomatic plants in Thailand are infected by *white leaf phytoplasma*, not by *leaf yellows phytoplasma. Australasian Plant Pathology* 42: 723–729.

Srivastava, S., Singh, V., Gupta, P.S., Sinha, O.K., and Baitha, A. 2006. Nested PCR assay for detection of sugarcane grassy shoot phytoplasma in the leafhopper vector Deltocephalus vulgaris: a first report. *Plant Pathology* 55(1): 25–28.

Tamura, K., Nei, M., and Kumar, S. 2004. Prospects for inferring very large phylogenies by using the neighbor-joining method. *Proceedings of the National Academy of Sciences, USA*, 101: 11030–11035

Tiwari, A.K., Vishwakarma, S.K., and Rao, G.P. 2012. Increasing incidence of sugarcane grassy shoot disease in Uttar Pradesh, India and its impact on yield and quality of sugarcane. *Phytopathogenic Mollicutes* 2(2):63–67.

Tomlinson, J.A., Boonham, N., and Dickinson, M. 2010. Development and evaluation of a one-hour DNA extraction and loop-mediated isothermal amplification assay for rapid detection of phytoplasmas. *Plant Pathology* 59(3): 465–471.

Tran-Nguyen, L., Blanche, K.R., Egan, B., and Gibb, K.S. 2000. 84 Diversity of phytoplasmas in northern Australian sugarcane and other grasses. *Plant Pathology* 49: 666–679.

Vega, J., Scagliusi, S.M., and Ulian, E.C. 1997. Sugarcane yellow leaf disease in Brazil: evidence for association with *a Luteovirus. Plant Disease* 81: 21–26.

Viswanathan, R. 1997. Detection of phytoplasmas associated with grassy shoot disease of sugarcane by ELISA techniques. *Z. PflKrankh. PflSchutz* 104: 9–16.

Viswanathan, R. 2001. Serodiagnosis of phytoplasmas causing grassy shoot disease in sugarcane. In: Rao, G.P., Ford, R.E., Tosir, M., Teakle, D.S. (eds.) *Sugarcane Pathology. Volume II: Virus and Phytoplasma Diseases*. Science Publishers. Inc, Enfield (NH), USA, p. 209–220.

Viswanathan, R., 2002. Sugarcane yellow leaf syndrome in India: incidence and effect on yield parameters. *Sugar Cane International* 20(5): 17–23.

Viswanathan, R. and G.P. Rao. 2011. Disease scenario and management of major sugarcane diseases in India. *Sugar Tech* 13(4): 336–353.

Viswanathan, R., Chinnaraja, C., Malathi, P., Gomathi, R., Rakkiyappan, P., Neelamathi, D., and Ravichandran, V. 2014. Impact of *Sugarcane yellow leaf virus* (SCYLV) infection on physiological efficiency and growth parameters of sugarcane under tropical climatic conditions in India. *Acta physiologiae plantarum* 36: 1805–1822.

Viswanathan, R., Malathi, P., and Neelamathi, D. 2018. Enhancing sugarcane yield per hectare through improved virus-free seed nursery programme. *Icar News* 24: 4–5.

Viswanathan, R., Padmanaban, P., Mohanraj, D., Sundar, A. Ramesh, and Premachandran, M.N. 1999. Suspected yellow leaf syndrome in sugarcane. *Sugarcane Breeding Institute Newsletter* 18(3): 2–3.

Vu, N.T., Pardo, J.M., Alvarez, E., Le, H.H., Wyckhuys, K., Nguyen. K.L., and Le, D.T. 2016. Establishment of a loop-mediated isothermal amplification (LAMP) assay for the detection of phytoplasma-associated cassava witches' broom disease. *Applied Biological Chemistry* 59(2): 151–156.

Wang, Q., Liu, Y., Xie, Y., and You, M. 2006. Cryotherapy of potato shoot tips for efficient elimination of Potato leaf roll virus (PLRV) and Potato virus Y (PVY). *Potato Research* 49: 119–129.

Wang, Q.C., Bart, P., Engelmann, F., Lambardi, M., and Valkonen, J.P.T. 2009 Cryotherapy of shoot tips: a technique for pathogen eradication to produce healthy planting material and for cryopreservation of healthy plant genetic resources. *Annals of Applied Biology* 154: 351–363.

Wang, Q.C., Munir, M., Li, P., Gafny, R., Sela, I., and Tanne, E. 2003. Elimination of grapevine virus A (GVA) by cryopreservation of in vitro-grown shoot tips of Vitis vinifera L. *Plant Science* 165: 321–327.

Wang, Q.C. and Valkonen, J.P.T. 2008. Efficient elimination of sweetpotato little leaf phytoplasma from sweetpotato by cryotherapy of shoot tips. *Plant Pathology* 57: 338–347.

Wang, X.Y., Li, W.F., Huang, Y.K., Zhang, R.Y., Luo, Z.M., Shan, H.L., Yin, J., Shen, K., Jia, Y.M., and Bai, Z.G. 2014. Analyses of the 16S-23S intergenic region of the phytoplasma causing the sugarcane white leaf disease in Yunnan Province, China. *Tropical Plant Pathology* 39: 184–188.

Wilson, M.R. and Turner, J.A. 2010. Leafhopper, Planthopper and Psyllid vectors of plant disease. Amgueddfa Cymru–National Museum Wales. Vectors, March 29, Available Online: http://naturalhistory.museumwales.ac.uk/

Wongkaew, P., Hanboonsong, Y., Sirithorn, P., Choosai, C., Boonkrong S., Tinnangwattana, T., Kitchareonpanya, R., and Damak, S. 1997. Differentiation of phytoplasmas associated with sugarcane and gramineous weed white leaf disease and sugarcane grassy shoot by RFLP analysis and sequencing. *Theoretical and Applied Genetics* 95: 660–663.

Yadav, A., Thorat, V., Deokule, S., Shouche, Y., and Prasad, D.T. 2017. New subgroup 16SrXI-F phytoplasma strain associated with sugarcane grassy shoot (SCGS) disease in India. *International Journal of Systematic and Evolutionary Microbiology* 67(2): 374–378.

14 The Improvement of Sugarcane (*Saccharum officinarum L.*) for Sugar, Ethanol and Biofuel Production Through Innovative Biotechnology: A Perspective View on its Scope, Importance & Challenges

Pramod N. Nair
Department of Forestry, Fiji National University, Fiji

H.K. Sachan
Department of Crop Science, Fiji National University, Fiji

CONTENTS

14.1 INTRODUCTION

Sugarcane is a grass native to Asia and grown primarily in the tropics and subtropics, where it has grown for over 4,000 years. It is referred to as a group of lofty grass species of the tropical region and is used for sugar production (Paterson et al. 2013). This parched crop required a lot of water for its survival. Its ability to remain in the soil for one year makes a major impact on many environmentally receptive regions where the sugarcane grows. Due to sugarcane's ability to withstand in soil for a year or more, the planting of sugarcane throughout the world has had an important impact on the biodiversity of plant species. Sugar is prevalent in the modern diet and is increasingly a source of ethanol and bioplastics. Sugarcane was one of the grass species that motivated humans to build up sugar-production technology (Goldstein and Mintz 2015), where they crushed and boiled for the extraction

DOI: 10.1201/9781003292425-14

of sugarcane juice (Gopal 1964). While considering the development of technology worldwide, sugar production technology has not altered much to date, but sugarcane production exceeds other food crops, such as wheat, rice, and maize, etc. (FAO 2017).

Even though humans have been using sugarcane, the scientific community has given the least attention to this perennial tropical grass, and its physiological and biochemical characteristics were not studied in detail until the C4 cycle of photosynthesis was revealed (Hatch et al. 1967; Hatch 2005). Sugarcane genomic studies lagged behind other grass species, such as rice, wheat sorghum, etc. Then, after years, the sugarcane genome, a polyploid genome, was discovered to be very complex among all plant species (Thirugnanasambandam et al. 2018). Currently, sugarcane research has attained its peak stage because it contemplates using biofuel crops as a replacement for oil (Souza et al. 2011; Hoang et al. 2015a). The very complex genome in sugarcane was a barrier to the scientific community for genetic analysis and sequencing until recenlty, but through recent innovations, scientists have overcome the technological difficulties linked with sugar, bioenergy, and biomaterial production from sugarcane using advanced genetic perspectives (Hotta et al. 2010; Furtado et al. 2014). Scortecci et al. (2012) revealed that the sub-genomic and sequencing studies in the important traits like fiber or sugar will help execute the genomic-assisted breeding, especially for the end-product-specific genotypes. The above methods are required and helpful to alleviate some of the nastiest effects of climate change and guarantee feasible sugarcane production in the future (Abberton et al. 2016).

The sugarcane yield decreased worldwide due to continuous monoculture cropping, which may lose its ability to fight against soil pathogens and other nematode attacks (Stirling et al. 2001). Extreme weather can also have significant impacts on sugarcane yield. Researchers in a country like Fiji, where 5.2M tons of production were shown in 1994, showed significantly reduced production in subsequent years (Gawander 2007). Due to the above condition, it is very important to advance good management practices for the development of genetically improved cultivars to enhance the productivity of sugarcane worldwide. It is reported by Stevenson (1965) and Hatch et al. (1967) that in the late 1800s, the first sugarcane breeding programs were established in Java and Barbados. Up to the earlier stages of the 20th century, *S. officinarum,* or noble cane sugar from Papa New Guinea, were commonly used for industrial bulk production (Yadav et al. 2020). Yadav et al. (2020) said that the earlier genetic improvement practices of sugarcane were classified into three stages: intercrossing, interspecific hybridization, and backcrossing between two species, such as *S. officinarum* and *S. spontaneum.* The above genetic-improvement practices were helpful to increase the yield, disease resistance, and ratooning ability of sugarcane. Thirugnanasambandam et al. (2018) revealed that the sugarcane species is an apolyploid genome, and other scientists reported that the existing cultivars may be formed from hybridization between these two species, *S. officinarum* and *S. spontaneum* (Daniels and Roach 1987; Moore et al. 2013). The above assumptions were supported by the whole-genome sequencing studies in sugarcane, which found that the hybrid's genome size is 10 Gb (D'Hont and Glaszmann 2001). Molecular evidence suggests that polyploidy genomes can make dynamic changes in DNA sequences and gene expression and this reflects in epigenetic changes in homologous genes in chromosomes due to inter-genomic interactions (McClintock 1984). The polyploid species are evolutionary successful and have a higher phenotypic freshness than other diploid species or their parents (Ramsey and Schemske 2002). While considering other factors of sugarcane polyploidy genome, this increase in the capacity for phenotypic variation capacity may be caused due to the control of more alleles in the species (Birchler et al. 2005).

14.2 SCOPE AND IMPORTANCE OF BIOTECHNOLOGY APPLICATIONS IN THE IMPROVEMENT OF SUGARCANE FOR SUGAR PRODUCTION

As a result of biotechnology applications worldwide, genetically modified soybean, corn, canola, and cotton have been grown commercially for more than 20 years. Genetically modified (GM) tomato, rice, papaya, sugar beet, pumpkin, potato, alfalfa, and aubergine are also used by farmers worldwide. Globally, sugarcane is grown in tropical and subtropical countries and contributes 70%

of sugar production. The most important sugar cane producing countries are Brazil, India, China, and Thailand, and among those countries, Brazil contributes 40% of the world's sugarcane. In recent years, the cultivation of genetically modified Bt sugarcane began in Brazil, at industry named CTC (Centro de TecnologiaCanavieira), and was approved by the government for cultivation in 2017; later, it was developed for commercial cultivation (Pocket 45 ISAAA).

A genetically modified glyphosate-resistant sugar beet was developed by Monsanto and KWS SAAT SE & Co in 2000; this variety in was commercialized in 2007. This genetically modified, U.S.-grown sugar beet increased to just about 95% (Dillen et al. 2013). America approved the GM sugar beets, and people are consuming GM sugar beets because the research studies revealed that the sugar from GM sugar beets is molecularly identical to typically grown sugar beets and has the same nutritional value (Klein et al. 1998; Oguchi et al. 2009). Genetic manipulation is being conducted to increase the sucrose content of sugarcane. For conducting this experimental work, the scientist should understand many biochemical reactions, especially the enzymatic reactions in sugar for the accumulation of sugar production and its storage in sugarcane stumps (Figure 14.1). The above-mentioned key enzymatic reactions were regulated by using genetic-engineering mechanisms to increase sugar production (Patrick 2013; Groenewald and Botha 2008). Brazil and Indonesia are the two countries that approved genetically modified sugarcane for commercial cultivation; in 2018, Canada also approved the sale of genetically modified sugar.

The C4 plants including sugarcane do not have a photorespiration process. This different mechanism of photosynthesis is linked with the increase in the production of sugar (Figure 14.2). This trait is very important for tropical and subtropical countries where the rate of photosynthesis increases with increased light and temperature. In sugarcane like C4 plants, chlorophyll-containing photosynthetic tissues are arranged in a bundle sheath and mesophyll cells. Phosphoenolpyruvate carboxylase (PEPC) receives carbon dioxide molecules and initiates the carbon-fixation process in the mesophyll cells. PEPC catalyzes and adds one molecule of CO2 to phosphoenol pyruvate (PEP) and forms a four-carbon compound called oxaloacetate. In the same mesophyll cells, oxaloacetate converts to another four-carbon molecule called malate. This malate has the ability to move to a bundle sheath through a small channel called plasmodesmata and then enter the Calvin cycle and produce a new molecule called pyruvate and produce glucose (Du et al. 2000).

Globally, the attention in sugarcane production has increased considerably in the last 30 years. Even more, production is expected to increase due to biotechnology research advancements. Sugarcane's polyploidy genome, fine gene pool, composite metabolic reactions, long breeding cycle, rare flowering, and complex environmental interactions make this crop more complex, and therefore, harder to achieve progress in natural genetic improvement. Therefore, a better alternative is to adopt biotechnology tools and produce GM sugarcane that contains resistant genes to protect from biotic and abiotic stresses; in the future, this use may lead to an increase in the production of sugarcane globally. The recent studies conducted in India revealed that sugarcane is sensitive to salt, drought, and biotic stress. Therefore, it leads to a significant reduction in the yield (Nasir et al. 2000). Biochemically, it is proven that sugarcane is a glucophyte, so it grows healthily in soils with a low content of sodium salts and is very sensitive to salt (Suprasanna 2010). Transgenic sugarcane production is a good alternative to overcome all the above stress because it can integrate desired genes and make up resistance and yield traits in a short period.

The response of sugarcane to plant-tissue culture is remarkable, and tissue culture is finding a more important and integral role in plant genetic improvement. The best way to ensure the production of disease-free and resistant sugarcane for commercial cultivation is plant-tissue culture. It is required to propagate the plants in a colossal amount *in vitro,* and this micropropagation increases the sugarcane productivity globally. This technology can also be utilized for supplying new varieties of sugarcane to farmers for cultivation within a short time (Salokhe 2021). The sugarcane germplasm conservation worldwide has many resistant and non-resistant varieties; currently, sugarcane breeders are in a strong position to do breeding and release it to farmers and industry because they can achieve this through molecular breeding. This molecular breeding, combined

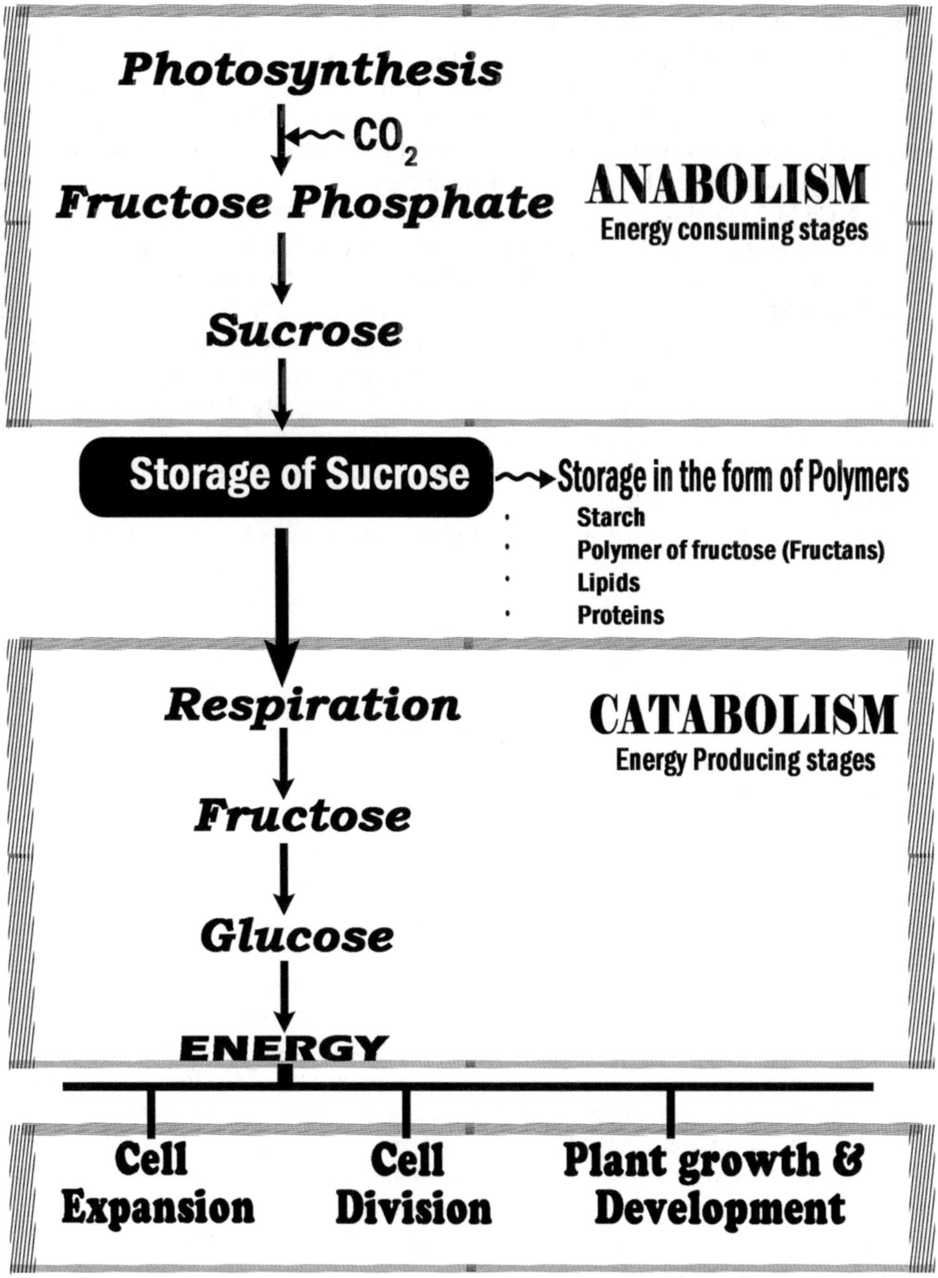

FIGURE 14.1 Sugar accumulation, storage, and its usage in sugarcane.

with *in vitro* propagation, makes the task easy to attain in a short period. Another method for genetic transformation through *in vitro* culture is somatic embryos and can serve as an ideal material forin *vitro* propagation and plant regeneration. Sugarcane plantlets derived from somatic embryogenesis show a uniform growth pattern (Suprasanna 2010). The scope and importance of genetic engineering of the sugarcane cultivars can be useful in the future for producing medicinal compounds, artificial sweeteners, nutraceuticals, useful pigments, and enzymes.

With the advent of recombinant DNA technology or genetic engineering, it is now possible to clone or integrate the desired gene to plants that, coupled with *in vitro* propagation systems, have

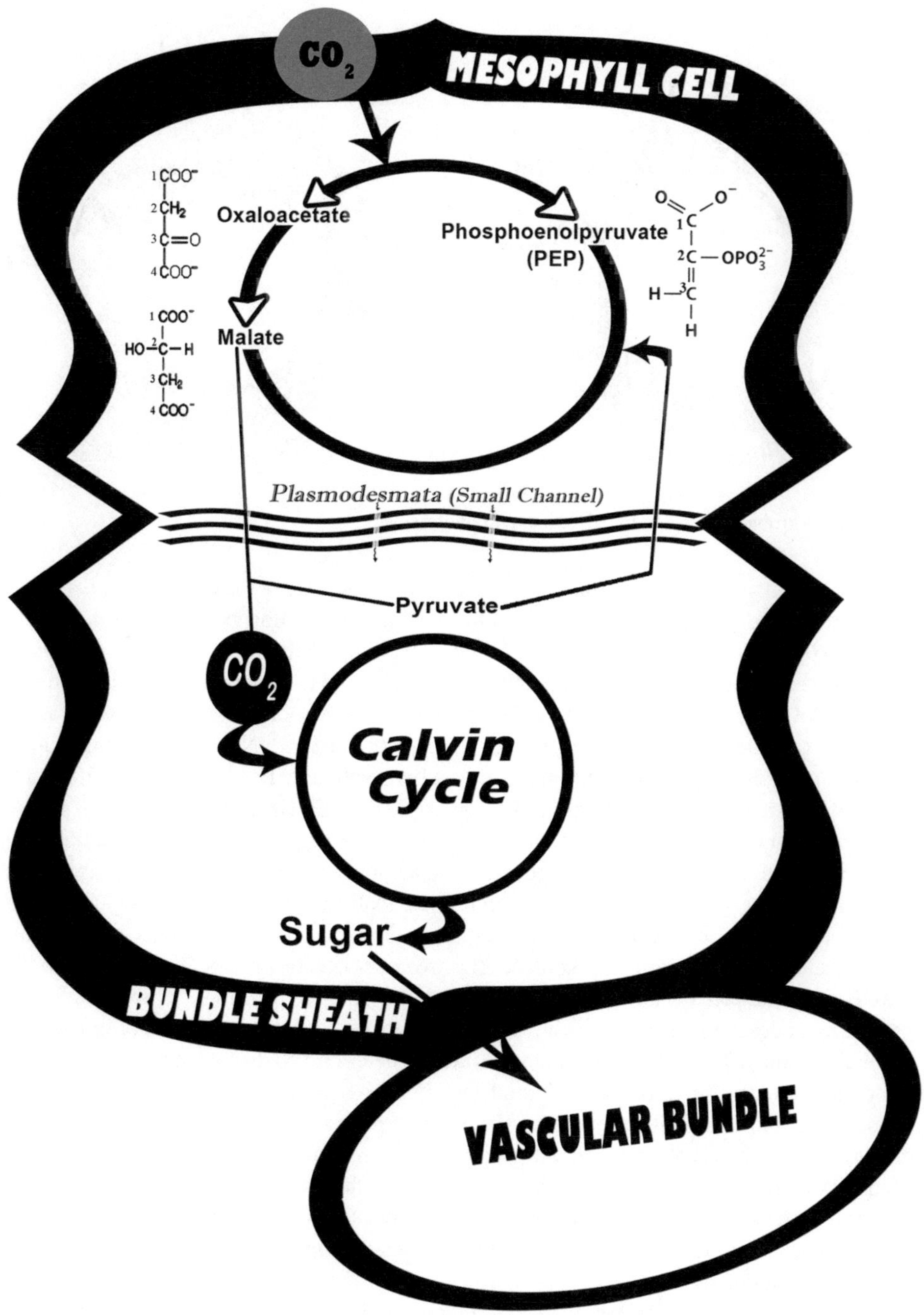

FIGURE 14.2 Mechanism of Photosynthesis in C4 plant sugarcane.

made a huge success in *Agrobacterium*-mediated gene transfer (Figure 14.3). The major work was the transfer of insecticidal crystal protein of *Bacillus thuringiensis* to sugarcane to fight against larvae of cane borers (Schuler et al. 1998).

Another method of transformation is the fusion of protoplast and the joining of naked DNA in presence of polyethylene glycol, PEG in short (Dinesh et al. 2017). PEG is found as a hydrated polymer; it removes water molecules between them and attracts them between plasma membrane,

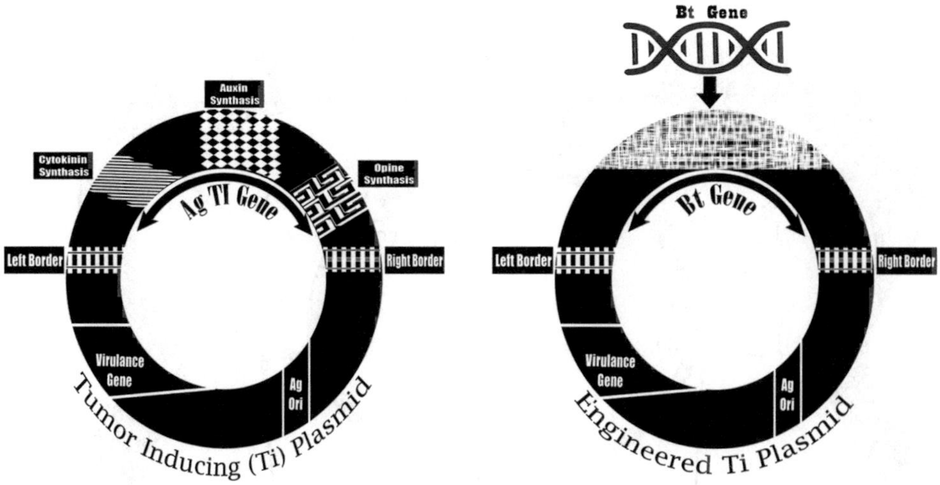

FIGURE 14.3 Agrobacterium mediated transfer in sugarcane.

then inducing the fusion between lipid-bilayers of protoplast cells because the cell walls are enzymatically degraded (Figure 14.4). The plasmodesmatal connection between the adjacent cells enlarge due to the removal of the cell wall; they allow the entry of cell organelles and other biomolecules, including the genes (Chaudary et al. 2018). The importance of this technique in sugarcane is that for crop improvement, it is extremely efficient in this species (Aftab and Iqbal 2001). Silicon carbide fibers (Sic fibers) were used to transform DNA by vortexing vigorously without any specialized types of equipment. Sic fibers help the DNA to penetrate the plasma membrane of sugarcane cells and transfer the foreign DNA (El-Tayeb NSM 2008).

Other than the above techniques, there are three common practices in the transformation of genes in sugarcane: particle-bombardment gene gun, microinjection, and electroporation (Figure 14.5). The particle-bombardment gene gun is another technique that can be useful for sugarcane for the transformation and improvement of the species. This biolistic process for gene construction and transfer is a valuable application in sugarcane improvement because it can be used for the visualization and quantification of the newly expressed gene and its functions in specific tissues of sugarcane. Electroporation and electrofusion are the two techniques that have a major impact on gene transfer in all organisms, including sugarcane. Arceniba et al. (1995) discovered an efficient method in commercial varieties, such as PoJ 2878 and Ja 605 based on electroporation from embryogenic calli, which was biochemically proven by the GUS reporter assay system and DNA sequence identification by southern blotting. The advantage of electroporation is that all the cells may have equivalent physiological characters after gene transfer, and the efficiency of the DNA shift is found high, always based on the electric field intensity and the concentration of the electroporation medium. Several experiments and research proved earlier that the electroporation methods are useful for sugarcane transformation.

The gene transformation process by the particle-gene bombardment gun method was used in callus tissues (Franks and Birch 1991). The Cry 1Ab gene was expressed efficiently against borer diseases in sugarcane plants (Barga et al. 2003). Several transgenic sugarcane lines have been developed, expressing desired genes against borers. Mainly, Cry1Ab, aprotinin gene, and Cry 1Aa3 were evaluated, and expression is thoroughly proven by serological and molecular techniques (Arvinth et al. 2009; Christy et al. 2009; Kalunke et al. 2009). Basnayake et al. (2011) reported the controlled tissue culture conditioning a different set of Australian sugarcane cultivars for effective genetic transformation using the particle-bombardment technique. The Chitinase gene is an alternative means to prevent fungal damage, which shows resistance against fungal disease in sugarcane cultivars in Thailand (Khamrit et al. 2012).

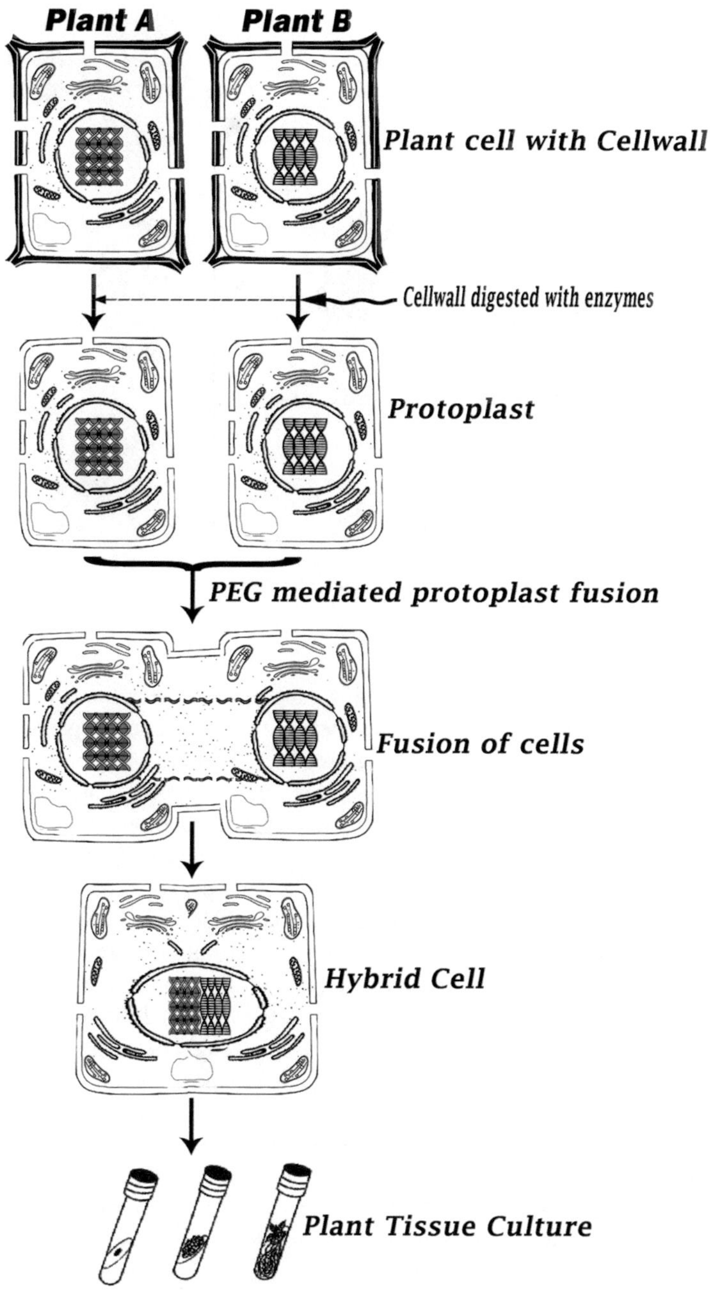

FIGURE 14.4 PEG-mediated protoplast fusion.

The current trends in sugarcane breeding programs are mainly focused on developing cultivars for sugar production and making resistant varieties. The existing sugarcane breeding programs follow the earlier hybridization and continue the same (Berding et al. 2007; Yadav et al. 2020). These limitations are due to the lack of genetic tools and modern tools for transformation and identification of desired genes. Next, they will study the expression of those genes in the transgenic cultivars (Botha andMoore 2013). Molecular breeding in sugarcane includes the collection of germplasm, identification of genes, transformation, evaluation, and selection. The ultimate goal of molecular phenotyping or breeding is to discover desired genes that could be transformed to improve quality traits in

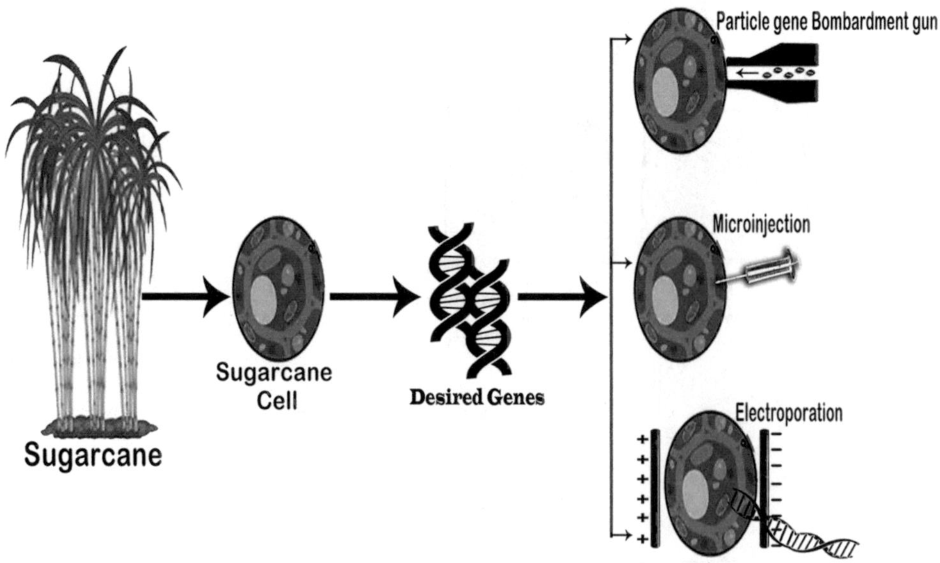

FIGURE 14.5 Techniques for transfer of genes.

existing varieties using recombinant DNA technology and plant tissue culture. Through this method, it is possible to integrate certain genes of different species, which is new to sugarcane germplasm and is considered a major advantage to the sugarcane breeding programs.

Sugars consist of pure sucrose, and the impurities present in sugar are ashes, soil dust, and moisture. Modern technologies can identify the DNA and the proteins in the sugar, and the transgene is also considered as another impurity by certain agencies. However, it is found negligible and harmless to people, and no regulatory agency in the world considers them relevant for the determination of the quality of sugar. Genetically modified sugarcane should be considered identical to sugar produced from conventional sugarcane, but it is referred to as genetically modified sugarcane because it contains approximately 40,000 genes (Nascimento et al. 2019) in the diploid chromosome. Two or three genes are manipulated by transformation, nearly identical to the parent, but this cane is referred to as genetically modified sugar because it is completely distinct from the other conventional varieties with certain characteristics.

Sugar is one of the important ingredients in human food and the food industry; therefore, research and experiments will continue to increase the production of sugarcane by molecular breeding, genetic transformation, and tissue-culture technology. As per the earlier studies, the global sugar production increased to 181 metric tons for the 2019–2020 harvest (Lajolo et al. 2020). The above research clearly indicates that sugar consumption and the demand is increased worldwide, and also points toward the importance of genetic improvement of sugarcane to attain its demand. The adoption of GM sugarcane has a positive impact and gives scope for satisfying the restrictive environmental concerns of the farmers and consumers.

14.3 SCOPE AND IMPORTANCE OF BIOTECHNOLOGY APPLICATIONS IN THE IMPROVEMENT OF SUGARCANE FOR ETHANOL AND BIOFUEL PRODUCTION

Biofuels are gaining worldwide attention as fossil fuels are reducing drastically. Sugarcane is one of the crops that has a great potential for the production of ethanol as biofuel; the carbohydrate can convert to ethanol as favorable energy. Sugarcane ethanol is a clean, affordable, and low-carbon biofuel; sugarcane ethanol has emerged as a leading renewable fuel for the transportation sector.

Brazil is the first country to start biofuel production from sugarcane and has encouraged the upgrade of the technology to the industrial level for the production of an alternative fuel with less pollution as a natural alternative to fossil fuels (Dias De Oliveira et al. 2005). For the production of ethanol as a sugarcane biofuel, genetic and genomic researches are relevant and important. Currently, the genetic tools and researches are well developed, and the researchers are studying the complex genome of sugarcane's ethanol-forming mechanism. These studies may be effective for the transformation of sugarcane biomass to biofuel production through fermentation and will also help to do the genetic improvement of this grass as a bioenergy crop. At the University of Illinois, a group of researchers has proved that genetically modified sugarcane can produce more sugar, which could be used for ethanol production by fermentation methods. The plant biomass includes the cell wall components such as cellulose, hemicellulose, and lignin, which can be altered to biofuels as an alternative to fossil fuels. These biofuels are not only renewable sources of energy but also reduce air pollution (Matsuoka et al. 2009). The production of ethanol from sugarcane is considered a successful second-generation fuel with low production cost (Yuan et al. 2008; Botha 2009).

The general process of ethanol production from the sugarcane is simple because it mainly consists of sugar in the form of disaccharide (sucrose), which is readily fermented into ethanol by yeast (*Saccharomyces cerevisiae*). The process starts when sugarcane stalks are crushed to extract sugar juice and then delivered to a fermentation tank where the yeast fermentation reaction occurs to generate ethanol (Huang et al. 2020). The ethanol produced from sugarcane is an alcohol-based fuel produced by the fermentation of sugarcane juice and molasses (Figure 14.6). It is a very pure form of ethanol, and so it never causes any pollution if used as a fuel. After fermentation, the leftover contains a solid residue and a liquid residue. The solid residue is bagasse or waste fibrous residue that can also be used to generate heat/electricity, and this bagasse contains 45–50% moisture. The liquid residue is the vinasse or the fermentation broth that contains 5–12% ethanol, which can now become beer. This beer is brought back to the distillation column, and ethanol is recovered at the bottom of the distillation unit with a purity of 92–95%. Then, removal of residual water is carried out by a further process, and finally, a fuel-grade 100% anhydrous ethanol is produced (Rodrigues and Hu 2017).

The world is improving technology every day, and scientists are determined to help mankind by fulfilling their needs and uplifting living standards through the genetic improvement of plant

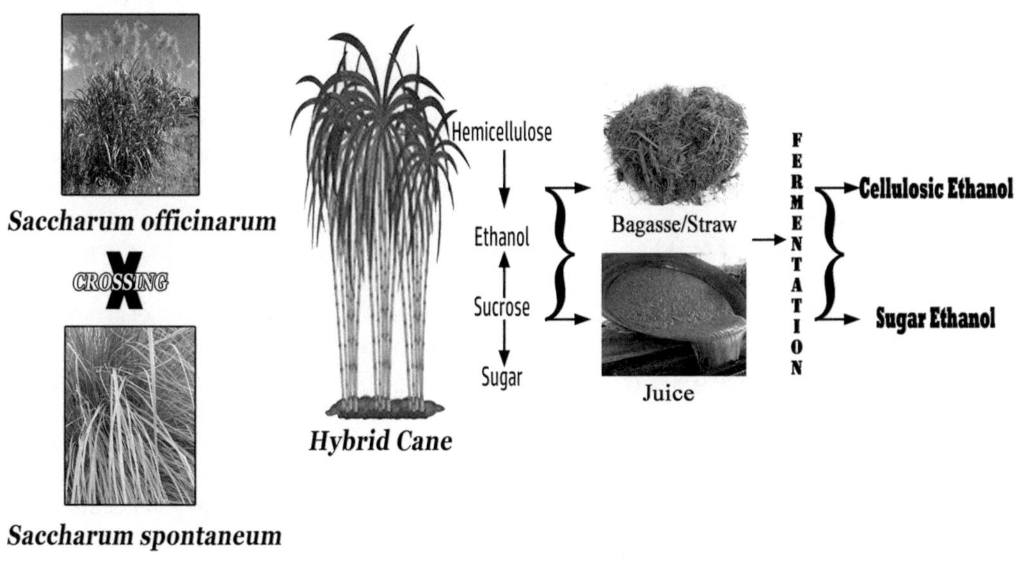

FIGURE 14.6 Fermentation in sugarcane.

genomes, especially in the agricultural sector by making transgenic crops or GM crops. The demand for biological products is increasing every day; likewise, genetically modified sugarcane can be utilized for enhancing the production of sucrose and ethanol (Arruda 2012). Genetic manipulation of growth hormones and biomass pathways of the cell wall components make sugarcane an efficient C4 grass in biomass accumulation; biofuels can be produced widely from its soluble sugar and other lignocellulosic components (Tew and Cobill 2008; Macrelli et al. 2014; Furtado et al. 2014). India and America are the two countries that hold the largest germplasm collections of the *Saccharum* species, and these germplasms can be potentially exploited for breeding purposes to improve sugarcane for the exploitation of advanced biofuel characters (Todd et al. 2014).

The wild varieties of any species show more genetic variations; likewise, wild sugarcane also shows more genetic variability than the domesticated cane. Sugarcane parental species and their diversity in genome content have been exploited by the usage of molecular markers and make it easy to understand the complex genome structure of sugarcane (Hoang Nam et al. 2015). DNA-based molecular markers could be a useful tactic in the case of sugarcane research to study complex hybrids and to do genetic manipulation for biomass production (Henry et al. 2012). The genetically modified sugarcane-producing hydrolytic enzymes may produce ethanol because the currently available grasses do not require much chemical treatment to break down composite cell wall polymers such as lignin, cellulose, and hemicelluloses into sugars and then fermented to produce ethanol. Most biotechnology researches are focusing to isolate the gene from bacteria and fungi-producing enzymes that degrade the plant cell wall.

As a result of the above researches, the scientist has identified three enzymes that can degrade the cellulose to glucose: endo-glucanases, cellobiohydrolases, and β-glucosidases. Recently, a synthetically designed gene of endo-glucanases (EG) enzymes derived from the bacteria has been introduced in the sugarcane. This EG catalysis of the internal portions of the cellulose polymer creates smaller chains and is easy for biofuel production (Willis et al. 2016). The caffeic acid O-methyltransferase (COMT) degrades lignin compounds in the normal sugarcane varieties; 6–12 percentages were observed in transgenic sugarcane plants. This reduction of lignin improves approximately 20% the sugar production, followed by ethanol production in fermentation (Jung et al. 2013). These types of recent discoveries gained a name to sugarcane as "energy cane" or biorefinery, especially focusing on high fiber and high biomass varieties for the production of ethanol and other biological products (Figure 14.7) (Landell and Bressiani 2010; Knoll et al. 2013). Due to the high fiber and biomass gifts, these sugarcane varieties have an excellent tillering and rationing ability also (Tew and Cobill 2008).

The breeding in sugarcane is challenging due to its complex polyploid genome. Each cultivar of sugarcane has a sole set of chromosomes (Grivet and Arruda 2002); therefore, separate alleles at different locus make it challenging to reveal the characteristics of sugarcane in conventional breeding programs (Matsuoka et al. 2009). It is very important in the sugarcane industrial development for the long-term conventional breeding process to be a swing to molecular breeding and the sugarcane tissue culture for release as commercial lines or cultivars. When conventional breeding swings to molecular breeding, the research should more focus on improving the important traits to increase the yield of sugar, bioenergy, and biochemical products, resistance to biotic andabiotic stress, etc. (Waclawovsky et al. 2010). Conventional sugarcane breeding depends on the same family and genetic constituents, but in advanced technology, the cross transformation may be possible with closely related species or else from the wild varieties (Tabasum et al. 2010). The intimately linked species such as *Miscanthus* (silvergrass) and *Erianthus* (sweetcane) have already been used for the genetic improvement of sugarcane, and the results are good, not only in the increase of biomass production but also in the resistance characteristics (De Siqueira Ferreira et al. 2013). Marker-assisted selection in the sugarcane breeding for the precision in gene identification and transformation, DNA sequencing, phenotyping, genotyping, and quantity trait locus (QTL) mapping to improve breeding programs and is relevant to the commercial production of biofuels (Lam et al. 2009; Berkman et al. 2012).

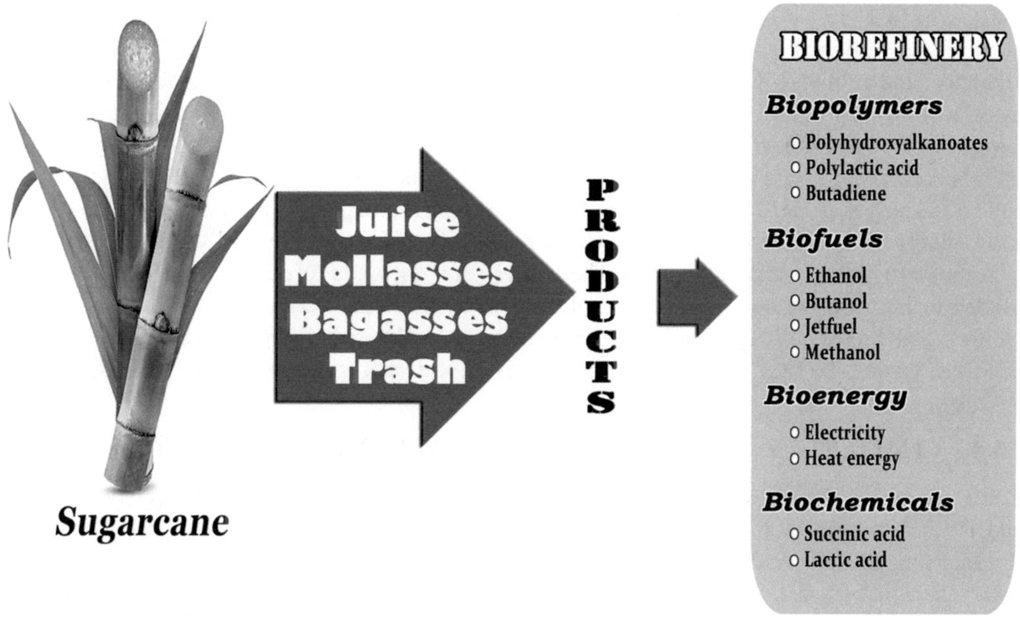

FIGURE 14.7 Sugarcane as energy cane or biorefinery.

Sugarcane plants are utilized to produce sugar and molasses fractions for alcoholic fermentation. Conventional breeding is too slow and has been done for several years to retain the sugar content in all hybrids. Approximately 80%–90% of the genes in currently cultivated sugarcane varieties originated from the primitive cane species, such as *S. officinarum and S. Spontanum.* Any crossbreeding is done with a thought to produce superior varieties compared to the parents, and for doing so, the most difficult task is the process of selecting the parents. In current stages, the selection is marker-assisted selection or molecular selection leads to saving much in costs and time for the completion of breeding and the release of new breeds. The plant cell walls are very thick and need cellulose-degrading enzymes to the extraction of more ethanol during fermentation. Now through genetic modification, sugarcane is produced with in-built cell wall-degrading enzymes. Alfalfa is a genetically modified variety of sugarcane that stopped the lignin biosynthesis; almost twice the production of sugar is extracted from it. This research is successful and beneficial for the biomass production in sugarcane (Chen and Dixon 2007).

Water is the main limiting factor in sugarcane production in almost all sugar-producing tropical and non-tropical countries. In many regions, India included, drought tolerance is under development in sugarcane in Brazil, Australia, and Mauritius using genetic modification (Singh et al. 2006). A sugar named trehalose is protecting the cell structures of several organisms from damage due to dehydration. The gene responsible for this sugar is trehalose synthase, present in *Grifolafrondosa*, a mushroom species; this gene is identified as a recent discovery to recover sugarcane from drought, and the presence of this enzyme helps to produce more concentration of sugar in cane varieties (Zhang et al. 2006). Another research study about the salt resistance in sugarcane is continuing; the salt-tolerance genes are identified from natural salt-tolerant plant species *Sueada salsa*. These genes are used in sugarcane for genetic modification and the development of transgenic salt-resistant sugarcane (Zhang etal. 2006; Li et al. 2007; Wong et al. 2007). Antarctic hair grass (Deschampsiaantarctica) from the antarctic region is a grass species like sugarcane that can withstand –30°C; the gene from this grass may utilize to produce the genetically modified sugarcane (John et al. 2009).

High-precision genetically modified sugarcane increases the scope and could now provide an interesting new solution to stress-related problems in the future. In GM sugarcane, the resistance, fuel productivity, and eco-efficiencies are increased and currently classified as bioenergy plants. The varieties of sugarcane combine the genes of trehalose synthase, drought, and salt tolerance, and cellulose-degrading capacity may be frequently cultivated in sugarcane production areas worldwide (Tammisola 2010). Fuel energy such as fossil fuel or biofuel is an inevitable necessity of human life. Demand for it will be increasing, not decreasing. The availability of fossil fuels is decreasing every day, and the price of fossil fuels is increasing worldwide. So the world is enthusiastically waiting for an alternative fuel to meet the world's fuel demand. Sugarcane ethanol and biofuels are eco-friendly and sustainable energy sources from nature, and sugarcane is a major contributor to biofuel, next to sorghum. There is a high scope to do research, and efforts are mainly needed to increase the commercial biofuel production from sugarcane. Recent advancements in biology and biotechnology, combined with other technological disciplines, have opened new horizons, paving the way to develop sugarcane as a future energy crop in place of fossil energy.

14.4 CHALLENGES IN THE GENETIC IMPROVEMENT OF SUGARCANE

Conventional breeding was the first step of genetic modification, and this method was reported years ago. Consecutive research and developments in science and technology made us stand on a platform for upraising GM crop cultivation worldwide. The challenges in the production of GM crops commercially start with its safety for human use. Repeated studies on GM crops by the scientific community have declared them as safe for human consumption, and based on those reports, the regulatory authorities are releasing the GM crops, which are safe to the ecosystem and human health (Lang and Hallman 2005). Trust from the consumers is very important, and achieving that trust is challenging among all producers of GM sugar. The trust should achieve from three parties, such as farmers, merchants, and the public. The above parties consider several factors, including trust, beliefs, risk, and benefits of the developed crops. Therefore, rejection and acceptance of genetically engineered plants are based on different arguments from different people. Due to the political interference, opinions from the experts of GM, a scientist from research institutes and universities, medical professionals, and media are not unique.

The different opinions from different people interfere with the people's decisions in the selection of GM plants. These opinions point toward the safety of GM food and are found as an important impediment to the adoption of the technology (Lang and Hallman 2005). Like the existing food allergies of human beings due to several crops, GM crops may also cause some allergies to a few people, but not to all (Grice et al. 2003). Due to a lack of knowledge about GM crops, the people's attitude toward genetically modified plants is depressing, with the same in the case of GM sugarcane production. So, based on biased public opinion, it is currently challenging to GM sugarcane production and marketing success. Government initiatives are important to give awareness to the people and tackle the situation, encouraging the GM crops in the market with GM-labeling schemes. So the consumers get their choice to select the GM crops, and thereby possibly to improve the market of GM foods (Lucht 2015).

An important challenge for GM crops, including sugarcane, is based on ecological grounds; it is revealed based on studies that GM crops may affect the biodiversity of natural ecosystems and crops. A common belief is that the modified genes in plants are neither transported nor retained for a long time because, like other genes, the transformed genes also undergo natural variation in the later stages. Based on the biodiversity and ecosystem, GM field trial regulations exist in several countries for the release of GM crops, especially in commercial production. While doing the laboratory studies, the safety protocols of GM lines are followed strictly, again the same to be repeated for commercial cultivation for assessment is another challenge for the commercial production (Tammisola 2010). This difficulty is identified in isomaltulose sugarcane trials in Australia, and finally, it was the most promising experimental line and started commercial cultivation of GM sugarcane (Basnayake et al. 2011).

Sugarcane biomass becomes an emerging biomass stock worldwide. As mentioned earlier in this article, the complex polyploid genome, narrow gene pool, breeding cycle, and sugarcane flowering is a challenge for biomass improvement. The extraction, isolation, and transformation of the desired gene from different species for biofuel or biomass production is another challenging task for the scientist. Sugarcane breeders are trying to enhance the biomass yield by increasing the cell wall components and enhancing the biofuel conversion efficiency. So, the incorporation of different genes with different traits in one crop such as sugarcane is very challenging because of its complex polyploidy genome (Li et al. 2007). In future, the breeding activities should focus on the high biomass yield and quality. Many species databases are available for the improvement of sugarcane at the molecular level, so it is challenging to identify the specific gene from the databases that suits sugarcane, especially the cell-wall degrading enzyme and genes involved to improve the quality and yield of biomass produced from sugarcane.

In the coming future, GM plants will be common and commercially available in all places. The advantage of this technology is that, to date, no negative impact to human health or the environment has been reported. The world population is increasing and demands of sugar, food, energy, and healthcare items are also increasing; therefore, new and useful technologies of genetically modified and transgenic organisms will be supporting mankind to meet the future demand of the above-mentioned items. Sugarcane is an economically important crop and is cultivated for sugar and ethanol, and the current varieties are mostly derived from *S. officinarum and S. spontaneum* hybrid. These hybrids hold very complex polyploidy genome and create an immense challenge in modern marker-assisted selection and molecular breeding. Several methodologies are adopted by the researchers to overcome the complex polyploidy of sugar cane to make a good genetically modified variety (Joyce et al. 2010).

14.5 CONCLUSION

The interest in sugarcane cultivation is significantly increased worldwide due to its sustainable energy production ability. The breeding practices in the sugarcane were successful practice and obtained high yield in the sugar production. When comparing sugarcane with other crops, breeding practices are limited. In addition to breeding, the newly introduced biotechnology approaches also influence the further increase of sugarcane yield, but the approaches were not so successful when compared to other crops for a certain period. Certain constraints in sugarcane, such as narrow gene pool, complex polyploidy genome, poor fertility, and long breeding, make this crop a potential candidate for molecular breeding. As time passes, with the development of scientific research and technology, the molecular breeding and genetically modified sugarcane, which contains specific genes to increase biotic and abiotic stress, was successful and released to the market for commercial production. Recent transgenic approaches and more studies in this grass genome made it possible to sequence the whole genome of this crop, develop important agronomic traits, and increase the sugar, bioethanol, and biofuel production. All the efforts made in the above sugarcane research directed this crop to be named "energy cane" or "biorefinery" for its valuable products. With the rapid advancements in science and technology and emerging biotechnology efforts, we can expect more useful sugarcane research for the improvement of this crop in the future.

REFERENCES

Abberton M., Batley J., Bentley A., Bryant J., Cai H., Cockram J., Costa de Oliveira, A., Cseke L.J., Dempewolf H., De Pace C., and Edwards D. (2016). Global agricultural intensification during climate change: a role for genomics. *Plant Biotechnology Journal*, 14: 1095–1098.

Aftab F. and Iqbal J. (2001). PEG mediated somatic hybridization studies in sugarcane (Saccharum species Hybrid cvs Col-54 and CP-43/33). *Pakistan Journal of Botany*, 33: 233–238.

Arencibia A., Molina P.R., de la Riva G., and Selman-Housein G. (1995). Production of transgenic sugarcane (*Saccharumofficinarum* L.) plants by intact cell electroporation. *Plant cell reports*, 14:305–309.

Arruda P. (2012). Genetically modified sugarcane for bioenergy generation. *Current opinion in biotechnology*, 23:315–322.

Arvinth S., Selvakesavan R.K., Subramonian N., and Premachandran M.N. (2009). Transmission and expression of transgenes in progeny of sugarcane clones with cry1Ab and aprotinin genes. *Sugar Tech* 11: 292–295.

Barga D.P.V., Arrigoni E.D.B., Silvia-Filho M.C., and Ulian E.C. (2003). Expression of the CRY1 Ab protein in genetically modified sugarcane for the control if Diatraeasaccharalis (Lepidoptera Crambedae). *Journal of New Seeds*, 5: 209–221.

Basnayake S.W.V., Moyle R., and Birch R.G. (2011). Embryogenic callus proliferation and regeneration conditions for genetic transformation of diverse sugarcane cultivars. *Plant Cell Reports*, 30: 339–348.

Berding N., Hogarth M., and Cox M. (2007). "Plant Improvement of Sugarcane," In *Sugarcane*, ed. G. James. Blackwell Publishing Ltd, Oxford: 20–53.

Berkman P.J., Lai K., Lorenc M.T., and Edwards D. (2012). Next-generation sequencing applications for wheat crop improvement. *American Journal of Botany*, 99: 365–371.

Birchler J.A., Riddle N.C., Auger D.L., and Veitia R.A. (2005). Dosage balance in gene regulation: biological implications. *Trends in Genetics* 21: 219–226.

Botha F. (2009). "Energy yield and cost in a sugarcane biomass system," in: Proceedings Australian Society Sugarcane Technologists: 1–9.

Botha F.C. and Moore P.H. (2013). Biomass and bioenergy. Sugarcane: Physiology, biochemistry, and functional biology: 521–540.

Chaudhary K. (2018). CRISPR/Cas13a targeting of RNA virus in plants. *Plant cell reports*, 37(12): 1707–1712.

Chen F. and Dixon R.A. (2007) Lignin modification improves fermentable sugar yields for biofuel production. *Nature Biotechnology* 25:759–761.

Christy L.A., Arvinth S., Saravanakumar M., Kanchana M., Mukunthan N., Srikanth J., Thomas G., and Subramonian N. (2009). Engineering sugarcane cultivars with bovine pancreatic trypsin inhibitor (aprotinin) gene for protection against top borer (Scirpophagaexcerptalis Walker). *Plant Cell Reports*, 28:175–184.

D'Hont A., Souza G.M., Menossi M., Vincentz M., Van-Sluys M.-A., Glaszmann J.C., and Ulian E. (2008). "Sugarcane: a major source of sweetness, alcohol, and bio-energy," in *Genomics of Tropical Crop Plants*. Springer, New York, NY: 483–513.

Daniels J. and Roach B.T. (1987). Taxonomy and evolution. Sugarcane improvement through breeding/edited by Don. J. Heinz.

De Siqueira Ferreira S., Nishiyama M.Y., Jr., Paterson A.H., and Souza G.M. (2013). Biofuel and energy crops: high-yield Saccharinae take center stage in the post-genomics era. *Genome Biology* 14: 1–12.

D'Hont, and Glaszmann, J.C. (2001). Sugarcane genome analysis with molecular markers: a first decade of research. In International Society of Sugar Cane Technologists. Proceedings of the XXIV Congress, Brisbane, Australia, 17-21 September 2001. Australian Society of Sugar Cane Technologists, 2:556–559

Dias De Oliveira M.E., Vaughan B.E., and Rykiel E.J. (2005). Ethanol as fuel: energy, carbon dioxide balances, and ecological footprint. *Bioscience* 55: 593–602.

Dillen K., Demont M., Tillie P., and Rodriguez Cerezo E. (2013). Bred for Europe but grown in America: the case of GM sugar beet. *New Biotechnology*, 30(2): 131–135.

Dinesh I.M., Maharana M., and Swapan T.K. (2017) *Scope of Genetic Transformation in Sugarcane: A Review Genomics and Applied Biology*, 8: 1–7.

Du Y.C., Nose A., Kondo A., and Wasano, K. (2000). Diurnal changes in photosynthesis in sugarcane leaves: II. Enzyme activities and metabolite levels relating to sucrose and starch metabolism. *Plant Production Science*, 3: 9–16.

El-Tayeb NSM (2008). A study on the potential sugarcane fibers/ polyester composite for tribological applications. *Wear*, 265: 223–235.

FAO (2017). Food and Agriculture Data. Available at: http://www.fao.org/faostat/en/#home

Franks T. and Birch R.G. (1991). Transgenic Gene transfer into intact sugarcane cells using microprojectile bombardment. *Australian Journal of plant physiology*, 18: 471–480.

Furtado A., Lupoi J.S., Hoang N.V., Healey A., Singh S., Simmons B.A., and Henry, R.J. (2014). Modifying plants for biofuel and biomaterial production. *Plant biotechnology journal*, 12: 1246–1258.

Gawander, J. (2007). The impact of climate change on sugar cane production in Fiji. *World Meteorological Organization Bulletin*, 56: 34–39.

Goldstein, D. and Mintz, S. (2015). *The Oxford Companion to Sugar and Sweets*. Oxford University Press, Oxford

Gopal, L. (1964). Sugar-making in ancient India. *Journal of the Economic and Social History of the Orient*, 7: 57–72.

Grice J., Wegener M.K., Romanach L.M., Paton S., Bonaventura P. and Garrad, S. (2003). Genetically modified sugarcane: a case for alternate products. *AgBioForum*, 6: 162–168.

Grivet L. and Arruda P. (2002). Sugarcane genomics: depicting the complex genome of an important tropical crop. *Current Opinion in Plant Biology*, 5:122–127.

Groenewald J.H. and Botha F.C. (2008) Down-regulation of pyrophosphate: fructose 6-phosphate 1-phosphotransferase (PFP) activity in sugarcane enhances sucrose accumulation in immature internodes. *Transgenic Research*, 17: 85–92.

Hatch M., Slack C., and Johnson H.S. (1967). Further studies on a new pathway of photosynthetic carbon dioxide fixation in sugar-cane and its occurrence in other plant species. *Biochemical. Journal*, 102: 417–422.

Hatch M.D. (2005). "C4 photosynthesis: discovery and resolution," in *Discoveries in Photosynthesis*, eds. Govindjee, J.T. Beatty, H. Gest and J.F. Allen, Springer, Dordrecht: 875–880.

Henry R.J., Edwards M., Waters D.L., Gopala Krishnan S., Bundock P., Sexton T.R., Masouleh A.K., Nock C.J., and Pattemore J. (2012). Application of large-scale sequencing to marker discovery in plants. *Journal of biosciences*, 37: 829–841.

Hoang N.V., Furtado A., Botha F.C., Simmons B.A., and Henry, R.J. (2015a). Potential for genetic improvement of sugarcane as a source of biomass for biofuels. *Frontiers in Bioengineering and Biotechnology*, 3:182.

Hoang Nam V., Furtado Agnelo, Botha Frederik C., Simmons Blake A., Henry Robert J. (2015) Potential for Genetic Improvement of Sugarcane as a Source of Biomass for Biofuels. *Frontiers in Bioengineering and Biotechnology*: 182.

Hotta C., Lembke C., Domingues D., Ochoa E., Cruz G.Q., Melotto-Passarin D., Marconi T.G., Santos M.O., Mollinari M., Margarido G.R., and Crivellari A.C. (2010). The biotechnology roadmap for sugarcane improvement. *Tropical Plant Biology*, 3: 75–87.

Huang J., Khan M.T., Perecin D., Coelho S.T., and Zhang M., (2020). Sugarcane for bioethanol production: potential of bagasse in Chinese perspective. *Renewable and Sustainable Energy Reviews*, 133: 110296.

John U.P., Polotnianka R.M., Sivakumaran K.A., Chew O., MacKin L., Kuiper M.J., Talbot J.P., Nugent G.D., Mautord J., Schrauf G.E., and Spangenberg G.C. (2009). Ice recrystallization inhibition proteins (IRIPs) and freeze tolerance in the cryophilic Antarctic hair grass Deschampsia antarctica E. Desv. *Plant, cell & environment*, 32: 336–348.

Joyce P., Kuwahata M., Turner N., and Lakshmanan P. (2010). Selection system and co-cultivation medium are important determinants of Agrobacterium-mediated transformation of sugarcane. *Plant Cell Reports*, 29: 173–183.

Jung J.H., Vermerris W., Gallo M., Fedenko J.R., Erickson J.E. and Altpeter F. (2013). RNA interference suppression of lignin biosynthesis increases fermentable sugar yields for biofuel production from field-grown sugarcane. *Plant biotechnology journal*, 11(6), pp. 709–716.

Kalunke R.M., Kolge A.M., Babu K.H., and Prasad D.T. (2009). Agrobacterium mediated transformation of sugarcane for borer resistance using cry 1Aa3 gene and one-step regeneration of transgenic plants. *Sugar Tech* 11: 355–359.

Khamrit R., Jaisil P., and Bunnag P. (2012). Callus induction, regeneration and transformation of sugarcane (*Saccharumofficinarum* L.) with chitinase gene using particle bombardment. *African Journal of Biotechnology* 11: 6612–6618.

Klein J., Altenbuchner J., and Mattes R. (1998). Nucleic acid and proteinelimination during the sugar manufacturing process of conventionaland transgenic sugar beets. *Journal of Biotechnology*, 60(3): 145–153.

Knoll J.E., Anderson W.F., Richard Jr, E.P., Doran-Peterson J., Baldwin B., Hale A.L. and Viator R.P. (2013). Harvest date effects on biomass quality and ethanol yield of new energycane (*Saccharum* hyb.) genotypes in the Southeast USA. *Biomass and Bioenergy*, 56:147–156.

Lajolo F.M., Yokoyama S.M. and CheavegattiGianotto, A. (2020). Sugar derived from genetically modified sugarcane. *Food Science and Technology*, 41:1–7.

Lam E., Shine J., Da Silva J., Lawton M., Bonos S., Calvino M., Carrer, H., Silva-Filho, M.C., Glynn, N., Helsel, Z., and Ma, J. (2009). Improving sugarcane for biofuel: engineering for an even better feed-stock. *Global Change Biology and Bioenergy*, 1:251–255.

Landell M.G.A. and Bressiani J.A. (2010). *Genetic Improvement, Characterization and Management Varietal*. In: Dinardo-Miranda, L.L., Vasconcelos, A.C.M., Landell, M.G.A. (Eds.), Agronomic Institute, Campinas, Sugarcane.

Lang J.T. and Hallman W.K., (2005). Who does the public trust? The case of genetically modified food in the United States. *Risk Analysis: An International Journal*, 25: 1241–1252.

Li J., Jiang G., Huang P., Ma J., and Zhang F. (2007). Over expression of the Na+ /H+ antiporter gene from Suaeda salsa confers cold and salt tolerance to transgenic Arabidopsis thaliana. *Plant Cell Tissue and Organ Culture*, 90 (41-8): 144.

Liu S., Liu Y., Yang X., Tong C., Edwards D., Parkin I.A., Zhao M., Ma J., Yu J., Huang S., and Wang X. (2007). The Brassica oleracea genome reveals the asymmetrical evolution of polyploid genomes. *Nature communications*, 5:1–11.

Lucht J.M. (2015). Public acceptance of plant biotechnology and GM crops. *Viruses*, 7:4254–4281.

Macrelli S., Galbe M., and Wallberg O. (2014). Effects of production and market factors on ethanol profitability for an integrated first and second generation ethanol plant using the whole sugarcane as feedstock. *Biotechnology for biofuels*, 7(1):1–16.

Matsuoka S., Ferro J., and Arruda P. (2009). The Brazilian experience of sugarcane ethanol industry. *In Vitro Cellular & Developmental Biology* 45: 372–381.

McClintock B. (1984). The significance of responses of the genome to challenge. *Science*, 226: 792–801.

Moore P.H., Paterson A.H., and Tew T. (2013). Sugarcane: the crop, the plant, and domestication. *Sugarcane: physiology, biochemistry, and functional biology*: 1–17.doi.org/10.1002/9781118771280.ch1

Nascimento L.C., Yanagui K., Jose J., Camargo E.L., Grassi M.C.B., Cunha C.P., Bressiani J.A., Carvalho G.M., Carvalho C.R., Prado P.F., and Mieczkowski P. (2019). Unraveling the complex genome of *Saccharumspontaneum* using Polyploid Gene Assembler. *DNA Research*, 26: 205–216.

Nasir N.M., Qureshi R.H., and Aslam M. (2000), Effect of salinity on emergence of sugarcane lines. *Pakistan Sugar Journal*, 15: 12–14.

Oguchi T., Onishi M., Chikagawa Y., Kodama T., Suzuki E., KasaharaM., Akiyama H., Teshima R., Futo S., Hino A., Furui S., and KittaK. (2009). Investigation of residual DNAs in sugar from sugar beet(Beta vulgaris L.). food hygiene and safety science. *ShokuhinEiseigakuZasshi*, 50(1): 41–46.

Patrick J.W. (2013). Metabolic engineering of sugars and simple sugar derivatives in plants. *Plant Biotechnology Journal*, 11: 142–156.

Paterson A.H., Moore P.H., and Tew T.L. (2013). The Gene Pool of Saccharum Species and Their Improvement. In: *Genomics of the Saccharinae. Plant Genetics and Genomics: Crops and Models*. Paterson A. (eds), Springer, New York, NY, vol 11.

Pocket K. No. 45 Biotechnology for Sugarcane (2021). International Service for the Acquisition of Agri-biotech Applications (ISAAA) (https://www.isaaa.org/resources/publications/pocketk/document/Doc-Pocket%20K45.pdf)

Ramsey J. and Schemske D.W. (2002). Neopolyploidy in flowering plants. *Annual Review of Ecology, Evolution, and Systematics*, 33: 589–639.

Rodrigues Reis C.E. and Hu B. (2017). Vinasse from sugarcane ethanol production: better treatment or better utilization? *Frontiers in Energy Research*, 5: 7.

Salokhe S. (2021). Development of an efficient protocol for production of healthy sugarcane seed cane through Meristem culture. *Journal of Agriculture and Food Research*, 4: 100126.

Schuler T.H., Poppy G.M., Kerry B.R., and Denholm I. (1998) Insect resistant transgenic plants. *Trends in biotechnology*, 16: 168–175.

Scortecci K.C., Creste S., Calsa Jr, T., Xavier M.A., Landell M.G., Figueira, A., and Benedito V.A. (2012). Challenges, opportunities and recent advances in sugarcane breeding. *Plant breeding*, 1: 267–296.

Singh R.D., Singh P.N., and Kumar A. (2006). Evaluation of sugarcane (*Saccharum officinarum* L.) genotypes under variable water regimes. *Indian Journal of Crop Science* 1: 142–145.

Souza G.M., Berges H., Bocs S., Casu R., D'hont A., Ferreira J.E., Henry R., Ming R., Potier B., Van Sluys M.A. and Vincentz M. (2011). The sugarcane genome challenge: strategies for sequencing a highly complex genome. *Tropical Plant Biology*, 4:145–156.

Stevenson G.C. (1965). *Genetics and Breeding of Sugar Cane*. Longmans, London, UK.

Stirling G., Blair B., Pattemore J., Garside, A., and Bell M. (2001). Changes in nematode populations on sugarcanefollowing fallow, fumigation and crop rotation, and implications for the role of nematodes in yield decline. *Australasian Plant Pathology*, 30(4): 323–335.

Suprasanna P. (2010). Biotechnological interventions in sugarcane improvement: Strategies, methods and progress. *Nuclear Agric. & Biotech. Div., BARC Newsletter*, 316:47–53.

Tabasum S., Khan F.A., Nawaz S., Iqbal M.Z., and Saeed A. (2010). DNA profiling of sugarcane genotypes using randomly amplified polymorphic DNA. *Genetics and Molecular Research*, 9: 471–483.

Tammisola J., (2010). Towards much more efficient biofuel crops— can sugarcane pave the way? *GM Crops*, 1: 181–198.

Tew T., and Cobill R. (2008). "Genetic improvement of sugarcane (*Saccharum* spp.) as an energy crop," In *Genetic Improvement of Bioenergy Crops*, ed.Vermerris W.. Springer, New York, NY: 273–294.

Thirugnanasambandam, P.P., Hoang, N.V., and Henry, R.J. (2018). The challenge of analyzing the sugarcane genome. *Frontiers in plant science*, 9: 616.

Todd J., Wang J., Glaz B., Sood S., Ayala-Silva T., Nayak S.N., Glynn N.C., Gutierrez O.A., Kuhn D.N., Tahir M., and Comstock J.C. (2014). Phenotypic characterization of the Miami World Collection of sugarcane (*Saccharum* spp.) and related grasses for selecting a representative core. *Genetic Resources and Crop Evolution*, 61(8):1581–1596.

Waclawovsky A.J., Sato P.M., Lembke C.G., Moore P.H., and Souza G. (2010). Sugarcane for bioenergy production: an assessment of yield and regulation of sucrose content. *Plant Biotechnology Journal*, 8: 263–276.

Willis J.D., Mazarei M., and Stewart Jr, C.N. (2016). Transgenic plant-produced hydrolytic enzymes and the potential of insect gut-derived hydrolases for biofuels. *Frontiers in plant science* 7:675.

Wong Y.Y., Ho C.L., Nguyen P.D., Teo S.S., Harikrishna J.A., Rahim R.A., and Wong M.C. (2007). Isolation of salinity tolerant genes from the mangrove plant, Bruguieracylindrica by using suppression subtractive hybridization (SSH) and bacterial functional screening. *Aquatic botany*, 86:117–122.

Yadav S., Jackson P., Wei X., Ross E.M., Aitken K., DeomanoE., Atkin F., Hayes B.J., and Voss-Fels K.P. (2020). Accelerating genetic gain in sugarcane breeding using genomic selection. *Agronomy* 10:585.

Yuan J.S., Tiller K.H., Al-Ahmad H., Stewart N.R., and Stewart C.N. (2008). Plants to power: bioenergy to fuel the future. *Trends in Plant Science* 13: 421–429.

Zhang S.Z., Yang B.P., Feng C.L., Chen R.K., Luo J.P., Cai W.W., and Liu F.H. (2006). Expression of the Grifolafrondosatrehalose synthase gene and improvement of drought-tolerance in sugarcane (*Saccharumofficinarum* L.). *Journal of Integrative Plant Biology*, 48:453–459.

15 Industrial Biotechnology Applied to Sugarcane

Manuel Díaz de los Ríos and Georgina Michelena Álvarez
Cuban Research Institute of Sugar Cane By-products, Havana, Cuba

CONTENTS

15.1 INTRODUCTION

Sugarcane, among commercial crops, has excellent qualities due to its capacity to generate green mass composed of sugars and lignocellulosic substances; all of them are raw materials for a wide range of productions (Srivastava, 2020; Singh, 2020). This graminea, together with sugar, gives rise to several by-products that originate during harvesting and processing and constitute raw materials for biotechnological productions.

The use of efficient, flexible, and simplified technological schemes, which allow for greater production alternatives, reduced processing time, and lower energy consumption, broaden the availability and options of raw materials to obtain derivatives and make it possible to direct the production according to the prices of the products and the marketing strategy established by the factory.

To produce sugar from sugarcane, a series of mechanical operations are implemented, such as cane preparation, grinding and fiber breaking, to facilitate juice extraction; these steps are followed by classical chemical engineering unit operations, such as clarification, filtration, evaporation, crystallization and drying, among others aimed at guaranteeing the basic services of the process. However, the diversification of sugar production, through the transformation of its co-products and residues into products with higher added value, is mainly based on their biotechnological transformation, as illustrated in Figure 15.1.

As a result of the production of sugar, bagasse, molasses, and filter mud are obtained as by-products. Depending on the technology available in each sugar factory, some streams of low-quality

DOI: 10.1201/9781003292425-15

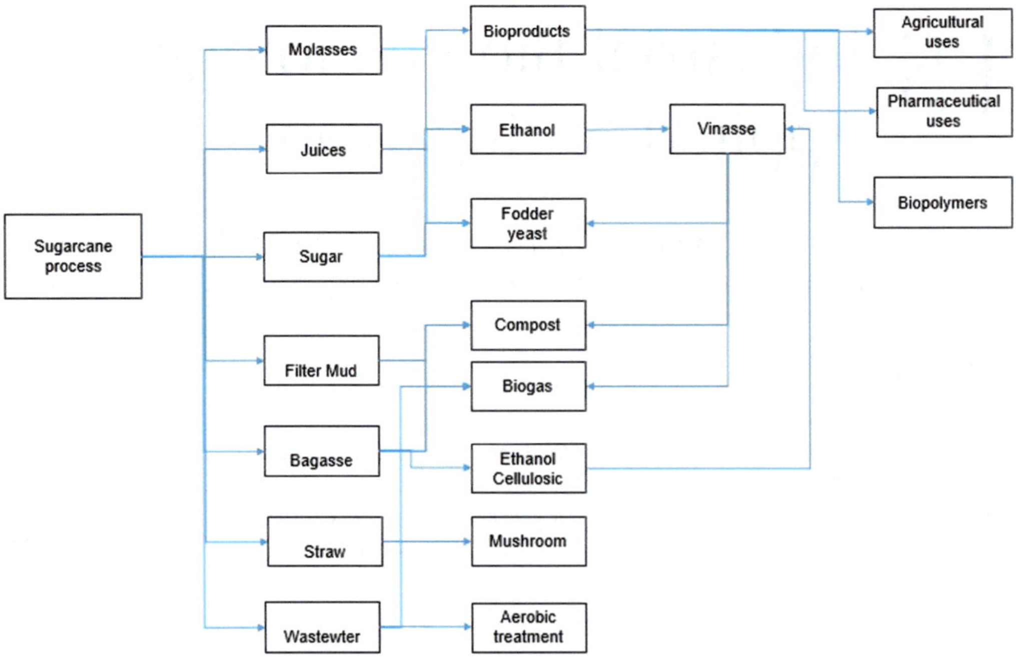

FIGURE 15.1 Biotechnological processes for sugarcane co-products and wastewater transformation in added value products.

sugary juices are also used to obtain derivatives. As less valued "waste," although not without utility, are cane straw, bagasse, and ashes resulting from the bagasse burning in the boilers.

As illustrated in Figure 15.1, sugar, molasses, and juices constitute excellent substrates for the production of ethanol, fodder yeast for animal feed, and to obtain various bio-products of interest for the pharmaceutical industry, agriculture, and other industrial uses. The bagasse can be transformed by chemical or enzymatic hydrolysis into ethanol, while its residual liquid or vinasse, as it is also called, can be used in the production of fodder yeast or anaerobic digestion to obtain biogas.

Other by-products of the process, such as filter cake, bagasse, and cane straw can also be used in the biogas production. A common practice in the sector is the use of filter cake in the production of organic fertilizer or compost, where stillage and, to a lesser extent, bagasse and cane straw are also used. The wastewater from the mill receives, on many occasions, a treatment that generates higher-value products.

The diagram shows that most of the derivatives of the sugar process are based on biotechnological transformations, widely extended internationally. The philosophy of biotechnological diversification has become the concept of "biorefineries" (Formann et al., 2020; Puglia et al., 2021).

15.2 FRUCTOOLIGOSACCHARIDES (FOS) PRODUCTION

The technology for making this prebiotic in the form of syrup is based on the enzymatic transformation of sucrose into fructooligosaccharides (FOS). At present, greater attention is paid to the potential of certain foods to promote health, improve physical well-being, and reduce the risk of diseases. Functional foods, including prebiotics (Srivastava, 2020; Singh, 2020) and fructooligosaccharides (FOS), stand out as the most interesting since they meet all the classification and food-safety criteria (Slavin, 2003, Davani-Davari et al., 2019).

These are oligofructose that have a low degree of polymerization (less than nine fructose units) and, due to their structure, represent a linear chain of fructose with β(2→1) linkage; they are not attacked by the digestive enzymes of the stomach or small intestine. In this way, they reach the large intestine, promoting the growth of the intestinal probiotic flora, functioning as a metabolic substrate (Davani-Davari et al., 2019). FOS are widely used as sweetener additives because they provide approximately one-third of the sweetness of sucrose and are low in calories (1.5 kcal/g) compared to digestible carbohydrates (4 kcal/g) (Roberfroid, 1999; Biedrzycka and Bielecka, 2004).

Currently, the industrial production of FOS is carried out using two strategies: partial degradation of inulin or enzymatic synthesis from sucrose, with the use of enzymes with high transfructosylase activity. Both technologies give rise to a mixture of FOS whose main components are 1-kestose, nystose, fructose, glucose ,and sucrose, the first two components being those of greatest commercial interest due to their double effect as a hypocaloric sweetener and a prebiotic supplement, as well as for their possible use for consumption by diabetics.

The technology for making this prebiotic in the form of syrup is essentially based on the enzymatic transformation of sucrose into fructooligosaccharides (FOS). The process consists of the use of a biocatalyst composed of 1-SST (1-sucrose: sucrose fructosyltransferase) which is a recombinant free enzyme expressed in *Pichia pastoris.*

During the product synthesis stage, a sucrose liquor at 50° Brix is put in contact with the biocatalyst in the catalytic hydrogenation reactor to obtain sorbitol from glucose. The reaction mixture is kept with stirring under controlled conditions, at a temperature of 45° C and at pH 5.5–6.0. Once the FOS mixture has been obtained in the synthesis stage, it is necessary to purify the product, in such a way that it reaches physicochemical parameters similar to commercial products and that also allows its conservation.

For this, it is demineralized using three columns in series: cationic and anionic resins and mixed beds. Subsequently, it is treated with a mixture of activated carbon and filter earth and passed through a pressure filter with vertical plates, achieving established quality parameters of pH = 5–7, brix ≥ 45°, cero color and conductivity ≤ 2 mS/cm. It is then concentrated by evaporation at 60°C until reaching values of 75%–78% soluble solids by means of a "falling film" evaporator. Finally, the syrup obtained is packaged in high-density polyethylene tanks and kept at room temperature in a cool place until use.

15.3 ETHANOL PRODUCTION

Cane molasses, and especially the final molasses, has been the main raw material for ethanol production. Molasses constitutes a highly heterogeneous substrate and can change composition, even for the same producer, within a given season. The composition of the final molasses is not uniform, and different factors can influence it. Among the most important are: the variety of cane, type of soil, harvesting methods, and the specifications of the particular process applied to each factory.

Obtaining ethyl alcohol or ethanol from sugars has been known for centuries and has been used by man in the production of alcoholic beverages since ancient times; it is, therefore, the oldest biotechnological derivative of sugarcane. It is obtained by the fermentation of the sugars present in molasses and juices, for which the international information available on its properties, ways of obtaining, and applications are considerably extensive.

In 2020, ethanol production in the world exceeded 25 billion gallons, where Brazil ranked second in world production, whose production, unlike that originating in the United States, is from sugarcane. In Latin America, Brazil, Colombia, and Ecuador (Coral-Medina and Magalhaes, 2020) stand as the largest producers of ethanol from sugar cane. It is estimated that this raw material reaches 30%–37% of world production.

A ton of final molasses can produce about 69 gallons of ethanol, while a ton of raw sugar could yield 135 gallons of ethanol.

The biotechnological production of ethanol is produced by fermentation of hexoses with yeasts of the type *Saccharomyces cerevisiae,* according to the following general expressions:

$$C_2H_{12}O_6 + O_2 \rightarrow CH_{1.83}O_{0.56}N_{0.17} + CO_2 + H_2O$$

$$C_2H_{12}O_6 \rightarrow 2C_2H_5OH + 2CO_2$$

The alcoholic fermentation mechanism consists of a first step, where the *invertase* enzyme breaks the sucrose molecule into glucose and fructose, for later, through a glycolysis stage, the glucose molecule is divided into two pyruvate molecules, according to the following equation:

$$C_6H_{12}O_6 + 2ADP + 2P_I + 2NAD^+ \rightarrow 2CH_3COCOO^- + 2ATP + 2NADH + 2H_2O + 2H^+$$

Finally, the pyruvate is converted to ethanol and the oxidized NAD^+ is regenerated.

$$CH_3COCOO^- + H^+ \rightarrow CH_3CHO + CO_2$$

In the industrial process, initially, through an aerobic stage, the growth and propagation of yeast occur, and later in the anaerobic stage, ethanol is obtained.

The production of carbon dioxide, of great interest to the food industry, as well as the availability of biomass with high protein content (38%–40%) for animal feed, are important byproducts of alcoholic fermentation.

During the fermentation process of molasses, other organic components are also obtained; these are secondary metabolites of the process, such as ethyl acetate, isopropyl alcohol, amyl alcohol, and isobutanol, among others, which are separated during the fractional distillation process of ethanol. These fractions are commonly referred to as head alcohol, rich in ethyl acetate, and fusel oil, which contain products of interest, such as solvents and additives for the chemical and pharmaceutical industries. Today they constitute chemical products resulting from green technologies, for which they acquire great importance in the context of a renewable and sustainable policy.

The following table provides the composition of these compounds in the stream called fusel oil, according to various authors. The variability of the results depends on several factors, such as the composition of the molasses, the fermentation process itself, and the characteristics of the fractional distillation technology used. Some industrial facilities wash the fusel with water to recover the ethanol retained by this stream. (Table 15.1)

TABLE 15.1
Fusel Oil Composition Reported by Several Authors

Component	% w/w Ferreira, M.C. et al	% w/w Montoya, N. et al	% w/w Patil, A.G. et al
isoamyl alcohol	55.70	64.68–80.74	60.00
active amyl alcohol	12.07	–	–
isobutanol	6.71	5.87–14.23	1.83
butanol	0.57	1.31–1.51	4.53
pentanol	0.03	–	
1-propanol	0.94	0–4.76	3.20
2-propanol	–	–	19.00
ethanol	8.66	1.40–6.26	1.23
methanol	0.02	–	–
water	15.31	5.52–12.30	5.93
Isopropyl alcohol	–	0–1.69	–
Others	–	–	3.62

Amyl alcohol and other components also represent raw materials for the third generation of bioproducts, such as the enzymatic esterification of isoamyl alcohol, obtained from fusel oil, with acetic acid and the use of immobilized lipase from *Candida antárctica* (Güvenç et al., 2007).

15.4 PRODUCTION OF FODDER YEAST OR TORULA YEAST (*CANDIDA UTILIS YEAST*)

Under the name of forage yeasts, those that are easily propagated are studied, especially to be used in animal feed, with a view to taking advantage of the nutritional components that constitute it, especially proteins and vitamins. The preferred yeast strains for industrial production are: *Candida* and *Sacharomyces*, including the species *Candida utilis, C. pulcherrima, C. reukauffi, Saccharomyces cerevisiae, S. carlsbergensis* and *S. fragilis*. The choice of a strain so that it can be developed industrially must consider the following factors: assimilation capacity, high yields, high development speed, biochemical and cultivation stability, easy recovery, high nutritional value, pleasant taste, and good appearance.

World experience has proven the possibility of using sulphite liquors, molasses, wood hydrolysates and cellulosic residues, hydrocarbons, starch residues, whey, and citrus molasses as a carbon source for this production. Among the usable microorganisms, the *Candida utilis* strain has been shown to be favored, for its ability to assimilate not only hexoses and pentoses but also other nonsugar organic compounds, such as organic acids, alcohols, and aldehydes. However, fodder yeast is not produced exclusively for these purposes, but also that which is obtained as a by-product of other fermentation processes, such as ethyl alcohol (distillery yeast) and beer (brewer's yeast); these have a nutritional value a little lower than that of forage use.

In tropical countries with a large sugarcane industry, the production of fodder yeast, from final molasses is particularly attractive, especially if one considers the impossibility of having low-cost vegetable proteins, such as soy, motivated by climatic characteristics. In addition, it allows independence from the foreign market, is a local production of medium technological complexity, and contributes to the diversification of the sugar cane industry. The production of fodder yeast, from molasses, is characterized by being a process in which this compound provides carbohydrates as a source of energy and growth factors in the form of vitamins and minerals. This production is carried out through five basic process units: preparation of raw and auxiliary materials, yeast propagation, yeast separation or recovery, concentration and drying of yeast, packaging and storage.

Although the nutritional properties of unicellular protein have been known for centuries, it was not until the Second World War that it became an industrial production in Germany, with the purpose of contributing to the feeding of the troops. Subsequently, its production was directed, in many countries, toward obtaining animal feed, although it has been affected by competition from various sources of vegetable protein, such as soybean meal and corn. The production of unicellular protein as a source for animal feeding is carried out by propagation in continuous or discontinuous fermentation systems of yeasts of the *Candida* type, mainly various species of *Candida utilis* yeasts. Its propagation requires a medium rich in a carbon source, such as cane molasses and juices, abundant in sugars, which are supplemented with nitrogen, phosphorus, and potassium (NPK). The cost of these raw materials, together with the high energy demand required by aeration, concentration, and drying operations, determine their economic feasibility.

Although it is still used by some countries in the intensive feeding of various species for human consumption, its high costs have directed its use toward the production of food for affective animals and the obtaining of food supplements or amino acid concentrates for agriculture.

Obtaining forage yeast with *Candida utilis* NRRL Y-660 with the use of sugarcane molasses as a carbon source has been extensively investigated by Almazan, 1981 and summarized by Díaz de los Ríos et al., 2020. However, the use of other cheaper carbon sources, such as wastewater from various agricultural productions, has attracted the attention of various authors. Their production has been investigated in substrates such as wastewater from the pulp and paper industry (Streit

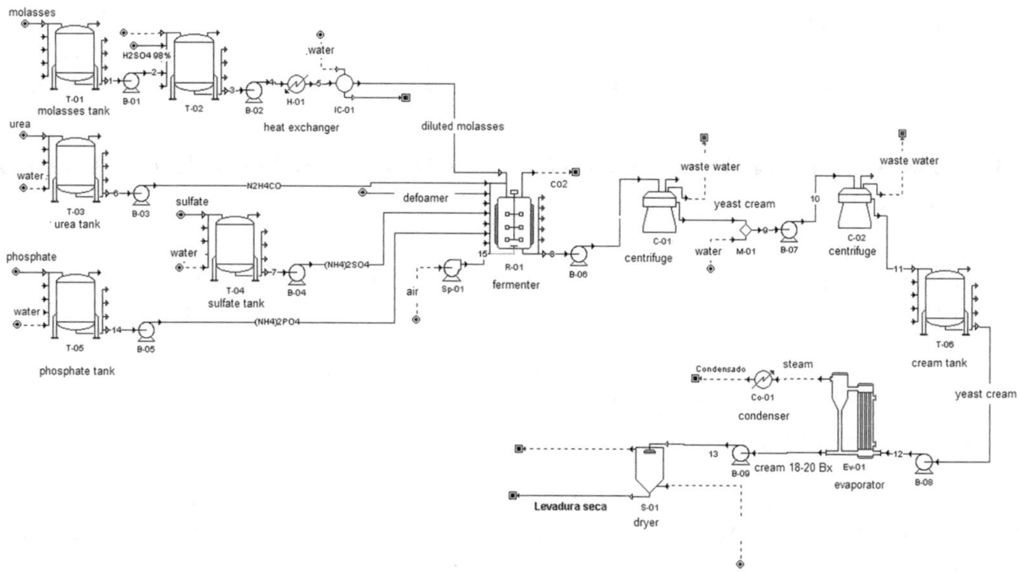

FIGURE 15.2 Technological process for obtaining torula yeast from sugarcane molasses (Garrido et al., 2010).

et al., 1987), rice production residues (Rajoka et al., 2006), fruit extracts (Munawar et al., 2010; Hamad et al., 2016), wastewater from potato processing (Kurcz, 2016), and waste from pineapple production (Rosma and Cheong, 2007), among others.

However, the use of vinasses resulting from ethanol production has been widely studied; see the work developed by Otero Rambla et al., 2007. Other authors have also ventured into the use of stillage for this production (Rocha de Meira Pires et al., 2018). The growth in the concentration of biomass in this medium reports, as an additional benefit, a reduction in the organic load of the wastewater (COD), for which this technology becomes an alternative for the waste treatment in ethanol industry. Studies on the feeding of pigs, substituting up to 66% of soybean meal for yeast produced from vinasse, did not show significant differences in the weight gain of the animals during the evaluation period (Mora et al., 2012).

The forage yeast-production process can be summarized in the following steps: the molasses is pasteurized and diluted to reach concentrations of 20–25 g/l of sugars to obtain yeast concentrations at the exit of the fermenter between 8–25 g/l. Additionally, the salts that supply the nitrogen and phosphorus necessary for the yeast growth are prepared. After propagation, the yeasts are separated in two stages of centrifugation, with an intermediate wash step for the reduction of ashes and reach concentrations of 120–150g/l. The concentration is continuous up to 20–22% in falling-film evaporators that include a thermolysis step to break up yeast and increase digestibility. Finally, the product is dried to a humidity less than 8% and packed in 25 kg multilayer bags. The process flow diagram is illustrated in Figure 15.2.

15.4.1 Obtaining Biogas from Sugar Agribusiness Waste

The development of the sugar industry and its derivatives from sugar cane is one of the sources with the greatest impact on pollution due to the generation of high volumes of highly polluting liquid waste. This has led to increasingly severe restrictions regarding the discharge of industrial effluents, so they must be treated properly to reduce environmental pollution.

Several treatment methods for the wastewater of this industry can be listed, which, depending on their efficiency, will have a higher or lower investment and operating cost.

The use of treatment systems based on the combination of reception ponds, biogas production, and then the effluent for irrigation is a convenient way, considering that these technloogies are easy to operate and maintain, thus managing the residuals economically and efficient (Budiyono and Sumardiono, 2014).

It is recognized that the anaerobic digestion (AD) process is an interesting alternative for vinasse treatment because it reduces the content of organic matter, recovers energy in the form of biogas, and obtains biofertilizers (Cabrera et al., 2017), and greenhouse gas (GHG) emissions can be reduced (Wang, 2014).

Different studies refer to the use of anaerobic reactors that ensure efficient elimination of the organic matter contained in this waste. Among them is the hybrid anaerobic reactor (RAH) (Díaz Iglesias et al. 2016; Moraes, 2017), which results from the combination of a sludge blanket reactor (UASB) in its lower part and an anaerobic filter in its upper part, thus replacing the gas-solid-liquid separation system carried by the UASB by packaging material. Its main benefit is its filtration effect. Furthermore, it reduces the risk of biomass washing from the reactor and improves gas-solid-liquid separation (Lorenzo et al. 2014).

On the other hand, natural and modified zeolites are known to have been widely used in the treatment of industrial effluents (Montalvo et al., 2012; Wang et al., 2011). It has also been used as biocatalytic support, in fixed and fluidized bed configurations, to retain high biomass concentrations and operate with significantly low HRT (Lin et al., 2013). Its use positively influences the microbial and enzymatic transformations of various substances, such as ammonium, sulfur, carbohydrates, proteins, and phenolic compounds (Andreza et al., 2020); all of this, coupled with increases in the efficiency values of COD removal of the treated waste; the performance of the biogasification occurs during the digestion process (Sunil et al., 2007).

The principles of anaerobic digestion and the design of biodigesters have been described in detail by Chernicharo, de Lemus, 2007, while the mathematical model for the detailed description of the process is known as ADM1 (Batstone et al., 2002). This model has recently been limited to the production of hydrogen from biogas (Couto et al., 2020).

Biogas consists mainly of a mixture of CO_2 and methane, where the latter occupies 55%–65% v/v (Belhadj et al., 2013). The potential of the various by-products and residues from the production of sugar and its derivatives for obtaining biogas has been determined by various authors, as is summarized in Table 15.2.

In the case of anaerobic digestion of stillage, due to the great variability of this raw material, inherent to the characteristics of the distillery (ethanol from molasses, autonomous for ethanol from juices, production of anhydrous alcohol, brandy, various qualities of ethanol, among others),

TABLE 15.2
Biogas/Methane Potential of Different Sugarcane Coproducts and Wastes

	Bagasse	Filter mud	Straw	Vinasse	Reference
Biogas (ml/g VS)	236–326	245–281	199–252	246–302	Janke, L. et al.
Biogas (ml/g VS)	259	–	274	–	López-Dávila, E. et al.
Methane (ml/g VS)	76	–	–	–	Talha, Z. et al.
Methane (ml/g COD removed)	–	–	–	344	Harada, H. et al.
Methane (ml/g COD removed)	–	250	–	–	Sánchez, E. et al.
Methane (ml/g COD)	–	–	–	228–339	Soares-Santos et al.
Methane (ml/g VS removed) with pretreatment	–	–	344–360	–	Sánchez-Herrera et al.

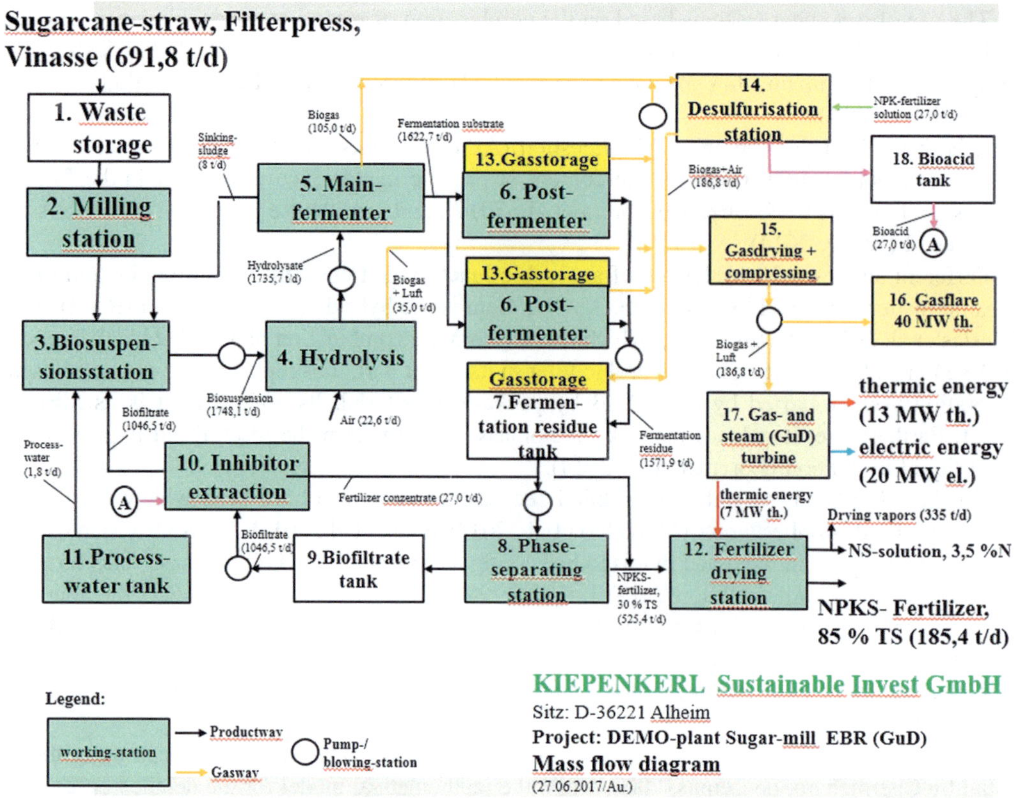

FIGURE 15.3 Biogas and organic fertilizer plant block diagram from sugar industry waste processing (Courtesy of KIEPENKERL Sustainable Invest GmbH).

there is evidence of different kinetic behavior in the generation of biogas, which determines the reactor configurations and modes of operation. However, Soares-Santos et al., 2019 do not determine variations in the methane yield of vinasse and point out that it reaches between 328 and 339 NmL-CH^4 g^{-1}COD.

In sugar mills, the generation of energy is carried out by burning bagasse in high-pressure boilers, a paradigm difficult to modify. Its conversion into biogas through co-digestion, together with the remaining residues, would constitute a new type of biorefinery, which has been conceptually evaluated by technology suppliers, as illustrated in Figure 15.3. Processing 592.5 t/h of waste (bagasse and filter cake) with prior biodigestion, they reach a generation power of 18 kWh/h and about 15 t/d of organic biofertilizers. Other residues, such as cane straw and stillage, can also be incorporated.

15.5 ORGANIC BIOFERTILIZER OR COMPOST

Organic matter residual from crops and agro-industrial productions undergoes decomposition because of the action of a varied consortium of microorganisms and insects. This natural degradation process takes place spontaneously in nature, characterized by its low speeds, and its occurrence on the surface of soils, at room temperature and preferably under aerobic conditions.

In the case of sugarcane, the policy of retaining a large part of the cane straw in the soil during harvesting is intended to supply the soil with the organic matter that the plant requires for its development; however, sugarcane straw, bagasse and distillery stillage are used in many mills, together with filter cake, to produce an organic biofertilizer or compost. An extensive bibliographic review on the subject and its positive effects for the management of residues in the sugar agroindustry has been presented by Saranraj, and Stella, 2014.

Through the development of the composting process, it is possible to considerably accelerate the biodegradation and biotransformation of the organic material, obtaining products with more favorable agrobiological properties than those of the starting material. The compost production process goes through five fundamental stages:

- Preparation: Mix of raw materials trying to achieve a particle size between 1.3–5 cm, an initial carbon/nitrogen ratio (C:N) around 25:1 and initial humidity around 65%. The control of some of these indicators is feasible by means of an adequate filter cake: bagasse balance (Meunchang et al., 2005) for the adjustment of the C:N ratio, while the filter cake: stillage mixtures, with the gradual addition of the latter, help the control of the humidity of the compost and the temperature (Sarangi et al., 2008).
- Mesophilic decomposition: The degradation of sugars and amino acids occurs due to the action of bacteria such as *Bacillus* and *Thermus*. There is an increase in temperature up to 40°C and a decrease in pH.
- Thermophilic decomposition: Degradation of hemicelluloses, waxes and polymers occurs through the action of fungi from the group of *Actinomycetes* (*Micromonospora, Streptomyces* and *Actinomyces*) and a rise in temperature that can reach 60–70°C is evidenced. These process conditions require that the process be constantly monitored with regard to humidity and temperature control, which has led to the development of various aerators for compost piles and systems for restoring humidity by adding water or stillage. The addition of vinasse allows the incorporation of potassium into the biofertilizer.
- Mesophilic cooling decomposition: The degradation of cellulose and lignin by bacteria and fungi occurs and the temperature is stabilized at about 40°C.
- Maturation: The humus stabilizes at room temperature and oxygen consumption decreases.

It is important to highlight that many technologies incorporate the addition of various biological inoculum to the process as an alternative to reduce composting times and increase the efficiency in the degradation of organic matter.

15.6 CELLULOSIC ETHANOL

The growth of the world ethanol market is based on increasing its use as an alternative fuel, considering the fermentation of lignocellulosic materials that do not compete with food production; these are: wood and forest residues, recycled paper and waste from the paper industry, sugarcane bagasse and agricultural wastes, as well as solid urban wastes. This process involves several stages of pretreatment, enzymatic hydrolysis, and fermentation, where the cellulolytic complex consisting of endoglucanases, cellobiohydrolases, and beta-glucosidase participates. Each stage has specific functions and synergistic action that produces the degradation of cellulose into glucose. Currently, the trend is to integrate the enzymatic process and the fermentation of glucose and xylose sugars in a single reactor using genetically modified microorganisms.

The production of cellulosic ethanol is carried out according to two platforms: biochemistry (hydrolysis on pretreated lignocellulosic material, followed by fermentation of the sugars generated) and thermochemistry (gasification) by transforming the lignocellulosic material into "synthesis gas" consisting of CO, CO_2 and H_2. These gases can be converted to ethanol by fermentation or thermochemical catalysis.

Fermentation of six-carbon sugars has traditionally been done with the yeast *Saccharomyces cerevisiae*. Due to the complex nature of lignocellulosic biomass, glucose and xylose-sugar-type sugars are present in hydrolysates, so it is necessary that microorganisms have the ability to use all of these sugars to increase the economic competitiveness of the process. In recent years, the genetically modified microorganisms used for this production have shown significant progress.

On the other hand, the ethanol production process based on the thermochemical platform (gasification) establishes two methods of obtaining: thermochemical and biological system, in which the carbon of the lignocellulosic biomass is gasified and the "synthesis gas" (H_2, CO_2, CO) bubbles in a special fermenter containing a microorganism (*Clostidium bacterium*) for its conversion into cellulosic ethanol. Using the thermochemical synthesis method, the gasified biomass and the generated "synthesis gas" pass to a reactor containing a catalyst for its conversion to cellulosic ethanol.

In recent years, the importance of lignocellulosic residues has been emphasized as the most promising way to promote the development of the fuel ethanol program and boost the alcohol, pharmaceutical, and cosmetic industries.

The efforts to develop a bagasse pretreatment that makes enzymatic hydrolysis viable have not yet been able to generate a technology that allows glucose to be produced at a cost low enough to be used in industrial processes. Still, with the most efficient pretreatment processes, the quantity of enzymes required in the hydrolysis to convert cellulose to glucose is high, and the alternatives proposed, including hydrolysis and alcohol production simultaneously, have not solved the problem yet, given the difference between the optimal working temperatures of enzymes and yeasts.

Sugarcane bagasse is composed of approximately 41%–44% cellulose, 25%–27% hemicelluloses, 20%–22% lignin, and 8%–10% of other components, such as ash (Singh, 2020).

Research into the processes of converting cellulose to glucose and other monomers has taken two main approaches: acid and enzymatic hydrolysis. One of the oldest methods of converting cellulose to glucose is acid hydrolysis. Unlike this process, enzymatic hydrolysis is more specific, with a higher yield, and avoids the use of chemicals difficult to remove.

The enzyme usually used in enzymatic hydrolysis is cellulase, an extract from microorganisms, which constitutes a complex, formed by several enzymes with specific functions that lead to the cellulose degradation into glucose. The cellulolytic complex contains the enzymes endoglucanases, cellobiohydrolases, and beta-glucosidase. The synergistic action of these three enzymes on the cellulolytic complex completely hydrolyzes the cellulose into glucose. Cellobiose, a product formed in the process, has an inhibitory effect, which is why cellobiase is required for its degradation and it is present in small concentrations.

The basic treatment of enzymatic hydrolysis is carried out by mixing the substrate with water to achieve a pulp of 5%–12% by weight of cellulose, which reacts with the cellulolytic complex. Typically, hydrolysis is carried out in a period of 24 to 150 hours, at temperatures close to 50°C and pH 5.0. At the end of hydrolysis, glucose, which is soluble in water, is in the liquid, while unconverted cellulose, lignin, and other compounds remain in suspension. The glucose liquor is recovered by filtration and washing.

However, cellulase cannot fully hydrolyze naturally occurring cellulose. The enzymatic degradation of lignocellulosic materials is governed by the structural features of cellulose, of strong crystalline structure, attached with hemicellulose and lignin, and the activity of the cellulolytic complex, under conditions not well known and by the inhibitory effect of cellobiose and glucose. To overcome this resistance, pretreatments are used. Different approaches have been used for the pretreatment of lignocellulosic materials: biological, physical, physicochemical, and chemical.

Several alternatives have been used to try to reduce the hydrolysis phase cost. Attempts have been made to add less cellulase to the system, but the marked decrease in the amount of glucose produced makes this process unacceptable. The variant of increasing the contact time of the enzyme in the material also leads to uneconomical processes since the completion of the stage may take weeks.

The strategy of reducing the amount of cellulase enzyme in the process by making simultaneous hydrolysis and fermentation of the glucose in the same reactor allows the reduction of the inhibition effects, which allows working with lower enzyme loads and increasing the speed of hydrolysis.

However, this process is not commercially viable yet because the yeasts work with an optimum temperature much lower than the optimum of the enzyme. Another alternative studied is to use the recycling of enzymes to reduce the load of fresh ones.

The dominant approach today is to seek to produce microorganisms that generate a greater number of enzymes to reduce their cost. It has been possible to considerably reduce the cost of commercial cellulases due to the production of strains of microorganisms through genetic engineering, aimed at the massive production of this enzyme.

The fermentation process constitutes another challenge that affects the economy of the process. In this sense, work is being done on: increasing the tolerance of microorganisms to inhibitors produced in pretreatments, increasing the performance of the process with modified microorganisms for the transformation of all sugars (Michelena et al., 2009), introducing operons that encode enzymes for the assimilation of xylose and the pentose pathway in *Z. mobilis*, using recombinant strains of E. coli with *Z. mobilis* genes for more efficient conversion of pyruvate to ethanol, introducing plasmids with xylose reductase and xylitol genes dehydrogenase of *P. stipitis* in *Saccharomyces ssp* for glucose and xylose cofermentation, simultaneous saccharification and co-fermentation (SCFS) (Olofsson et al., 2008), and purification of hydrolytic liquors to facilitate their fermentability (Ziolkowska, 2020).

15.7 EDIBLE MUSHROOMS

One of the important directions in the fight against hunger is the production of unconventional proteins such as that of unicellular microorganisms (Rajoka et al., 2012), especially that of higher fungi (Zenebe et al., 2016). *Pleurotus ostreatus* is one of the most widely produced species in the world; it is attracting growing general interest due to the nutritional quality of its biomass and the metabolites it contains and its recognized antioxidant and antitumor activity (Thanh et al., 2020).

The technological process begins when the dry sugar cane residues are mechanically processed until a suitable chopping and shredding are achieved. Subsequently, they are treated by means of a process with steam or hot water at 90 °C for 2 hours. Optionally a pasteurization of the substrate could be employed. They are then conditioned to room temperature, including disinfection with applied fungicidal solutions between 0.02 and 0.05%, to reduce the risks of contamination.

Inoculation of the substrate uses inocula prepared from strains propagated in medium 3% malt sown in different supports (seeds of wheat, millet, ground corn cobs, and even the fine sugar cane residues themselves, thoroughly crushed, giving rise to the first-generation inoculum. Subsequently, from this, commercial or second-generation flasks are inoculated.

Once inoculated, the substrate is placed in PVC bags of such dimensions that it avoids an excessive increase in temperature inside; the bags are placed on structural supports for the growth or colonization stage, which is carried out at dark and at a temperature of 26 to 28°C, depending on the strain used. During this period, the mycelium gradually colonizes the packed substrate for about 10 or 12 days, after which cuts are made in the bags and environmental conditions are radically changed.

At this stage, lighting, adequate ventilation, and a high humidity content are necessary in the growing rooms. About 18–19 days of being inoculated the substrate begin the fruiting of the fungi, and after 22 or 23 days, the first fruits are collected. Usually, three crops are harvested as average in each cycle of approximately 40 days. The increase in yields has been a factor to be achieved using different ways that have been studied: supplementation of substrates with nitrogen, use of phytohormones in substrates, mixes with substrates richer in nitrogen, and use of growth-stimulating products.

To obtain good mushroom harvests, it is necessary to have a series of cares that prevent the entry of pests and diseases.

15.8 PRODUCTION OF BIOFERTILIZERS, BIOPESTICIDES, AND BIOSTIMULATORS OF PLANT GROWTH FROM FERMENTATION PROCESSES USING SUGAR SUBSTRATES

Problems that in the past could be solved under relatively favorable auspices in terms of land availability, irrigation and fertilization possibilities, and plant protection based on the use of chemicals have proven unsustainable, with negative implications for the environment. Today, it is recognized, without a doubt, that the solution to the world's food supply, without damage to the environment, would be inconceivable and unfeasible without biotechnology.

It is known that during growth, the plant interacts closely with the microorganisms in its surrounding environment. The interaction between microorganisms and plants can be beneficial, neutral, or even harmful, depending on the plant in question and the environment in which it grows.

There are three classes of microorganisms beneficial to plants:

1. The first class is formed by those microorganisms that can increase the supplement to the plant of essential mineral nutrients for growth, such as nitrogen and phosphorus: these are the microorganisms in the soil responsible for biofertilization.
2. The second class of microorganisms comprises those that stimulate the growth or action of pathogenic organisms to plants. This form of disease prevention is sometimes called biocontrol.
3. The third class of microorganisms beneficial to plants includes those that are directly responsible for the biological growth of the plant, for example by the production of phytohormones in the rhizosphere.

These three classes of interactions open opportunities for biotechnological applications. The microorganisms and their metabolites for these applications are produced from fermentation, where the carbon source is usually molasses or cane juice, supplemented with a nitrogen source, often derived from yeast, and a phosphorous source. The production of bioproducts for agriculture is considered an alternative to diversify sugarcane and its industrial process.

15.8.1 Biofertilization

Microorganisms that form symbiotic associations with plant roots are particularly important for their role in plant nutrition. The lack of appropriate sources of nitrogen and phosphorus are, after the lack of available water, the second and third limiting factors, respectively, in plant growth.

Almost all plants can live in symbiosis either with bacteria that nodule or with *Mycorhizae*, which are capable of providing the plant with sources of nitrogen and phosphorus, respectively. In the roots of certain plants, the bacteria *Rhizobium* and *Bradyrhyzobium* form nodules in which atmospheric nitrogen is fixed and converted to ammonia, a source of nitrogen that can be used by plants. In return, the plants provide the microorganisms with the carbon necessary for their growth.

The commercial production of *Rhizobium* inoculum has been used for almost a century to ensure that legume crops have a sufficient number of host bacteria, with excellent nitrogen-fixing capacity.

The *Azospirillum* bacteria (Dalla Santa et al., 2004; Sangoquiza Caiza et al., 2018) are associated with monocots, such as wheat and corn. Some of these species make it possible to reduce the consumption of fertilizers to levels of 25–30%.

Other known bacteria are also associated with nitrogen fixation from the atmosphere, such as *Azotobacter* and others, and *Acetobacter* specifically related to sugarcane.

Mycorrhizae are highly specialized associations between roots and certain soil fungi that occur in almost all existing plant species. Its primary effect is to increase plant growth through a greater supplement of mineral nutrients from the soil to the root. Almost all mycorrhizae increase the assimilation of phosphorus; some also favor the entry of other nutrients such as ammonium and trace elements.

Some bacteria, fungi, and actinomycetes are capable of solubilizing phosphorus from the soil, which is hardly assimilated by the plant. Some examples can be mentioned: *Bacillus meghaterium* and *Pseudomonas fluorecens* among bacteria, fungi such as *Aspergillus niger* and *Penicillum lilacilum,* and other actinomycetes such as *Streptomyces* sp.

The biofertilizer *Azospirillum* allows the substitution of 25% of the nitrogen fertilizer in rice and the increase of the yield between 5% and 15%. Results of the use of an *Azospirillium* strain isolated in Cuban soils are also reported, which has had very promising results in sugarcane with yield increases between 17% and 50%, depending on the variety of sugarcane and the type of soil (Acebo et al., 2007).

The use of *Pseudomonas* strains with phosphorus solubilizing characteristics isolated from Cuban soils, allows reducing the necessary phosphorus between 50% and 100% in different crops, stimulating yields between 5% and 25%.

15.8.2 Biocontrol

Some bacteria and fungi produce different types of antibiotics providing a form of biological control. The production of Chetomin by *Chaetomiun Globosum* and Pyrrolnitrin by *Pseudomonas fluorescent* can be mentioned as an example (Caulier et al., 2018).

Many bacteria such as *Pseudomonas fluorescent* secrete substances with an extreme affinity for iron and are called siderophores. These compounds are very powerful in sequestering low amounts of iron, limiting the growth of microorganisms that are harmful to the plant (Michelena et al., 2004).

Parasitism is another alternative to biocontrol. In this case, the biocontrol agent produces an extracellular enzyme that degrades the fungal cell wall. An example of this effect is the chitin produced by the soil bacterium *Serratia marcescens* against the pathogenic fungus of the bean *Sclerotium rolfsií.*

One of the most important biocontrol mechanisms is the one that occurs in *Bacillus thurigiensis* (BT). BT species produce a crystalline protein (endotoxin) that inhibits the activity of insects such as Lepidoptera, Coleoptera, mosquito larvae, flies, etc. Biocontrol of insects and nematodes harmful to plants with BT has been widely used commercially since its approval in 1961 by the EPA.

Other types of biocontrol strategies, such as the production of volatile substances, competition for nutrients, induced resistance, and others have also been reported.

15.9 DIRECT GROWTH PROMOTION OF THE PLANTS

Plant growth and the differentiation process are regulated by the interaction of different endogenous hormones in plants. Plant growth regulators, also known as phytohormones, are chemical substances that participate in the control of normal plant development or a delicate balance between growth and the factors that coordinate it.

Currently, four types of plant hormones are recognized: auxins, gibberellins, cytokinins, and inhibitors. The hormonal properties of ethylene have also been described. Auxins is a generic term for a grade of compounds that are characterized by inducing cell lengthening, cell division, and root initiation.

Many auxins exist naturally, others are produced synthetically and some are also synthesized by some microorganisms such as: *Azospirillum*, *Azotobacter*, *Pseudomonas*, *Bacillus*, *Rhizobium*, etc., as is the case with indoleacetic acid (IAA).

The effect of auxins, in general, is to stimulate cell elongation or promote its depression, depending on the concentration used. They can form tumors that disorganize the anatomy of the organs and can cause death, as occurs with auxin herbicides (2–4 D acid). Effects on apical dominance and tropisms due to AIA polarization, also called auxin polar movement, are also reported.

IAA exerts a positive action on the roots; the initiation of the lateral root is promoted by the action of auxins, which cause the reduction of wall pressure and induce the synthesis of specific enzymes, which entails to the increase of plasticity of the cell wall, favoring germination (Mohite, 2013).

Since the second half of the 1980s, it is reported on the evaluation of the microorganisms of the rhizosphere for the production of IAA since, naturally, these microorganisms live in various forms of association with plants, increasing the speed of growth of the latter, their performance, and productivity. In many cases, its direct or indirect action is to produce certain hormones that favor the development of the plant.

IAA production has been identified in bacteria of the genus *Azotobacter*; in *Pseudomonas* strains, its production occurs simultaneously with its own biological control. The bacteria of the genus *Rhizobium* responsible for nodulation in legumes is one of the most studied genera in the production of IAA: Bacillus soil bacteria produce growth-promoting compounds such as auxins (IAA), cytokinins, and gibberellins; different strains of *Arthobacter sp.* isolated from aquatic plants produce IAA in the culture medium when tritcophane is added; some bacteria with nitrogen-fixation capacity, such as *Azospirillum* and *Streptomyces,* have been reported. More recently, evidence has emerged about the ability of *Bradyrhzobium* strains to synthesize IAA; Indolic compounds, especially IAA, have been identified in *Flavobacterium* isolated from the rhizoplane of sugarcane; the ability of *Mycorrhizogenic fungi* in the production of auxins is also recognized.

Many of the microorganisms that form symbiotic or parasitic associations with higher plants produce large amounts of cytokinins, which are considered one of the most important classes of plant growth hormones produced by microorganisms. Cytokinins have the ability to stimulate cell division and also tend to cause expansion of cells in all directions, rather than elongation.

It is known that many microorganisms are capable of excreting quantities of cytokinins. Among them *Corynebacterium fasciens*, *Agrobacterium tumefaciens*, *Rhizobium sp*., *Pseudomonas savastanoi*, *Azotobacter chroococcum,* and others. Investigations of the microbiological production of cytokinins are concentrated in plant pathogens, but as with auxins, there are no specific reports of cytokinin isolation productions.

Gibberellins, especially gibberellic acid (GA), are another growth regulator in plants produced by microorganisms. Gibberellins have found wide use in horticulture and agriculture in general due to their ability to increase vegetative growth, accelerate the speed of seed germination, increase agricultural yields, marked elongation of stems, promote seeds in fruits, and induction of flowering in plants, among others. Another widespread use of GA is in the barley malting process, increasing malt yield and reducing malting time.

For this production, the *Gibberella fujikuroi* mushroom continues to be the most productive system since no other microorganism is capable of producing gibberellic acid or similar substances at commercially feasible levels.

In addition to the selection of the strain, the aspects of genetic improvement of the wild strain, the development of culture media, the use of appropriate culture techniques, and downstream techniques, are important requirements to achieve the economic feasibility of gibberellins production. It is also a well-established fact that the development of the separation stages involves large expenses that can reach up to more than 35% of the cost of the product.

After the fermentation, the cell-free extract is subjected to unit recovery operations that involve absorption into ion-exchange resins, metal hydroxides, diatomaceous earth, etc. Alternatively, GA3 is extracted with solvents and then purified by liquid-liquid extractions and is subsequently vacuum dried, where a crystalline and amorphous powder is obtained. Jasmonic acid (JA), another bio-stimulator, was initially isolated from the fungal filtrate of *Lasiodiplodia theobromae* in 1971 and reported as a plant growth inhibitor, although its production is also reported in *Botryodiplodia theobromae* and *Gibberella fujikuroi.*

JA and related compounds are widely distributed among higher plants and play a role in regulating their development, inhibiting their growth, and promoting leaf senescence. They have been reported as potent inducers of the expression of some genes, such as the protease inhibitor gene in potatoes and the vegetative protein genes in soybean plants (Laredo-Alcalá et al., 2016).

Among its different beneficial effects is the stimulation of various physiological processes, such as stomatal closure, scissor rolling, tuber formation, breaking the seed lethargy, respiration and ethylene biosynthesis. On the other hand, there are countless inhibitory effects, among which are: photosynthetic activity, embryogenesis, root growth, and opening of the pulvulus.

A small group of microorganisms capable of synthesizing jasmonates is reported: *Lasiodiplodia theobromae* and *Gibberella fujikuroi*, although studies carried out with superior fungi of 23 different genera evidenced the genera *Collybia, Miscena* and *Coprinus* as producers (Eng et al., 2021).

Michelena et al., 2001, Eng et al., 2008 and Laredo-Alcalá et al., 2016 have carried out studies on the selection of strains and fermentation conditions. Currently having the definition of the form of inoculation, carbon and nitrogen sources, C/N ratio, fermentation systems, stirring-aeration conditions, and the methods separation and purification.

The uses of jasmonic acid in agriculture have been diverse, both in the treatment of seeds and *in vitro* cultivation and exogenous application in crops; adding to this, more recently, its fungicidal properties in local and systematic treatments.

15.10 CONCLUSIONS

Diversification, through the integral use of sugar cane, applying biotechnological processes, with flexible production schemes, is a viable economic alternative, in which the production of alcohol, biogas, and compost must be present, although other derivatives of more complex processes, such as the production of yeast for animal feed, fructooligosaccharides and other bioproducts depend on technological availability and local markets.

The productive experiences in most of the derivatives obtained by biotechnological processes show that the schemes that consider these technologies increase the added value, generate higher profits that justify investment in this field.

REFERENCES

Acebo, Y., Rives, N., Heydrich, M. y, and Hernández, A. 2007. Efecto promotor del crecimiento vegetal de cepas de Azospirillum sp. En el cultivo del arroz, *Cultivos Tropicales* 28 (3): 29–32.

Almazán del Olmo, O. 1981. Producción de levaduras forrajeras a partir de subproductos de las industrias azucarera y alcoholera, Tesis presentada en opción del grado de doctor en ciencias biológicas, CNIC, La Habana, Cuba.

Andreza, A. Longati, Anderson, R. A. Lino, Roberto, C. Giordano, Felipe, F. Furlan, and Antonio, J. G. Cruz. 2020. Biogas Production from Anaerobic Digestion of Vinasse in Sugarcane Biorefinery: A Techno-economic and Environmental Analysis, *Waste and Biomass Valorization* 11: 4573–4591, 10.1007/s12649-019-00811-w

Batstone, D.J., Keller, J., Angelidaki, I., Kalyuzhnyi, S.V., Pavlostathis, G., Rozzi, A., Sanders, W.T.M., Siegrist, H., and Vavilin, V.A. The IWA Anaerobic Digestion Model No 1 (ADM1). 2002. *Water Science and Technology*, vol 45, 10, 65-73, IWA Publishing.

Belhadj, S., Karouach, F., El Bari H., and Joute, Y. 2013. The biogas production from mesophilic anaerobic digestion of vinasse. IOSR Journal Of Environmental Science, *Toxicology And Food Technology (IOSR-JESTFT)* e-ISSN: 2319-2402, p- ISSN: 2319-2399. 5(6) (Sep. - Oct.): 72–77.

Biedrzycka, E. and Bielecka, M. 2004. Prebiotic effectiveness of fructans of different degrees of polymerization. *Trends Food Sci Technol* 15: 170–175.

Budiyono, B. and Sumardiono, S. 2014. Effect of total solid content to biogas production rate from vinasse. *Int. J. Eng.* 27: 177.

Cabrera Díaz, A., Urbano Quintero, M., Pereda Reyes, I., Díaz Marrero, M.A., and Chávez Fando, I. 2017. Digestión anaerobia de vinaza pura cubana en un sistema combinado UASB-filtro anaerobio Ingeniería Hidráulica y Ambiental vol.38 no.3.

Caulier, S., Gillis, A., Colau, G., Licciardi, F., Liépin, M., Desoignies, N., Modrie, P., Legrève, A., Mahillon, J., and Bragard, C. 2018. Versatile Antagonistic Activities of Soil-Borne Bacillus spp. and Pseudomonas spp. against Phytophthora infestans and Other Potato Pathogens. *Front. Microbiol.* 10.3389/fmicb.2018.00143

Chacón Villalobos, A. 2006. Perspectivas agroindustriales actuales de los oligofructosacáridos (FOS). *Agronomía Mesoamericana* 17(2): 265–286.

Chernicharo, C.A. de Lemus. 2007. Biological Wastewater treatment in warm climate region. Anaerobic reactors, vol 4, Federal University of Minas Gerais, Brazil, IWA Publishing.

Coral Medina, J.D. and Magalhaes Jr, A.I. 2020. Ethanol Production, Current Facts, Future Scenarios, and Techno-Economic Assessment of Different Biorefinery Configurations. 10.5772/intechopen.95081

Couto, P.T., Eng, F., Naessens, W., Nopens,I., M. Zaiat, M., and Ribeiro, R. 2020. Modelling sugarcane vinasse processing in an acidogenic reactor to produce hydrogen with an ADM1-based model. *International Journal of Hydrogen Energy* 45(11): 6217–6230, 10.1016/j.ijhydene.2019.12.206

Dalla Santa, O.R., Fernández Hernández, R., Michelena Alvarez, G.L., Ronzelli Junior, and Ricardo Soccol, O.C. 2004. Azospirillum sp. inoculation in wheat, barley and oats seeds greenhouse experiments Agriculture, Agribusiness and Biotechnology, *Braz. Arch. Biol. Technol.* 47 (6), 10.1590/S1516-89132 004000600002

Davani-Davari, D. Manica Negahdaripour, Iman Karimzadeh, Mostafa Seifan, Milad Mohkam, Seyed Jalil Masoumi, Aydin Berenjian, and Younes Ghasemi (2019). Prebiotics: Definition, types, sources, mechanisms, and clinical applications, *Foods* 8(3): 92. 10.3390/foods8030092

Díaz Iglesia, J. A., Miranda Caballero, A., and Almirall Romero, E.J. 2016.Metodología para obtención de biogás a partir de residuos de cosechas del arroz utilizando como inóculo aguas residuales.

Díaz de los Ríos, M., Saura Laria, G. y, and Pérez Bermúdez, I. 2020. Capitulo 4: La producción de levadura *candida utilis* (levadura torula), *Resultados de los Institutos Cubanos de Investigación. Desarrollo e Innovación en las tecnología sobre azúcar y derivados*, ISBN 978-959-7165-60-6, 338-351.

Díaz Iglesias, J.A., Miranda Caballero, A. y, and Almirall Romero, E.J. 2016. Metodología para obtención de biogás a partir de residuos de cosechas del arroz utilizando como inóculo aguas residuales. http://www.ciget.pinar.cu/ojs/index.php/publicaciones/article/view/181/531

Eng, F., Gutiérrez, M., y, and Favela, E. 2008. Estudio del efecto de la relación carbono: nitrógeno, el tipo de inóculo y la adición de extracto de levadura en la producción de ácido jasmónico con *Botryodiplodia theobromae* Pat. cepa RC1. *Rev. Iberoam. Micol.* 25: 188–192.

Eng, F., Marin, J.E., Zienkiewicz, K., Gutiérrez-Rojas, M., Favela-Torres, E., and Feussne, I. 2021. Jasmonic acid biosynthesis by fungi: derivatives, first evidence on biochemical pathways and culture conditions for production. *Peer J.* 9: e10873.

Ferreira, M.C., Meirelles, A.J.A., and Batista, E.A.C. 2013. Study of the fusel oil distillation process, *Ind. Eng. Chem. Res.* 52: 2336–2351, 10.1021/ie300665z

Formann, S., Hahn, A., Janke, L., Stinner, W., Sträuber, H., Logroño, W., and Nikolausz M. Nov 2020. Beyond Sugar and Ethanol production: Value generation opportunities through sugarcane residues, *Frontiers in Energy Research* 8: 20202, article 579577. 10.3389/fenrg.2020.579577

Garrido, N., Navarro, H., de los Ríos, M.D. y, and Pérez, I. 2010. Evaluación de alternativas de producción de levadura forrajera a partir de vinazas mediante simulación con SuperPro Designer, *ICIDCA sobre los derivados de la caña de azúcar* 44(2): 34–40

Güvenç, A., Kapucu, N., Kapucu, H., Aydoğan, O., and Mehmetoğlu, ü. 2007. Enzymatic esterification of isoamyl alcohol obtained from fusel oil: Optimization by response surface methodology, *Enzyme and Microbial Technology* 40, 778–785. 10.1016/j.enzmictec.2006.06.010

Hamad, S.H., Aleid, S.M., Aljasass, F.M., and Yousif, M. 2016. Production of Single Cell Protein from *Candida utilis* 70163 Using Date Fruit Syrup as Substrate, *Jokull Journal* 66(3): 58–75.

Harada, H., Uemura, S., Chen, A.-C., and Jayadevan, J. 1996. Anaerobic treatment of a recalcitrant distillery wastewater by a thermophilic UASB reactor. *Bioresource Technol* 55, 215–250.

Janke, L., Leite, A., Nikolausz, M., Thomas Schmidt, T., Liebetrau, J., Nelles, M., and Stinner, W. 2015. Biogas production from sugarcane waste: Assessment on kinetic challenges for process designing, *Int. J. Mol. Sci.* 16: 20685–20703. 10.3390/ijms160920685

Kurcz, A., Błazejak, S., Kot, A.M., Anna Bzducha-Wrobel, A., and Kieliszek, M. 2018. Application of industrial wastes for the production of microbial single-cell protein by fodder yeast *candida utilis*, *Waste Biomass Valor* 9: 57–64. 10.1007/s12649-016-9782-z

Laredo-Alcalá, E.I., Michelena-Álvarez, G., Iliná, A., Borjas-Ramos, J., Hernández-Castillo, F.D. y, and Martínez-Hernández, J.L. 2016. Uso de *Botryodiplodia theobromae* como productor de ácido jasmónico en dos sistemas de fermentación. *Rev. Mex. Mic*, 44, 1–9.

Lin, L., Wan, Ch., Liu, X., Lei, Z., Lee, D.J., Zhang, Y., Tay, J.H., and Zhang, Z. 2013. Anaerobic digestion of swine manure under natural zeolite addition: VFA evolution, cation variation, and related microbial diversity. *Applied Microbiology and Biotechnology* 97: 10575–10583.

López González, L.M., Pereda Reyes, I., and Osvaldo Romero Romero, O. 2017. Anaerobic co-digestion of sugarcane press mud with vinasse on methane yield, *Waste Management* 68: 139–145, 10.1016/j.wasman.2017.07.016

López-Dávila, E., Jiménez, J., Cisneros-Ortiz, M.E., and Romero-Romero, O. 2013. Potencial de biogás de la fermentación anaerobia de residuos sólidos de la agroindustria azucarera, utilizando lodo anaerobio de estiércol porcino, *CentroAzucar* 40(1): 1–7.

Lorenzo, Y., Valdés, A., Domenech, F., Rojas, L., and Eng, F. 2014. Cálculos técnicos en el diseño de una planta de biogás. Caso de estudio "Tratamiento de vinazas de destilerías en reactores UASB", *ICIDCA sobre los derivados de la caña de azúcar* 48 (2): 29–34.

Meunchang, S., Supamard Panichsakpatana, S., and Weaver, R.W. 2005. Co-composting of filter cake and bagasse; by-products from a sugar mill, *Bioresource Technology* 96: 437–442, 10.1016/j.biortech.2004.05.024

Michelena, G., Carrera, E., Bell, A., Altuna, B., yand Almeida, G. 2001. Proceso de separación y decoloración del ácido jasmónico a partir de *Botryodiplodia theobromae* cepa 715. *Rev. Tecnol. Ciencia, Educación* 16: 12–19.

Michelena, G., Frias, A., Dorta, M. y, and Díaz de Villegas, M.E. 2004. Metabolitos como agentes de biocontrol a partir de *Pseudomonas sp.*: una reseña. *Agroenfoque, Año* XV(1151).

Michelena, G., Bueno, G., Eng, F. Viñals, M., Álvarez, A., Bell, A., Carrera, E. y, and Manganelly, E. 2009. Etanol celulósico: una alternativa energética sustentable sin competencia con los alimentos ICIDCA. *Sobre los Derivados de la Caña de Azúcar* 1: 10 – 16.

Mohite, B. 2013. Isolation and characterization of indole acetic acid (IAA) producing bacteria from rhizospheric soil and its effect on plant growth, *Journal of Soil Science and Plant Nutrition* 13(3): 638–649.

Montalvo, S., Guerrero, L., Borja, R., Sánchez, E., Milán, Z., Cortés, I., and de la Rubia, M.A. 2012. Application of natural zeolites in anaerobic digestion processes: a review. *Applied Clay Science* 58: 125–133.

Montoya, N., Durán, J. Córdoba, F., Gil, I., Trujillo, C., and Rodríguez, G. 2016. Colombian fusel oil, *Ingeniería e Investigación* 36 (2): 21–27, 10.15446/ing.investig.v36n2.52369

Mora, L.M., Lezcano, P., Hidalgo, K., and Rodríguez, B. 2012. Torula yeast (*Candida utilis*) on distiller's vinasse in growing pig diets, *Cuban Journal of Agricultural Science* 46(1): 63–65.

Moraes, B.S., Petersen, S.O., Zaiat, M., Sommer, S.G., and Triolo, J.M. 2017. Reduction in greenhouse gas emissions from vinasse through anaerobic Digestion. *Applied Energy* 189: 21–30.

Munawar, R.A., Irfan, M., M.Nadeem, M., Syed, Q.A., and Siddique, Z.H. 2010. Biosynthesis of single cell biomass of *Candida utilis* by submerged fermentation, *Pakistan Journal of Science*, 62(1): 1–5.

Olofsson, K., Bertilsson, M., and Lidén, G.A. 2008. Short review on SSf- an interesting process option for ethanol production from lignocellulosic feedstocks. *Biotechnol Biofuels Univ. de California* 1 (1): 7.

Otero Rambla, M.A., Saura Laria, G., Martínez Valdivielso, J.A., and Almazán del Olmo, O.A. 2007. Fodder yeast production: a new approach for distillery vinasse treatment, *Proc. Int. Soc. Sugar Cane Technol.* 26: 1127–1133.

Patil, A.G., Koolwal, S.M., and Butala, H.D. 2002. Fusel oil: Composition, removal and potencial utilization. *Int. Sugar J.* 104: 51–58.

Puglia, D., Daniela Pezzolla, D., Gigliotti, G., Torre, L., Bartucca, M.L., and Del Buono, D. 2021. The opportunity of valorizing agricultural waste, through its conversion into biostimulants, biofertilizers, and biopolymers, *Sustainability* 13: 2710. 10.3390/su13052710

Rajoka, M.I., Ahmed, S., Hashmi, A.S., and Athar, M. 2012. Production of microbial biomass protein from mixed substrates by sequential culture fermentation of *Candida utilis* and *Brevibacterium lactofermentum*. *Ann Microbiol* 62: 1173–1179.

Rajoka, M.I., Khan, M.S., Jabbar, M.A., Awan, M.S., and Hashmi, A.S. 2006. Kinetics of batch single cell protein production from rice polishings with *Candida utilis* in continuously aerated tank reactors, *Bioresource Technology* 97: 1934–1941.

Roberfroid, M. 1999. Caloric value of inulin and oligofructose. *J Nutr* 1999(129): 1436–1437.

Rocha de Meira Pires, J., Grazziotti, P.H., Villela Dessimoni Pinto, N.A., Barbosa dos Reis, A., Fernandes Reis, M.L., Barbosa Borges, C.L., and Lee Nelson, D. 2018. Optimization of protein production by *candida utilis* in industrial vinasse with applicability in food, *Australian Journal of Basic and Applied Sciences* 12(1): 13–26, 10.22587/ajbas.2018.12.3.2

Rosma, A. and Cheong, M.W. 2007. Effects of nitrogen supplementation on yeast (*Candida utilis*) biomass production by using pineapple (Ananas comosus) waste extracted medium, *Malaysian Journal of Microbiology* 3(1): 19–26.

Rosma, A. and Cheong, M.W., 2007. Effects of nitrogen supplementation on yeast (candida utilis) biomass production by using pineapple (ananas comosus) waste extracted medium malaysian, *Journal of Microbiology* 3(1): 19–26 10.21161/mjm.0030

Sánchez, E., Borja, R., and Lopez, M. 1996. Determination of the kinetic constants of anaerobic digestion of sugar mill- mud –waste (SMMW), *Bioresource Technol.* 56: 245–249.

Sanchez-Herrera, D., Sanchez, O., Houbron, E., Rustrian, E., Toledano, T., Tapia-Tussell, F.R., and Alzate-Gaviria, L. 2018. Biomethane potential from sugarcane straw in Veracruz, Mexico: combined liquid hot water pretreatment and enzymatic or biological hydrolysis, *Revista Mexicana de Ingeniería Química* 17(3): 1105–1120.

Sangoquiza Caiza, C.A., Viera Tamayo, Y. y, and Yánez Guzmán, C.F. 2018. Respuesta biológica de aislados de Azospirillum spp. frente a diferentes tipos de estrés, *Centro Agricola*, 45(1): 40–46.

Sarangi, B.K., Mudliar, S.N., Bhatt, P., Kalve, S., Chakrabarti, T., and Pandey, R.A. 2008. Compost from sugarmill pressmud and distillery spentwash for sustainable agriculture, *Enviromental Biotechnology Division*, National Enviromental Engineering Research Institute (NEERI), Nehru Marg, Nagpur-440020, India, Global Science Books.

Saranraj, P. and Stella, D. 2014. Composting of sugar mill wastes: a review, *World Applied Sciences Journal* 31(12): 2029–2044. 1010.5829/idosi.wasj.2014.31.12.546

Singh, A. (2020). Benefits of crop diversification in Fiji's sugarcane farming 7(1): 65–80. 10.1002/app5.291

Slavin, J. 2003. Impact of the proposed definition of dietary fiber on nutrients data bases, *J Food Com Anal 2003* 16: 287–291.

Soares Santos, P., Zaiat, M., Oller do Nascimento, C.A., and Tadeu Fuess, L. 2019. Does sugarcane vinasse composition variability affect the bioenergy yield in anaerobic systems? A dual kinetic-energetic assessment, *Journal of Cleaner Production* 240: 118005, 10.1016/j.jclepro.2019.118005

Srivastava, S. (2020). Diversification of sugar and sugarcane industry: agro-industrial alternatives, *Sugar and Sugar Derivatives: Changing Consumer Preferences*: 151–169 10.1007/978-981-15-6663-9_10

Streit, J.D., Prior, B.A., and Kilian S.G. 1987. Substrate utilization by yeast in a spent sulphite liquor permeate, *Water SA* 13(3): 145–150.

Sunil, K.G., Gupta, S.K., and Singh, G. 2007. Biodegradation of distillery spent wash in anaerobic hybrid reactor. *Water Research*, 41: 721–730.

Talha, Z., Ding, W., Mehryar, E., Hassan, M., and Bi, J. 2016. Alkaline pretreatment of sugarcane bagasse and filter mud codigested to improve biomethane production, *BioMed Research International*, Article ID 8650597, 10 pages, 10.1155/2016/8650597

Thanh, M., Nguyen, Senaratne L., and Ranamukhaarachchi. 2020. Effect of different culture media, grain sources and alternate substrates on the mycelial growth of pleurotus eryngii and pleurotus ostreatus *Pak J Biol Sci* 23(3): 223–230. 10.3923/pjbs.2020.223.230.

Wang, Q.H., Yang, Y.N., Yu, C., Huang, H., Kim, M., Feng, C.P., and Zhang, Z.Y. 2011. Study on a fixed zeolite bioreactor for anaerobic digestion of ammonium-rich swine wastes. *Bioresource Technology* 102(14): 7064–7068.

Wang, J. 2014. Decentralized biogas technology of anaerobic digestion and farm ecosystem: opportunities and challenges *Front. Energy Res.* 26. 10.3389/fenrg.2014.00010

Zenebe, G., Gorems, W., Birhanu, G., and Zewdie S., 2016, Growth and yield performance of Pleurotus ostreatus (Jacq. Fr.) Kumm (oyster mushroom) on different substrates, *AMB Express* 6(1): 87, 10.1186/s13568-016-0265-1. Epub 2016 Oct 4.

Ziolkowska, J.R. 2020. Chapter 1 - Biofuels technologies: An overview of feedstocks, processes, and technologies in Biofuels for a More Sustainable Future, Pages 1–19.